LINEAR ALGEBRA
With Geometric Applications

UNIVERSITÉ DU QUÉBEC À HULL
BIBLIOTHÈQUE

MONOGRAPHS AND TEXTBOOKS IN PURE AND APPLIED MATHEMATICS

1. K. Yano. Integral Formulas in Riemannian Geometry (1970)
2. S. Kobayashi. Hyperbolic Manifolds and Holomorphic Mappings (1970)
3. V. S. Vladimirov. Equations of Mathematical Physics (A. Jeffrey, editor; A. Littlewood, translator) (1970)
4. B. N. Pshenichnyi. Necessary Conditions for an Extremum (L. Neustadt, translation editor; K. Makowski, translator) (1971)
5. L. Narici, E. Beckenstein, and G. Bachman. Functional Analysis and Valuation Theory (1971)
6. D. S. Passman. Infinite Group Rings (1971)
7. L. Dornhoff. Group Representation Theory (in two parts). Part A: Ordinary Representation Theory. Part B: Modular Representation Theory (1971, 1972)
8. W. Boothby and G. L. Weiss (eds.). Symmetric Spaces: Short Courses Presented at Washington University (1972)
9. Y. Matsushima. Differentiable Manifolds (E. T. Kobayashi, translator) (1972)
10. L. E. Ward, Jr. Topology: An Outline for a First Course (1972)
11. A. Babakhanian. Cohomological Methods in Group Theory (1972)
12. R. Gilmer. Multiplicative Ideal Theory (1972)
13. J. Yeh. Stochastic Processes and the Wiener Integral (1973)
14. J. Barros-Neto. Introduction to the Theory of Distributions (1973)
15. R. Larsen. Functional Analysis: An Introduction (1973)
16. K. Yano and S. Ishihara. Tangent and Cotangent Bundles: Differential Geometry (1973)
17. C. Procesi. Rings with Polynomial Identities (1973)
18. R. Hermann. Geometry, Physics, and Systems (1973)
19. N. R. Wallach. Harmonic Analysis on Homogeneous Spaces (1973)
20. J. Dieudonné. Introduction to the Theory of Formal Groups (1973)
21. I. Vaisman. Cohomology and Differential Forms (1973)
22. B.-Y. Chen. Geometry of Submanifolds (1973)
23. M. Marcus. Finite Dimensional Multilinear Algebra (in two parts) (1973, 1975)
24. R. Larsen. Banach Algebras: An Introduction (1973)
25. R. O. Kujala and A. L. Vitter (eds). Value Distribution Theory: Part A; Part B. Deficit and Bezout Estimates by Wilhelm Stoll (1973)
26. K. B. Stolarsky. Algebraic Numbers and Diophantine Approximation (1974)
27. A. R. Magid. The Separable Galois Theory of Commutative Rings (1974)
28. B. R. McDonald. Finite Rings with Identity (1974)
29. I. Satake. Linear Algebra (S. Koh, T. Akiba, and S. Ihara, translators) (1975)
30. J. S. Golan. Localization of Noncommutative Rings (1975)
31. G. Klambauer. Mathematical Analysis (1975)
32. M. K. Agoston. Algebraic Topology: A First Course (1976)
33. K. R. Goodearl. Ring Theory: Nonsingular Rings and Modules (1976)
34. L. E. Mansfield. Linear Algebra with Geometric Applications (1976)

LINEAR ALGEBRA
With Geometric Applications

LARRY E. MANSFIELD

Queens College of the City
University of New York
Flushing, New York

MARCEL DEKKER, INC. New York and Basel

MARCEL DEKKER, INC.
270 Madison Avenue, New York, New York 10016

LIBRARY OF CONGRESS CATALOG CARD NUMBER: 75–10345
ISBN: 0–8247–6321–1

Current printing (last digit):
10 9 8 7 6 5 4 3 2 1

PRINTED IN THE UNITED STATES OF AMERICA

Contents

Appendix: Determinants

Preface

Until recently an introduction to linear algebra was devoted primarily to solving systems of linear equations and to the evaluation of determinants. But now a more theoretical approach is usually taken and linear algebra is to a large extent the study of an abstract mathematical object called a vector space. This modern approach encompasses the former, but it has the advantage of a much wider applicability, for it is possible to apply conclusions derived from the study of an abstract system to diverse problems arising in various branches of mathematics and the sciences.

Linear Algebra with Geometric Applications was developed as a text for a sophomore level, introductory course from dittoed material used by several classes. Very little mathematical background is assumed aside from that obtained in the usual high school algebra and geometry courses. Although a few examples are drawn from the calculus, they are not essential and may be skipped if one is unfamiliar with the ideas. This means that very little mathematical sophistication is required. However, a major objective of the text is to develop one's mathematical maturity and convey a sense of what constitutes modern mathematics. This can be accomplished by determining how one goes about solving problems and what constitutes a proof, while mastering computational techniques and the underlying concepts. The study of linear algebra is well suited to this task for it is based on the simple arithmetic properties of addition and multiplication.

Although linear algebra is grounded in arithmetic, so many new concepts must be introduced that the underlying simplicity can be obscured by terminology. Therefore every effort has been made to introduce new terms only when necessary and then only with sufficient motivation. For example, systems of linear equations are not considered until it is clear how they arise, matrix multiplication is not defined until one sees how it will be used, and complex scalars are not introduced until they are actually needed. In addition, examples are presented with each new term. These examples are usually either algebraic or geometric in nature. Heavy reliance is placed on geometric examples because geometric ideas are familiar and they provide good interpretations of linear algebraic concepts. Examples employing polynomials or functions are also easily understood and they supply nongeometric interpretations. Occasionally examples are drawn from other fields to suggest the range of possible application, but this is not done often because it is difficult to clarify a new concept while motivating and solving problems in another field.

The first seven chapters follow a natural development begining with an algebraic approach to geometry and ending with an algebraic analysis of second degree curves and surfaces. Chapter 8 develops canonical forms for matrices under similarity and might be covered at any point after Chapter 5. It is by far the most difficult chapter in the book. The appendix on determinants refers to concepts found in Chapters 4 and 6, but it could be taken up when determinants are introduced in Chapter 3.

Importance of Problems The role of problems in the study of mathematics cannot be overemphasized. They should not be regarded simply as hurdles to be overcome in assignments and tests. Rather they are the means to understanding the material being presented and to appreciating how ideas can be used. Once the role of problems is understood, it will be seen that the first place to look for problems is not necessarily in problem sets. It is important to be able to find and solve problems while reading the text. For example, when a new concept is introduced, ask yourself what it really means; look for an example in which the property is not present as well as one in which it is, and then note the differences. Numerical examples can be made from almost any abstract expression. Whenever an abstract expression from the text or one of your own seems unclear, replace the symbols with particular numerical expressions. This usually transforms the abstraction into an exercise in arithmetic or a system of linear equations. The next place to look for problems is in worked out examples and proved theorems. In fact, the best way to understand either is by working through the computation or deduction on paper as you read, filling in any steps that may have been omitted. Most of our theorems will have fairly simple proofs which can be constructed with little more than a good understanding of what is being claimed and the knowledge of how each term is defined. This does not mean that you should be able to prove each theorem when you first encounter it, however the attempt to construct a proof will usually aid in understanding the given proof. The problems at the end of each section should be considered next. Solve enough of the computational problems to master the computational techniques, and work on as many of the remaining problems as possible. At the very least, read each problem and determine exactly what is being claimed. Finally you should often try to gain an overview of what you are doing; set yourself the problem of determining how and why a particular concept or technique has come about. In other words, ask yourself what has been achieved, what terms had to be introduced, and what facts were required. This is a good time to see if you can prove some of the essential facts or outline a proof of the main result.

At times you will not find a solution immediately, but simply attempting to set up an example, prove a theorem, or solve a problem can be very useful. Such an attempt can point out that a term or concept is not well understood and thus lead to further examination of some idea. Such an examination will often provide the basis for finding a solution, but even if it does not, it should lead to a fuller understanding of some aspect of linear algebra.

Because problems are so important, an extensive solution section is provided for the problem sets. It contains full answers for all computational problems and some theoretical problems. However, when a problem requires a proof, the actual development of the argument is one objective of the problem. Therefore you will often find a suggestion as to how to begin rather than a complete solution. The way in which such a solution begins is very important; too often an assumption is made at the begining of an argument which amounts to assuming what is to be proved, or a hypothesis is either misused or omitted entirely. One should keep in mind that a proof is viewed in its entirety, so that an argument which begins incorrectly cannot become a proof no matter what is claimed in the last line about having solved the problem. A given suggestion or proof should be used as a last resort, for once you see a completed argument you can no longer create it yourself; creating a proof not only extends your knowledge, but it amounts to participating in the development of linear algebra.

Acknowledgments At this point I would like to acknowledge the invaluable assistance I have received from the many students who worked through my original lecture notes. Their observations when answers did not check or when arguments were not clear have lead to many changes and revisions. I would also like to thank Professor Robert B. Gardner for his many helpful comments and suggestions.

LINEAR ALGEBRA
With Geometric Applications

1

A Geometric Model

Before begining an abstract study of vector spaces, it is helpful to have a concrete example to use as a guide. Therefore we will begin by defining a particular vector space and after examining a few of its properties, we will see how it may be used in the study of plane geometry.

§1. The Field of Real Numbers

Our study of linear algebra is based on the arithmetic properties of real numbers, and several important terms are derived directly from these properties. Therefore we begin by examining the basic properties of arithmetic. The set of all real numbers will be denoted by R, and the symbol "$\in$" will mean "is a member of." Thus $\sqrt{2} \in R$ can be read as "$\sqrt{2}$ is a member of the real number system" or more simply as "$\sqrt{2}$ is a real number." Now if r, s, and t are any real numbers, then the following properties are satisfied:

Properties of addition:

$r + s \in R$ or R is *closed under addition*

$r + (s + t) = (r + s) + t$ or addition is *associative*

$r + s = s + r$ or addition is *commutative*

$r + 0 = r$ or 0 is an *additive identity*

For any $r \in R$, there is an *additive inverse* $-r \in R$ such that $r + (-r) = 0$.

Properties of multiplication:

$r \cdot s \in R$ or R is *closed under multiplication*

$r \cdot (s \cdot t) = (r \cdot s) \cdot t$ or multiplication is *associative*

$r \cdot s = s \cdot r$ or multiplication is *commutative*

$r \cdot 1 = r$ or 1 is a *multiplicative identity*

For any $r \in R$, $r \neq 0$, there is a *multiplicative inverse* $r^{-1} \in R$ such that $r \cdot (r^{-1}) = 1$.

The final property states that multiplication distributes over addition and ties the two operations together:

$$r \cdot (s + t) = r \cdot s + r \cdot t \qquad \text{a } \textit{distributive law.}$$

This is a rather special list of properties. On one hand, none of the properties can be derived from the others, while on the other, many properties of real numbers are omitted. For example, it does not contain properties of order or the fact that every real number can be expressed as a decimal. Only certain properties of the real number system have been included, and many other mathematical systems share them. Thus if r, s, and t are thought of as complex numbers and R is replaced by C, representing the set of all complex numbers, then all the above properties are still valid. In general, an algebraic system satisfying all the preceding properties is called a *field*. The real number system and the complex number system are two different fields, and there are many others. However, we will consider only the field of real numbers in the first five chapters.

Addition and multiplication are *binary operations*, that is they are only defined for two elements. This explains the need for associative laws. For if addition were not associative, then $r + (s + t)$ need not equal $(r + s) + t$ and $r + s + t$ would be undefined. The field properties listed above may seem obvious, but it is not to difficult to find binary operations that violate any or all of them.

One phrase in the preceding list which will appear repeatedly is the statement that a set is closed under an operation. The statement is defined for a set of numbers and the operations of addition and multiplication as follows:

Definition Let S be a set of real numbers. S is *closed under addition* if $r + t \in S$ for all $r, t \in S$. S is *closed under multiplication* if $r \cdot t \in S$ for all $r, t \in S$.

For example, if S is the set containing only the numbers 1, 3, 4, then S is not closed under either addition or multiplication. For $3 + 4 = 7 \notin S$ and $3 \cdot 4 = 12 \notin S$, yet both 3 and 4 are in S. As another example, the set of all odd integers is closed under multiplication but is not closed under addition.

Some notation is useful when working with sets. When the elements are easily listed, the set will be denoted by writing the elements within brackets. Therefore $\{1, 3, 4\}$ denotes the set containing only the numbers 1, 3, and 4. For larger sets, a set-building notation is used which denotes an arbitrary member of the set and states the conditions which must be satisfied by any member of the set. This notation is $\{\cdots \mid \cdots\}$ and may be read as "the set of all $\cdots$ such that $\cdots$." Thus the set of odd integers could be written as:

$\{x \mid x \text{ is an odd integer}\}$, "the set of all x such that x is an odd integer." Or it could be written as $\{2n + 1 \mid n \text{ is an integer}\}$.

Problems

1. Write out the following notations in words:
   a. $7 \in R$. b. $\sqrt{-6} \notin R$. c. $\{0, 5\}$. d. $\{x \mid x \in R, x < 0\}$.
   e. $\{x \in R \mid x^2 = -1\}$.
2. a. Show by example that the set of odd integers is not closed under addition.
   b. Prove that the set of odd integers is closed under multiplication.
3. Determine if the following sets are closed under addition or multiplication:
   a. $\{1, -1\}$. b. $\{5\}$. c. $\{x \in R \mid x < 0\}$. d. $\{2n \mid n \text{ is an integer}\}$.
   e. $\{x \in R \mid x \geq 0\}$.
4. Using the property of addition as a guide, give a formal definition of what it means to say that "addition of real numbers is commutative."
5. A distributive law is included in the properties of the real number system. State another distributive law which holds and explain why it was not included.

§2. The Vector Space $\mathscr{V}_2$

It would be possible to begin a study of linear algebra with a formal definition of an abstract vector space. However, it is more fruitful to consider an example of a particular vector space first. The formal definition will essentially be a selection of properties possessed by the example. The idea is the same as that used in defining a field by selecting certain properties of real numbers. The mathematical problem is to select enough properties to give the essential character of the example while at the same time not taking so many that there are few examples that share them. This procedure obviously cannot be carried out with only one example in hand, but even with several examples the resulting definition might appear arbitrary. Therefore one should not expect the example to point directly to the definition of an abstract vector space, rather it should provide a first place to interpret abstract concepts.

As with the real number system, a vector space will be more than just a collection of elements; it will also include the algebraic structure imposed by operations on the elements. Therefore to define the vector space $\mathscr{V}_2$, both its elements and its operations must be given.

The elements of $\mathscr{V}_2$ are defined to be all ordered pairs of real numbers and are called *vectors*. The operations of $\mathscr{V}_2$ are addition and scalar multiplication as defined below:

Vector addition: The sum of two vectors (a_1, b_1) and (a_2, b_2) is defined by: $(a_1, b_1) + (a_2, b_2) = (a_1 + a_2, b_1 + b_2)$.

For example, $(2, -5) + (4, 7) = (2 + 4, -5 + 7) = (6, 2)$.

Scalar multiplication: For any real number r, called a *scalar*, and any vector (a, b) in $\mathscr{V}_2$, $r(a, b)$ is a scalar multiple and is defined by $r(a, b) = (ra, rb)$.

For example, $5(3, -4) = (15, -20)$.

Now the set of all ordered pairs of real numbers together with the operations of vector addition and scalar multiplication forms the vector space $\mathscr{V}_2$. The numbers a and b in the vector (a, b) are called the *components* of the vector. Since vectors are ordered pairs, two vectors (a, b) and (c, d) are *equal* if their corresponding components are equal, that is if $a = c$ and $b = d$.

One point that should be made immediately is that the term "vector" may be applied to many different objects, so in other situations the term may apply to something quite different from an ordered pair of real numbers. In this regard it is commonly said that a vector has magnitude and direction, but this is not true for vectors in $\mathscr{V}_2$.

The strong similarity between the vectors of $\mathscr{V}_2$ and the names for points in the Cartesian plane will be utilized in time. However, these are quite different mathematical objects, for $\mathscr{V}_2$ has no geometric properties and the Cartesian plane does not have algebraic properties. Before relating the vector space $\mathscr{V}_2$ with geometry, much can be said of its algebraic structure.

Theorem 1.1 (Basic properties of addition in $\mathscr{V}_2$) If U, V, and W are any vectors in $\mathscr{V}_2$, then

1. $U + V \in \mathscr{V}_2$
2. $U + (V + W) = (U + V) + W$
3. $U + V = V + U$
4. $U + (0, 0) = U$
5. For any vector $U \in \mathscr{V}_2$, there exists a vector $-U \in \mathscr{V}_2$ such that $U + (-U) = (0, 0)$.

Proof Each of these is easily proved using the definition of addition in $\mathscr{V}_2$ and the properties of addition for real numbers. For example, to prove part 3, let $U = (a, b)$ and $V = (c, d)$ where $a, b, c, d \in R$, then

$$\begin{aligned} U + V &= (a, b) + (c, d) \\ &= (a + c, b + d) && \text{Definition of vector addition} \\ &= (c + a, d + b) && \text{Addition in } R \text{ is commutative} \\ &= (c, d) + (a, b) && \text{Definition of addition in } \mathscr{V}_2 \\ &= V + U. \end{aligned}$$

The proof of 2 is similiar, using the fact that addition in R is associative, and 4 follows from the fact that zero is an additive identity in R. Using the above notation, $U + V = (a + c, b + d)$ and $a + c, b + d \in R$ since R is closed under addition. Therefore $U + V \in \mathscr{V}_2$ if $U, V \in \mathscr{V}_2$ and 1 holds. Part 5 follows from the fact that every real number has an additive inverse, thus if $U = (a, b)$

$$U + (-a, -b) = (a, b) + (-a, -b) = (a - a, b - b) = (0, 0)$$

and $(-a, -b)$ can be called $-U$.

Each property in Theorem 1.1 arises from a property of addition in R and gives rise to similar terminology. (1) states that $\mathscr{V}_2$ is *closed under addition.* From (2) and (3) we say that addition in $\mathscr{V}_2$ is *associative* and *commutative*, respectively. The fourth property shows that the vector (0, 0) is an *identity for addition* in $\mathscr{V}_2$. Therefore (0, 0) is called the *zero vector* of the vector space $\mathscr{V}_2$ and it will be denoted by $\mathbf{0}$. Finally the fifth property states that every vector U has an *additive inverse* denoted by $-U$.

Other properties of addition and a list of basic properties for scalar multiplication can be found in the problems below.

Problems

1. Find the following vector sums:

   a. $(2, -5) + (3, 2)$. c. $(2, -3) + (-2, 3)$.
   b. $(-6, 1) + (5, -1)$. d. $(1/2, 1/3) + (-1/4, 2)$.

2. Find the following scalar multiples:

   a. $\frac{1}{2}(4, -5)$. b. $0(2, 6)$. c. $3(2, -1/3)$. d. $(-1)(3, -6)$.

3. Solve the following equations for the vector U:

a. $U + (2, -3) = (4, 7)$. c. $(-5, 1) + U = (0, 0)$.
b. $3U + (2, 1) = (1, 0)$. d. $2U + (-4)(3, -1) = (1, 6)$.

4. Show that for all vectors $U \in \mathscr{V}_2$, $U + \mathbf{0} = U$.

5. Prove that vector addition in $\mathscr{V}_2$ is associative; give a reason for each step.

6. Suppose an operation were defined on pairs of vectors from $\mathscr{V}_2$ by the formula $(a, b) \circ (c, d) = ac + bd$. Would $\mathscr{V}_2$ be closed under this operation?

7. The following is a proof of the fact that the additive identity of $\mathscr{V}_2$ is unique, that is, there is only one vector that is an additive identity. Find the reasons for the six indicated steps.

Suppose there is a vector $W \in \mathscr{V}_2$ such that $U + W = U$ for all $U \in \mathscr{V}_2$. Then, since $\mathbf{0}$ is an additive identity, it must be shown that $W = \mathbf{0}$.

Let $U = (a, b)$ and $W = (x, y)$	a. ?
then $U + W = (a, b) + (x, y)$	
$= (a + x, b + y)$	b. ?
but $U + W = (a, b)$	c. ?
therefore $a + x = a$ and $b + y = b$	d. ?
so $x = 0$ and $y = 0$	e. ?
and $W = \mathbf{0}$.	f. ?

8. Following the pattern in problem 7, prove that each vector in $\mathscr{V}_2$ has a unique additive inverse.

9. Prove that the following properties of scalar multiplication hold for any vectors $U, V \in \mathscr{V}_2$ and any scalars $r, s \in R$:

a. $rU \in \mathscr{V}_2$
b. $r(U + V) = rU + rV$
c. $(r + s)U = rU + sU$
d. $(rs)U = r(sU)$
e. $1U = U$ where 1 is the number 1.

10. Show that, for all $U \in \mathscr{V}_2$: a. $0U = \mathbf{0}$. b. $-U = (-1)U$.

11. Addition in R and therefore in $\mathscr{V}_2$ is both associative and commutative, but not all operations have these properties. Show by examples that if subtraction is viewed as an operation on real numbers, then it is neither commutative nor associative. Do the same for division on the set of all positive real numbers.

§3. Geometric Representation of Vectors of $\mathscr{V}_2$

The definition of an abstract vector space is to be based on the example provided by $\mathscr{V}_2$, and we have now obtained all the properties necessary for the definition. However, aside from the fact that $\mathscr{V}_2$ has a rather simple definition, the preceding discussion gives no indication as to why one would want to study it, let alone something more abstract. Therefore the remainder of this chapter is devoted to examing one of the applications for linear algebra, by drawing the connection between linear algebra and Euclidean geometry. We will first find that the Cartesian plane serves as a good model for algebraic concepts, and then begin to see how algebraic techniques can be used to solve geometric problems.

Let E^2 denote the Cartesian plane, that is the Euclidean plane with a Cartesian coordinate system. Every point of the plane E^2 is named by an ordered pair of numbers, the coordinates of the point, and thus can be used to represent a vector of $\mathscr{V}_2$ pictorially. That is, for every vector $U = (a, b)$, there is a point in E^2 with Cartesian coordinates a and b that can be used as a picture of U. And conversely, for every point with coordinates (x, y) in the plane, there is a vector in $\mathscr{V}_2$ which has x and y as components. Now if the vectors of $\mathscr{V}_2$ are represented as points in E^2, how are the operations of vector addition and scalar multiplication represented?

Suppose $U = (a, b)$ and $V = (c, d)$ are two vectors in $\mathscr{V}_2$, then $U + V = (a + c, b + d)$, and Figure 1 gives a picture of these three vectors in E^2. A little plane geometry shows that when the four vectors $\mathbf{0}$, U, V, and $U + V$ are viewed as points in E^2, they lie at the vertices of a parallelogram, as in Figure 2. Thus the sum of two vectors U and V can be pictured as the fourth vertex of the parallelogram having the two lines from the origin to U and V as two sides.

To see how scalar multiples are represented, let $U = (a, b)$ and $r \in R$, then $rU = (ra, rb)$. If $a \neq 0$, then the components of the scalar multiple rU satisfy the equation $rb = (b/a)ra$. That is, if $rU = (x, y)$, then the components

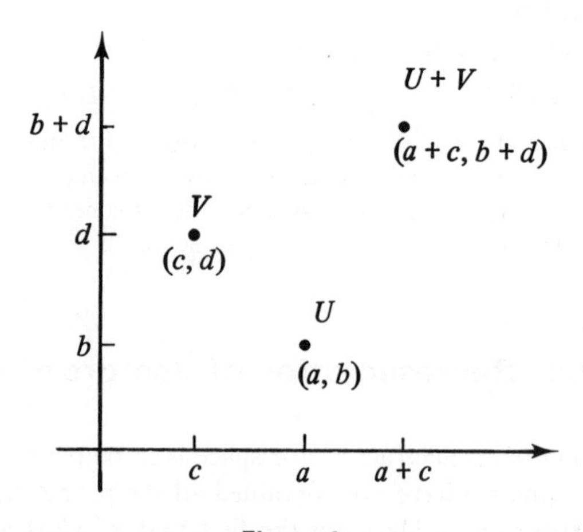

Figure 1

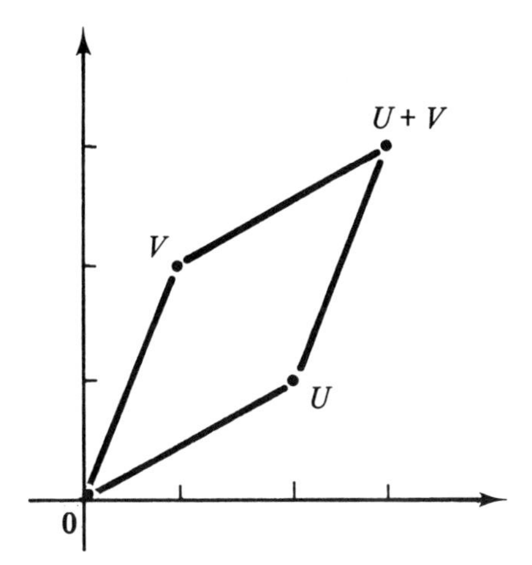

Figure 2

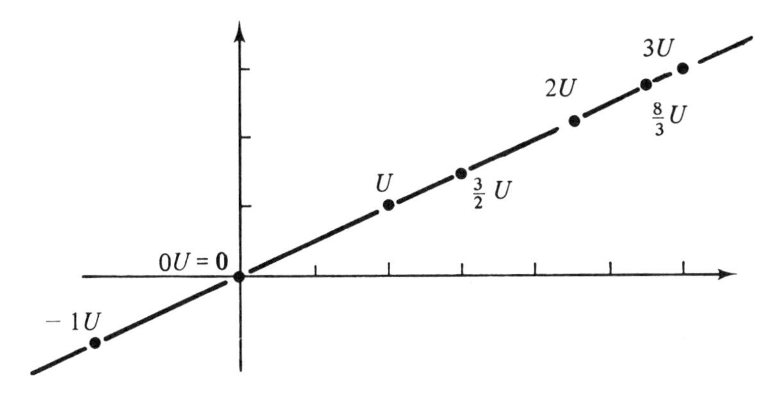

Figure 3

satisfy the equation of the line given by $y = (b/a)x$. Conversely, any point on the line with equation $y = (b/a)x$ has coordinates $(x, (b/a)x)$, and the vector $(x, (b/a)x)$ equals $(x/a)(a, b)$ which is a scalar multiple of U. In Figure 3 several scalar multiples of the vector $U = (2, 1)$ are shown on the line which represents all the scalar multiples of (2, 1). If $a = 0$ and $b \neq 0$, then the set of all multiples of U is represented by the vertical axis in E^2. If $U = \mathbf{0}$, then all multiples are again $\mathbf{0}$. Therefore in general, the set of all scalar multiples of a nonzero vector U are represented by the points on the line in E^2 through the origin and the point representing the vector U.

This interpretation of scalar multiplication provides vector equations for lines through the origin. Suppose ℓ is a line in E^2 passing through the origin and U is any nonzero vector represented by a point on ℓ. Then identifying points and vectors, every point on ℓ is tU for some scalar t. Letting P denote this variable point, $P = tU$, $t \in R$ is a vector equation for ℓ, and the variable t is called a *parameter*.

Example 1 Find a vector equation for the line ℓ with Cartesian equation $y = 4x$.

ℓ passes through the origin and the point (1, 4). Therefore $P = t(1, 4)$, $t \in R$ is a vector equation for ℓ. If the variable point P is called (x, y), then $(x, y) = t(1, 4)$ yields $x = t$, $y = 4t$ and eliminating the parameter t gives $y = 4x$. Actually there are many vector equations for ℓ. Since $(-3, -12)$ is on ℓ, $P = s(-3, -12)$, $s \in R$ is another vector equation for ℓ. But $s(-3, -12) = (-3s)(1, 4)$, so the two equations have the same graph.

Using the geometric interpretation of vector addition, we can write a vector equation for any line in the plane. Given a line ℓ through the origin, suppose we wish to find a vector equation for the line m parallel to ℓ passing through the point V. If U is a nonzero vector on ℓ, then tU is on ℓ for each $t \in R$. Now $\mathbf{0}$, tU, V, and $tU + V$ are vertices of a parallelogram, Figure 4, and m is parallel to ℓ. Therefore $tU + V$ is on m for each value of t. In fact every point on m can be expressed as $tU + V$ for some t, so $P = tU + V$, $t \in R$ is a vector equation for the line m.

Example 2 Find a vector equation for the line m passing through the point (1, 3) and parallel to the line ℓ with Cartesian equation $x = 2y$.

The point (2, 1) is on ℓ so $P = t(2, 1) + (1, 3)$, $t \in R$ is a vector equation

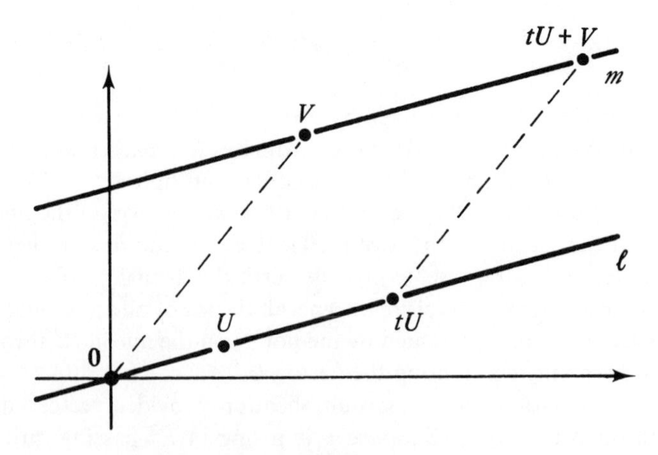

Figure 4

for m. Setting $P = (x, y)$ and eliminating the parameter t from $x = 2t + 1$ and $y = t + 3$ gives the Cartesian equation $x = 2y - 5$. The same Cartesian equation for m can be obtained using the point-slope formula.

Notice the use of the vector U in the equation $P = tU + V$ for m and $P = tU$ for ℓ, in Example 2. U determines the direction of m and ℓ, or rather its use results from the fact that they are parallel. Measurement of direction for lines is a relative matter, for example slope only measures direction relative to given directions, whereas the fact that lines are parallel is independent of coordinate axes. For the vectors in $\mathscr{V}_2$, direction is best left as an undefined term. However, each nonzero vector of $\mathscr{V}_2$ can be used to express the direction of a class of parallel lines in E^2.

Definition If m is a line with vector equation $P = tU + V$, $t \in R$, then any nonzero scalar multiple of U gives *direction numbers* for m.

Example 3 The line ℓ with Cartesian equation $y = 3x$ has $P = t(1, 3)$ as a vector equation. So $(1, 3)$, $2(1, 3) = (2, 6)$, $-3(1, 3) = (-3, -9)$ are all direction numbers for ℓ. Notice that $3/1 = 6/2 = -9/-3 = 3$ is the slope of ℓ.

In general, if a line has slope p, then it has direction numbers $(1, p)$.

Example 4 Find a vector equation for the line m with Cartesian equation $y = 3x + 5$.

m is parallel to $y = 3x$, so $(1, 3)$ gives direction numbers for m. It passes through the point $(-1, 2)$, so $P = t(1, 3) + (-1, 2)$, $t \in R$ is a vector equation for m. To check the result, let $P = (x, y)$, then $x = t - 1$ and $y = 3t + 2$. Eliminating t yields $y = 3x + 5$.

Aside from providing a parameterization for lines, this vector approach will generalize easily to yield equations for lines in 3-space. In contrast, the approach to lines in the plane found in analytic geometry does not generalize to lines in 3-space.

There are two possible points of confusion in this representation of vectors with points. The first is in the nature of coordinates. To say that a point P has coordinates (x, y) is not to say that P equals (x, y). An object is not equal to its name. However, a statement such as "consider the point with coordinates (x, y)" is often shortened to "consider the point (x, y)." But such a simplification in terminology should not be made until it is clear that points and coordinates are different. The second point of confusion is between vectors and points. There is often the feeling that "a vector really is a point isn't it?" The answer is emphatically no. If this is not clear, consider the situation with a number line such as a coordinate axis. When the identification is made between real numbers and points on a line, no one goes away saying "but of course, numbers are really points!"

Problems

1. For each of the vectors $(3, -4)$ $(0, -1/2)$ and $(2, 3)$ as U;

a. Plot the points in E^2 which represent U, $-U$, $\frac{1}{2}U$, $0U$, and $2U$.
b. Without using a formula, find a Cartesian equation for the line representing all scalar multiples of U.

2. Plot the points in E^2 which represent the vectors $\mathbf{0}$, U, V, and $U + V$ and sketch the parallelogram they determine when

a. $U = (1, 4)$; $V = (3, 2)$. b. $U = (1, 1)$; $V = (-1, 1)$.

3. How are the points representing U, V, and $U + V$ related if U is a scalar multiple of V and $V \neq \mathbf{0}$?

4. Find a vector equation for the lines with the following Cartesian equations:

a. $y = x$. b. $y = 5x$. c. $2x - 3y = 0$. d. $y = 0$.
e. $y = 5x - 6$. f. $x = 4$. g. $2x - 4y + 3 = 0$.

5. Find direction numbers for each of the lines in problem 4.

6. Find a vector equation for the line through

a. $(5, -2)$ and parallel to $3x = 2y$.
b. $(4, 0)$ and parallel to $y = x$.
c. $(1, 2)$ and parallel to $x = 5$.

7. Find Cartesian equations for the lines with the given vector equations.

a. $P = t(6, 2) + (7, 3)$, $t \in R$. b. $P = t(5, 5) + (5, 5)$, $t \in R$.
c. $P = t(2, 8) + (1, 2)$, $t \in R$.

8. Show that if $P = tU + V$, $t \in R$ and $P = sA + B$, $s \in R$ are two vector equations for the same line, then U is a scalar multiple of A.

§4. The Plane Viewed as a Vector Space

Pictures of vectors as points in the Cartesian plane contain Euclidean properties which are lacking in $\mathscr{V}_2$. For example, the distance from the origin to the point (a, b) in E^2 is $\sqrt{a^2 + b^2}$ but this number does not represent anything in $\mathscr{V}_2$. Similarly angles can be measured in E^2 while angle has no meaning in $\mathscr{V}_2$. In order to bring the geometry of E^2 into the vector space, we could define the length of a vector (a, b) to be $\sqrt{a^2 + b^2}$, but this definition would come from the plane rather than the vector space. The algebraic approach is to add to the algebraic structure of the vector space so that the properties of length and angle in the plane will represent properties of vectors. The following operation will do just that.

Definition Let $U = (a, b)$ and $V = (c, d)$; then the *dot product* of U and V is $U \circ V = (a, b) \circ (c, d) = ac + bd$.

For example,

$$(2, -6) \circ (2, 1) = 2 \cdot 2 + (-6) \cdot 1 = -2;$$

$$(3, -4) \circ (-4, -3) = -12 + 12 = 0 \qquad \text{and} \qquad (1, 1) \circ (1, 1) = 2.$$

Notice that the dot product of two vectors is a scalar or number, *not* a vector. Therefore the dot product is not a multiplication of vectors in the usual sense of the word; for when two things are multiplied, the product should be the same type of object.

The algebraic system consisting of all ordered pairs of real numbers together with the operations of addition, scalar multiplication, and the dot product is not really $\mathscr{V}_2$. This new space will be denoted by $\mathscr{E}_2$. The vector space $\mathscr{E}_2$ differs from $\mathscr{V}_2$ only in that it possesses the dot product and $\mathscr{V}_2$ does not. The similarity between the notations E^2 and $\mathscr{E}_2$ is not accidental; it will soon become evident that $\mathscr{E}_2$ is essentially the Cartesian plane together with the algebraic structure of a vector space.

Notice that for any vector $U = (a, b)$ in $\mathscr{E}_2$, $U \circ U = a^2 + b^2$. This is the square of what we thought of calling the length of U.

Definition The *length* or *norm* of a vector $U \in \mathscr{E}_2$ denoted $\|U\|$, is defined by $\|U\| = \sqrt{U \circ U}$.

The notation $\|U\|$ may be read "the length of U." In terms of components, if $U = (a, b)$ then $\|(a, b)\| = \sqrt{a^2 + b^2}$. For example

$$\|(2, -1)\| = \sqrt{4 + 1} = \sqrt{5}, \quad \|(3/5, 4/5)\| = 1, \quad \text{and} \quad \|\mathbf{0}\| = 0.$$

Now with the length of a vector $U = (a, b)$ equal to the distance between the origin and the point with coordinates (a, b) in E^2, the line segment between the origin and (a, b) could be used to represent both the vector U and its

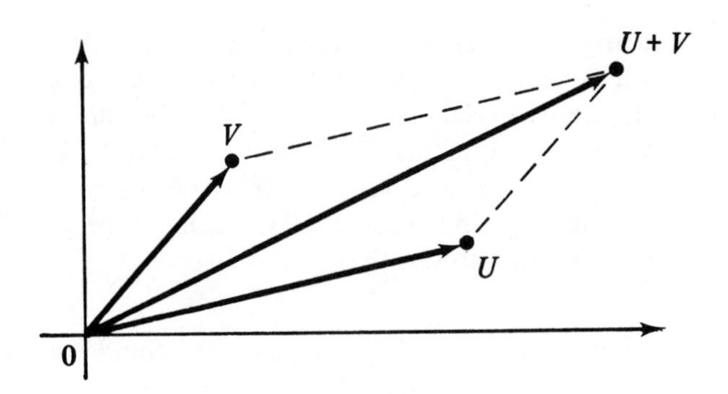

Figure 5

length. But each nonzero vector of $\mathscr{V}_2$ and hence $\mathscr{E}_2$ determines a direction in the Cartesian plane. Therefore the vector $U = (a, b)$ can be represented pictorially by an arrow from the origin to the point with coordinates (a, b). Using this representation for vectors from $\mathscr{E}_2$, the picture for addition of vectors in Figure 2 can be redrawn. In Figure 5, the sum $U + V$ is represented by an arrow which is the diagonal of the parallelogram determined by $\mathbf{0}$, U, V, and $U + V$. This figure illustrates the *parallelogram rule of addition* which states that the sum of two vectors, viewed as arrows, is the diagonal (at the origin) of the parallelogram they determine.

An arrow is not necessarily the best way to represent a vector from $\mathscr{E}_2$. Viewing vectors as arrows would not improve the picture of all the scalar multiples of a vector. The choice of which representation to use depends on the situation.

An arrow which is not based at the origin may be viewed as representing a vector. This arises from a representation of the difference of two vectors and leads to many geometric applications. For $U, V \in \mathscr{E}_2$, $U - V = U + (-V)$, so that viewing U, V, $-V$, and $U - V$ as arrows in E^2 gives a picture such as that in Figure 6. If these four vectors are represented by points and a few lines are added, then two congruent triangles with corresponding sides parallel are obtained, as shown in Figure 7. Therefore the line segment from V to U has the same length and is parallel to the line from the origin to $U - V$. In other words, the arrow from V to U has the same length and direction as the arrow representing the difference $U - V$. Therefore the arrow from V to U can also be used to represent $U - V$, as shown in Figure 8. This representation can be quite confusing if one forgets that the arrow is only a picture for an ordered pair of numbers; it is not the vector itself. For example, when $U = (4, 1)$ and $V = (3, 7)$, the difference $U - V = (1, -6)$ is an ordered pair. The simplest representation of the vector $(1, -6)$ is either as a point or as an arrow at the origin ending at the point with coordinates $(1, -6)$. There is nothing in the

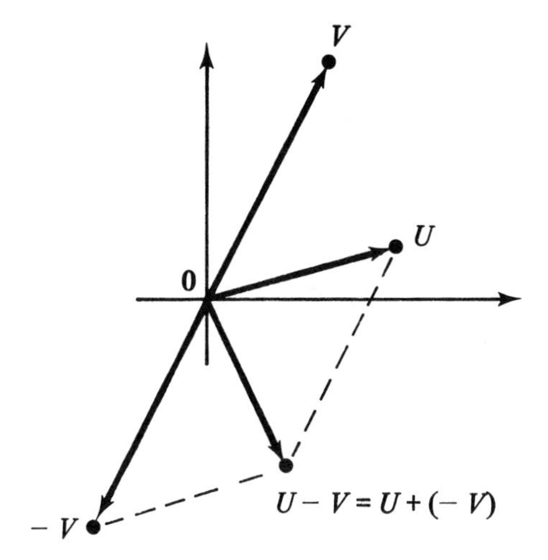

Figure 6

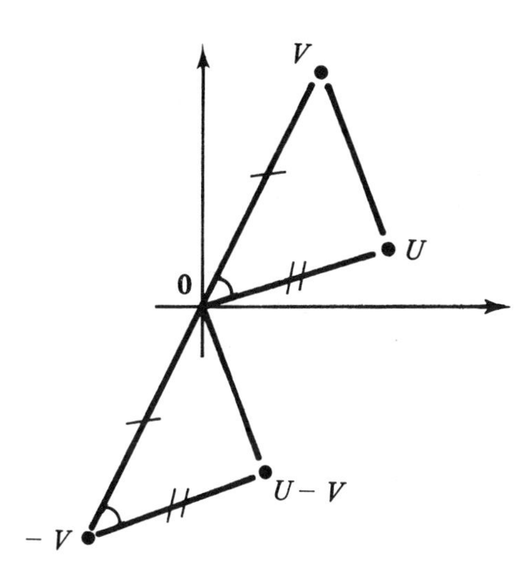

Figure 7

vector $(1, -6)$ which indicates that it should be viewed as some other arrow.

A vector equation for the line determined by two points can be obtained using the above representation of a vector difference. Suppose A and B are two points on a line ℓ in E^2. Viewing these points as vectors, the difference $B - A$ may be represented by an arrow from A to B; see Figure 9. Or $B - A$ is on the line parallel to ℓ which passes through the origin. Therefore the

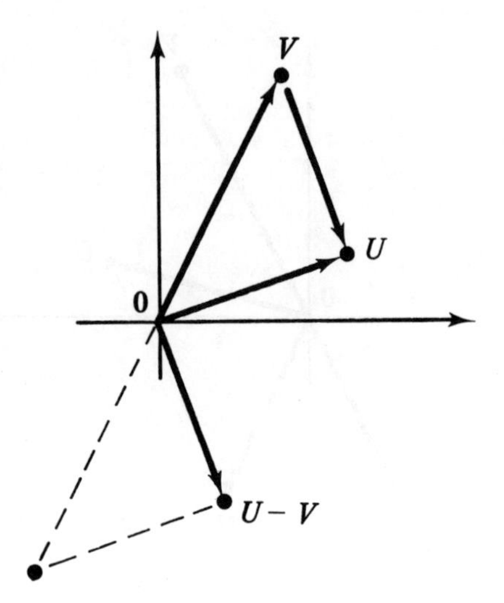

Figure 8

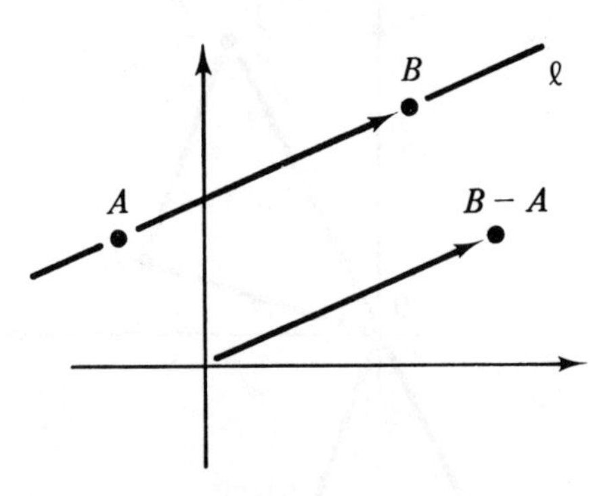

Figure 9

difference $B - A$ gives direction numbers for the line ℓ, and a vector equation for ℓ is $P = t(B - A) + A$, $t \in R$.

Example 1 Find a vector equation for the line ℓ through $(-2, 3)$ and $(4, 6)$.

ℓ has direction numbers $(6, 3) = (4, 6) - (-2, 3)$, and $(-2, 3)$ is a point on the line. Thus $P = t(6, 3) + (-2, 3)$, $t \in R$ is a vector equation for ℓ. Several other equations could be obtained for ℓ, such as $P = s(-6, -3) + (4, 6)$, $s \in R$. These two equations have the same graph, so for each particular value of t, there is a value of s which gives the same point. In this case this occurs when $s = 1 - t$.

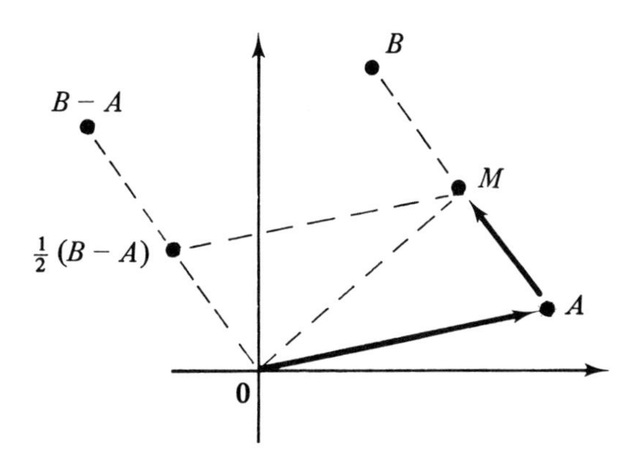

Figure 10

Example 2 Find the midpoint M of the line segment joining the two points A and B in the plane E^2.

If A, B, and M are viewed as vectors, then $B - A$ can be thought of as an arrow from A to B. M is in the middle of this arrow, but $\frac{1}{2}(B - A)$ is not M. Rather, $\frac{1}{2}(B - A)$ can be represented by the arrow from A to M, as in Figure 10. Therefore $\frac{1}{2}(B - A) = M - A$, or A must be added to $\frac{1}{2}(B - A)$ to obtain M; and $M = A + \frac{1}{2}(B - A) = \frac{1}{2}(A + B)$.

$M = \frac{1}{2}(A + B)$ is a simple vector formula for the midpoint of a line which does not explicitly involve the coordinates of points. One of the main strengths of the vector approach to geometry is the ability to express relationships in a form free of coordinates. On the other hand, the vector formula can easily be converted to the usual expression for the coordinates of a midpoint in analytic geometry. For if the point A has coordinates (x_1, y_1) and B has coordinates (x_2, y_2), then $M = \frac{1}{2}(A + B) = (\frac{1}{2}(x_1 + x_2), \frac{1}{2}(y_1 + y_2))$. The vector form of the midpoint formula is used in the next example to obtain a short proof of a geometric theorem.

Example 3 Prove that the line joining the midpoints of two sides of a triangle is parallel to and half the length of the third side.

Suppose A, B, and C are the vertices of a triangle, M is the midpoint of side AB and N is the midpoint of side BC, as in Figure 11. Then it is necessary to show that the line MN is half the length and parallel to AC. In terms of vectors this is equivalent to showing that $N - M = \frac{1}{2}(C - A)$. But we know that $M = \frac{1}{2}(A + B)$ and $N = \frac{1}{2}(B + C)$, so

$$\begin{aligned} N - M &= \tfrac{1}{2}(B + C) - \tfrac{1}{2}(A + B) = \tfrac{1}{2}B + \tfrac{1}{2}C - \tfrac{1}{2}A - \tfrac{1}{2}B \\ &= \tfrac{1}{2}(C - A). \end{aligned}$$

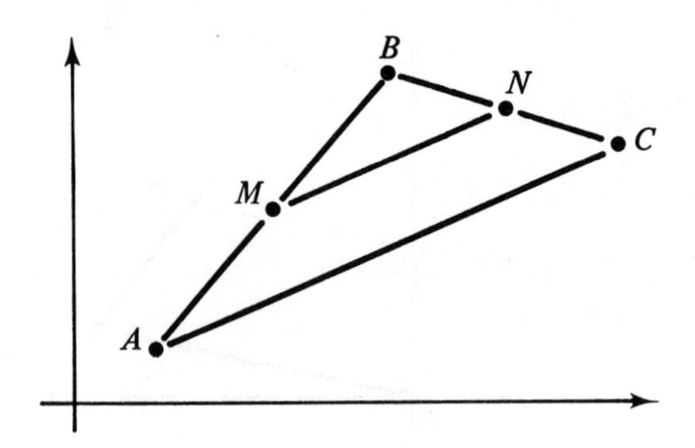

Figure 11

Problems

1. Compute the following dot products: a. $(2, -1)\circ(3, 4)$. b. $(0, 0)\circ(1, 7)$. c. $(2, -3)\circ(3, 2)$. d. $(5, -1)\circ(-4, 3)$.
2. Find the lengths of the following vectors from $\mathscr{E}_2$:

   a. $(1, 0)$. b. $(1/\sqrt{2}, 1/\sqrt{2})$. c. $(-4, 3)$. d. $(1, -3)$.
3. Show that $\|rU\| = |r|\|U\|$ for any vector $U \in \mathscr{E}_2$ and any scalar r.
4. Let $U, V, W \in \mathscr{E}_2$. a. What is the meaning of $U\circ(V\circ W)$? b. Show that $U\circ(V + W) = U\circ V + U\circ W$.
5. Use arrows from the origin in E^2 to represent the vectors U, V, $U + V$, $-V$, and $U - V = U + (-V)$ when

   a. $U = (2, 4)$, $V = (-2, 3)$. b. $U = (5, 1)$, $V = (2, -4)$.
6. Find a vector equation for the line through the given pair of points.

   a. $(2, -5)$, $(-1, 4)$. b. $(4, 7)$, $(10, 7)$. c. $(6, 0)$, $(3, 7)$.
7. Find the midpoint of each line segment joining the pairs of points in problem 6.
8. Let A and B be two points. Without using components, find a vector formula for the point Q which is on the line segment between A and B and one-fifth the distance to A from B.
9. Without using components, prove that the midpoints of the sides of any quadrilateral are the vertices of a parallelogram.
10. Let A, B, and C be the vertices of a triangle. Without using components, show that the point on a median which is 2/3 of the distance from the vertex to the opposite side can be expressed as $\frac{1}{3}(A + B + C)$. Conclude from this that the medians of a triangle are concurrent.
11. Prove without using components that the diagonals of a parallelogram bisect each other. Suggestion: Show that the midpoints of the diagonals coincide.

§5. Angle Measurement in $\mathscr{E}_2$

The dot product was used to define length for vectors in $\mathscr{E}_2$ after first finding what the length should be in order to have geometric meaning. The definition of angle can be motivated in the same way. We need the properties listed below but a detailed discussion of them will be postponed to Chapter 6.

Theorem 1.2 Let U, V, and $W \in \mathscr{E}_2$, and $r \in R$, then

1. $U \circ V = V \circ U$
2. $U \circ (V + W) = U \circ V + U \circ W$
3. $(rU) \circ V = r(U \circ V)$
4. $U \circ V \geq 0$ and $U \circ U = 0$ if and only if $U = \mathbf{0}$.

To motivate the definition of an angle in $\mathscr{E}_2$, suppose U and V are two vectors such that one is not a scalar multiple of the other. Representing U, V, and $U - V$ with arrows in E^2 yields a triangle as in Figure 12. Then the angle θ between the arrows representing U and V is a well-defined angle inside the triangle. Since the three sides of the triangle can be expressed in terms of the given vectors, the law of cosines has the form

$$\|U - V\|^2 = \|U\|^2 + \|V\|^2 - 2\|U\|\|V\| \cos \theta.$$

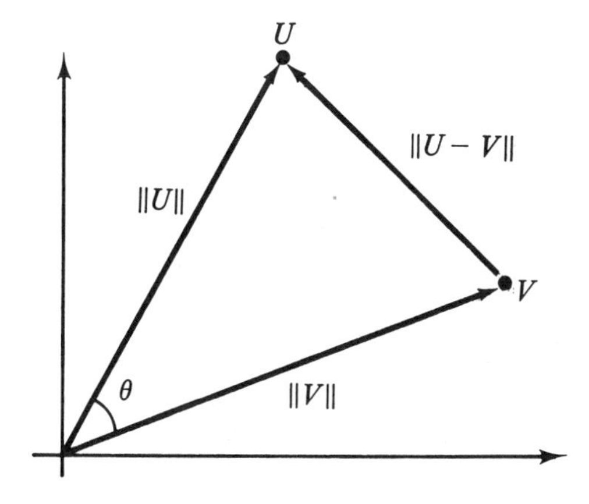

Figure 12

Using Theorem 1.2 and the definition of length,

$$\begin{aligned}\|U - V\|^2 = (U - V) \circ (U - V) &= U \circ U - U \circ V - V \circ U + V \circ V \\ &= \|U\|^2 - 2U \circ V + \|V\|^2.\end{aligned}$$

Therefore the law of cosines gives us the relation $U \circ V = \|U\|\|V\| \cos \theta$. This relation ties together the angle between the arrows and the vectors they represent, making it clear how angles should be defined in $\mathscr{E}_2$.

Definition If U and V are nonzero vectors in $\mathscr{E}_2$, the *angle* between them is the angle θ which satisfies $0 \leq \theta \leq 180°$ and

$$\cos \theta = \frac{U \circ V}{\|U\|\|V\|}.$$

If U or V is the zero vector, then the *angle* between them is taken to be zero.

With this definition, angles in $\mathscr{E}_2$ will have the usual geometric interpretation when the vectors of $\mathscr{E}_2$ are represented by arrows in E^2. The second part of the definition insures that an angle is defined for any pair of vectors, and we can write $U \circ V = \|U\|\|V\| \cos \theta$ for all vectors U, V from $\mathscr{E}_2$.

Example 1 If $U = (1, 2)$ and $V = (-1, 3)$, then the angle between U and V satisfies

$$\cos \theta = \frac{(1, 2) \circ (-1, 3)}{\|(1, 2)\|\|(-1, 3)\|} = \frac{1}{\sqrt{2}},$$

and $\theta = 45°$.

Example 2 If $U = (1, 1)$, $V = (-1, 2)$ and θ is the angle between U and V, then $\cos \theta = 1/\sqrt{10}$ or $\theta = \cos^{-1} 1/\sqrt{10}$. It would be possible to obtain a decimal approximation for θ if required.

Example 3 When $U = (4, 6)$ and $V = (-3, 2)$, $\cos \theta = 0$. Therefore the angle between U and V is $90°$, and if U and V are represented by arrows in E^2, then the arrows are perpendicular.

Definition Two nonzero vectors in $\mathscr{E}_2$ are *perpendicular* or *orthogonal* if the angle between them is 90°. The zero vector is taken to be *perpendicular* or *orthogonal* to every vector.

The zero vector is defined to be perpendicular to every vector for convenience, for then it is not necessary to continually refer to it as a special case. Although the definition of perpendicular has a good geometric interpretation, it is easier to use the following algebraic condition.

Theorem 1.3 Two vectors U and V in $\mathscr{E}_2$ are perpendicular if and only if $U \circ V = 0$.

Proof ($\Rightarrow$)* Suppose U and V are perpendicular. Then the angle θ between U and V is 90° and $\cos\theta = 0$, or either U or V is the zero vector and $\|U\|$ or $\|V\|$ is 0. In each case $U \circ V = \|U\|\|V\| \cos\theta = 0$.

($\Leftarrow$) Suppose $U \circ V = 0$. Then since $U \circ V = \|U\|\|V\| \cos\theta$, one of the three factors is 0. If $\|U\|$ or $\|V\|$ is 0, then U or V is the zero vector so U and V are perpendicular. If $\cos\theta = 0$, then $\theta = 90°$ and U is again perpendicular to V.

Example 4 Let ℓ be the line given by $P = t(2, 5) + (7, 1)$, $t \in R$. Find a vector equation for the line m passing through $(1, 0)$ which is perpendicular to ℓ.

ℓ has direction numbers $(2, 5)$, therefore the direction numbers (a, b) for m must satisfy $0 = (2, 5) \circ (a, b) = 2a + 5b$. There are many solutions for this equation besides $a = b = 0$. Taking $(a, b) = (5, -2)$ yields the equation $P = t(5, -2) + (1, 0)$, $t \in R$, for the line m.

Problems

1. Find the angles between the following pairs of vectors.
   a. $(1, 5), (-10, 2)$. b. $(1, -1), (-1, 3)$. c. $(1, \sqrt{3}), (\sqrt{6}, \sqrt{2})$.
2. Find a vector equation for the line passing through the given point and perpendicular to the given line.
   a. $(0, 0)$, $P = t(1, -3)$, $t \in R$. b. $(5, 4)$, $P = t(4, 4)$, $t \in R$.
   c. $(2, 1)$, $P = t(0, 7) + (4, 2)$, $t \in R$.

*The statement "P if and only if Q" contains two assertions; first "P only if Q" or "P implies Q" which may be denoted by $P \Rightarrow Q$ and second "P if Q" or "P is implied by Q" which may be denoted by $P \Leftarrow Q$. Each assertion is the *converse* of the other. An assertion and its converse are independent statements, therefore a theorem containing an if and only if statement requires two separate proofs.

3. Show that, in $\mathscr{E}_2$, the vectors (a, b) and $(b, -a)$ are perpendicular.
4. The proof of Theorem 1.3 uses the fact that if $U \in \mathscr{E}_2$ and $\|U\| = 0$, then $U = \mathbf{0}$. Show that this is true.
5. Give a vector proof of the Pythagorean theorem without the use of components. (The proof is simplest if the vertex with the right angle represents the zero vector of $\mathscr{E}_2$.)
6. Show that the diagonals of a rhombus (a parallelogram with equal sides) are perpendicular without using components.
7. Without using components, prove that any triangle inscribed in a circle with one side a diameter is a right triangle having the diameter as hypotenuse. (Suppose the center of the circle represents the zero vector of $\mathscr{E}_2$.)

§6. Generalization of $\mathscr{V}_2$ and $\mathscr{E}_2$

At this time the examples provided by $\mathscr{V}_2$ and $\mathscr{E}_2$ are sufficient to begin a general discussion of vector spaces. But before leaving the geometric discussion, a few ideas concerning 3-dimensional Euclidean geometry should be covered. The vector techniques developed for two dimensions generalize to n dimensions as easily as to three. Thus $\mathscr{V}_2$ can be generalized in one process to an infinite number of vector spaces.

Definition An *ordered n-tuple* of numbers has the form $(a_1, \ldots, a_n)$ where $a_1, \ldots, a_n \in R$ and $(a_1, \ldots, a_n) = (b_1, \ldots, b_n)$ if and only if $a_1 = b_1, \ldots, a_n = b_n$.

Thus when $n = 1$, the ordered n-tuple is essentially a real number; when $n = 2$, it is an ordered pair; and when $n = 3$, it is an ordered triple.

Definition For each positive integer n, the *vector space* $\mathscr{V}_n$ consists of all ordered n-tuples of real numbers, called *vectors*, together with the operation of *vector addition*:

$$(a_1, \ldots, a_n) + (b_1, \ldots, b_n) = (a_1 + b_1, \ldots, a_n + b_n)$$

and the operation of *scalar multiplication*:

$$r(a_1, \ldots, a_n) = (ra_1, \ldots, ra_n) \qquad \text{for any} \quad r \in R.$$

This definition gives an infinite number of vector spaces, one for each positive integer. Thus in the vector space $\mathscr{V}_4$,

$$(-2, 0, 5, 3) + (8, -3, -6, 1) = (6, -3, -1, 4).$$

And in $\mathscr{V}_6$,

$$2(0, 1, 0, -7, 3, 4) = (0, 2, 0, -14, 6, 8).$$

The numbers in an ordered n-tuple from $\mathscr{V}_n$ will be called *components*. Thus 4 is the second component of the vector $(3, 4, \sqrt{2}, 0, 1)$ from $\mathscr{V}_5$.

Since the vectors of $\mathscr{V}_3$ are ordered triples, they can be represented by points in Cartesian 3-space. Such a space has three mutually perpendicular coordinate axes with each pair of axes determining a coordinate plane. A point P is given coordinates (a, b, c) if a, b, and c are the directed distances from the three coordinate planes to P. If a rectangular box is drawn at the origin with height, width, and depth equal to $|c|$, $|b|$, and $|a|$, then P is at the vertex furthest from the origin, as in Figure 13. Now P can be used as a pictorial representation of the vector in $\mathscr{V}_3$ with components a, b, and c.

Since we live in a 3-dimensional world, we cannot draw such pictures for vectors from $\mathscr{V}_n$ when $n > 3$. This does not affect the usefulness of these spaces, but at first you may wish to think of $\mathscr{V}_n$ as either $\mathscr{V}_2$ or $\mathscr{V}_3$.

The vector notation used in the previous sections is free of components. That is, capital letters were used to denote vectors rather than ordered pairs. For example, the midpoint formula was written as $M = \frac{1}{2}(A + B)$ and the equation of a line as $P = tU + V$, $t \in R$. Now that there are other vector

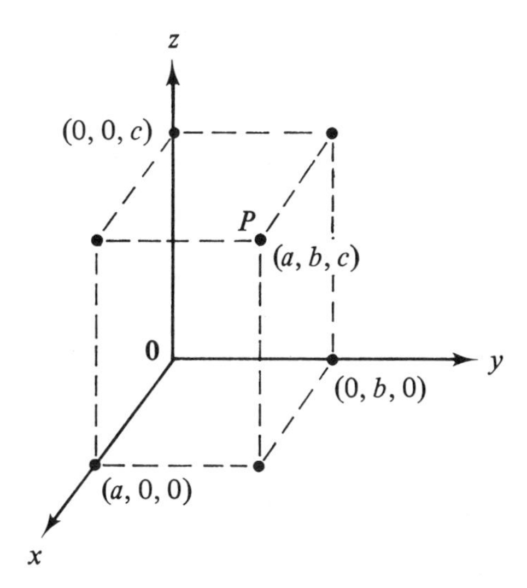

Figure 13

spaces, the question arises, when can these vectors be interpreted as coming from $\mathscr{V}_n$? The answer is that all the general statements about $\mathscr{V}_2$ carry over to $\mathscr{V}_n$. Thus it can be shown that addition in $\mathscr{V}_n$ is associative and commutative or that $r(U + V) = rU + rV$ for all $U, V \in \mathscr{V}_n$ and $r \in R$. Everything carries over directly to $\mathscr{V}_n$.

The properties of a line are the same in all dimensions. A line is either determined by two points or there is a unique line parallel to a given line through a given point. Therefore, the various vector statements about lines in the plane can be regarded as statements about lines in higher dimensions. This means that the set of all scalar multiples of a nonzero vector from $\mathscr{V}_n$ can be pictured as a line through the origin in n-dimensional Cartesian space.

Example 1 The three coordinate axes of 3-space have vector equations $P = (t, 0, 0)$, $P = (0, t, 0)$, and $P = (0, 0, t)$ as t runs through the real numbers. The line passing through the point with coordinates $(1, 3, -2)$ and the origin has the vector equation $P = t(1, 3, -2)$, $t \in R$.

In particular, each nonzero vector in $\mathscr{V}_n$ gives *direction numbers* for a set of parallel lines in n-dimensional Cartesian space. Notice that the measurement of slope used in the plane is a ratio requiring only two dimensions and cannot be applied to lines in three or more dimensions.

Example 2 The graph of the vector equation $P = t(1, 3, -2) + (4, 0, 1)$, $t \in R$ is the line with direction numbers $(1, 3, -2)$, passing through the point with coordinates $(4, 0, 1)$. It is also parallel to the line considered in Example 1.

Euclidean properties, that is, length and angle, are again introduced using the dot product.

Definition The *dot product* of the n-tuples $U = (a_1, \ldots, a_n)$ and $V = (b_1, \ldots, b_n)$ is $U \circ V = a_1b_1 + \cdots + a_nb_n$. The algebraic system consisting of all ordered n-tuples, vector addition, scalar multiplication, and the dot product will be denoted by $\mathscr{E}_n$.

The space $\mathscr{E}_2$ was obtained from $\mathscr{V}_2$ using the plane E^2 as a guide, and 3-dimensional geometric ideas will provide a basis for viewing properties of $\mathscr{E}_3$. But for higher dimensions there is little to build on. Therefore, for dimensions larger than three the procedure can be reversed, and properties of n-dimensional Cartesian space can be found as geometric properties of $\mathscr{E}_n$. That is, we can gain geometric insights by studying the vector spaces $\mathscr{E}_n$ when $n > 3$.

As with $\mathscr{V}_2$ and the spaces $\mathscr{V}_n$, the properties of $\mathscr{E}_2$ carry over directly to $\mathscr{E}_n$. The terms length, angle, and perpendicular are defined in just the same way. Thus the vectors $(4, 1, 0, 2)$ and $(-1, 2, 7, 1)$ are perpendicular in $\mathscr{E}_4$ because

$$(4, 1, 0, 2) \circ (-1, 2, 7, 1) = -4 + 2 + 0 + 2 = 0.$$

Arrows, triangles, and parallelograms look the same in two, three, or higher dimensions, so the geometric results obtained in the plane also hold in each of the other spaces.

Example 3 Find a vector equation for the line ℓ in 3-space passing through the points with coordinates $(2, 5, -4)$ and $(7, -1, 2)$.

Direction numbers for ℓ are given by $(7, -1, 2) - (2, 5, -4) = (5, -6, 6)$ and ℓ passes through the point $(7, -1, 2)$. Therefore, $P = t(5, -6, 6) + (7, -1, 2)$, $t \in R$ is a vector equation for ℓ. As in the 2-dimensional case, there are many other equations for ℓ. But notice that setting $P = (x, y, z)$ yields three parametric equations

$$x = 5t + 7, \qquad y = -6t - 1, \qquad z = 6t + 2, \qquad t \in R.$$

It is not possible to eliminate t from these three equations and obtain a single Cartesian equation. In fact three equations are obtained by eliminating t, which may be written in the form

$$\frac{x-7}{5} = \frac{y+1}{-6} = \frac{z-2}{6}.$$

These are called *symmetric equations* for ℓ. They are the equations of three planes that intersect in ℓ.

Next in complexity after points and lines in 3-space come the planes. We will use the following characterization of a plane in 3-space: A plane π is determined by a normal direction N and a point A, that is, π consists of all lines through A perpendicular to N. This is a perfect situation for the vector space approach. If P is any point on π, then viewing the points as vectors in $\mathscr{E}_3$, $P - A$ is perpendicular to N. This gives rise to a simple vector equation, for P is on the plane π if and only if $(P - A) \circ N = 0$. The direction N is

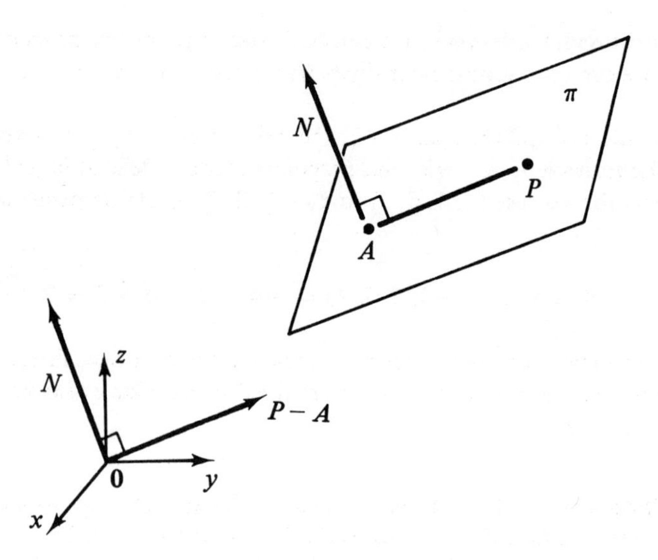

Figure 14

called a *normal* to the plane, and it makes a nice picture if N is represented by an arrow perpendicular to π, as in Figure 14. However, it must be remembered that N is an ordered triple not an arrow.

Example 4 Find a vector equation and a Cartesian equation for the plane passing through the point $A = (1, -2, 3)$ with normal $N = (4, -1, 2)$.

A vector equation for this plane is

$$[P - (1, -2, 3)] \circ (4, -1, 2) = 0.$$

Now suppose the arbitrary point P has coordinates (x, y, z). Then the vector equation

$$[(x, y, z) - (1, -2, 3)] \circ (4, -1, 2) = 0$$

becomes

$$(x - 1)4 + (y + 2)(-1) + (z - 3)2 = 0$$

or

$$4x - y + 2z = 12.$$

That is, the Cartesian equation of this plane is a first degree or linear equation in the coordinates.

This example suggests that the graph of a first degree equation in 3-space is a plane not a line.

Theorem 1.4 The Cartesian equation of a plane in 3-space is a linear equation. Conversely, the locus of points (x, y, z) satisfying the linear equation $ax + by + cz = d$ is a plane with normal (a, b, c).

Proof The first part is left to the reader, see problem 7 at the end of this section. For the converse, suppose $A = (x_0, y_0, z_0)$ is a point on the locus, then $ax_0 + by_0 + cz_0 = d$. If $P = (x, y, z)$ is a general point on the locus, then $ax + by + cz = d$, and putting the equations together gives

$$ax + by + cz = ax_0 + by_0 + cz_0$$

or

$$a(x - x_0) + b(y - y_0) + c(z - z_0) = 0$$

or

$$(a, b, c) \circ [(x, y, z) - (x_0, y_0, z_0)] = 0,$$

which is $N \circ (P - A) = 0$ when $N = (a, b, c)$. That is, P satisfies the equation of the plane passing through the point A with normal N.

The point of Theorem 1.4 is that planes not lines have linear Cartesian equations in $\mathscr{E}_3$. Since lines in the plane have linear equations, it is often assumed that lines in 3-space must also have linear equations. But a linear equation should not be associated with a line, rather it is associated with the figure determined by a point and a normal direction.

Definition A *hyperplane* in n-space is the set of all points P satisfying the vector equation $(P - A) \circ N = 0$, where $P, A, N \in \mathscr{E}_n$ and $N \neq \mathbf{0}$.

For $n = 3$ a hyperplane is simply a plane in $\mathscr{E}_3$.

Every hyperplane has a linear Cartesian equation and conversely. For P, N, A in the equation $(P - A) \circ N = 0$, let $P = (x_1, \ldots, x_n)$, $N = (a_1, \ldots, a_n)$, and $A \circ N = d$. Then $(P - A) \circ N = 0$ or $P \circ N = A \circ N$ becomes $a_1x_1 + \cdots + a_nx_n = d$, a general linear equation in the n coordinates of P. The converse follows just as in the proof of Theorem 1.4.

This means that a line in the plane E^2 can be regarded as a hyperplane.

For example, consider the line ℓ in E^2 with equation $y = mx + b$.

$$\begin{aligned} 0 = y - mx - b &= (x, y) \circ (-m, 1) - b \\ &= (x, y) \circ (-m, 1) - (0, b) \circ (-m, 1) \\ &= [(x, y) - (0, b)] \circ (-m, 1) = (P - A) \circ N. \end{aligned}$$

Here $A = (0, b)$ is a point on ℓ and $N = (-m, 1)$ is the direction perpendicular to ℓ.

Problems

1. Perform the indicated operations and determine the vector space from which the vectors come.

a. $(2, -1, 1) + (6, 2, 5)$.
b. $4(2, -3, 1, 0) + 3(-1, 4, 2, 6)$.
c. $(2, 4, -1)\circ(3, -3, -6)$.
d. $(1, 3, 2, 0, 1) - (1, 4, 3, 6, 5)$.
e. $(2/3, -1/3, 2/3)\circ(2/3, -1/3, 2/3)$.
f. $(3, 1, -4, 2)\circ(-1, 1, 3, 4)$.

2. Prove that $U + V = V + U$ when $U, V \in \mathscr{V}_3$.

3. Prove that $(r + s)U = rU + sU$ when $r, s \in R$ and $U \in \mathscr{V}_n$.

4. Find the lengths of the following vectors from $\mathscr{E}_3$:

a. $(1/2, 1/2, 1/2)$. b. $(1, 3, -2)$. c. $(1/3, 2/3, -2/3)$.

5. What is the angle between a diagonal of a cube in 3-space and an edge of the cube?

6. Find a vector equation for the line passing through the following pairs of points:

a. $(7, 2, -4), (2, -1, 4)$.
b. $(-4, 3, 0), (2, 1, 7)$.
c. $(1, 3, 0, 2), (0, -1, 2, 0)$.
d. $(3, 7, -1, -5, 0), (1, -1, 4, 2, 5)$.

7. Show that the Cartesian equation of a plane in 3-space is linear.

8. Use the corner of a room, the floor and two walls, to visulaize the three coordinate planes in Cartesian 3-space. Determine where the points with coordinates $(2, 1, 4)$, $(2, -1, 4)$, and $(2, 1, -4)$ would be located if the coordinates are given in feet. Then determine where the graphs with the following equations would be located.

a. $P = (0, 0, t)$, $t \in R$. b. $z = 0$. c. $z = 4$. d. $x = 3$.
e. $P = t(0, 5, -5) + (0, 3, 0)$, $t \in R$. f. $y = -1$. g. $y = x$.
h. $P = t(1, 1, 0)$, $t \in R$. i. $x + y + z = 3$.

9. Find vector and Cartesian equations for the hyperplane with given normal N passing through the given point A.

a. $N = (0, 0, 1)$, $A = (1, 2, 0)$.
b. $N = (-3, 2, 1)$, $A = (4, 0, 8)$.
c. $N = (2, -3)$, $A = (1, 1)$.
d. $N = (1, 0)$, $A = (-7, 0)$.
e. $N = (3, 1, 0, 2)$, $A = (2, 1, 1, 0)$.

10. What is the graph of the Cartesian equation $y = 5$:

a. in E^2? b. in 3-space? c. in 4-space?

11. Determine if the following pairs of lines intersect. That is, are there values for the parameters t and s which give the same point?

a. $P = t(2, 1, -5)$, $P = s(4, 3, 0)$, $t, s \in R$
b. $P = t(1, -1, 2)$, $P = s(1, 3, -2) + (4, 4, 0)$, $t, s \in R$.
c. $P = t(1, 3, 0) + (2, 0, 1)$, $P = s(4, -1, 3) + (1, 1, 2)$, $t, s \in R$.
d. $P = t(1, 0, 3) + (4, -5, -6)$, $P = s(-3, 2, 6) + (0, 2, 0)$, $t, s \in R$.
e. $P = t(1, -2, -1) + (3, -3, -1)$, $P = s(3, 0, -4) + (4, 1, -3)$, $t, s \in R$.

12. Write equations in the form $(P - A) \circ N = 0$ for the lines in the plane with the following Cartesian or vector equations:

a. $y = 4x - 3$. b. $y = 7$. c. $P = t(3, 2) + (6, 1)$, $t \in R$.
d. $P = t(1, 3) + (1, 1)$, $t \in R$. e. $P = (t, t)$, $t \in R$.

13. A vector $U \in \mathscr{E}_n$ is a *unit vector* if $\|U\| = 1$.

a. Show that for any nonzero vector $U \in \mathscr{E}_n$, $U/\|U\|$ is a unit vector.
b. Obtain unit vectors from (2, 2, 1), (3, 4), (1, 1, 1).

14. Let U be direction numbers for a line ℓ in 3-space. Show that if U is a unit vector then its three components are the cosines of the angles between U and the three positive coordinate axes. (The components of U are called *direction cosines* of the line ℓ.)

15. Find direction cosines for the line through
a. (2, 4, 6) and (3, 2, 4). b. (2, 3, 5) and (3, 3, 6). c. (0, 4, 1) and (0, 9, 1).

16. Three vertices of the box in Figure 13 are not labeled. What are the coordinates of these three points?

17. How might the vectors of $\mathscr{V}_1$ be represented geometrically?

2

Real Vector Spaces

The space $\mathscr{V}_2$ was introduced to illustrate the essential properties to be included in the definition of an abstract vector space. Since this space itself appeared abstract and uninteresting, the dot product was added yielding $\mathscr{E}_2$ and a new approach to Euclidean geometry. We now return to the simple properties of $\mathscr{V}_2$ which lead to many different spaces and ideas; it will be some time before we again find the need to add more algebraic structure.

§1. Definition and Examples

Definition A *real vector space* $\mathscr{V}$ is a set of elements $U, V, W, \ldots$, called *vectors*, together with the following algebraic structure:

Addition: The *sum* $U + V$ is defined for every pair of vectors and satisfies

1. $U + V \in \mathscr{V}$ (Closure)
2. $U + (V + W) = (U + V) + W$ (Associative Law)
3. $U + V = V + U$ (Commutative Law)
4. There exists a *zero vector* $\mathbf{0} \in \mathscr{V}$ such that $U + \mathbf{0} = U$ for all $U \in \mathscr{V}$.
5. For every vector $U \in \mathscr{V}$ there is an *additive inverse* $-U \in \mathscr{V}$ such that $U + (-U) = \mathbf{0}$.

Scalar multiplication: The scalar multiple rU is defined for every scalar $r \in R$ and every vector $U \in \mathscr{V}$, and satisfies

1. $rU \in \mathscr{V}$ (Closure)
2. $r(U + V) = rU + rV$
   $(r + s)U = rU + sU$ (Distributive Laws)
3. $(rs)U = r(sU)$ (Associative Law)
4. $1U = U$, 1 the multiplicative identity in R.

The word "real" in the term real vector space indicates that the scalars are real numbers. It would be possible to choose the scalars from some other field such as the complex number system, in which case the system would be called a "complex vector space." One advantage of using the term scalar instead of real number is that the study of real vector spaces can easily be generalized to complex vector spaces by simply changing the meaning of the term scalar. However, at this point little would be gained by allowing the scalars to come from some other field, and since only real vector spaces will be under consideration until Chapter 6, the adjective real will be dropped for convenience.

The definition of a vector space is abstract in that it says nothing about what vectors are; it only lists a few basic properties. The fact that $\mathcal{V}_2$ is a vector space, as defined above, is contained in a theorem and a problem found in the preceding chapter. But many other algebraic systems also satisfy the definition.

Example 1 The n-tuple spaces $\mathcal{V}_n$ are all vector spaces as defined above. The proof of this is very similar to the proof for $\mathcal{V}_2$, the only difference being in the number of components to a vector. Since $\mathcal{E}_n$ is obtained from $\mathcal{V}_n$ by adding the dot product, $\mathcal{E}_n$ is a vector space for each positive integer n.

Example 2 Consider the set of all arrows or directed line segments $\overrightarrow{OP}$ from the origin O to the point P in the plane E^2. Define addition of arrows with the parallelogram rule for addition. To define scalar multiplication, suppose the length of $\overrightarrow{OP}$ is x. For $r \in R$, let $r\overrightarrow{OP}$ be

i. the arrow in the same direction as OP with length rx if $r > 0$;
ii. the origin if $r = 0$;
iii. the arrow in the opposite direction from $\overrightarrow{OP}$ with length $-rx$ if $r < 0$.

Using a little plane geometry one can show that the set of all arrows together with these two operations satisfies the definition of a vector space. Notice that this vector space is neither $\mathcal{V}_2$ nor $\mathcal{E}_2$.

Example 3 The real number system R is a vector space. In fact, the conditions in the definition of a vector space become the field properties of R when U, V, and W are taken to be real numbers. Thus a real number can be a vector.

Example 4 Let $R[t]$ denote the set of all polynomials in t with real coefficients, for example, $2/3 + 5t - 4t^5 \in R[t]$. With the usual algebraic rules for adding polynomials and multiplying a polynomial by a real number,

$R[t]$ is a vector space. To prove this it is necessary to prove that all nine properties of addition and scalar multiplication hold in $R[t]$. For example, to prove commutativity of addition it is necessary to show that $P + Q = Q + P$ when

$$P = a_0 + a_1 t + \cdots + a_n t^n \text{ and } Q = b_0 + b_1 t + \cdots + b_m t^m, \quad a_i, b_j \in R.$$

This follows from the fact that addition in R is commutative, for the term $(a_k + b_k)t^k$ in the sum $P + Q$ becomes $(b_k + a_k)t^k$, a term in $Q + P$. The zero vector of $R[t]$ is the number 0, since the elements of $R[t]$ are not polynomial equations, a polynomial such as $3 + t$ cannot be zero.

This example begins to show how general the term "vector" is, for in this space polynomials are vectors. Thus $5 - 7t^3 + 2t^6$ is a vector.

Example 5 Consider the set of all complex numbers $a + bi$, with $a, b \in R$, $i = \sqrt{-1}$ and equality given by $a + bi = c + di$ if $a = c$, $b = d$. If addition is defined in the usual way,

$$(a + bi) + (c + di) = a + c + (b + d)i$$

and scalar multiplication is defined by

$$r(a + bi) = ra + (rb)i \qquad \text{for} \quad r \in R,$$

then the complex numbers form a (real) vector space, see problem 3 at the end of this section.

Example 6 Let $\mathcal{F}$ denote the set of all continuous functions with domain $[0, 1]$ and range in R. To make $\mathcal{F}$ into a vector space it is necessary to define addition and scalar multiplication and show that all the nine properties hold. For f and g in $\mathcal{F}$ define $f + g$ by

$$(f + g)(x) = f(x) + g(x) \qquad \text{for all} \quad x \in [0, 1].$$

It is proved in elementary calculus that the sum of continuous functions is continuous, therefore $\mathcal{F}$ is closed under addition. Commutativity follows from commutativity of addition in R for if $f, g \in \mathcal{F}$,

$$\begin{aligned} (f + g)(x) &= f(x) + g(x) && \text{Definition of addition in } \mathcal{F} \\ &= g(x) + f(x) && \text{Addition is commutative in } R \\ &= (g + f)(x) && \text{Definition of addition in } \mathcal{F}. \end{aligned}$$

Now two functions in $\mathscr{F}$ are equal if and only if they give the same value for each value of $x \in [0, 1]$, therefore $f + g = g + f$. Associativity of addition follows from the fact that addition in R is associative in the same way.

Let $\mathbf{0}$ denote the zero function in $\mathscr{F}$, that is $\mathbf{0}(x) = 0$ for all $x \in [0, 1]$. Then $\mathbf{0}$ is the zero vector of $\mathscr{F}$, for given any $f \in \mathscr{F}$,

$$(f + \mathbf{0})(x) = f(x) + \mathbf{0}(x) = f(x) + 0 = f(x),$$

therefore $f + \mathbf{0} = f$. Finally if $-f$ is defined by $(-f)(x) = -[f(x)]$, one can show that $-f$ is the additive inverse of f.

Scalar multiplication is defined in $\mathscr{F}$ by

$$(rf)(x) = r[f(x)] \qquad \text{for any} \quad f \in \mathscr{F} \quad \text{and} \quad r \in R.$$

Aside from closure, the properties of scalar multiplication which must be satisfied in $\mathscr{F}$, if it is to be a vector space, are proved using field properties of the real number system. See problem 5 at the end of this section.

Therefore $\mathscr{F}$ is a vector space, and such continuous functions as $x^2 + 4x$, e^x, $\sin x$ and $\sqrt{x^3 + 1}$, $x \in [0, 1]$, are vectors.

Vector spaces similar to $\mathscr{F}$ can be obtained in endless variety. For example, the condition that the functions be continuous can be dropped or changed, say to differentiable, or the domain can be changed. In fact, other vector spaces, besides R, could be used for the domain or range, and in time such vector spaces of functions will become our main concern.

Example 7 A set containing only one element, say Z, can be made into a vector space by defining $Z + Z = Z$ and $rZ = Z$ for all $r \in R$. The conditions in the definition of a vector space are easily satisfied. Since every vector space must have a zero vector, Z is that vector, and this simple vector space is called a *zero vector space*.

Example 8 Consider the set of all sequences $\{a_n\}$ of real numbers which converge to zero, that is $\lim_{n \to \infty} a_n = 0$. With addition and scalar multiplication defined by

$$\{a_n\} + \{b_n\} = \{a_n + b_n\} \qquad \text{and} \qquad r\{a_n\} = \{ra_n\}.$$

this set becomes a vector space. Thus the sequences $\{1/n\}$ or $1, 1/2, 1/3, 1/4, \ldots$ and $\{(-1/2)^n\}$ or $-1/2, 1/4, -1/8, 1/16, \ldots$ can be regarded as vectors.

Other vector spaces can be constructed similar to this one. For example, the set of all convergent infinite series could be used. In which case $\sum_{n=1}^{\infty} (-1)^n \, 1/n = -1 + 1/2 - 1/3 + 1/4 - 1/5 + \cdots$ would be a vector.

The above examples should give an idea of the diversity of vector spaces. It is because of this diversity that an abstract vector space will be studied rather than a particular example. For the proof that a certain condition holds for the ordered pairs of $\mathscr{V}_2$ cannot be used to claim that the condition holds for the functions of $\mathscr{F}$ or the polynomials of $R[t]$. On the other hand, a condition shown to hold in an abstract vector space $\mathscr{V}$, also holds in $\mathscr{V}_2$, $\mathscr{F}$, and $R[t]$, as well as in vector spaces constructed after the proof is completed. But since the vector space $\mathscr{V}$ is abstract, nothing can be assumed about it without justification. Just how little is known about $\mathscr{V}$ is indicated by the fact that it must be assumed that $1U = U$ for all $U \in \mathscr{V}$. The vectors of $\mathscr{V}$ cannot be described. In fact, the term "vector" now only means "member of a vector space" and a vector might be anything from a number to a function.

It should be noticed that two symbols appear repeatedly in the above examples. The zero vector was always denoted by $\mathbf{0}$, even though it might be the number 0, an n-tuple of zeros, or a zero function. Although it is possible to denote the zero vector of a space $\mathscr{V}$ by $\mathbf{0}_{\mathscr{V}}$, the meaning of $\mathbf{0}$ is usually clear from the context in which it is used. The other symbol which can mean many things is the plus sign. Addition is a different operation in each vector space. The precise meaning of a plus sign must also be determined from the context.

Not all the examples presented above will be of equal interest. Aside from vector spaces to be obtained later, we will be interested primarily in $\mathscr{V}_n$ and $\mathscr{E}_n$, with many references to $R[t]$ and $\mathscr{F}$. An examination of the examples and the problems below will show that a system forms a vector space by virtue of properties of real numbers, or more generally, because the real number system is itself a vector space. Future examples of vector spaces will also be based on the properties of R or some other vector space.

Problems

1. Find the sum of the following pairs of vectors.

a. $(2, -4, 0, 2, 1), (5, 2, -4, 2, -9) \in \mathscr{V}_5$.
b. $2 + 6t^2 - 3t^4, 2t + 3t^4 \in R[t]$.
c. $2 - i, 4i - 7 \in C$.
d. $\cos x, 4 - \sqrt{x} \in \mathscr{F}$.

2. Find the additive inverse for each vector in problem 1.

3. Show that the complex number system is a vector space, as claimed in Example 5.

4. Verify that scalar multiplication in $R[t]$ satisfies the conditions required if $R[t]$ is to be a vector space.

5. a. In Example 6, fill in the reasons in the chain of equalities yielding $f + \mathbf{0} = f$.
   b. Show that $f + (-f) = \mathbf{0}$, $f \in \mathscr{F}$, with reasons.
   c. Verify that scalar multiplication in $\mathscr{F}$ satisfies the conditions necessary for $\mathscr{F}$ to be a vector space.

6. What is the zero vector in a. $\mathscr{V}_3$? b. $\mathscr{F}$? c. C? d. $R[t]$? e. $\mathscr{V}_5$? f. the space of sequences which converge to zero? g. the space of arrows in E^2 starting at the origin?

7. Each of the following systems fails to be a vector space. In each case find all the conditions in the definition of a vector space which fail to hold.
   a. The set of all ordered pairs of real numbers with addition defined as in $\mathscr{V}_2$ and scalar multiplication given by $r(a, b) = (ra, b)$.
   b. As in a except scalar multiplication is $r(a, b) = (ra, 0)$.
   c. The set of all ordered pairs with scalar multiplication as in $\mathscr{V}_2$ and addition given by $(a, b) + (c, d) = (a - c, b - d)$.
   d. The set of all functions f from R to R such that $f(0) = 1$, together with the operations of $\mathscr{F}$. ($\cos x$ and $x^3 + 3x + 1$ are in this set.)

8. A real 2×2 (read "two by two") matrix is defined to be an array of four numbers in the form $\begin{pmatrix} a & b \\ c & d \end{pmatrix}$, with $a, b, c, d \in R$. Using the vector spaces $\mathscr{V}_n$ as a guide, define addition and scalar multiplication for 2×2 matrices and prove that the resulting system is a vector space.

§2. Subspaces

In the study of an abstract algebraic system such as a vector space, subsystems which share the properties of the system are of major importance. The consideration of such subsystems in vector spaces will lead to the central concepts of linear algebra. An analogy can be drawn with geometry and the importance of lines in the plane or lines and planes in 3-space. Before stating the definition of a subspace, a little terminology concerning sets is needed.

Definition Let A and B be sets. A is a *subset* of B if every element of A is in B, that is, if $x \in A$ implies $x \in B$. If A is a subset of B, write $A \subset B$ for "A is contained in B" or $B \supset A$ for "B contains A."

For example, the set of all integers is a subset of R and $\{t^4, 3t + 2\} \subset R[t]$. In general a set contains two simple subsets, itself and the "empty set," the set with no elements.

Definition Two sets A and B are *equal* if $A \subset B$ and $B \subset A$.

Definition A subset $\mathscr{S}$ of a vector space $\mathscr{V}$ is a *subspace* of $\mathscr{V}$ if $\mathscr{S}$ is a vector space relative to the operations of $\mathscr{V}$. (*Note*: $\mathscr{S}$ is a Captal script S.)

Every vector space $\mathscr{V}$ contains two simple subspaces. Namely, $\mathscr{V}$ itself, since every set is a subset of itself, and $\{\mathbf{0}\}$, the set containing only the zero vector of $\mathscr{V}$. $\{\mathbf{0}\}$ is easily seen to be a vector space in its own right and is therefore a subspace of $\mathscr{V}$. $\{\mathbf{0}\}$ is called the *zero subspace* of $\mathscr{V}$ and should not be confused with the zero vector space defined in the previous section. These two subspaces are called *trivial* or *improper* subspaces of $\mathscr{V}$, and all other subspaces are called *nontrivial* or *proper* subspaces of $\mathscr{V}$. Thus $\mathscr{S}$ is a proper subspace of $\mathscr{V}$ if it contains some nonzero vectors but not all the vectors of $\mathscr{V}$.

Example 1 Let $\mathscr{S} = \{(0, r, 0) | r \in R\}$. Show that $\mathscr{S}$ is a proper subspace of $\mathscr{V}_3$.

The elements of $\mathscr{S}$ are ordered triples, so $\mathscr{S}$ is a subset of $\mathscr{V}_3$. It is not difficult to show, using the operations of $\mathscr{V}_3$, that $\mathscr{S}$ satisfies all nine conditions in the definition of a vector space. For example, suppose $U, V \in \mathscr{S}$, then there exist $r, s \in R$ such that $U = (0, r, 0)$, $V = (0, s, 0)$, and $U + V = (0, r + s, 0)$. Since R is closed under addition, $r + s \in R$ and the sum $U + V$ is in $\mathscr{S}$. That is, $\mathscr{S}$ is closed under addition. Since $U, V \in \mathscr{V}_3$ and addition in $\mathscr{V}_3$ is commutative, $U + V = V + U$ in $\mathscr{S}$. The remaining properties are proved similarly.

$\mathscr{S}$ is a proper subspace of $\mathscr{V}_3$ for the vector $(0, -5, 0) \in \mathscr{S}$ implying that $\mathscr{S} \neq \{\mathbf{0}\}$ and $(0, 0, 7) \notin \mathscr{S}$ but $(0, 0, 7) \in \mathscr{V}_3$ so $\mathscr{S} \neq \mathscr{V}_3$.

Applying the definition directly is not the best way to determine if a subset is a subspace. However, a few basic facts are needed before obtaining a simpler method.

Theorem 2.1 Let $\mathscr{V}$ be a vector space. Then for all vectors $U \in \mathscr{V}$,

1. $-U$ and $\mathbf{0}$ are unique.
2. $0U = \mathbf{0}$ and $r\mathbf{0} = \mathbf{0}$, for all $r \in R$.
3. If $rU = \mathbf{0}$, then either $r = 0$ or $U = \mathbf{0}$.
4. $-U = (-1)U$.

Proof 1. To show that $\mathbf{0}$ is unique, suppose there is a vector X in $\mathscr{V}$ such that $U + X = U$ for all $U \in \mathscr{V}$. Consider $X + \mathbf{0}$. By the definition of $\mathbf{0}$, $X + \mathbf{0} = X$. But by the commutativity of addition, $X + \mathbf{0} = \mathbf{0} + X$ and $\mathbf{0} + X = \mathbf{0}$ by assumption. Therefore $X = \mathbf{0}$.

To show that the additive inverse of U is unique, suppose $U + Y = \mathbf{0}$ for some $Y \in \mathscr{V}$. Then

$$-U = -U + \mathbf{0} = -U + (U + Y) = ((-U) + U) + Y$$
$$= (U + (-U)) + Y = \mathbf{0} + Y = Y + \mathbf{0} = Y.$$

2. To show that $0U = \mathbf{0}$ note that $0U = (0 + 0)U = 0U + 0U$. Therefore

$$\mathbf{0} = 0U - 0U = (0U + 0U) - 0U = 0U + (0U - 0U) = 0U + \mathbf{0} = 0U.$$

Can you find a similar proof for the statement $r\mathbf{0} = \mathbf{0}$?

3. Left to the reader.

4. To prove that $-U = (-1)U$, it need only be shown that $(-1)U$ acts like the additive inverse of U since by part 1 the additive inverse of U is unique. But $U + (-1)U = 1U + (-1)U = (1 + (-1))U = 0U = \mathbf{0}$. Therefore $(-1)U$ is the additive inverse of U, $-U$.

Parts of Theorem 2.1 were obtained for $\mathscr{V}_2$ in problems 7, 8, and 10 on page 7, however, the theorem is stated for an abstract vector space. Therefore these are properties of any algebraic system satisfying the definition of a vector space be it $\mathscr{V}_n$, $R[t]$, or a system to be defined at some future time.

Definition Let S be a subset of a vector space $\mathscr{V}$. S is *closed under addition* if $U + V \in S$ for all $U, V \in S$, where $+$ is the addition in $\mathscr{V}$. S is *closed under scalar multiplication* if $rU \in S$ for all $U \in S$ and $r \in R$ where scalar multiplication is as defined in $\mathscr{V}$.

Example 2 Show that $S = \{(a, 0) | a \in R, a > 0\} \subset \mathscr{V}_2$ is closed under addition but not scalar multiplication.

Let $U, V \in S$, then $U = (x, 0)$, $V = (y, 0)$ for some $x > 0$ and $y > 0$. The sum $x + y > 0$ so $(x + y, 0) \in S$. But $(x + y, 0) = U + V$, therefore, S is closed under addition. However, $(-1)U = (-x, 0)$ and $-x < 0$. Therefore the scalar multiple $(-1)U \notin S$, and S is not closed under scalar multiplication.

The following theorem shows that it is only necessary to verify three conditions to prove that a subset of a vector space is a subspace.

Theorem 2.2 $\mathscr{S}$ is a subspace of a vector space $\mathscr{V}$ if the following conditions hold:

1. $\mathscr{S}$ is a nonempty subset of $\mathscr{V}$.
2. $\mathscr{S}$ is closed under addition.
3. $\mathscr{S}$ is closed under scalar multiplication.

Proof Since it is given that $\mathscr{S}$ is closed under addition and scalar multiplication, it must be shown that $\mathscr{S}$ satisfies the remaining seven conditions in the definition of a vector space. Most of these follow because $\mathscr{S}$ is a subset of $\mathscr{V}$. For example, if $U, V \in \mathscr{S}$, then $U, V \in \mathscr{V}$ and since $\mathscr{V}$ is a vector space, $U + V = V + U$, therefore, addition is commutative in $\mathscr{S}$. Similarly, associativity of addition and the remaining properties of scalar multiplication hold in $\mathscr{S}$. So it only remains to show that $\mathscr{S}$ has a zero vector and that each element of $\mathscr{S}$ has an additive inverse in $\mathscr{S}$.

By part 1 there exists a vector $V \in \mathscr{S}$ and by part 3, $0V \in \mathscr{S}$, but $0V = \mathbf{0}$ by Theorem 2.1, so $\mathbf{0} \in \mathscr{S}$. Now suppose U is any vector in $\mathscr{S}$. By 3, $(-1)U \in \mathscr{S}$, and $(-1)U = -U$ by Theorem 2.1, so $-U \in \mathscr{S}$. Therefore $\mathscr{S}$ is a vector space and hence a subspace of $\mathscr{V}$.

Example 3 For a positive integer n, let $R_n[t]$ denote the set of all polynomials in t of degree less than n. $R_n[t]$ is a subset of $R[t]$, and since $t^{n-1} \in R_n[t]$, it is nonempty. Addition and scalar multiplication cannot increase the degree of polynomials, so $R_n[t]$ is closed under the operations of $R[t]$. Therefore, for each positive integer n, $R_n[t]$ is a subspace of $R[t]$.

Example 4 Show that $\mathscr{W} = \{f \in \mathscr{F} | f(1/3) = 0\}$ is a subspace of the vector space of functions $\mathscr{F}$.

$\mathscr{W}$ is nonempty since the function $f(x) = 18x^2 - 2$ is a member. To prove that $\mathscr{W}$ is closed under the operations of $\mathscr{F}$, let $f, g \in \mathscr{W}$. Then $f(1/3) = g(1/3) = 0$ so $(f + g)(1/3) = f(1/3) + g(1/3) = 0 + 0 = 0$. Therefore $f + g \in \mathscr{W}$, and $\mathscr{W}$ is closed under addition. For any $r \in R$, $(rf)(1/3) = r(f(1/3)) = r0 = 0$, so $rf \in \mathscr{W}$ and $\mathscr{W}$ is closed under scalar multiplication. Thus $\mathscr{W}$ satisfies the hypotheses of Theorem 2.2 and is a subspace of $\mathscr{F}$. In this example it should be clear that the choice of 1/3 was arbitrary while the choice of $0 = f(1/3)$ was not.

Example 5 Show that $\mathscr{S} = \{f \in \mathscr{F} | f \text{ is differentiable and } f' = f\}$ is a subspace of $\mathscr{F}$. Note that $f \in \mathscr{S}$ if and only if $y = f(x)$ is a solution of the differential equation $dy/dx = y$.

The zero function clearly satisfies $dy/dx = y$, so $\mathscr{S}$ is nonempty. If $f, g \in \mathscr{S}$, then $(f + g)' = f' + g'$ so $f' = f$ and $g' = g$ implies $(f + g)' = f + g$. That is, $\mathscr{S}$ is closed under addition. Similarly for any scalar r, $(rf)' = rf'$, so that if $f' = f$, then $(rf)' = rf$ and $\mathscr{S}$ is closed under scalar multiplication also. Therefore, $\mathscr{S}$ is a subspace of $\mathscr{F}$, in fact, $\mathscr{S} = \{ce^x | c \in R\}$.

It is not always the case that the solutions of a differential equation form a vector space, but when they do all the theory of vector spaces can be applied.

Example 6 If $\mathscr{E}_n$ is viewed as Euclidean n-space, then lines through the origin are subspaces of $\mathscr{E}_n$.

A line through the origin is given by all scalar multiples of some nonzero vector $U \in \mathscr{E}_n$. Therefore the claim is that $\{rU | r \in R\}$ is a subspace of $\mathscr{E}_n$. The set is nonempty because it contains U. Since $rU + sU = (r + s)U$ and $r(sU) = (rs)U$ for all $r, s \in R$, the set is closed under the operations of $\mathscr{E}_n$. Therefore each line through the origin is a subspace of $\mathscr{E}_n$.

Example 7 $\mathscr{T} = \{a(2, 1, 3) + b(1, 4, 1) + c(1, -3, 2) | a, b, c \in R\}$ is a proper subspace of $\mathscr{V}_3$.

$\mathscr{T} \subset \mathscr{V}_3$ since $\mathscr{V}_3$ is closed under addition and scalar multiplication. $\mathscr{T}$ is nonempty for $(2, 1, 3) \in \mathscr{T}$ when $a = 1, b = c = 0$. For closure suppose $U, V \in \mathscr{T}$, then there exist scalars such that

$$U = a_1(2, 1, 3) + b_1(1, 4, 1) + c_1(1, -3, 2)$$

and

$$V = a_2(2, 1, 3) + b_2(1, 4, 1) + c_2(1, -3, 2).$$

So

$$U + V = [a_1 + a_2](2, 1, 3) + [b_1 + b_2](1, 4, 1) + [c_1 + c_2](1, -3, 2)$$

and

$$rU = [ra_1](2, 1, 3) + [rb_1](1, 4, 1) + [rc_1](1, -3, 2).$$

Therefore the vectors $U + V$ and rU satisfy the condition for membership in $\mathscr{T}$, and $\mathscr{T}$ is closed under addition and scalar multiplication. Thus $\mathscr{T}$ is a subspace of $\mathscr{V}_3$.

Since $\mathscr{T} \neq \{\mathbf{0}\}$, $\mathscr{T}$ is a proper subspace of $\mathscr{V}_3$ if not every ordered triple is in $\mathscr{T}$. That is, given $(r, s, t) \in \mathscr{V}_3$, do there exist scalars a, b, c such that

$$a(2, 1, 3) + b(1, 4, 2) + c(1, -3, 2) = (r, s, t)?$$

This vector equation yields the following system of linear equations:

$$\begin{aligned} 2a + b + c &= r \\ a + 4b - 3c &= s \\ 3a + b + 2c &= t. \end{aligned}$$

If c is eliminated from the first pair and last pair of equations, then

$$\begin{aligned} 7a + 7b &= s + 3r \\ 11a + 11b &= 2s + 3t \end{aligned}$$

is obtained, and for some choices of r, s, t these equations have no solution. In particular, if $r = 1$, $s = t = 0$, no values for a and b can satisfy the equations. Therefore $(1, 0, 0) \notin \mathscr{T}$, and $\mathscr{T}$ is a proper subspace of $\mathscr{V}_3$.

Example 8 $\mathscr{U} = \{a(2, -5) + b(3, -7) \mid a, b \in R\}$ is the trivial subspace $\mathscr{V}_2$ of the vector space $\mathscr{V}_2$.

The proof that $\mathscr{U}$ is a subspace of $\mathscr{V}_2$ follows the pattern in Example 7. Assuming this is done, $\mathscr{U} \subset \mathscr{V}_2$ since $\mathscr{U}$ is a subspace of $\mathscr{V}_2$. Therefore to show that $\mathscr{U} = \mathscr{V}_2$, it is necessary to prove that $\mathscr{V}_2 \subset \mathscr{U}$. If $(x, y) \in \mathscr{V}_2$, then it is necessary to find scalars a and b such that $(x, y) = a(2, -5) + b(3, -7)$ or $x = 2a + 3b$ and $y = -5a - 7b$. These equations are easily solved for a and b in terms of x and y giving

$$(x, y) = [-7x - 3y](2, -5) + [5x + 2y](3, -7).$$

Therefore $(x, y) \in \mathscr{U}$. So $\mathscr{V}_2 \subset \mathscr{U}$ and $\mathscr{U} = \mathscr{V}_2$.

Problems

1. Show that the following sets are subspaces of the indicated spaces.
   a. $\{(a, b, 0) \mid a, b \in R\} \subset \mathscr{V}_3$.
   b. $\{(x, y, z) \mid x = 3y, x + y = z\} \subset \mathscr{V}_3$.
   c. $\{at^3 + bt \mid a, b \in R\} \subset R[t]$.
   d. $\{f \mid f(x) = a \sin x + b \cos x, a, b \in R\} \subset \mathscr{F}$.

2. Show that a plane through the origin in 3-space represents a subspace of $\mathscr{E}_3$.

3. Prove that if a subspace $\mathscr{S}$ of $\mathscr{V}_3$ contains the vectors (1, 1, 1), (0, 1, 1), and (0, 0, 1), then $\mathscr{S} = \mathscr{V}_3$.

4. Suppose a subspace $\mathscr{S}$ of $\mathscr{V}_3$ contains the following vectors, determine in each case if $\mathscr{S}$ can be a proper subspace of $\mathscr{V}_3$.
   a. (1, 0, 0), (0, 1, 0), and (0, 0, 1).
   b. (2, 1, 0) and (3, 0, 1).
   c. (1, 0, 2), (4, 0, −5), (0, 0, 1), and (6, 0, 0).
   d. (1, 1, 0), (1, 0, 1), and (0, 1, 1).

5. Write an arbitrary element from the following sets.
   a. $\{(x, y, z) | x + 4y - z = 0\}$.
   b. $\{(a, b, c, d) | 3a = 2c, b = 5d\}$.
   c. $\{(a, b, c) | a + b - c = 0 \text{ and } 3a - b + 5c = 0\}$.
   d. $\{(x, y, z) | x + 3y - 2z = 0 \text{ and } 2x + 7y - 6z = 0\}$.

6. Determine if the following are subspaces.
   a. $\{P \in R[t] | \text{degree of } P \text{ is } 3\} \subset R[t]$.
   b. $\{(a, b) | a \text{ and } b \text{ are integers}\} \subset \mathscr{V}_2$.
   c. $\{(2a + 3b, 4a + 6b) | a, b \in R\} \subset \mathscr{V}_2$.
   d. $\{(x, y) | y = x^2\} \subset \mathscr{V}_2$.
   e. $\{f \in \mathscr{F} | f(x) = ax^2 \text{ or } f(x) = ax^3 \text{ for some } a \in R\} \subset \mathscr{F}$.
   f. $\{(x, y, z) | x + y + z = 1\} \subset \mathscr{V}_3$.

7. Why is it impossible for a line that does not pass through the origin of the Cartesian plane to represent a subspace of $\mathscr{V}_2$?

8. Prove that the vector spaces $\mathscr{V}_1$ and R have no proper subspaces.

9. Determine if the following are subspaces of $\mathscr{F}$.
   a. $\mathscr{S} = \{f | f' = 0\}$. What is the set $\mathscr{S}$?
   b. $\mathscr{S} = \{f | f' + 3f + 5 = 0\}$.
   c. $\mathscr{S} = \{f | f'' + 4f' - 7f = 0\}$.
   d. $\mathscr{S} = \{f | [f']^2 = f\}$.

10. Suppose $\mathscr{S}$ and $\mathscr{T}$ are subspaces of a vector space $\mathscr{V}$. Define the *sum* of $\mathscr{S}$ and $\mathscr{T}$ by

$$\mathscr{S} + \mathscr{T} = \{U + V | U \in \mathscr{S}, V \in \mathscr{T}\},$$

the *intersection* of $\mathscr{S}$ and $\mathscr{T}$ by

$$\mathscr{S} \cap \mathscr{T} = \{U | U \in \mathscr{S} \text{ and } U \in \mathscr{T}\},$$

and the *union* of $\mathscr{S}$ and $\mathscr{T}$ by

$$\mathscr{S} \cup \mathscr{T} = \{U | U \in \mathscr{S} \text{ or } U \in \mathscr{T}\}.$$

   a. Prove that the sum $\mathscr{S} + \mathscr{T}$ is a subspace of $\mathscr{V}$.
   b. Prove that the intersection $\mathscr{S} \cap \mathscr{T}$ is a subspace of $\mathscr{V}$.
   c. Show that in general the union $\mathscr{S} \cup \mathscr{T}$ satisfies only two of the conditions in Theorem 2.1.
   d. Find a necessary condition for $\mathscr{S} \cup \mathscr{T}$ to define a subspace of $\mathscr{V}$.

11. For the proof of Theorem 2.1, fill in the reasons for the chain of equalities showing that

a. $-U = Y$ in part 1. b. $\mathbf{0} = 0U$ in part 2. c. $U = \mathbf{0}$ in part 4.

12. Prove that $rU = \mathbf{0}$ implies $r = 0$ or $U = \mathbf{0}$.

13. Suppose $\mathscr{V}$ is a vector space containing a nonzero vector. Prove that $\mathscr{V}$ has an infinite number of vectors.

14. Suppose Z is an element of a vector space $\mathscr{V}$ such that for some $U \in \mathscr{V}$, $U + Z = U$. Prove that $Z = \mathbf{0}$.

§3. Linear Dependence

The previous problem set touches on several ideas that are of fundamental importance to linear algebra. In this section we will introduce terms for these ideas and begin to look at their consequences.

Definition If $U_1, \ldots, U_k$ are vectors from a vector space and $r_1, \ldots, r_k$ are scalars, then the expression $r_1U_1 + \cdots + r_kU_k$ is a *linear combination* of the k vectors.

The expressions $2(2, 1) + 3(3, -5) - \frac{1}{2}(6, 2) = (10, -14)$ and $5(2, 1) - (6, 2) = (4, 3)$ are linear combinations of the three vectors $(2, 1)$, $(3, -5)$ and $(6, 2)$ from $\mathscr{V}_2$. In fact, every vector in $\mathscr{V}_2$ can be expressed as a linear combination of these vectors, one way to do this is

$$(a, b) = [17a + 3b](2, 1) + a(3, -5) - [6a + b](6, 2).$$

As further examples, the sets $\mathscr{T}$ and $\mathscr{U}$ in Examples 7 and 8 of the previous section consist of all possible linear combinations of three and two vectors, respectively.

Definition Let S be a subset of a vector space $\mathscr{V}$. The set of all possible linear combinations of elements in S is called the *span* of S and is denoted by $\mathscr{L}(S)$. If S is the empty subset, $\mathscr{L}(S)$ contains only the zero vector of $\mathscr{V}$.

From the above discussion, it can be said that $\mathscr{V}_2$ is the span of $(2, 1)$, $(3, -5)$, and $(6, 2)$, or $\mathscr{V}_2 = \mathscr{L}\{(2, 1), (3, -5), (6, 2)\}$.

The span of a single vector is simply the set of all scalar multiples of the vector. Thus if U is a nonzero vector in $\mathscr{V}_2$, $\mathscr{L}\{U\}$ can be thought of as the line in the plane through U and the origin.

Example 1 Suppose $U, V \in \mathscr{V}_3$ are nonzero. Determine the possible geometric interpretations of $\mathscr{L}\{U, V\}$.

If one vector is a scalar multiple of the other, say $U = rV$, then any linear combination $aU + bV$ becomes $a(rV) + bV = (ar + b)V$. Therefore $\mathscr{L}\{U, V\} = \{tV | t \in R\}$, a line through the origin.

If neither vector is a scalar multiple of the other, the scalar multiples of U and V determine two lines through the origin. A linear combination of U and V, $aU + bV$, is the sum of a point from each line, which by the parallelogram rule for addition is a point on the plane containing the two lines (see Figure 1). Therefore $\mathscr{L}\{U, V\}$ is the plane determined by U, V, and the origin.

This example provides a geometric argument for the fact that the span of two vectors from $\mathscr{V}_3$ cannot equal all of $\mathscr{V}_3$.

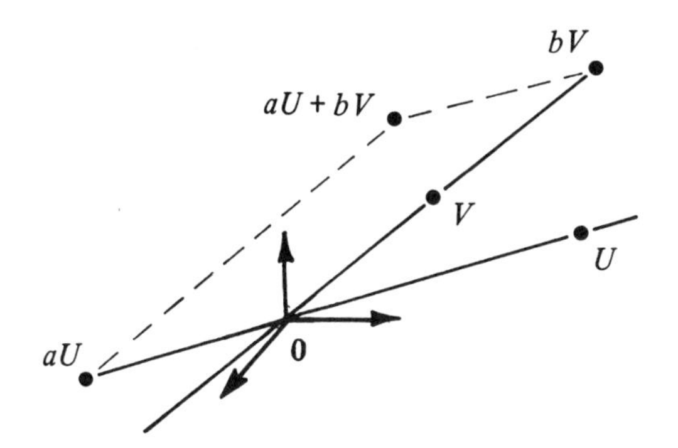

Figure 1

Example 2 Show that the span of $\{(2, -1, 5), (0, 3, -1)\}$ contains the vector $(3, 6, 5)$ but not $(2, 5, 0)$.

$(3, 6, 5) \in \mathscr{L}\{(2, -1, 5), (0, 3, -1)\}$ if there exist scalars x and y such that $(3, 6, 5) = x(2, -1, 5) + y(0, 3, -1)$. This vector equation yields the linear equations $3 = 2x$, $6 = -x + 3y$, and $5 = 5x - y$, which are satisfied by $x = 3/2$ and $y = 5/2$. Therefore $(3, 6, 5)$ is in the span.

The vector $(2, 5, 0)$ is in the span if there exist $x, y \in R$ such that $(2, 5, 0) = x(2, -1, 5) + y(0, 3, -1)$. This equation yields $2 = 2x$, $5 = -x + 3y$, and $0 = 5x - y$. With $x = 1$, the other two equations become $y = 2$ and $y = 5$, therefore no solution exists. That is, $(2, 5, 0) \notin \mathscr{L}\{(2, -1, 5), (0, 3, -1)\}$.

Example 3 $\mathscr{L}\{1, t, t^2, t^3, \ldots, t^n, \ldots\} = R[t]$.

In this case the set is infinite, but linear combinations are defined only for a finite number of vectors. Therefore $\mathscr{L}\{1, t, t^2, \ldots\}$ is the set of all finite linear combinations of the vectors $1, t, t^2, t^3, \ldots, t^n, \ldots$. This is precisely what polynomials are.

Theorem 2.3 If S is a subset of a vector space $\mathscr{V}$, then the span of S is a subspace of $\mathscr{V}$.

Proof The proof that $\mathscr{L}(S)$ is nonempty and closed under addition is left to the reader. To show $\mathscr{L}(S)$ is closed under scalar multiplication, let $U \in \mathscr{L}(S)$, then by the definition of span there exist vectors $V_1, \ldots, V_p \in S$ and scalars $r_1, \ldots, r_p$ such that $U = r_1V_1 + \cdots + r_pV_p$. So

$$\begin{aligned} rU &= r(r_1V_1 + \cdots + r_pV_p) \\ &= (rr_1)V_1 + \cdots + (rr_p)V_p. \end{aligned}$$

Therefore rU is a linear combination of $V_1, \ldots, V_p$, or $rU \in \mathscr{L}(S)$.

Definition Let S be a subset of a vector space and $\mathscr{U}$ the span of S. The vector space $\mathscr{U}$ is said to be *spanned* or *generated* by S and the elements of S *span* $\mathscr{U}$ or are *generators* of $\mathscr{U}$.

Therefore two of the above results can be restated as, $R[t]$ is spanned by the vectors $1, t, t^2, \ldots$ and $\mathscr{V}_2$ is spanned by $(2, 1)$, $(3, -5)$, and $(6, 2)$. Every vector space spans itself, that is $\mathscr{L}(\mathscr{V}) = \mathscr{V}$. Therefore every vector space is spanned by some set. But the space itself is not a very interesting spanning set. An "interesting" spanning set would just span in the sense that if one vector were removed, it would no longer span.

Consider the vector space $\mathscr{T}$ introduced in Example 7 of the previous section; $\mathscr{T} = \mathscr{L}\{(2, 1, 3), (1, 4, 1), (1, -3, 2)\}$. Although $\mathscr{T}$ is spanned by the three vectors, it can be spanned by any two of them. For example, $(2, 1, 3) = (1, 4, 1) + (1, -3, 2)$, so if $U \in \mathscr{T}$, then

$$\begin{aligned} U &= a(2, 1, 3) + b(1, 4, 1) + c(1, -3, 2) \\ &= a[(1, 4, 1) + (1, -3, 2)] + b(1, 4, 1) + c(1, -3, 2) \\ &= [a + b](1, 4, 1) + [a + c](1, -3, 2). \end{aligned}$$

Thus $U \in \mathscr{L}\{(1, 4, 1), (1, -3, 2)\}$ or $\mathscr{T} \subset \mathscr{L}\{(1, 4, 1), (1, -3, 2)\}$. The other containment is immediate so $\mathscr{T} = \mathscr{L}\{(1, 4, 1), (1, -3, 2)\}$.

Definition A vector U is *linearly dependent* on the vectors $U_1, \ldots, U_k$ if $U \in \mathscr{L}\{U_1, \ldots, U_k\}$, that is, if there are scalars $a_1, \ldots, a_k$ such that $U = a_1U_1 + \cdots + a_kU_k$.

Thus the vector (2, 1, 3) is linearly dependent on (1, 4, 1) and (1, −3, 2) from the above considerations.

If a vector U is linearly dependent on V, then U is a scalar multiple of V. Thus if $U, V \in \mathscr{V}_2$ and U is linearly dependent on V, then as points in the plane, U and V are collinear with the origin. If $U, V, W \in \mathscr{V}_3$ are nonzero and W is linearly dependent on U and V, then either U, V, and W are collinear with the origin, or W is on the plane determined by U, V, and the origin.

Example 4 Show that the vector $8 - t + 7t^2$ is linearly dependent on $2 - t + 3t^2$ and $1 + t - t^2$.

That is find scalars a and b such that

$$8 - t + 7t^2 = a(2 - t + 3t^2) + b(1 + t - t^2)$$

or that

$$(8 - 2a - b) + (-1 + a - b)t + (7 - 3a + b)t^2 = \mathbf{0}.$$

All the coefficients in the zero polynomial are zero, so a and b satisfy

$$\begin{aligned} 8 - 2a - b &= 0 \\ -1 + a - b &= 0 \\ 7 - 3a + b &= 0 \end{aligned} \qquad \text{or} \qquad \begin{aligned} 2a + b &= 8 \\ -a + b &= -1 \\ 3a - b &= 7. \end{aligned}$$

(That is, polynomials are equal if their coefficients are equal.) These equations are satisfied by $a = 3$, $b = 2$, so that

$$8 - t + 7t^2 = 3(2 - t + 3t^2) + 2(1 + t - t^2)$$

is the desired linear combination.

If U is linearly dependent on the vectors $U_1, \ldots, U_k$, then $U = a_1U_1 + \cdots + a_kU_k$ for some scalars $a_1, \ldots, a_k$. Therefore $(-1)U + a_1U_1 + \cdots + a_kU_k = \mathbf{0}$. That is, $\mathbf{0}$ is a linear combination of the $k + 1$ vectors, with at least one of the scalars nonzero. (Although -1 is not zero, all the other scalars $a_1, \ldots, a_k$ could equal zero.)

Definition The finite set of vectors $\{V_1, \ldots, V_h\}$ is *linearly dependent* if there exist scalars $b_1, \ldots, b_h$, not all zero, such that $b_1V_1 + \cdots + b_hV_h = \mathbf{0}$. An infinite set is *linearly dependent* if it contains a finite linearly dependent subset.

The phrase "not all zero" causes much confusion. It does not mean "all not zero" and in some situations zero must be used as a coefficient. For example, the set $\{(1, 2), (2, 4), (1, 0)\}$ is linearly dependent since $2(1, 2) + (-1)(2, 4) + 0(1, 0) = \mathbf{0}$.

Every set containing the zero vector is linearly dependent; even the set $\{\mathbf{0}\}$ for $1(\mathbf{0}) = \mathbf{0}$ is a linear combination satisfying the conditions of the definition.

Example 5 Determine if the set $\{(2, 5), (-1, 1), (-4, -3)\}$ is linearly dependent.

Suppose

$$x(2, 5) + y(-1, 1) + z(-4, -3) = \mathbf{0} = (0, 0).$$

This vector equation yields the system of linear equations

$$2x - y - 4z = 0 \qquad \text{and} \qquad 5x + y - 3z = 0.$$

If $x = y = z = 0$ is the only solution, the vectors are not linearly dependent. But there are many other solutions, for example, if $x = 1$, then $y = -2$, $z = 1$. Therefore $(2, 5) - 2(-1, 1) + (-4, -3) = \mathbf{0}$, and the vectors are linearly dependent.

Example 6 Remove one vector from the above set and determine if $\{(2, 5), (-1, 1)\}$ is linearly dependent.

That is, are there scalars x, y not both zero, such that

$$x(2, 5) + y(-1, 1) = (0, 0)?$$

This vector equation yields $2x - y = 0$ and $5x + y = 0$, which together have only the solution $x = y = 0$. Therefore the set is not linearly dependent.

Definition A set of vectors is *linearly independent* if it is not linearly dependent.

Thus the set in Example 6 is linearly independent.

From our discussion of linear dependence and geometry we can say that a set of two vectors in $\mathscr{V}_2$ is linearly independent if as points they are not collinear with the origin or equivalently if they span $\mathscr{V}_2$. A set of three vectors in $\mathscr{V}_3$ is linearly independent if as points in 3-space they are not coplanar with the origin. We will see that in such a case the three vectors span $\mathscr{V}_3$.

Theorem 2.4 The set of vectors $\{W_1, \ldots, W_n\}$ is linearly independent provided $t_1 W_1 + \cdots + t_n W_n = \mathbf{0}$ implies $t_1 = \cdots = t_n = 0$, that is, provided the only linear combination yielding the zero vector is the one with all coefficients equal to zero.

Proof This follows directly from the definition of linear dependence for sets of vectors.

Example 7 Determine if the set of vectors $\{\sin x, \cos x\}$ is linearly independent in the vector space $\mathscr{F}$.

Suppose $a(\sin x) + b(\cos x) = \mathbf{0}(x)$ for some scalars $a, b \in R$. Then $a(\sin x) + b(\cos x) = 0$ for every value of x in the interval $[0, 1]$. But when $x = 0$, $\sin x = 0$, so $b = 0$ and the equation becomes $a(\sin x) = 0$ for all $x \in [0, 1]$. Now $\sin 1 \neq 0$ implies $a = 0$. Therefore, the only linear combination of these functions yielding the zero function is $0(\sin x) + 0(\cos x)$ and the set is linearly independent.

Example 8 $\{1, t, t^2, \ldots, t^n, \ldots\}$ is linearly independent in $R[t]$.

This is immediate, for a (finite) linear combination of these vectors is a polynomial with each term involving a different power of t. Such a polynomial is zero only if all the coefficients are zero.

We have now encountered several terms using the word linear; linear equation, linear combination, linear dependence, linear independence, and thus we begin to see why the study of vector spaces is called linear algebra. The concepts introduced in this section are basic to both the theory and the applications of linear algebra. Therefore it is important to gain a good understanding of their meaning.

Problems

1. Determine which of the following vectors are in the span of the set $\{(-2, 1, -9), (2, -4, 6)\}$. a. $(0, 3, 3)$. b. $(1, 0, 1)$. c. $(0, 0, 0)$. d. $(1, 0, 0)$. e. $(1, -5, 0)$. f. $(1, -4, 1)$.

2. Determine which of the following vectors are members of $\mathscr{L}\{t^3 - t + 1,\ 3t^2 + 2t,\ t^3\}$. a. t^2. b. $t - 1$. c. $5t^3 + 6t^2 + 4$. d. $t^3 + t^2 + t + 1$. e. $t^3 + 3t^2 + 3t - 1$.

3. Which of the following sets span $\mathscr{V}_3$?
   a. $\{(1, 1, 0), (0, 1, 1), (1, 1, 1)\}$.
   b. $\{(1, 0, 0), (4, 2, 0), (0, 1, 0), (0, 0, 0), (0, 0, 1)\}$.
   c. $\{(1, 1, 0), (2, 0, -1), (4, 4, 0), (5, 3, -1)\}$.
   d. $\{(1, 1, 1), (1, 2, 3)\}$.

4. Write out how the following statements are read.
   a. $U \in \mathscr{L}\{V, W\}$. b. $\mathscr{S} \subset \mathscr{L}\{U, V\}$. c. $\mathscr{S} = \mathscr{L}\{U, V\}$. d. $\mathscr{L}\{U_1, \ldots, U_k\}$.

5. a. Prove that no finite set can span $R[t]$.
   b. Use part a to show that no finite set can span $\mathscr{F}$.

6. Why is it unnecessary to write linear combinations with parentheses, such as $(\cdots((r_1U_1 + r_2U_2) + r_3U_3) + \cdots + r_kU_k)$?

7. Show that $\{(a, b), (c, d)\}$ is linearly dependent in $\mathscr{V}_2$ if and only if $ad - bc = 0$.

8. Under what conditions on a and b are the following sets linearly independent in $\mathscr{V}_3$?
   a. $\{(1, a, 2), (1, 1, b)\}$. b. $\{(1, 2, 0), (a, b, 2), (1, 0, 1)\}$.
   c. $\{(a, 2, 3), (0, 4, 6), (0, 0, b)\}$.

9. a. Write the statement "some vector in $\{U_1, \ldots, U_n\}$ is a linear combination of the other vectors" in terms of the vectors and scalars.
   b. Show that the statement in a implies $\{U_1, \ldots, U_n\}$ is linearly dependent.
   c. Why is the statement in part a not a satisfactory definition of linear dependence?

10. Are the following sets linearly independent or dependent?
    a. $\{2 - 3i, 1 + i\}$ in C.
    b. $\{\cos^2 x, 17, \sin^2 x\}$ in $\mathscr{F}$.
    c. $\{4t, t^2 + t, 3t^2 - t\}$ in $R[t]$.
    d. $\{3x^2/(x + 1), -x/(2x + 2), 4x\}$ in $\mathscr{F}$.
    e. $\{e^x, e^{2x}, e^{3x}\}$ in $\mathscr{F}$.
    f. $\{t + 3t^4, t^4 - 5, 6t\}$ in $R[t]$.
    g. $\{\tan x, \sin x\}$ in $\mathscr{F}$.

11. Let $S = \{(2, -4), (1, 3), (-6, -3)\} \subset \mathscr{V}_2$.
    a. Show that S is linearly dependent.
    b. Show that the vector $(5, 0)$ can be expressed as a linear combination of the vectors in S in an infinite number of ways. Do the same for the vector $(0, -20)$.
    c. Show that two vectors from S, say the first two, span $\mathscr{V}_2$.
    d. Show that $\mathscr{L}(S)$ is the trivial subspace $\mathscr{V}_2$ of $\mathscr{V}_2$.

12. Let $S = \{(6, 2, -3), (-2, -4, 1), (4, -7, -2)\} \subset \mathscr{V}_3$.
    a. Show that S is linearly dependent.
    b. Show that $\mathscr{L}(S)$ is a proper subspace of $\mathscr{V}_3$. Suggestion: consider the vector $(0, 0, 1)$.

c. Show that each of the following vectors can be expressed in an infinite number of ways as a linear combination of the vectors in S: (0, 0, 0), (0, —1, 0), (2, 0, -1), and (6, 2, -3).
d. Prove that $\mathscr{L}(S) \subset \mathscr{L}\{(0, 1, 0), (2, 0, -1)\}$.
e. Show that $\mathscr{L}(S) = \mathscr{L}\{(0, 1, 0), (2, 0, -1)\}$.

13*. Suppose $\{U_1, \ldots, U_k\}$ is linearly independent and $V \notin \mathscr{L}\{U_1, \ldots, U_k\}$. Prove that $\{V, U_1, \ldots, U_k\}$ is linearly independent.

14*. Suppose $\mathscr{V} = \mathscr{L}\{U_1, \ldots, U_k\}$ and $U_1 \in \mathscr{L}\{U_2, \ldots, U_k\}$. Prove that $\mathscr{V} = \mathscr{L}\{U_2, \ldots, U_k\}$.

15*. Suppose every vector in a vector space $\mathscr{V}$ has a unique expression as a linear combination of the vectors $U_1, \ldots, U_n$. Prove that the set $\{U_1, \ldots, U_n\}$ is linearly independent.

16. The fact that $r(U_1 + \cdots + U_k) = rU_1 + \cdots + rU_k$ for any k vectors in $\mathscr{V}$ and any $r \in R$ was used in the proof of Theorem 2.3. Use induction to prove this fact.

§4. Bases and Dimension

The coordinate axes in the Cartesian plane E^2 can be viewed as all scalar multiples of the vectors (1, 0) and (0, 1) from $\mathscr{V}_2$. Therefore (1, 0) and (0, 1) span the plane or rather the vector space $\mathscr{V}_2$. A set of generators for a vector space can be as important as the coordinate system in E^2. However, if a set of generators is linearly dependent, then a smaller set will also span. Therefore we shall look for linearly independent sets that span a vector space.

Definition A *basis* for a vector space $\mathscr{V}$ is a linearly independent, ordered set that spans $\mathscr{V}$.

Example 1 The set $\{(1, 0), (0, 1)\}$ is a basis for $\mathscr{V}_2$.

If $x(1, 0) + y(0, 1) = \mathbf{0}$, then $(x, y) = \mathbf{0} = (0, 0)$ and $x = y = 0$. Therefore the set is linearly independent. And every vector $(a, b) \in \mathscr{V}_2$ can be written as $(a, b) = a(1, 0) + b(0, 1)$, so the set spans $\mathscr{V}_2$. The set $\{(1, 0), (0, 1)\}$ will be called the *standard basis* for $\mathscr{V}_2$.

Example 2 It has already been determined in previous examples that

*You should not go on to the next section section before working on problems 13, 14, and 15. If the problems are not clear, start by looking at numerical examples from $\mathscr{V}_2$ or $\mathscr{V}_3$.

$\{1, t, t^2, \ldots, t^n, \ldots\}$ spans $R[t]$ and is linearly independent. Therefore this infinite set is a basis for $R[t]$.

Example 3 Construct a basis for $\mathscr{V}_3$.

We can start with any nonzero vector in $\mathscr{V}_3$, say $(1, -4, 2)$. The set $\{(1, -4, 2)\}$ is linearly independent, but since its span is $\{r(1, -4, 2) | r \in R\}$, it does not span $\mathscr{V}_3$. Since the vector $(1, -4, 3)$ is clearly not in $\mathscr{L}\{(1, -4, 2)\}$, the set $\{(1, -4, 2), (1, -4, 3)\}$ is linearly independent (problem 13, Section 3). Does this set span $\mathscr{V}_3$? That is, given any vector (a, b, c), are there real numbers x, y such that $(a, b, c) = x(1, -4, 2) + y(1, -4, 3)$? If so then x and y must satisfy $a = x + y$, $b = -4x - 4y$, and $c = 2x + 3y$. But these equations have a solution only if $a = -4b$. Therefore,

$$\mathscr{L}\{(1, -4, 2), (1, -4, 3)\} = \{(a, b, c) | a = -4b\}.$$

This is not $\mathscr{V}_3$, so any vector (a, b, c) with $a \neq -4b$ can be added to give a linearly independent set of three vectors. Choose the vector $(-1, 3, 5)$ arbitrarily, then $\{(1, -4, 2), (1, -4, 3), (-1, 3, 5)\}$ is linearly independent. This set spans $\mathscr{V}_3$ if it is possible to find $x, y, z \in R$ such that

$$(a, b, c) = x(1, -4, 2) + y(1, -4, 3) + z(-1, 3, 5)$$

for any values of a, b, c, that is, for any $(a, b, c) \in \mathscr{V}_3$. A solution is given by $x = -29a - 8b - c$, $y = 26a + 7b + c$, and $z = -4a - b$. Therefore $\{(1, -4, 2), (1, -4, 3), (-1, 3, 5)\}$ is a basis for $\mathscr{V}_3$.

The above construction can be viewed geometrically. First an arbitrary nonzero vector was chosen. The second vector was chosen from all vectors not on the line determined by the first vector and the origin, that is, the line given by $P = t(1, -4, 2)$, $t \in R$. Then the third vector was chosen from all vectors not on the plane determined by the first two and the origin, that is, the plane with equation $x + 4y = 0$. Obviously it is possible to obtain many bases for $\mathscr{V}_3$ with no vectors in common. The important point to be established is that every basis for $\mathscr{V}_3$ has the same number of vectors.

Consider for a moment the vector space $\mathscr{F}$. If it has a basis, it cannot be obtained by inspection as with $\mathscr{V}_2$ or $R[t]$. And since a basis for $\mathscr{F}$ must be infinite (problem 5, Section 3), it is not possible to build a basis as in Example 3. But there is a theorem which states that every vector space has a basis. The proof of this theorem is beyond the scope of this course, but it means that $\mathscr{F}$ has a basis even though we cannot write one out. Most of our study will concern vector spaces with finite bases, but the spaces $R[t]$ and $\mathscr{F}$ should be kept in mind to understand what is meant by, "let $\mathscr{V}$ be a vector space."

The next theorem will be used to prove that if a vector space $\mathscr{V}$ has a finite basis, then all bases for $\mathscr{V}$ have the same number of vectors.

Theorem 2.5 If $\{W_1, \ldots, W_k\}$ is a linearly dependent ordered set of nonzero vectors, then $W_j \in \mathscr{L}\{W_1, \ldots, W_{j-1}\}$ for some j, $2 \leq j \leq k$. That is, one of the vectors is a linear combination of the vectors preceding it.

Proof Since $W_1 \neq \mathbf{0}$, the set $\{W_1\}$ is linearly independent. Now sets can be built from $\{W_1\}$ by adding one vector at a time from $W_2, \ldots, W_k$ until a linearly dependent set is obtained. Since $\{W_1, \ldots, W_k\}$ is linearly dependent, such a set will be obtained after adding at most $k - 1$ vectors. Thus there is an index j, $2 \leq j \leq k$, such that $\{W_1, \ldots, W_{j-1}\}$ is linearly independent and $\{W_1, \ldots, W_{j-1}, W_j\}$ is linearly dependent. Then there exist scalars $a_1, \ldots, a_j$, not all zero, such that

$$a_1 W_1 + \cdots + a_{j-1} W_{j-1} + a_j W_j = \mathbf{0}.$$

If $a_j \neq 0$, we are done, for then

$$W_j = -\frac{1}{a_j}(a_1 W_1 + \cdots + a_{j-1} W_{j-1}).$$

But if $a_j = 0$, then $a_1 W_1 + \cdots + a_{j-1} W_{j-1} = \mathbf{0}$ with not all the coefficients equal to zero. That is, $\{W_1, \ldots, W_{j-1}\}$ is linearly dependent, contradicting the choice of j. Therefore $a_j \neq 0$ as desired, and the proof is complete.

Two false conclusions are often drawn from this theorem. It does not say that every linearly dependent set contains a vector which is a linear combination of the others or of those preceding it. For example, consider the linearly dependent sets $\{\mathbf{0}\}$ and $\{\mathbf{0}, t\} \subset R[t]$. It also does not say that the last vector of a linearly dependent set is a linear combination of the vectors preceding it; consider the linearly dependent set $\{t, 3t, t^2\} \subset R[t]$.

Theorem 2.6 If a vector space $\mathscr{V}$ has a finite basis, then every basis has the same number of vectors.

Proof Suppose $B_1 = \{U_1, \ldots, U_n\}$ and $B_2 = \{V_1, \ldots, V_m\}$ are bases for $\mathscr{V}$. We will first show that $n \leq m$ by replacing the vectors in B_2

with vectors from B_1. Each step of the argument will use the fact that B_2 spans $\mathscr{V}$ and B_1 is linearly independent.

To begin $U_1 \in \mathscr{V}$ and $\mathscr{V} = \mathscr{L}\{V_1, \ldots, V_m\}$, therefore the ordered set $\{U_1, V_1, \ldots, V_m\}$ is linearly dependent. All the vectors in this set are nonzero (Why?), so by Theorem 2.5 some vector, say V_j, is a linear combination of $U_1, V_1, \ldots, V_{j-1}$, and the set

$$\{U_1, V_1, \ldots, V_{j-1}, V_{j+1}, \ldots, V_m\}$$

spans $\mathscr{V}$ (problem 14, Section 3). Now renumber the vectors in this set so that V_1 no longer appears. Then $\{U_1, \ldots, U_n\}$ is linearly independent and $\{U_1, V_2, \ldots, V_m\}$ is a spanning set for $\mathscr{V}$.

This argument can be repeated until the first n vectors of B_2 have been replaced by the vectors of B_1. A typical argument is as follows: Suppose k of the V's have been replaced to obtain (with relabeling) $\{U_1, \ldots, U_k, V_{k+1}, \ldots, V_m\}$. This set spans $\mathscr{V}$ and $U_{k+1} \in \mathscr{V}$, so $\{U_1, \ldots, U_k, U_{k+1}, V_{k+1}, \ldots, V_m\}$ is linearly dependent. Therefore one of the vectors in this ordered set is a linear combination of the preceding vectors. This vector cannot be one of the U's since $\{U_1, \ldots, U_{k+1}\} \subset B_1$, which is linearly independent, so it is one of the V's. (Note that the set is linearly dependent, so there must be at least one of the V's left.) Removing this vector and relabeling the remaining V's yields the set $\{U_1, \ldots, U_{k+1}, V_{k+2}, \ldots, V_m\}$, a spanning set for $\mathscr{V}$. Thus for each vector from B_1, there must be a vector to remove from B_2, that is, $n \leq m$.

Now interchanging the roles of B_1 and B_2, and using the fact that B_1 spans $\mathscr{V}$ and B_2 is linearly independent yields $m \leq n$. So $n = m$.

Theorem 2.6 permits the following definition.

Definition If a vector space $\mathscr{V}$ has a basis containing n vectors, then the *dimension* of $\mathscr{V}$ is n; write $\dim \mathscr{V} = n$. If $\mathscr{V} = \{\mathbf{0}\}$, then $\dim \mathscr{V} = 0$. In either case, $\mathscr{V}$ is said to be *finite dimensional*. $\mathscr{V}$ is *infinite dimensional* if it is not finite dimensional.

From the above examples, $\dim \mathscr{V}_2 = 2$ and $\dim \mathscr{V}_3 = 3$, whereas $R[t]$ and $\mathscr{F}$ are infinite dimensional.

Example 4 The dimension of $\mathscr{V}_n$ is n.

Consider the vectors $E_1 = (1, 0, \ldots, 0)$, $E_2 = (0, 1, 0, \ldots, 0), \ldots,$ $E_n = (0, \ldots, 0, 1)$ in $\mathscr{V}_n$. These n vectors are linearly independent, for if

$x_1E_1 + x_2E_2 + \cdots + x_nE_n = \mathbf{0}$, then $(x_1, x_2, \ldots, x_n) = (0, 0, \ldots, 0)$ and $x_1 = x_2 = \cdots = x_n = 0$.

$\mathscr{V}_n$ is spanned by these n vectors, for if $U \in \mathscr{V}_n$, then $U = (a_1, a_2, \ldots, a_n)$ for some $a_1, a_2, \ldots, a_n \in R$. But

$$a_1E_1 + a_2E_2 + \cdots + a_nE_n = (a_1, a_2, \ldots, a_n),$$

so $\mathscr{V}_n = \mathscr{L}\{E_1, E_2, \ldots, E_n\}$. Thus $\{E_1, \ldots, E_n\}$ is a basis for $\mathscr{V}_n$, and dim $\mathscr{V}_n = n$. This basis will be called the *standard basis* for $\mathscr{V}_n$ and will be denoted by $\{E_i\}$. In $\mathscr{V}_2$, $\{E_i\} = \{(1, 0), (0, 1)\}$ and in $\mathscr{V}_3$, $\{E_i\} = \{(1, 0, 0), (0, 1, 0), (0, 0, 1)\}$.

Example 5 The set $\{1, t, t^2, \ldots, t^{n-1}\}$ can easily be shown to be a basis for $R_n[t]$, the set of all polynomials in t of degree less than n. Therefore dim $R_n[t] = n$.

Several corollaries follow from Theorem 2.6.

Corollary 2.7 If dim $\mathscr{V} = n$ and S is a set of n vectors that span $\mathscr{V}$, then S is a basis for $\mathscr{V}$.

Example 6 Show that $\{1 + t + t^2, 1 + t, 1\}$ is a basis for $R_3[t]$.

It can be determined by inspection that this set spans $R_3[t]$, for

$$a + bt + ct^2 = c(1 + t + t^2) + (b - c)(1 + t) + (a - b)1.$$

The set has 3 vectors and dim $R_3[t] = 3$, so by Corollary 2.7, the set is a basis for $R_3[t]$.

Corollary 2.8 If dim $\mathscr{V} = n$ and S is a linearly independent subset of n vectors, then S is a basis for $\mathscr{V}$.

Example 7 The set $\{(a, b), (c, d)\}$ is linearly independent in $\mathscr{V}_2$ if $ad - bc \neq 0$ (problem 7, Section 3). Dim $\mathscr{V}_2 = 2$, so by Corollary 2.8, $\{(a, b), (c, d)\}$ is a basis for $\mathscr{V}_2$ if $ad - bc \neq 0$.

Corollary 2.9 If dim $\mathscr{V} = n$, then any subset of $\mathscr{V}$ containing more than n vectors is linearly dependent.

With this corollary, several previous problems can be solved by inspection. For example, a set such as $\{(1, 5), (2, 7), (4, 0)\}$ must be linearly dependent in $\mathscr{V}_2$.

Theorem 2.10 Let $\mathscr{S}$ be a proper subspace of a finite-dimensional vector space $\mathscr{V}$. Then dim $\mathscr{S} <$ dim $\mathscr{V}$.

Proof Suppose dim $\mathscr{V} = n$, then no linearly independent subset of $\mathscr{S}$ can contain more than n vectors by Corollary 2.9. Hence dim $\mathscr{S} \leq$ dim $\mathscr{V}$. But if $\mathscr{S}$ has a linearly independent subset of n vectors, then by Corollary 2.8, it is a basis for $\mathscr{V}$ and $\mathscr{S} = \mathscr{V}$. Since $\mathscr{S}$ is a proper subspace in $\mathscr{V}$, a linearly independent subset in $\mathscr{S}$ cannot have more than $n - 1$ vectors, that is dim $\mathscr{S} <$ dim $\mathscr{V}$.

Example 8 The dimension of $R[t]$ is infinite, so Theorem 2.10 does not apply. In fact, $\mathscr{L}\{1, t^2, t^4, \ldots, t^{2n}, \ldots\}$ is a proper subspace of $R[t]$, yet it is also infinite dimensional.

Definition Let dim $\mathscr{V} = n$ and suppose S is a linearly independent subset of $\mathscr{V}$ containing k vectors, $k < n$. If B is a basis for $\mathscr{V}$ which contains the vectors in S, then B is an *extension* of S.

Theorem 2.11 Any linearly independent subset of a finite-dimensional vector space can be extended to a basis of the vector space.

Proof Suppose dim $\mathscr{V} = n$, and $S = \{U_1, \ldots, U_k\}$ is linearly independent in $\mathscr{V}$, with $k < n$. Since $k < n$, Corollary 2.7 implies that S does not span $\mathscr{V}$. Therefore there exists a vector U_{k+1} in $\mathscr{V}$ but not in the span of S. Now $\{U_1, \ldots, U_k, U_{k+1}\}$ is linearly independent (see problem 13, Section 3) and it contains $k + 1$ vectors. If $k + 1 = n$, this is a basis for $\mathscr{V}$ by Corollary 2.8. If $k + 1 < n$, then this procedure can be repeated, yielding a linearly independent set containing $k + 2$ vectors. When $n - k$ vectors have been added in this fashion, a basis will be obtained for $\mathscr{V}$ (by Corollary 2.8), which is an extension of S.

In general, a given basis for a vector space $\mathscr{V}$ does not contain a basis for a particular subspace $\mathscr{S}$. But Theorem 2.11 shows that if $\mathscr{V}$ is finite dimensional, then it is always possible to extend a basis for $\mathscr{S}$ to a basis for $\mathscr{V}$.

Example 9 Extend the linearly independent set $\{1 + t - 2t^2, 2t + 3t^2\}$ to a basis for $R_3[t]$.

That is, find a vector in $R_3[t]$ which is not in the space $\mathscr{S} = \mathscr{L}\{1 + t - 2t^2, 2t + 3t^2\}$. This could be done by trial and error but instead we will characterize the set $\mathscr{S}$. A vector $a + bt + ct^2$ is in $\mathscr{S}$ if there exist real numbers x and y such that $a + bt + ct^2 = x(1 + t - 2t^2) + y(2t + 3t^2)$. Therefore the equations $a = x$, $b = x + 2y$, and $c = -2x + 3y$ must have a solution for all $a, b, c \in R$. This is possible only if $b = a + 2y$ and $c = -2a + 3y$ or equivalently if $7a - 3b + 2c = 0$. So $\mathscr{S} = \{a + bt + ct^2 | 7a - 3b + 2c = 0\}$, and any vector $a + bt + ct^2$ such that $7a - 3b + 2c \neq 0$ may be added. Using $1 + t + t^2$, $\{1 + t - 2t^2, 2t + 3t^2, 1 + t + t^2\}$ is a basis for $R_3[t]$ by Corollary 2.8, and it is an extension of the given linearly independent set.

Problems

1. Show that the vector space of complex numbers C has dimension 2.
2. Show that any nonzero real number constitutes a basis for R when viewed as a vector space.
3. Use Theorem 2.10 to prove that neither $\mathscr{V}_1$ nor R has a proper subspace.
4. Determine if the following are bases:
   a. $\{(2, 1, -5), (7, 4, 3)\}$ for $\mathscr{V}_3$.
   b. $\{2 + 5i, 1 + 3i\}$ for C.
   c. $\{t, 3t^2 + 1, 4t + 2, 1 - 5t\}$ for $R_3[t]$.
   d. $\{(10, -15), (6, -9)\}$ for $\mathscr{V}_2$.
   e. $\{t^3, 4t^2 - 6t + 5, 2t^3 - 3t^2 + 5t - 9\}$ for $R_4[t]$.
   f. $\{(2, 1), (-3, 4)\}$ for $\mathscr{V}_2$.
5. Determine which of the following statements are true. If false, find a counterexample.
   a. Every linearly independent set is the basis of some vector space.
   b. The set $\{U_1, \ldots, U_n\}$ is linearly independent if $a_1U_1 + \cdots + a_nU_n = \mathbf{0}$ and $a_1 = \cdots = a_n = 0$.
   c. The dimension of a subspace is less that the dimension of the space.
6. Prove: a. Corollary 2.7 b. Corollary 2.8 c. Corollary 2.9.
7. Determine the dimension of the following vector spaces:
   a. $\mathscr{L}\{(2, -3, 1), (1, 4, 2), (5, -2, 4)\} \subset \mathscr{V}_3$.
   b. $\mathscr{L}\{e^x, e^{2x}, e^{3x}\} \subset \mathscr{F}$.
   c. $\mathscr{L}\{t - 1, 1 - t^2, t^2 - t\} \subset R_3[t]$.
8. Show that the vector space of 2×2 matrices as defined in problem 8, page 37, is four dimensional.

9. Extend the following sets to bases for the indicated vector space:

a. $\{(4, -7)\}$ for $\mathscr{V}_2$.
b. $\{(2, -1, 3), (4, 1, 1)\}$ for $\mathscr{V}_3$.
c. $\{t, t^2 + 4\}$ for $R_3[t]$.
d. $\{t - 1, t^2 + 5\}$ for $R_3[t]$.

10. Let $\mathscr{S}$ and $\mathscr{T}$ be subspaces of a finite-dimensional vector space and prove that $\dim(\mathscr{S} + \mathscr{T}) = \dim \mathscr{S} + \dim \mathscr{T} - \dim(\mathscr{S} \cap \mathscr{T})$. Begin by choosing a basis for the subspace $\mathscr{S} \cap \mathscr{T}$.

11. Use problem 10 to prove that two planes through the origin in 3-space intersect in at least a line through the origin.

§5. Coordinates

Up to this point there has been no mention of coordinates with regard to vector spaces. Care has been taken to call the numbers in an n-tuple vector components rather than coordinates as suggested by our experience with analytic geometry. And although coordinates have been used in the Cartesian plane and 3-space, they have not been considered on the same level as the discussion of vector spaces, for they are defined in terms of distances from lines and planes which is not the aspect we will generalize to vector spaces. However, a discussion of coordinates in vector spaces has direct application to geometry and should even aid in understanding what coordinates mean there. The theoretical basis for the definition of coordinates is in the following fact.

Theorem 2.12 The set $\{U_1, \ldots, U_n\}$ is a basis for the vector space $\mathscr{V}$ if and only if every vector of $\mathscr{V}$ has a unique expression as a linear combination of the vectors $U_1, \ldots, U_n$.

Proof ($\Rightarrow$) Suppose $\{U_1, \ldots, U_n\}$ is a basis for the vector space $\mathscr{V}$ and $U \in \mathscr{V}$. Since a basis spans the space, there exist scalars such that $U = a_1U_1 + \cdots + a_nU_n$. To see that this expression is unique, suppose $U = b_1U_1 + \cdots + b_nU_n$. Then

$$\mathbf{0} = U - U = (a_1 - b_1)U_1 + \cdots + (a_n - b_n)U_n.$$

But a basis is linearly independent, so $a_1 = b_1, \ldots, a_n = b_n$, and there is only one expression for U as a linear compination of the vectors in a basis.

($\Leftarrow$) If every vector of $\mathscr{V}$ has a unique expression as a linear combination of the vectors in the set $\{U_1, \ldots, U_n\}$, then the set spans $\mathscr{V}$, and it is only necessary to prove that it is linearly independent. Therefore suppose there are scalars c_i such that $c_1U_1 + \cdots + c_nU_n = \mathbf{0}$. By uniqueness, there is only one such expression for the zero vector and $\mathbf{0} = 0U_1 + \cdots + 0U_n$. Therefore $c_1 = \cdots = c_n = 0$ and the set is linearly independent. Thus $\{U_1, \ldots, U_n\}$ is a basis for $\mathscr{V}$.

Definition Let $B = \{U_1, \ldots, U_n\}$ be a basis for a (finite-dimensional) vector space $\mathscr{V}$. For $W \in \mathscr{V}$ suppose $W = a_1U_1 + \cdots + a_nU_n$, then the scalars $a_1, \ldots, a_n$ are the *coordinates of W with respect to the basis B*. We will write $W{:}(a_1, \ldots, a_n)_B$ for the statement "W has coordinates $a_1, \ldots, a_n$ with respect to B."

Bases were defined as ordered sets so that this coordinate notation would be meaningful.

Example 1 Consider the basis $B = \{(1, 3), (-2, 4)\}$ for $\mathscr{V}_2$.

$(1, 3) = 1(1, 3) + 0(-2, 4)$, therefore the coordinates of $(1, 3)$ with respect to this basis are 1 and 0, or $(1, 3){:}(1, 0)_B$. Similarly $(-2, 4){:}(0, 1)_B$. To find the coordinates of a vector such as $(9, 7)$ with respect to B, it is necessary to solve the vector equation $(9, 7) = x(1, 3) + y(-2, 4)$ for x and y. The solution is $x = 5$ and $y = -2$, so the coordinates are given by $(9, 7){:}(5, -2)_B$.

Example 2 Consider the basis $B' = \{(1, 3), (0, 2)\}$ for $\mathscr{V}_2$.

For this basis $(1, 3){:}(1, 0)_{B'}$ and $(0, 2){:}(0, 1)_{B'}$ and the coordinates of any other vector with respect to B' can be found by inspection. That is, the first component of $(0, 2)$ is zero, so the first coordinate of a vector with respect to B' must be its first component. Thus $(9, 7){:}(9, -10)_{B'}$.

Take care to notice that a vector, even an ordered pair, does not have coordinates per se. Coordinates can be found only when given a basis, and our notation will always make the connection clear. This connection is usually omitted in analytic geometry, and while this could be done here, one can afford to be sloppy only when the ideas being passed over are clear. But this is not the case with coordinates. In fact, problems are caused by the feeling that, having used coordinates in analytic geometry, there is nothing new here. Actually coordinates can be used in geometry with little understanding and no problem arises until it becomes necessary to change to another coordinate system.

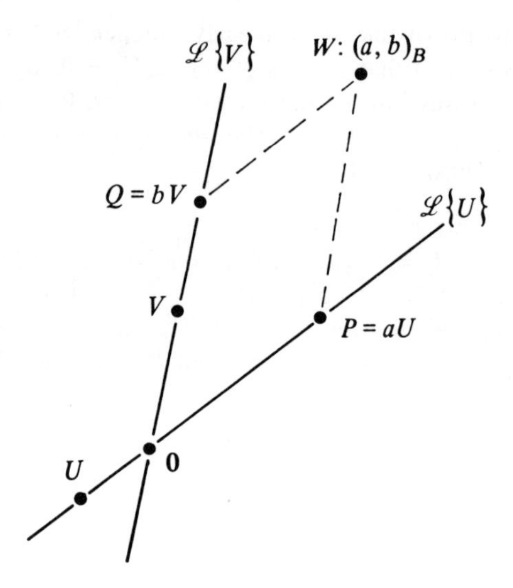

Figure 2

It is not difficult to represent a basis B for $\mathscr{V}_2$ and coordinates with respect to B geometrically. Suppose $B = \{U, V\}$, then B determines two lines in E^2 corresponding to $\mathscr{L}\{U\}$ and $\mathscr{L}\{V\}$; see Figure 2. These lines might be called "coordinate axes," and the coordinates of any vector W could be found using the parallelogram rule for addition. That is, if the line PW is parallel to $\mathscr{L}\{V\}$ and the line QW is parallel to $\mathscr{L}\{U\}$, then $P = aU$, $Q = bV$ implies that W:$(a, b)_B$.

Analytic geometry begins with a single coordinate system, but linear algebra is at the other extreme. We begin by considering all possible bases for a vector space, and the nature of the problem at hand will determine which should be used. It is not only important to be able to find coordinates with respect to any given basis, but the ease in handling coordinates will become one of the main strengths of linear algebra.

Example 3 It can be shown that $B = \{t^2 + 1, t + 1, t^2 + t\}$ is linearly independent and thus a basis for $R_3[t]$. The coordinates of the basis vectors with respect to B are easily found:

$$t^2 + 1{:}(1, 0, 0)_B, \quad t + 1{:}(0, 1, 0)_B, \quad \text{and} \quad t^2 + t{:}(0, 0, 1)_B.$$

To find the coordinates of a vector such as $2t$ with respect to B, consider the equation

$$x(t^2 + 1) + y(t + 1) + z(t^2 + 1) = 2t.$$

This equation has the solution $x = -1$, $y = z = 1$, so $2t:(-1, 1, 1)_B$. In the same way it can be shown that $8t^2 + 3t + 1:(3, -2, 5)_B$ and $2:(1, 1, -1)_B$.

Example 4 The coordinates of a vector in $\mathscr{V}_n$ with respect to the standard basis $\{E_i\}$ are the same as the components of the vector. Thus $(7, -1):(7, -1)_{\{E_i\}}$ and $(4, 8, 2):(4, 8, 2)_{\{E_i\}}$.

It is worthwhile repeating that a vector does not have coordinates, it only has coordinates with respect to a particular basis. Thus the numbers 4, 8, 2 are called the components of the vector (4, 8, 2) and not coordinates.

Example 5 $B = \{4 - i, i + 2\}$ is a basis for C. The coordinates of the basis vectors with respect to B are given by $4 - i:(1, 0)_B$ and $i + 2:(0, 1)_B$. Solving the equation $i = x(4 - i) + y(i + 2)$ for x and y gives $i:(-1/3, 2/3)_B$. Similarly $1:(1/6, 1/6)_B$ and $1 + i:(-1/6, 5/6)_B$.

In Example 3, coordinates of polynomial vectors from $R_3[t]$ are ordered triples, and in Example 5, coordinates of complex numbers from C are ordered pairs. So the process of obtaining coordinates associates vectors in an n-dimensional vector space with ordered n-tuples. What is more, the algebra of the vector space can be performed on the coordinates. Thus, in Example 5, if the coordinates of 1 and i with respect to B are added as elements of $\mathscr{V}_2$, the result is $(1/6, 1/6) + (-1/3, 2/3) = (-1/6, 5/6)$, which are the coordinates of the sum $1 + i$ with respect to B. n-tuples may be easier to handle than other vectors, such as polynomials, so that it will often be simpler to study an n-tuple space rather than some other n-dimensional vector space. The basis for doing this is given by the next theorem.

Theorem 2.13 Let $\dim \mathscr{V} = n$, B be a basis for $\mathscr{V}$, and $U, V \in \mathscr{V}$, $r \in R$. If $U:(a_1, \ldots, a_n)_B$ and $V:(b_1, \ldots, b_n)_B$, then

$$U + V:(a_1 + b_1, \ldots, a_n + b_n)_B \quad \text{and} \quad rU:(ra_1, \ldots, ra_n)_B.$$

That is, the coordinates of a sum are given by the sum of the two n-tuples of coordinates, and the coordinates of a scalar multiple are the scalar multiple of the n-tuple of coordinates.

Proof Suppose $B = \{W_1, \ldots, W_n\}$, then by assumption

$$U = a_1W_1 + \cdots + a_nW_n \quad \text{and} \quad V = b_1W_1 + \cdots + b_nW_n.$$

Therefore,

$$U + V = (a_1 + b_1)W_1 + \cdots + (a_n + b_n)W_n.$$

Thus the coordinates of $U + V$ with respect to B are $a_1 + b_1, \ldots, a_n + b_n$ as desired.

For rU,

$$rU = r(a_1 W_1 + \cdots + a_n W_n) = (ra_1)W_1 + \cdots + (ra_n)W_n$$

or $rU{:}(ra_1, \ldots, ra_n)_B$.

Example 6 $B = \{1, t, t^2, t^3\}$ is a basis for $R_4[t]$, for which coordinates can be found by inspection.

Let $U = t^3 - 3t + 7$ and $V = -t^3 + 2t^2 + 6t$. Then the sum $U + V = 2t^2 + 3t + 7$. Now $U{:}(7, -3, 0, 1)_B$, $V{:}(0, 6, 2, -1)_B$, $U + V{:}(7, 3, 2, 0)_B$ and in $\mathscr{V}_4$, $(7, -3, 0, 1) + (0, 6, 2, -1) = (7, 3, 2, 0)$. Thus addition of vectors in $R_4[t]$ can be performed by adding their coordinates with respect to B in $\mathscr{V}_4$.

In general, if $\dim \mathscr{V} = n$ and B is a basis for $\mathscr{V}$, then for each vector in $\mathscr{V}$ there is an n-tuple of coordinates with respect to B and conversely. Also the operations of addition and scalar multiplication in $\mathscr{V}$ can be performed on the coordinates as if they were vectors of $\mathscr{V}_n$. In other words, as sets the spaces $\mathscr{V}$ and $\mathscr{V}_n$ correspond and as algebraic systems they correspond. Such a relationship is called an "isomorphism." The definition of this correspondence is in four parts: the first two establish the correspondence between the sets and the last two state that the correspondence "preserves addition and scalar multiplication."

Definition A vector space $\mathscr{V}$ is *isomorphic* to the vector space $\mathscr{W}$, written $\mathscr{V} \cong \mathscr{W}$, if there exists a correspondence T from the elements of $\mathscr{V}$ to those of $\mathscr{W}$ which satisfies:

1. For each $V \in \mathscr{V}$, $T(V)$ is a unique vector of $\mathscr{W}$.
2. For each $W \in \mathscr{W}$, there is a unique vector $V \in \mathscr{V}$, such that $T(V) = W$.
3. For all $U, V \in \mathscr{V}$, $T(U) + T(V) = T(U + V)$.
4. For all $U \in \mathscr{V}$ and $r \in R$, $rT(U) = T(rU)$.

In Chapter 4 terminology will be adopted which will considerably shorten this definition. But for now it precisely states the nature of an isomorphism. The first two conditions insure that for every vector in $\mathscr{V}$ there corresponds one and only one vector in $\mathscr{W}$, and conversely. The last two conditions insure that the algebraic structure of $\mathscr{V}$ corresponds to the algebraic structure of $\mathscr{W}$. For example, $T(U) + T(V) = T(U + V)$ means that the result obtained when two vectors, U and V, are added in $\mathscr{V}$ and then the corresponding vector, $T(U + V)$ is found in $\mathscr{W}$, is the same as the result obtained when the vectors $T(U)$ and $T(V)$ corresponding to U and V are added in $\mathscr{W}$. A vector space is a set of vectors together with the operations of addition and scalar multiplication, and an isomorphism establishes a total correspondence between the vectors and the algebraic operations. Therefore, any algebraic result obtainable in $\mathscr{V}$ could be obtained in $\mathscr{W}$, and the two vector spaces are indistinguishable as algebraic systems. Of course the vectors in $\mathscr{V}$ and $\mathscr{W}$ may be quite different, as shown by the next example.

Example 7 The vector space C is isomorphic to $\mathscr{V}_2$.

We will use the basis $B = \{1, i\}$ for C to establish a correspondence T from C to $\mathscr{V}_2$. For each vector $U \in C$, if $U:(a, b)_B$, then set $T(U) = (a, b)$. Thus $T(4 - 7i) = (4, -7)$ and $T(i) = (0, 1)$.

T is a "transformation" or "mapping" which sends complex numbers to ordered pairs. Thus T satisfies the first condition in the definition of an isomorphism.

The second condition requires that no two complex numbers are sent by T to the same ordered pair and that every ordered pair corresponds to some complex number. To check the first part of this condition, suppose $U, V \in C$ are sent to the same ordered pair. That is, $T(U) = (a, b) = T(V)$. Then by the definition of T, $U = a + bi = V$, and only one vector corresponds to (a, b). For the second part, suppose $W \in \mathscr{V}_2$. If $W = (c, d)$, then $c + di \in C$ and $T(c + di) = W$. Thus T satisfies the second condition of the definition.

The third condition of the definition requires that T "preserve addition." Because of the way T is defined, this is established by Theorem 2.13. But to check it directly, suppose $U = a + bi$ and $V = c + di$, then $U + V = (a + c) + (b + d)i$. Therefore $T(U) = (a, b)$, $T(V) = (c, d)$, and $T(U + V) = (a + c, b + d)$, so that

$$T(U) + T(V) = (a, b) + (c, d) = (a + c, b + d) = T(U + V).$$

By a similar argument we can show that T satisfies the fourth condition in the definition of isomorphism and say that T "preserves scalar multiplication."

Thus C is isomorphic to $\mathscr{V}_2$ and as vector spaces they are indistinguishable; the difference in appearance of the vectors is of no algebraic significance. This example is merely a special case of a theorem which states that there is essentially only one vector space for each finite dimension.

Theorem 2.14 If dim $\mathscr{V} = n$, $n \neq 0$, then $\mathscr{V}$ is isomorphic to $\mathscr{V}_n$.

Proof We will use the association of vectors with coordinates to obtain a correspondence T from $\mathscr{V}$ to $\mathscr{V}_n$. Since $\mathscr{V}$ is a vector space, it has a basis B. Then if $U \in \mathscr{V}$, U: $(a_1, \ldots, a_n)_B$ and we define $T(U)$ by $T(U) = (a_1, \ldots, a_n)$. Thus for $U \in \mathscr{V}$, $T(U)$ is a uniquely determined ordered n-tuple in $\mathscr{V}_n$.

To see that T satisfies the second condition of the definition of an isomorphism, first suppose $T(U) = T(V)$ for some $U, V \in \mathscr{V}$. Then U and V have the same coordinates with respect to B. But coordinates uniquely determine a vector, so $U = V$. Now suppose $W \in \mathscr{V}_n$, say $W = (x_1, \ldots, x_n)$. Then there exists a vector $Z \in \mathscr{V}$ such that Z:$(x_1, \ldots, x_n)_B$. Therefore, $T(Z) = W$, and every n-tuple is associated with a vector from $\mathscr{V}$. Thus condition 2 is satisfied.

The third and fourth conditions follow at once from Theorem 2.13, or may be obtained directly as in Example 7. Therefore $\mathscr{V}$ is isomorphic to $\mathscr{V}_n$.

To say that two vector spaces are isomorphic is to say that one cannot be distinguished from the other by algebraic means. Thus there is essentially only one vector space for each finite dimension, and the study of finite-dimensional vector spaces could be carried out entirely with the spaces $\mathscr{V}_n$. However, this is not a practical approach. For example, the space of 2×2 matrices is isomorphic to $\mathscr{V}_4$ (by problem 8, Section 4), but there are many situations in which the arrangement of numbers in a matrix is too useful to be lost by writing the numbers in a 4-tuple. The same is true for the polynomial spaces $R_n[t]$. An operation such as differentiation has a simple definition for polynomials, but its meaning would be lost if defined in terms of n-tuples. So even though two vector spaces may be algebraically the same, the fact that their elements are different is sufficient to continue regarding them as distinct spaces.

Problems

1. Find the coordinates of $(2, -5)$ with respect to each of the following bases for $\mathscr{V}_2$:

   a. $\{(6, -5), (2, 5)\}$. b. $\{(3, 1), (-2, 6)\}$. c. $\{E_i\}$. d. $\{(1, 0), (2, -5)\}$.

2. Follow the same directions as in problem 1 for the vector $(1, 1)$.

3. Obtain by inspection coordinates of the following vectors with respect to the basis $B = \{(4, 1), (1, 0)\}$:

   a. $(1, 0)$. b. $(0, 1)$. c. $(3, -6)$. d. $(2, 8)$. e. $(7, -5)$.

4. Show that $B = \{t^3, t^3 + t, t^2 + 1, t + 1\}$ is a basis for $R_4[t]$ and find the coordinates of the following vectors with respect to B:
   a. $t^2 + 1$. b. t^3. c. 4 d. t^2. e. $t^3 - t^2$, f. $t^2 - t$.

5. Find the coordinates of the following vectors from C with respect to the basis $B = \{1 - 2i, i - 3\}$:
   a. $3 - i$. b. i. c. -5. d. $1 + 3i$. e. $3 + 4i$.

6. Find the coordinates of $(2, -3, 5)$ with respect to
   a. $B_1 = \{(1, 0, 0), (1, 1, 0), (1, 1, 1)\}$.
   b. $B_2 = \{(1, -1, 0), (-4, 6, -10), (-1, 3, -9)\}$.
   c. $B_3 = \{(1, 1, 0), (0, 1, 1), (1, 0, 1)\}$.
   d. $B_4 = \{(1, 0, 0), (0, 1, 0), (0, -5, 5)\}$.

7. Suppose B is a basis of $\mathscr{V}_2$ such that $(4, 1)$:$(3, 2)_B$ and $(1, 1)$:$(6, -3)_B$. Find the basis B.

8. Find B if $(1, 0)$:$(5, -8)_B$ and $(0, 1)$:$(-1, 2)_B$.

9. Suppose t^2:$(1, 0, 1)_B$, t:$(0, 1, 1)_B$, and 1:$(1, 1, 0)_B$. Find the basis B for $R_3[t]$.

10. Find the basis B for $R_2[t]$ if $10 - t$:$(4, -3)_B$ and $5t + 6$:$(1, 1)_B$.

11. Define a correspondence from $R[t]$ to C by $T(a + bt) = a + bi$. Prove that T is an isomorphism from $R_2[t]$ to C.

12. Prove that any two subspaces of $\mathscr{V}_3$ represented by lines through the origin are isomorphic.

13. Read the following statements.
    a. $(4, 7)$:$(1, 5)_B$. b. $2t^2 - 2$:$(-2, 1, 1)_{\{t+2, 3t^2, 4-t^2\}}$.

14. Show that the following maps satisfy all but one of the four conditions in the definition of an isomorphism.
    a. $T_1(a, b) = (a, b, 0)$; T_1 from $\mathscr{V}_2$ to $\mathscr{V}_3$.
    b. $T_2(a, b, c) = (a + b, 2a - c)$; T_2 from $\mathscr{V}_3$ to $\mathscr{V}_2$.

15. Let $T(a, b) = 2a + (a - b)t$.
    a. Find $T(1, 0)$ and $T(2, -4)$.
    b. Prove that T is an isomorphism from $\mathscr{V}_2$ to $R_2[t]$.

16. Prove that $\mathscr{V}_2$ and $\mathscr{V}_3$ are not isomorphic. (The idea used to obtain a correspondence in Example 7 cannot be applied here, however, this does not constitute a proof that no isomorphism exists.)

§6. Direct Sums

The sum of two subspaces $\mathscr{S}$ and $\mathscr{T}$ of a vector space $\mathscr{V}$ has been defined as $\mathscr{S} + \mathscr{T} = \{U + V \mid U \in \mathscr{S} \text{ and } V \in \mathscr{T}\}$. $\mathscr{S} + \mathscr{T}$ is again a subspace of $\mathscr{V}$ (problem 10 page 43), and the spaces $\mathscr{S}$ and $\mathscr{T}$ are called *summands* of the sum $\mathscr{S} + \mathscr{T}$.

Example 1 Let $\mathscr{S} = \mathscr{L}\{(1, 1, 0), (0, 1, 1)\}$ and $\mathscr{T} = \mathscr{L}\{(1, 0, 1), (1, 1, 1)\}$. Then the sum of $\mathscr{S}$ and $\mathscr{T}$ is given by

$$\mathscr{S} + \mathscr{T} = \mathscr{L}\{(1, 1, 0), (0, 1, 1), (1, 0, 1), (1, 1, 1)\}.$$

This sum equals $\mathscr{V}_3$ because any three of the vectors are linearly independent. If $\mathscr{S}$ and $\mathscr{T}$ are regarded as planes through the origin in 3-space, then $\mathscr{S} + \mathscr{T}$ expresses 3-space as the sum of two planes.

In Example 1, each vector $W \in \mathscr{V}_3$ can be expressed as a sum of a vector $U \in \mathscr{S}$ and a vector $V \in \mathscr{T}$. Figure 3 shows such a situation.

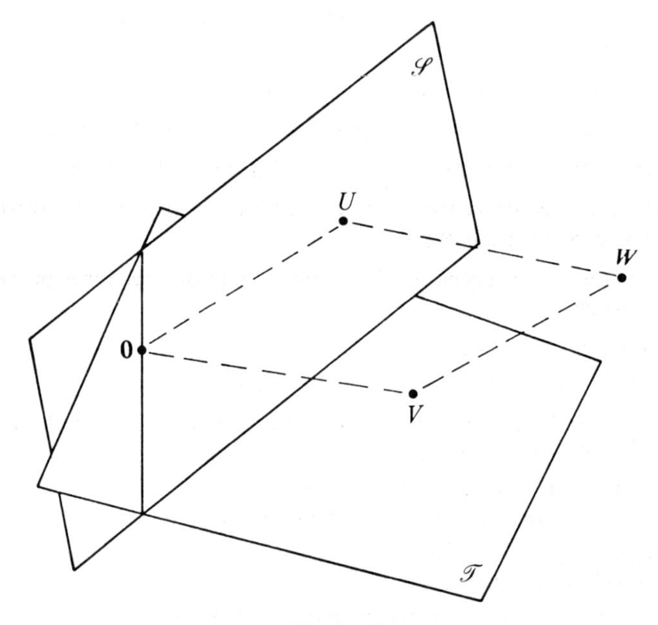

Figure 3

There is a close analogy between writing a vector space as the sum of subspaces and writing a vector as a linear combination of vectors. In the latter case, coordinates are obtained when there is a unique linear combination for each vector. For sums of subspaces, the most interesting use will again be when there are unique expressions for vectors. Suppose this is not the case for $\mathscr{V} = \mathscr{S} + \mathscr{T}$. That is, suppose there is a vector $W \in \mathscr{V}$ such that $W = U_1 + V_1$ and $W = U_2 + V_2$ with $U_1, U_2 \in \mathscr{S}$, $V_1, V_2 \in \mathscr{T}$, all distinct. Then $\mathbf{0} = W - W = U_1 + V_1 - (U_2 - V_2)$, implies $U_1 - U_2 = V_2 - V_1$. Call this vector Z, then $Z = U_1 - U_2 \in \mathscr{S}$ and $Z = V_2 - V_1 \in \mathscr{T}$, so $Z \in \mathscr{S} \cap \mathscr{T}$. By assumption, $U_1 \neq U_2$, so $Z \neq \mathbf{0}$ and $\mathscr{S} \cap \mathscr{T}$ is not the zero subspace. Therefore if a vector can be expressed as a sum of vectors from $\mathscr{S}$ and $\mathscr{T}$ in more then one way, then $\mathscr{S}$ and $\mathscr{T}$ have a nontrivial intersection.

Example 2 If $\mathscr{S}$ and $\mathscr{T}$ are the subspaces of $\mathscr{V}_3$ given in Example 1, then $\mathscr{S} \cap \mathscr{T} = \mathscr{L}\{(1, 2, 1)\} \neq \{\mathbf{0}\}$. Show that this implies that some vector $W \in \mathscr{V}_3$ does not have a unique expression as a sum of one vector from $\mathscr{S}$ and one vector from $\mathscr{T}$.

Choose $W = (1, -1, -1)$. Then $W = U_1 + V_1$ with $U_1 = (1, 0, -1) \in \mathscr{S}$ and $V_1 = (0, -1, 0) \in \mathscr{T}$. Now

$$\begin{aligned} W &= U_1 + V_1 + \mathbf{0} = U_1 + V_1 + (1, 2, 1) - (1, 2, 1) \\ &= [(1, 0, -1) + (1, 2, 1)] + [(0, -1, 0) - (1, 2, 1)] \\ &= (2, 2, 0) + (-1, -3, -1). \end{aligned}$$

Since $(1, 2, 1) \in \mathscr{S} \cap \mathscr{T}$ and $\mathscr{S}$, $\mathscr{T}$ are subspaces, $U_2 = (2, 2, 0) \in \mathscr{S}$ and $V_2 = (-1, -3, -1) \in \mathscr{T}$. Thus $W = U_2 + V_2$ is another expression for W. But $\mathscr{S} \cap \mathscr{T}$ contains an infinite number of vectors, so there are an infinite number of ways to express $(1, -1, -1)$ as a sum of vectors from $\mathscr{S}$ and $\mathscr{T}$. Moreover since $\mathscr{V}_3 = \mathscr{S} + \mathscr{T}$, any vector in $\mathscr{V}_3$ could have been taken for W.

Theorem 2.15 Let $\mathscr{S}$ and $\mathscr{T}$ be subspaces of a vector space $\mathscr{V}$ and $\mathscr{V} = \mathscr{S} + \mathscr{T}$. Every vector $W \in \mathscr{V}$ has a unique expression as a sum $W = U + V$ with $U \in \mathscr{S}$, $V \in \mathscr{T}$ if and only if $\mathscr{S} \cap \mathscr{T} = \{\mathbf{0}\}$.

Proof ($\Rightarrow$) Suppose $W \in \mathscr{V}$ and $W = U + V$, $U \in \mathscr{S}$, $V \in \mathscr{T}$ is unique. Let $Z \in \mathscr{S} \cap \mathscr{T}$. Then $U + Z \in \mathscr{S}$ and $V - Z \in \mathscr{T}$ since $\mathscr{S}$ and $\mathscr{T}$ are closed under addition. Now $W = (U + Z) + (V - Z)$, and by uniqueness $U = U + Z$ and $V = V - Z$, so $Z = \mathbf{0}$ and $\mathscr{S} \cap \mathscr{T} = \{\mathbf{0}\}$.

($\Leftarrow$) The contrapositive* of this statement was proved just before Example 2. That is, if there is not a unique expression, then $\mathscr{S} \cap \mathscr{T} \neq \{\mathbf{0}\}$.

Definition Let $\mathscr{S}$, $\mathscr{T}$ be subspaces of a vector space $\mathscr{V}$. The sum $\mathscr{S} + \mathscr{T}$ is said to be a *direct sum* and is written $\mathscr{S} \oplus \mathscr{T}$, if $\mathscr{S} \cap \mathscr{T} = \{\mathbf{0}\}$. If $\mathscr{V} =$

*The *contrapositive* of the statement "P implies Q" is the statement "not Q implies not P." A statement is equivalent to its contrapositive so that P implies Q may be proved by showing that the negation of Q implies the negation of P.

$\mathscr{S} \oplus \mathscr{T}$, then $\mathscr{S} \oplus \mathscr{T}$ is a *direct sum decomposition* of $\mathscr{V}$, and $\mathscr{S}$ and $\mathscr{T}$ may be called *direct summands* of $\mathscr{V}$.

Example 3 $B = \{(4, -7), (2, 5)\}$ is a basis for $\mathscr{V}_2$. Let $\mathscr{S} = \mathscr{L}\{(4, -7)\}$ and $\mathscr{T} = \mathscr{L}\{(2, 5)\}$. Consider the sum $\mathscr{S} + \mathscr{T}$.

Since B spans $\mathscr{V}_2$, $\mathscr{V}_2 = \mathscr{S} + \mathscr{T}$. Now suppose $Z \in \mathscr{S} \cap \mathscr{T}$, then $Z = a(4, -7)$ and $Z = b(2, 5)$ for some $a, b \in R$, and because B is linearly independent, $Z = \mathbf{0}$. Therefore, the sum $\mathscr{S} + \mathscr{T}$ is direct, and we can write $\mathscr{V}_2 = \mathscr{S} \oplus \mathscr{T}$.

Geometrically, Example 3 shows that the Cartesian plane E^2 may be expressed as the direct sum of two lines through the origin. It should be clear that the same result would be obtained using any two distinct lines through the origin in E^2. How does this relate to the idea of a coordinate system in E^2?

Example 4 Suppose $\mathscr{S} = \{(x, y, z) | x + y + z = 0\}$ and $\mathscr{T} = \mathscr{L}\{(3, 1, 4)\}$. The sum $\mathscr{S} + \mathscr{T}$ is direct, for if $(a, b, c) \in \mathscr{S} \cap \mathscr{T}$, then $a + b + c = 0$, and for some scalar t, $(a, b, c) = t(3, 1, 4)$. Eliminating a, b, c gives $3t + t + 4t = 0$ or $t = 0$, so $(a, b, c) = \mathbf{0}$ and $\mathscr{S} \cap \mathscr{T} = \{\mathbf{0}\}$. Thus we can write $\mathscr{S} \oplus \mathscr{T}$.

Now is $\mathscr{S} \oplus \mathscr{T}$ a direct sum decomposition of $\mathscr{V}_3$? That is, is $\mathscr{S} \oplus \mathscr{T} = \mathscr{V}_3$? Since $(1, -1, 0)$ and $(1, 0, -1)$ span $\mathscr{S}$, the direct sum is $\mathscr{L}\{(1, -1, 0), (1, 0, -1), (3, 1, 4)\}$. Since the three vectors in this set are linearly independent, $\mathscr{V}_3 = \mathscr{S} \oplus \mathscr{T}$.

Suppose one wished to find the expression for a vector, say $(8, 9, -1)$, using the sum $\mathscr{S} \oplus \mathscr{T}$. There must be scalars x, y, t such that $(8, 9, -1) = (x, y, -x - y) + t(3, 1, 4)$. This equation has the solution $(8, 9, -1) = (2, 7, -9) + (6, 2, 8)$ with $(2, 7, -9) \in \mathscr{S}$ and $(6, 2, 8) \in \mathscr{T}$ as desired.

The decomposition obtained in Example 4 expresses 3-space as the direct sum of a plane $\mathscr{S}$ and a line $\mathscr{T}$, not in $\mathscr{S}$. (Why must both the plane and the line pass through the origin?) Given a point W, the points $U \in \mathscr{S}$ and $V \in \mathscr{T}$ can be obtained geometrically. The point U is the intersection of $\mathscr{S}$ and the line through W parallel to the line $\mathscr{T}$. And V is the intersection of the line $\mathscr{T}$ and a line through W parallel to the plane $\mathscr{S}$. Then, as illustrated in Figure 4, $U + V = W$ by the parallelogram rule for addition.

It should be clear that 3-space can be expressed as the direct sum of any plane and a line not in the plane, if both pass through the origin.

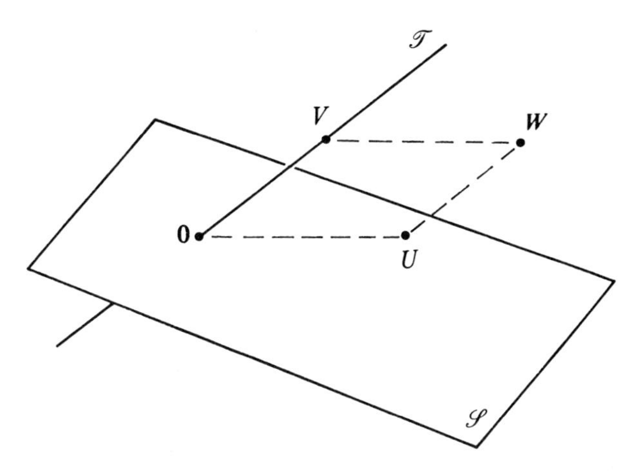

Figure 4

Example 5 Let $\mathscr{S}$ be as in Example 4 and $\mathscr{U} = \mathscr{L}\{(1, 0, 1)\}$. Since $(1, 0, 1) \notin \mathscr{S}$, $\mathscr{S} + \mathscr{T}$ is a direct sum. To see that $\mathscr{S} \oplus \mathscr{T} = \mathscr{V}_3$, consider the equation $(a, b, c) = (x, y, -x - y) + (z, 0, z)$. This has the solution $x = \frac{1}{2}(-a + b + c)$, $y = b$, $z = \frac{1}{2}(a + b + c)$ for any values of a, b, c and so $\mathscr{V}_3 \subset \mathscr{S} \oplus \mathscr{T}$. Therefore $\mathscr{S} \oplus \mathscr{T}$ is another direct sum decomposition of $\mathscr{V}_3$.

In terms of this direct sum, $(8, 9, -1) = (0, 9, -9) + (8, 0, 8)$ with $(0, 9, -9) \in \mathscr{S}$ and $(8, 0, 8) \in \mathscr{T}$.

Combining the results of Examples 4 and 5 gives $\mathscr{S} \oplus \mathscr{T} = \mathscr{S} \oplus \mathscr{U}$ but $\mathscr{T} \neq \mathscr{U}$, therefore there is no cancellation law for direct sums.

Theorem 2.16 If $\mathscr{S}$ and $\mathscr{T}$ are subspaces of a finite-dimensional vector space and $\mathscr{S} + \mathscr{T}$ is a direct sum, then

$$\dim(\mathscr{S} \oplus \mathscr{T}) = \dim \mathscr{S} + \dim \mathscr{T}.$$

Proof Suppose $\{U_1, \ldots, U_m\}$ and $\{V_1, \ldots, V_n\}$ are bases for $\mathscr{S}$ and $\mathscr{T}$, respectively. Then $\dim \mathscr{S} = m$, $\dim \mathscr{T} = n$, and the proof will be complete if it can be shown that $\{U_1, \ldots, U_m, V_1, \ldots, V_n\}$ is a basis for $\mathscr{S} \oplus \mathscr{T}$. It is easily seen that the set spans $\mathscr{S} \oplus \mathscr{T}$, therefore suppose

$$a_1U_1 + \cdots + a_mU_m + b_1V_1 + \cdots + b_nV_n = \mathbf{0}.$$

Let $Z = a_1U_1 + \cdots + a_mU_m$, so $Z \in \mathscr{S}$. But $Z = -b_1V_1 - \cdots - b_nV_n$, so $Z \in \mathscr{T}$ also. Thus $Z \in \mathscr{S} \cap \mathscr{T} = \{\mathbf{0}\}$ and $Z = \mathbf{0}$. Now the U's are linearly independent so $a_1 = \cdots = a_m = 0$, and the V's are linearly independent so $b_1 = \cdots = b_n = 0$. Therefore the set is linearly independent and a basis for $\mathscr{S} \oplus \mathscr{T}$. Since the basis contains $m + n$ vectors,

$$\dim(\mathscr{S} \oplus \mathscr{T}) = m + n = \dim \mathscr{S} + \dim \mathscr{T}.$$

(This result follows directly from the fact that $\dim(\mathscr{S} + \mathscr{T}) = \dim \mathscr{S} + \dim \mathscr{T} - \dim(\mathscr{S} \cap \mathscr{T})$, from problem 10, page 58.)

Unlike coordinates, it is not easy to see the value of a direct sum decomposition at this time. However, there are many situations in which it is advantageous to split a vector space up into subspaces via a direct sum.

Problems

1. Show that $\mathscr{L}\{(1, 7)\} + \mathscr{L}\{(2, 3)\}$ gives a direct sum decomposition of $\mathscr{V}_2$.
2. Show that the sum $\mathscr{L}\{(1, 4, -2)\} + \mathscr{L}\{(3, 0, 1)\}$ is direct. Does it give a direct sum decomposition of $\mathscr{V}_3$?
3. Let $\mathscr{S} = \mathscr{L}\{U, V\}$, $\mathscr{T} = \mathscr{L}\{W, X\}$ and determine if $\mathscr{S} + \mathscr{T}$ is a direct sum in each of the following cases:

   a. $U = (1, 0, 0)$, $V = (0, 1, 0)$, $W = (0, 0, 1)$, $X = (1, 1, 1)$.
   b. $U = (2, -1, 3)$, $V = (1, 2, 0)$, $W = (-1, 2, -1)$, $X = (2, -4, 2)$.
   c. $U = (1, 0, 0, 0)$, $V = (0, 1, 0, 0)$, $W = (0, 0, 1, 0)$, $X = (0, 0, 0, 1)$.
   d. $U = (3, -2, 1, 0)$, $V = (1, 3, 0, 4)$, $W = (4, 1, 1, 4)$, $X = (2, 1, 0, 1)$.

4. Let $N = (1, -1, 3)$, $\mathscr{S} = \{U \in \mathscr{E}_3 | U \circ N = 0\}$ and $\mathscr{T} = \mathscr{L}\{N\}$. Show that $\mathscr{E}_3$ is the direct sum of the plane $\mathscr{S}$ and the normal line $\mathscr{T}$.
5. In problem 4 each vector $W \in \mathscr{E}_3$ can be expressed as $W = V_1 + V_2$ with $V_1 \in \mathscr{S}$ and $V_2 \in \mathscr{T}$. Use the fact that W, V_1, V_2, and $\mathbf{0}$ determine a rectangle along with properties of the cosine function to show that

$$V_2 = \left(\frac{W \circ N}{N \circ N}\right)N \quad \text{and} \quad V_1 = W - V_2$$

6. For the following spaces, determine if $\mathscr{V}_3 = \mathscr{S} + \mathscr{T}$. When this is the case determine if the sum is direct.

   a. $\mathscr{S} = \{(x, y, z) | x = 2y\}$, $\mathscr{T} = \{(x, y, z) | z = 0\}$.
   b. $\mathscr{S} = \{(x, y, z) | x = y, z = 0\}$, $\mathscr{T} = \{(x, y, z) | x = 0, y = -z\}$.
   c. $\mathscr{S} = \mathscr{L}\{(4, 1, 3)\}$, $\mathscr{T} = \mathscr{L}\{(2, 1, 1), (1, 0, 2)\}$.
   d. $\mathscr{S} = \mathscr{L}\{(2, -1, 1)\}$, $\mathscr{T} = \mathscr{L}\{(1, 5, 3), (3, 4, 4)\}$.

7. a. Let $\mathscr{S}$ and $\mathscr{T}$ be as in problem 6c. Find $U \in \mathscr{S}$, $V \in \mathscr{T}$ such that $U + V = (-4, -1, 5)$.
   b. Let $\mathscr{S}$ and $\mathscr{T}$ be as in problem 3b. Find $U \in \mathscr{S}$ and $V \in \mathscr{T}$ such that $U + V = (0, -6, 5)$.

8. Let $\mathscr{S} = \{(a, b, c) | a + b - c = 0\}$ and $\mathscr{T} = \{(a, b, c) | 3a - b - c = 0\}$. Find $U \in \mathscr{S}$ and $V \in \mathscr{T}$ such that $U + V = (1, 1, 0)$.

9. Prove that if $\mathscr{S}$ is a proper subspace of a finite-dimensional vector space $\mathscr{V}$, then there exists a subspace $\mathscr{T}$ such that $\mathscr{V} = \mathscr{S} \oplus \mathscr{T}$.

10. Find a direct sum decomposition for $R[t]$.

Review Problems

1. Let $S = \{U_1, \ldots, U_n\}$. Determine if the following statements are true or false.
   a. If $a_1U_1 + \cdots + a_nU_n = \mathbf{0}$ when $a_1 = \cdots = a_n = 0$, then S is linearly independent.
   b. If $\mathscr{V} = \mathscr{L}(S)$, then $\mathscr{V} = a_1U_1 + \cdots + a_nU_n$ for some scalars $a_1, \ldots, a_n$.
   c. If $\mathscr{L}(S) = \mathscr{L}\{U_2, \ldots, U_n\}$, then $S = \{U_2, \ldots, U_n\}$.
   d. If S is linearly dependent, then $a_1U_1 + \cdots + a_nU_n = \mathbf{0}$ for some scalars $a_1, \ldots, a_n$.
   e. If S linearly dependent, then U_n is a linear combination of $U_1, \ldots, U_{n-1}$.
   f. If S spans the vector space $\mathscr{V}$, then $\dim \mathscr{V} \geq n$.

2. Suppose $\mathscr{V}$, $\mathscr{W}$ are vector spaces, U, V are vectors, and a, b are scalars. In our notation which of the following statements can be meaningful?

   a. $\{a, b\} \in R$.
   b. $(a, b) \in \mathscr{V}$.
   c. $a\mathscr{V}$.
   d. $\mathscr{V} + \mathscr{W}$.
   e. $\mathscr{V} \in \mathscr{W}$.
   f. $(a, b) \in R$.
   g. $\mathscr{W} \subset \mathscr{V}$.
   h. $a, b \in R$.
   i. $U + V$.
   j. $U \cap \mathscr{V}$.
   k. $U \in V$.
   l. $\{a\} \in R$.
   m. $U \subset \mathscr{V}$.
   n. $(U, V) \subset \mathscr{V}$.
   o. $\mathscr{V} \in \mathscr{V}$.
   p. $\mathscr{V} \cap \mathscr{W}$.
   q. $\mathscr{V} = aU + bV$.
   r. Ua.
   s. $\mathscr{V} = \{U, V\}$.
   t. $U \oplus V$.

3. Suppose $S = \{2t + 2, 3t + 1, 3 - 7t\} \subset R[t]$.
   a. Show that S is linearly dependent.
   b. Show that t may be expressed as a linear combination of the elements from S in an infinite number of ways.
   c. Show that $\mathscr{L}(S) = \mathscr{L}\{3t + 1, 3 - 7t\}$.

4. Suppose $\mathscr{V}$ is an n-dimensional vector space. Determine which of the following statements are true. If false, find a counterexample.
   a. Any n vectors span $\mathscr{V}$.
   b. Any $n + 1$ vectors are linearly dependent in $\mathscr{V}$.
   c. Any n vectors form a basis for $\mathscr{V}$.
   d. Any $n - 1$ vectors are linearly independent in $\mathscr{V}$.

e. Any basis for $\mathscr{V}$ contains n vectors.
f. Any $n + 1$ vectors span $\mathscr{V}$.
g. Any $n - 1$ vectors fail to span $\mathscr{V}$.

5. Suppose $\mathscr{S} = \{(x, y, z) | 2x - y + 4z = 0\}$ and $\mathscr{T} = \{(x, y, z) | 5y - 3z = 0\}$ are subspaces of $\mathscr{V}_3$.

   a. Show that the sum $\mathscr{S} + \mathscr{T}$ is not direct.
   b. Find $U \in \mathscr{S}$ and $V \in \mathscr{T}$ such that $U + V = (2, 9, 6)$.

6. Prove or disprove: There exists a vector space containing exactly two vectors.

7. Find a counterexample for each of the following false statements.

   a. If $\{U_1, \ldots, U_n\}$ is a basis for $\mathscr{V}$ and $\mathscr{S}$ is a k-dimensional subspace of $\mathscr{V}$, then $\{U_1, \ldots, U_k\}$ is a basis for $\mathscr{S}$.
   b. If $\{U_1, U_2, V_1, V_2\}$ is a basis for $\mathscr{S}$ and $\{U_1, U_2, W_1, W_2, W_3\}$ is a basis for $\mathscr{T}$, then $\{U_1, U_2\}$ is a basis for $\mathscr{S} \cap \mathscr{T}$.
   c. Any system of n linear equations in n unknowns has a solution.
   d. Any system of n linear equations in m unknowns has a solution provided $n \leq m$.
   e. If $\{U_1, \ldots, U_n\}$ is linearly independent in $\mathscr{V}$, then $\dim \mathscr{V} \leq n$.

3

Systems of Linear Equations

Many questions about vectors and vector spaces yield systems of linear equations. Such collections of first degree polynomial equations in several unknowns occur repeatedly in the problems and examples of the previous chapter. If one looks back over these situations, it will be noticed that the problem was not always to find values that satisfy all the equations simultaneously. Often the problem was to show that there are no solutions or that a system has many solutions. In addition the number of equations in a system was not always the same as the number of unknowns. However, even in the simple case of two equations in two unknowns it is possible to have no solution, a unique solution, or many solutions as illustrated by the following three systems:

$$\begin{aligned} 2x + 3y &= 4 \\ 4x + 6y &= 0, \end{aligned} \qquad \begin{aligned} 2x + 3y &= 4 \\ x - y &= 0, \end{aligned} \qquad \begin{aligned} 2x + 3y &= 4 \\ 4x + 6y &- 8. \end{aligned}$$

Systems of linear equations will continue to occur in our study of linear algebra, so a general procedure is needed to determine if a system has solutions and if so to find their values. Also, in the process of solving this problem, several important concepts will be obtained.

§1. Notation

If different letters are used to denote the various coefficients in a general system of n linear equations with m unknowns, we soon run out of symbols. Also, such notation would make it difficult to identify where particular coefficients occur in the system. Therefore a single symbol is used with a double indexing system. The first index or subscript designates the equation in which the coefficient appears and the second index designates the unknown for which it is the coefficient. Thus a system of n linear equations in m unknowns may be written as

$$\begin{aligned} a_{11}x_1 + a_{12}x_2 + \cdots + a_{1m}x_m &= b_1 \\ a_{21}x_1 + a_{22}x_2 + \cdots + a_{2m}x_m &= b_2 \\ &\vdots \\ a_{n1}x_1 + a_{n2}x_2 + \cdots + a_{nm}x_m &= b_n. \end{aligned} \tag{1}$$

With this notation a_{ij} is the coefficient of x_j, $1 \leq j \leq m$, in the ith equation, $1 \leq i \leq n$. If the system

$$\begin{aligned} 2x_1 + 3x_2 - 5x_3 &= 4 \\ 6x_2 + x_3 &= 7 \end{aligned}$$

were written with this notation, then $a_{11} = 2$, $a_{12} = 3$, $a_{13} = -5$, $a_{21} = 0$, $a_{22} = 6$, $a_{23} = 1$, $b_1 = 4$, and $b_2 = 7$. The double index notation is very useful and will be found throughout the remainder of the text.

The above system of linear equations may arise in several different ways. For example, let $U_1 = (a_{11}, \ldots, a_{n1})$, $U_2 = (a_{12}, \ldots, a_{n2}), \ldots, U_m = (a_{1m}, \ldots, a_{nm})$, and $W = (b_1, \ldots, b_n)$ be vectors from $\mathscr{V}_n$. Then the system of equations (1) arises in the following situations:

1. Determining if $W \in \mathscr{L}\{U_1, \ldots, U_m\}$. That is, do there exist scalars $x_1, \ldots, x_m$ such that $W = x_1U_1 + \cdots + x_mU_m$?
2. Determining if the set $\{U_1, \ldots, U_m\}$ is linearly independent. That is, do there exist scalars $x_1, \ldots, x_m$, not all zero, such that $x_1U_1 + \cdots + x_mU_m = \mathbf{0}$? In this case $W = \mathbf{0}$ so $b_1 = \cdots = b_n = 0$.
3. Determining the coordinates of W with respect to the basis $\{U_1, \ldots, U_n\}$. In this case $n = m$ and $W{:}(x_1, \ldots, x_n)_{\{U_1,\ldots,U_n\}}$.

There are many other ways in which the system (1) might arise, but these three are general enough to yield all possible systems of linear equations.

Once the doubly indexed notation for coefficients is adopted, it turns out that most of the remaining notation is superfluous. This is shown in the following example.

Example 1 The following is a system of two equations and an array of numbers derived from the coefficients and the constant terms.

$$\begin{aligned} x + 2y &= 1 \\ 2x + 3y &= 4 \end{aligned} \qquad \begin{matrix} 1 & 2 & 1 \\ 2 & 3 & 4 \end{matrix}$$

Using the technique of eliminating x to solve for y, one multiplies the first equation by -2 and adds the result to the second equation. Applying the same procedure to the array, we add -2 times the first row to the second row. This yields the following set of equations and corresponding array of numbers.

$$\begin{aligned} x + 2y &= 1 \\ -y &= 2 \end{aligned} \qquad \begin{matrix} 1 & 2 & 1 \\ 0 & -1 & 2 \end{matrix}$$

Now multiply the new second equation (row) by -1 to obtain the following equations and array.

$$\begin{aligned} x + 2y &= 1 \\ y &= -2 \end{aligned} \qquad \begin{matrix} 1 & 2 & 1 \\ 0 & 1 & -2 \end{matrix}$$

Finally multiply the new second equation (row) by -2 and add the result to the first.

$$\begin{aligned} x \quad &= 5 \\ y &= -2 \end{aligned} \qquad \begin{matrix} 1 & 0 & 5 \\ 0 & 1 & -2 \end{matrix}$$

The final array of numbers, if properly interpreted, gives the solution.

This example shows that the plus signs, the equal signs, and even the symbols for the unknowns are not needed to solve a system with an elimination technique. It might be argued that, in the course of eliminating a variable, one often separates the equations and would not continually write down both equations. Although such a procedure might be fine for two equations in two unknowns, it often leads to problems with larger systems having different numbers of unknowns and equations.

Definition An $n \times m$ *matrix*, read "n by m matrix," is a rectangular array of numbers with n *rows* and m *columns*. Let A be an $n \times m$ matrix with the element in the ith row and jth column denoted a_{ij}. Then

$$A = (a_{ij}) = \begin{pmatrix} a_{11} & a_{12} & \cdots & a_{1m} \\ a_{21} & a_{22} & \cdots & a_{2m} \\ & & \vdots & \\ a_{n1} & a_{n2} & \cdots & a_{nm} \end{pmatrix}.$$

The following are examples of a 2×3 matrix, a 3×1 matrix, and a 3×3 matrix, respectively:

$$\begin{pmatrix} 2 & 5 & 6 \\ 9 & 3 & 1 \end{pmatrix}, \quad \begin{pmatrix} 1 \\ 7 \\ 3 \end{pmatrix}, \quad \begin{pmatrix} 6 & -4 & 8 \\ 1 & 0 & 3 \\ 4 & 2 & 9 \end{pmatrix}.$$

Definition The $n \times m$ matrix $A = (a_{ij})$ is the *coefficient matrix* for the following system of n linear equations in m unknowns:

$$\begin{aligned} a_{11}x_1 + a_{12}x_2 + \cdots + a_{1m}x_m &= b_1 \\ &\vdots \\ a_{n1}x_1 + a_{n2}x_2 + \cdots + a_{nm}x_m &= b_n \end{aligned}$$

The *augmented (coefficient) matrix* for this system, denoted A^*, is the $n \times (m + 1)$ matrix obtained from A by adding the constant terms as the

$(m + 1)$st column. Thus A^* has the form

$$A^* = \begin{pmatrix} a_{11} & a_{12} & \cdots & a_{1m} & b_1 \\ & & \vdots & & \\ a_{n1} & a_{n2} & \cdots & a_{nm} & b_n \end{pmatrix}.$$

Example 2 If A and A^* denote the coefficient and augmented matrices of the system

$$\begin{aligned} 2x + 3y - 5z + w &= 4 \\ x + y \qquad - w &= 7 \\ 9x \qquad + 3z - w &= 8 \end{aligned}$$

then

$$A = \begin{pmatrix} 2 & 3 & -5 & 1 \\ 1 & 1 & 0 & -1 \\ 9 & 0 & 3 & -1 \end{pmatrix} \quad \text{and} \quad A^* = \begin{pmatrix} 2 & 3 & -5 & 1 & 4 \\ 1 & 1 & 0 & -1 & 7 \\ 9 & 0 & 3 & -1 & 8 \end{pmatrix}.$$

The array of numbers considered in Example 1 is the augmented matrix of the given system without the matrix notation. As shown in that example, the augmented matrix contains all the essential information of a system of linear equations.

With the next two definitions, matrices can be used to express any system of linear equations as a simple equation of matrices.

Definition Let $A = (a_{ij})$ be an $n \times m$ matrix and $B = (b_{hk})$ a $p \times q$ matrix, then $A = B$ if $n = p$, $m = q$, and $a_{ij} = b_{ij}$ for all i and j.

Definition Let $E = (e_{ij})$ be an $n \times m$ matrix and $F = (f_j)$ an $m \times 1$ matrix, then the *product* of E and F is the $n \times 1$ matrix $EF = (g_i)$ with $g_i = e_{i1}f_1 + e_{i2}f_2 + \cdots + e_{im}f_m$ for each i, $1 \leq i \leq n$.

For example,

$$\begin{pmatrix} 2 & 3 & 7 \\ 1 & 2 & 1 \end{pmatrix} \begin{pmatrix} 3 \\ 2 \\ -1 \end{pmatrix} = \begin{pmatrix} 2(3) + 3(2) + 7(-1) \\ 1(3) + 2(2) + 1(-1) \end{pmatrix} = \begin{pmatrix} 5 \\ 6 \end{pmatrix}$$

Now suppose $A = (a_{ij})$ is the $n \times m$ matrix of coefficients for the system (1) of n linear equations in m unknowns, let $X = (x_j)$ be an $m \times 1$ matrix of the unknowns and let $B = (b_i)$ be the $n \times 1$ matrix of constant terms. Then the equations in (1) can be written as a single *matrix equation* $AX = B$.

Example 3 Using a matrix equation to write a system of linear equations,

$$\begin{aligned} x_1 - 3x_2 + x_3 &= 0 \\ 4x_1 + 6x_2 - x_3 &= 7 \end{aligned}$$

becomes

$$\begin{pmatrix} 1 & -3 & 1 \\ 4 & 6 & -1 \end{pmatrix} \begin{pmatrix} x_1 \\ x_2 \\ x_3 \end{pmatrix} = \begin{pmatrix} 0 \\ 7 \end{pmatrix},$$

and

$$\begin{aligned} 2x + y &= 1 \\ x - 3y &= 2 \\ 5x - y &= 4 \end{aligned}$$

becomes

$$\begin{pmatrix} 2 & 1 \\ 1 & -3 \\ 5 & -1 \end{pmatrix} \begin{pmatrix} x \\ y \end{pmatrix} = \begin{pmatrix} 1 \\ 2 \\ 4 \end{pmatrix}.$$

Definition Suppose $AX = B$ is a system of n linear equations in the m unknowns $x_1, \ldots, x_m$. Then $x_1 = c_1, \ldots, x_m = c_m$ or $X = C = (c_j)$ is a *solution* of this system if $AC = B$.

For the second system in Example 3,

$$\begin{pmatrix} x \\ y \end{pmatrix} = \begin{pmatrix} 5/7 \\ -3/7 \end{pmatrix}$$

is a solution. That is,

$$\begin{pmatrix} 2 & 1 \\ 1 & -3 \\ 5 & -1 \end{pmatrix} \begin{pmatrix} 5/7 \\ -3/7 \end{pmatrix} = \begin{pmatrix} 1 \\ 2 \\ 4 \end{pmatrix}.$$

If a system of linear equations has a solution, it is said to be *consistent*. The problem of finding coordinates of a vector in $\mathscr{V}_n$ with respect to some basis always leads to a consistent system of linear equations. If a system of linear equations has no solution, it is *inconsistent*. An inconsistent system of linear equations is obtained when it is found that a vector from $\mathscr{V}_n$ is not in the span of a set.

If the constant terms in a system of linear equations are all zero, $AX = \mathbf{0}$. then there is always a solution, namely, the *trivial solution* $X = \mathbf{0}$ obtained by setting all the unknowns equal to zero. Such a system is called *homogeneous* because all the nonzero terms have the same degree. A homogeneous system of linear equations is obtained when determining if a subset of $\mathscr{V}_n$ is linearly dependent. Notice that in such a case, with $x_1U_1 + \cdots + x_mU_m = \mathbf{0}$, the question is not: Is there a solution? but rather: Is there a solution with not all the x's equal to zero? Such a solution, $X = C$ with $C \neq \mathbf{0}$, for a homogeneous system $AX = \mathbf{0}$ is called a *nontrivial solution*.

Example 4 Consider the homogeneous system of linear equations

$$\begin{aligned} 2x + 3y + z &= 0 \\ x - 6y - z &= 0 \end{aligned} \quad \text{or} \quad \begin{pmatrix} 2 & 3 & 1 \\ 1 & -6 & -1 \end{pmatrix} \begin{pmatrix} x \\ y \\ z \end{pmatrix} = \begin{pmatrix} 0 \\ 0 \end{pmatrix}.$$

As with all homogeneous systems, this system has the trivial solution $X = \mathbf{0}$ or $x = y = z = 0$. But it also has many nontrivial solutions. For example,

$$X = \begin{pmatrix} -1 \\ -1 \\ 5 \end{pmatrix} \quad \text{and} \quad X = \begin{pmatrix} 2 \\ 2 \\ -10 \end{pmatrix}$$

are both solutions.

If the above solutions are written as ordered triples in $\mathscr{V}_3$, then one is a scalar multiple of the other. Using the spaces $\mathscr{V}_n$ as a guide, it is a simple matter to turn the set of all $n \times m$ matrices, for any n and m, into a vector space.

Definition Let $\mathcal{M}_{n \times m} = \{A | A \text{ is an } n \times m \text{ matrix}\}$. If $A, B \in \mathcal{M}_{n \times m}$ with $A = (a_{ij})$ and $B = (b_{ij})$, set $A + B = (c_{ij})$ where $c_{ij} = a_{ij} + b_{ij}$ and $rA = (d_{ij})$ where $d_{ij} = ra_{ij}$, for any $r \in R$.

Example 5 In $\mathcal{M}_{2 \times 3}$,

$$\begin{pmatrix} 2 & 1 & 3 \\ 4 & 0 & 1 \end{pmatrix} + \begin{pmatrix} 3 & -4 & 2 \\ 1 & -1 & 5 \end{pmatrix} = \begin{pmatrix} 5 & -3 & 5 \\ 5 & -1 & 6 \end{pmatrix}$$

and

$$4\begin{pmatrix} 0 & 2 & 1 \\ 3 & 0 & 8 \end{pmatrix} = \begin{pmatrix} 0 & 8 & 4 \\ 12 & 0 & 32 \end{pmatrix}.$$

Theorem 3.1 The set $\mathcal{M}_{n \times m}$ of $n \times m$ matrices, together with the operations of addition and scalar multiplication as defined above, is a vector space.

Proof The proof is almost identical to the proof that $\mathscr{V}_n$ is a vector space and is left to the reader.

The definitions and theorems of Chapter 2 referred to an abstract vector space, therefore they apply to $\mathcal{M}_{n \times m}$ even though it was not in hand at the time. This means that we can speak of linearly dependent sets of $n \times m$ matrices or the dimension of $\mathcal{M}_{n \times m}$. (Can you find a simple basis for $\mathcal{M}_{n \times m}$?) In particular, the following theorem suggested by Example 4 is not difficult to prove.

Theorem 3.2 Let $AX = \mathbf{0}$ be a homogeneous system of n linear equation in m unknowns. Then the set of all solutions $\mathscr{S} = \{C | AC = \mathbf{0}\}$ is a subspace of the vector space $\mathcal{M}_{m \times 1}$.

Proof $\mathscr{S}$ is nonempty because $X = \mathbf{0}$ is a solution for $AX = \mathbf{0}$. We will show that $\mathscr{S}$ is closed under scalar multiplication and leave the proof that it is closed under addition as a problem.

Let $A = (a_{ij})$, $X = (x_j)$ and suppose $C = (c_j) \in \mathscr{S}$. Then $a_{i1}c_1 + \cdots + a_{im}c_m = 0$ for each i, $1 \le i \le n$. If $r \in R$, then using properties of the real numbers, we obtain

$$0 = r0 = r(a_{i1}c_1 + \cdots + a_{im}c_m) = a_{i1}(rc_1) + \cdots + a_{im}(rc_m).$$

That is, $rC = (rc_j)$ is a solution for $AX = \mathbf{0}$, and we have shown that $C \in \mathscr{S}$ implies $rC \in \mathscr{S}$ for any $r \in R$.

$\mathscr{S}$ is called the *solution space* of the homogeneous system $AX = \mathbf{0}$. Since the solution space is a subspace of an m-dimensional vector space, it has a finite basis. This means that a finite number of solutions generate all solutions by taking linear combinations.

Example 6 It can be shown that the solution space of the system

$$\begin{aligned} 3x + 3y + 6z - w &= 0 \\ x - y + 2z + w &= 0 \\ 5x + y + 10z + w &= 0 \end{aligned}$$

is

$$\mathscr{L}\left\{\begin{pmatrix} 3 \\ 0 \\ -1 \\ 0 \end{pmatrix}, \begin{pmatrix} 0 \\ 2 \\ 0 \\ 3 \end{pmatrix}\right\}.$$

Therefore every solution is a linear combination of two vectors, or every solution is of the form $x = 3a$, $y = 2b$, $z = -a$, $w = 3b$ for some $a, b \in R$.

Suppose $\mathscr{S}$ is the solution space of a homogeneous system of linear equations in m unknowns. Then the vectors of $\mathscr{S}$ are $m \times 1$ matrices. These matrices are like m-tuples written vertically and therefore might be called *column vectors*. In contrast, an element of $\mathscr{V}_n$ or $\mathscr{M}_{1 \times n}$ would be called a *row vector*. We have found that a plane in 3-space has a linear Cartesian equation. Thus the solution space of one homogeneous linear equation in three variables can be viewed as a plane in 3-space passing through the origin. In general, the column vectors of $\mathscr{S}$ and thus $\mathscr{S}$ itself can be represented by geometric points. Such graphs of solutions are considered in problem 14 below and in Section 4.

Problems

1. Show that each of the following questions leads to a system of linear equations.

a. Is $3t$ in the span of $\{1 + t, 3t^2 + t, 4t + 7\}$?
b. Is $2 + 3i \in \mathscr{L}\{6 + 9i, 7 - 2i\}$?

c. Is $(2, 1, -4) \in \mathscr{L}\{(1, 3, 6), (2, 1, 1)\}$?
d. What are the coordinates of $(2, 1, -4)$ with respect to the basis $\{(1, 3, 6), (2, 1, 1), (1, 0, 2)\}$?
e. What are the coordinates of $t^2 + 5$ with respect to the basis $\{3t - 2, t^2 + 4t - 1, t^2 - t + 6\}$?

2. For each of the following sets of vectors, show that a homogeneous system of linear equations arises in determining if the set is linearly dependent or independent.

a. $\{t^2 + 3t^3, t^2 - 5, 2t + t^2\}$ in $R_4[t]$.
b. $\{3 + 4i, 2i - 1, 6 - i\}$ in C.
c. $\{e^x, e^{2x}, e^{5x}\}$ in $\mathscr{F}$.
d. $\{(2, 1, -3, 4), (5, -1, -4, 7), (1, -3, 2, 1)\}$ in $\mathscr{V}_4$.

3. Using the double subscript notation a_{ij} for coefficients, what are the values of a_{11}, a_{23}, a_{13}, a_{24}, and a_{21} in the following system?

$$\begin{aligned} 2x_1 - 5x_2 + 6x_4 &= -6 \\ 3x_2 + x_3 - 7x_4 &= 8. \end{aligned}$$

4. Find the coefficient matrix and the augmented matrix for

a. the system of equations in problem 3.

b. $\begin{aligned} 2x - y &= 7 \\ x + y &= 3 \\ 5x - y &= 6. \end{aligned}$ c. $\begin{aligned} 2x_1 - 3x_2 + x_3 &= 0 \\ x_1 - 2x_3 &= 0. \end{aligned}$

5. Compute the following matrix products.

a. $\begin{pmatrix} 2 & 1 & 5 \\ 3 & 0 & 2 \end{pmatrix}\begin{pmatrix} 4 \\ 2 \\ -3 \end{pmatrix}$. b. $\begin{pmatrix} 2 & 1 \\ 3 & 4 \\ 7 & 1 \\ 5 & 3 \end{pmatrix}\begin{pmatrix} 2 \\ -4 \end{pmatrix}$. c. $\begin{pmatrix} 0 & 0 & 1 & 0 \\ 0 & 1 & 0 & 0 \end{pmatrix}\begin{pmatrix} 5 \\ 3 \\ 6 \\ 8 \end{pmatrix}$.

6. Write the systems in problem 4 as matrix equations.

7. Suppose the homogeneous system $AX = \mathbf{0}$ is obtained in determining if the set $S \subset \mathscr{V}_n$ is linearly dependent. Find S given that A is

a. $\begin{pmatrix} 5 & 2 & 7 \\ 1 & 3 & 8 \end{pmatrix}$. b. $\begin{pmatrix} 2 & 4 \\ 1 & 3 \end{pmatrix}$. c. $\begin{pmatrix} 2 & 5 & 7 & 8 \\ 1 & 3 & 0 & 2 \\ 2 & 1 & 0 & 6 \end{pmatrix}$. d. $\begin{pmatrix} 5 & 1 & 3 \\ 2 & 4 & 9 \\ 1 & 0 & 7 \\ 6 & 1 & 5 \end{pmatrix}$.

8. Suppose $AX = B$ is the system of linear equations obtained in determining if $W \in \mathscr{L}\{U_1, \ldots, U_m\}$. Given the augmented matrix A^*, find $W, U_1, \ldots, U_m \in \mathscr{V}_n$.

a. $A^* = \begin{pmatrix} 2 & -1 & 3 \\ 4 & 1 & 2 \end{pmatrix}$. b. $A^* = \begin{pmatrix} 1 & 5 & 0 \\ 2 & 3 & 0 \\ 4 & 9 & 0 \end{pmatrix}$. c. $A^* = \begin{pmatrix} 9 & 5 & 4 \\ 2 & 1 & 8 \\ 1 & 7 & 6 \\ 0 & 2 & 0 \end{pmatrix}$.

9. Give an argument involving the dimension of $\mathscr{V}_3$ which shows that not all systems of three linear equations in two unknowns can be consistent.

10. Perform the indicated operations.

a. $\begin{pmatrix} 2 & -3 & 7 \\ 1 & 4 & -5 \end{pmatrix} + \begin{pmatrix} 0 & 2 & 1 \\ 3 & 7 & 8 \end{pmatrix}$. b. $\begin{pmatrix} 4 \\ 1 \\ 3 \end{pmatrix} + \begin{pmatrix} -2 \\ 7 \\ -9 \end{pmatrix}$. c. $3\begin{pmatrix} 2 & 1 \\ 4 & 2 \\ 1 & 0 \end{pmatrix}$.

11. What is the zero vector of $\mathscr{M}_{3\times 2}$, $\mathscr{M}_{3\times 1}$, $\mathscr{M}_{1\times 4}$, and $\mathscr{M}_{2\times 2}$?

12. The homogeneous system $AX = \mathbf{0}$ of n linear equations in m unknowns has the trivial solution $X = \mathbf{0}$. What are the sizes of these two zero matrices?

13. Complete the proof of Theorem 3.2.

14. Suppose $AX = B$ not homogeneous. Let $S = \{C \in \mathscr{M}_{m\times 1} | AC = B\}$. Call S the solution set of the system.

a. Show that S is not a subspace of $\mathscr{M}_{m\times 1}$.
b. If $AX = B$ is a single linear equation in three unknowns, what is the geometric interpretation of S?
c. If $AX = B$ is two linear equations in three unknowns, what are the possible geometric interpretations of S?

§2. Row-equivalence and Rank

The technique of eliminating variables to solve a system of linear equations was not sufficient to handle all the systems that occurred in the previous chapter. However, it does contain the essential ideas necessary to determine if a system is consistent and, when it is, to find all of the solutions. In order to see this, we must first put the procedure we have been using to solve systems of linear equations on a firm basis.

Definition Two systems of linear equations are *equivalent* if every solution of one system is a solution of the other, and conversely.

That is, two systems with m unknowns, $A_1X = B_1$ and $A_2X = B_2$, are equivalent provided $A_1C = B_1$ if and only if $A_2C = B_2$ for $C \in \mathscr{M}_{m\times 1}$.

Example 1 The following two pairs of equivalent systems are given without justification at this time:

$$\begin{aligned} 2x + 3y &= -5 \\ 5x - 2y &= 16 \end{aligned} \quad \text{and} \quad \begin{aligned} x &= 2 \\ y &= -3, \end{aligned}$$

$$\begin{aligned} x - y + z &= 4 \\ 2x + y - 3z &= 2 \\ 5x + y - 5z &= 1 \end{aligned} \quad \text{and} \quad \begin{aligned} x - y + z &= 4 \\ 3y - 5z &= -6 \\ 0 &= -13. \end{aligned}$$

In the first case, the system is equivalent to equations which provide a unique solution. In the second, the system is equivalent to one that is obviously inconsistent.

Thus, determining if a system of linear equations is consistent and obtaining solutions amounts to finding an equivalent system in a particular form. The first fact to be established is that the technique of eliminating variables yields equivalent systems of equations.

Theorem 3.3 If one system of linear equations is obtained from another by a finite number of the following operations, then the two systems are equivalent.

Type I. Interchange two equations.
Type II. Multiply an equation by a nonzero scalar.
Type III. Add a scalar multiple of one equation to another equation.

Before proving the theorem consider some examples. The system

$$\begin{aligned} 2x - 5y + 4z &= 4 \\ 6x - 9y + 6z &= 7 \end{aligned}$$

becomes

$$\begin{aligned} 6x - 9y + 6z &= 7 \\ 2x - 5y + 4z &= 4 \end{aligned}$$

using an operation of type I. Multiplying the first equation of the original system by 1/2, an operation of type II yields

$$x - \tfrac{5}{2}y + 2z = 2$$
$$6x - 9y + 6z = 7.$$

Finally adding -3 times the first equation of the original system to the second equation, a type III operation, gives

$$2x - 5y + 4z = 4$$
$$6y - 6z = -5.$$

Proof It is sufficient to show that an equivalent system results when one operation of each type is performed, and it should be clear that an operation of type I yields an equivalent system. Therefore, suppose $A_1X = B_1$ is transformed to $A_2X = B_2$ with an operation of type II, say the kth equation of $A_2X = B_2$ is r times the kth equation of $A_1X = B_1$, $r \neq 0$. If $A_1 = (a_{ij})$, $X = (x_j)$, and $B_1 = (b_i)$, then the systems are identical except that the kth equation of the first system is $a_{k1}x_1 + \cdots + a_{km}x_m = b_k$ while the kth equation of the second is $ra_{k1}x_1 + \cdots + ra_{km}x_m = rb_k$.

Now if $X = C = (c_j)$ is a solution of $A_1X = B_1$, then $X = C$ is a solution of each equation in $A_2X = B_2$ except possibly the kth equation. But $a_{k1}c_1 + \cdots + a_{km}c_m = b_k$ so $ra_{k1}c_1 + \cdots + ra_{km}c_m = rb_k$ and $X = C$ is also a solution of this equation. Conversely, if $X = C$ is a solution of $A_2X = B_2$, then since $r \neq 0$, it can be divided out of the equality $ra_{k1}c_1 + \cdots + ra_{km}c_m = rb_k$ so that $X = C$ satisfies the kth equation of the first system. Therefore $X = C$ is a solution of $A_1X = B_1$, and the two systems are equivalent.

The proof that an operation of type III yields an equivalent system is left to the reader.

In Example 1 of the previous section, a system of linear equations was solved using an elimination technique. We now see that a sequence of operations of types II and III was used. At the same time the procedure was performed more simply on the augmented matrix. Combining this idea with the above theorem gives a general approach to the solution of systems of linear equations. When the word "equation" in the above three types of operations is replaced by "row" and the operations are performed on matrices they are called *elementary row operations*.

Example 2 Find the solution of

$$\begin{aligned} 2x - 3y &= 4 \\ x + 2y &= 7 \end{aligned}$$

by performing elementary row operations on the augmented matrix.

An arrow with I, II, or III below it will be used to indicate which type of elementary row operation has been used at each step.

$$\begin{pmatrix} 2 & -3 & 4 \\ 1 & 2 & 7 \end{pmatrix} \underset{\text{I}}{\rightarrow} \begin{pmatrix} 1 & 2 & 7 \\ 2 & -3 & 4 \end{pmatrix} \underset{\text{III}}{\rightarrow} \begin{pmatrix} 1 & 2 & 7 \\ 0 & -7 & -10 \end{pmatrix} \underset{\text{II}}{\rightarrow} \begin{pmatrix} 1 & 2 & 7 \\ 0 & 1 & 10/7 \end{pmatrix} \underset{\text{III}}{\rightarrow} \begin{pmatrix} 1 & 0 & 29/7 \\ 0 & 1 & 10/7 \end{pmatrix}.$$

First the two rows are interchanged, then -2 times the first row is added to the second row, next the second row is multiplied by $-1/7$, and finally the last row is multiplied by -2 and added to the first row. The resulting matrix is the augmented matrix for the system

$$\begin{aligned} x \quad &= 29/7 \\ y &= 10/7 \end{aligned}$$

which has $x = 29/7$, $y = 10/7$ as its only solution. Since this system is equivalent to the original system, the original system is consistent with the unique solution $x = 29/7$, $y = 10/7$.

Example 3 Show that

$$\begin{aligned} 2x + y &= 3 \\ 4x + 2y &= -1 \end{aligned}$$

is inconsistent.

Performing elementary row operations on the augmented matrix yields:

$$\begin{pmatrix} 2 & 1 & 3 \\ 4 & 2 & -1 \end{pmatrix} \underset{\text{II}}{\rightarrow} \begin{pmatrix} 1 & 1/2 & 3/2 \\ 4 & 2 & -1 \end{pmatrix} \underset{\text{III}}{\rightarrow} \begin{pmatrix} 1 & 1/2 & 3/2 \\ 0 & 0 & -7 \end{pmatrix} \underset{\text{II}}{\rightarrow} \begin{pmatrix} 1 & 1/2 & 3/2 \\ 0 & 0 & 1 \end{pmatrix}.$$

Since the equation $0 = 1$ is in a system equivalent to the given system, the given system is inconsistent.

Example 4 Find all solutions for

$$\begin{aligned} x - 2y + z &= 1 \\ 2x - 4y + 3z &= 0. \end{aligned}$$

An equivalent system can be obtained in which it is possible to read off all the solutions for this system.

$$\begin{pmatrix} 1 & -2 & 1 & 1 \\ 2 & -4 & 3 & 0 \end{pmatrix} \underset{\text{III}}{\rightarrow} \begin{pmatrix} 1 & -2 & 1 & 1 \\ 0 & 0 & 1 & -2 \end{pmatrix} \underset{\text{III}}{\rightarrow} \begin{pmatrix} 1 & -2 & 0 & 3 \\ 0 & 0 & 1 & -2 \end{pmatrix}.$$

Thus the system

$$\begin{aligned} x - 2y \quad &= \ \ 3 \\ z &= -2 \end{aligned}$$

is equivalent to the given system, and $x = 3 + 2t$, $y = t$, $z = -2$ is a solution for each value of the parameter t. Writing

$$(x, y, z) = t(2, 1, 0) + (3, 0, -2), \qquad t \in R,$$

shows that the two planes with the given equations intersect in a line.

Example 5 If consistent, solve the system

$$\begin{aligned} x + \ \ y - z &= 1 \\ 2x - 3y + z &= 4 \\ 4x - \ \ y - z &= 6 \end{aligned}$$

$$\begin{pmatrix} 1 & 1 & -1 & 1 \\ 2 & -3 & 1 & 4 \\ 4 & -1 & -1 & 6 \end{pmatrix} \underset{\text{III}}{\rightarrow} \begin{pmatrix} 1 & 1 & -1 & 1 \\ 0 & -5 & 3 & 2 \\ 4 & -1 & -1 & 6 \end{pmatrix} \underset{\text{III}}{\rightarrow} \begin{pmatrix} 1 & 1 & -1 & 1 \\ 0 & -5 & 3 & 2 \\ 0 & -5 & 3 & 2 \end{pmatrix}$$

$$\underset{\text{III}}{\rightarrow} \begin{pmatrix} 1 & 1 & -1 & 1 \\ 0 & -5 & 3 & 2 \\ 0 & 0 & 0 & 0 \end{pmatrix} \underset{\text{II}}{\rightarrow} \begin{pmatrix} 1 & 1 & -1 & 1 \\ 0 & 1 & -3/5 & -2/5 \\ 0 & 0 & 0 & 0 \end{pmatrix} \underset{\text{III}}{\rightarrow} \begin{pmatrix} 1 & 0 & -2/5 & 7/5 \\ 0 & 1 & -3/5 & -2/5 \\ 0 & 0 & 0 & 0 \end{pmatrix}.$$

After the first two steps, which eliminate x from the second and third equations, a system is obtained with the last two equations identical. Although the original system appeared to involve three conditions on x, y, and z, there are only two. All solutions for the given system are obtained from the final system of two equations by letting z be any real number. This gives $x = \frac{7}{5} + \frac{2}{5}t$, $y = -\frac{2}{5} + \frac{3}{5}t$, and $z = t$ with $t \in R$.

The above examples suggest that a complete answer to the problem of solving a system of linear equations is obtained by transforming the augmented matrix into a special form. This form isolates some or all of the unknowns. That is, certain unknowns appear only in one equation of the associated system of linear equations. Further, if an unknown, x_i, appears in only one equation, then none of the unknowns preceding it, x_k with $k < i$, are in that equation. Once such an equivalent system is found, it is clear if it is consistent. And if consistent, then all solutions can be read directly from it by assigning arbitrary values to the unknowns which are not isolated. The basic characteristics of this special form for the augmented matrix are as follows:

1. The first nonzero entry of each row should be one.
2. The column containing such a leading entry of 1, should have only 0's as its other entries.
3. The leading 1's should move to the right in succeeding rows. That is, the leading 1 of each row should be in a column to the right, of the column containing the leading 1 of the row above.
4. A row containing only 0's should come after all rows with nonzero entries.

Definition A matrix satisfying the four conditions above is in (row) *echelon form.*

The first and last matrices below are in echelon form, while the middle two are not.

$$\begin{pmatrix} 0 & 1 & 5 & 0 & 6 \\ 0 & 0 & 0 & 1 & 1 \\ 0 & 0 & 0 & 0 & 0 \end{pmatrix}, \begin{pmatrix} 1 & 2 & 3 \\ 0 & 1 & 4 \\ 0 & 0 & 1 \end{pmatrix}, \begin{pmatrix} 1 & 0 & 0 \\ 0 & 0 & 1 \\ 0 & 1 & 0 \end{pmatrix}, \begin{pmatrix} 1 & 0 & 2 & 0 & 2 & 0 \\ 0 & 1 & 1 & 0 & 1 & 0 \\ 0 & 0 & 0 & 1 & 3 & 0 \\ 0 & 0 & 0 & 0 & 0 & 1 \end{pmatrix}.$$

In the second matrix, two of the columns containing leading 1's, for the second and third rows, contain other nonzero entries. In the third matrix, the leading 1 in the third row is not in a column to the right of the column containing the leading 1 in the second row.

Definition Two $n \times m$ matrices are *row-equivalent* if there exists a finite sequence of elementary row operations that transforms one into the other.

Examples 2 through 5 suggest that any consistent system of linear equations might be solved by transforming the augmented matrix to echelon form.

This implies that it should be possible to transform any matrix to echelon form with elementary row operations. The proof that this is the case simply formalizes the procedure used in the examples.

Theorem 3.4 Given an $n \times m$ matrix A, there is a matrix in echelon form that is row-equivalent to A.

Proof If $A = \mathbf{0}$, then it is in echelon form. For $A \neq \mathbf{0}$, the proof is by induction on the number of rows in A. To begin, suppose the hth column of A is the first column with a nonzero element and $a_{kh} \neq 0$. Multiplying the kth row of A by $1/a_{kh}$ and interchanging the kth and 1st rows yields a matrix with 1 as the leading nonzero entry in its first row. Then with elementary row operations of type III, the other nonzero entries in the hth column can be replaced by 0's. Call the matrix obtained A_1, then the first row of A_1 conforms to the definition of echelon form and no column before the hth contains nonzero terms.

For the induction step, suppose the matrix $A_p = (b_{ij})$ has been obtained which conforms to the definition of echelon form in its first p rows, $1 \leq p < n$, and $b_{ij} = 0$ for $i > p$ and $j < q$, where $b_{pq} = 1$ is the leading one in the pth row. If there are no nonzero entries in the remaining rows, then A_p is in echelon form. Otherwise there exist indices r, s such that $b_{rs} \neq 0$ with $r > p$ (i.e., b_{rs} is in one of the remaining rows), and if $i > p$, $j < s$, then $b_{ij} = 0$ (i.e., b_{rs} is in the first column with nonzero entries below the pth row.) Now multiply the rth row of A_p by $1/b_{rs}$ and interchange the rth and $(p + 1)$st rows. This yields a matrix with 1 as the leading nonzero entry in the $(p + 1)$st row; by assumption, $s > q$, so this 1 is to the right of the leading 1 in the preceding row. Now use elementary row operations of type III to make all the other nonzero entries in the sth column equal to 0. If the matrix obtained is denoted by $A_{p+1} = (c_{ij})$, then the first $p + 1$ rows of A_{p+1} conform to the definition of a matrix in echelon form, and if $i > p + 1, j < s$, then $c_{ij} = 0$. Therefore the proof is complete by induction.

Example 6 The following sequence of elementary row operations illustrates the steps in the above proof.

$$\begin{pmatrix} 0 & 0 & 0 & 0 & 1 \\ 0 & 2 & 6 & 4 & 8 \\ 0 & 0 & 0 & 1 & 2 \end{pmatrix} \underset{\text{II}}{\rightarrow} \begin{pmatrix} 0 & 0 & 0 & 0 & 1 \\ 0 & 1 & 3 & 2 & 4 \\ 0 & 0 & 0 & 1 & 2 \end{pmatrix} \underset{\text{I}}{\rightarrow} \begin{pmatrix} 0 & 1 & 3 & 2 & 4 \\ 0 & 0 & 0 & 0 & 1 \\ 0 & 0 & 0 & 1 & 2 \end{pmatrix}$$

$$\underset{\text{I}}{\rightarrow} \begin{pmatrix} 0 & 1 & 3 & 2 & 4 \\ 0 & 0 & 0 & 1 & 2 \\ 0 & 0 & 0 & 0 & 1 \end{pmatrix} \underset{\text{III}}{\rightarrow} \begin{pmatrix} 0 & 1 & 3 & 0 & 0 \\ 0 & 0 & 0 & 1 & 2 \\ 0 & 0 & 0 & 0 & 1 \end{pmatrix} \underset{\text{III}}{\rightarrow} \begin{pmatrix} 0 & 1 & 3 & 0 & 0 \\ 0 & 0 & 0 & 1 & 0 \\ 0 & 0 & 0 & 0 & 1 \end{pmatrix}.$$

If the first matrix is the matrix A in the proof, then the third would be A_1, the fourth A_2, and the last A_3 which is in echelon form. In A the nonzero element denoted a_{kh} in the proof is 2 with $k = h = 2$. In A_1 the element b_{rs} is 1 with $r = 3$ and $s = 4$. Notice that $s = 4 > 2 = h$ as desired.

There are usually many sequences of elementary row operations which transform a given matrix into one in echelon form, and it is not at all clear that different sequences will not lead to different echelon forms. The proof that there is in fact only one echelon form is suggested by the relationship between an echelon form and the solution of a system of linear equations. However, a formal proof of this will be omitted, see problem 14.

Theorem 3.5 Each $n \times m$ matrix is row-equivalent to only one matrix in echelon form.

Because of the last two theorems it is possible to refer to "the echelon form" of a given matrix. Further the number of nonzero rows in the echelon form of a matrix A is determined only by A and not the procedure used to obtain the echelon form.

Definition The *rank* of a matrix is the number of rows with nonzero entries in its (row) echelon form.

Thus, from Example 4,

$$\operatorname{rank}\begin{pmatrix} 1 & -2 & 1 & 1 \\ 2 & -4 & 3 & 0 \end{pmatrix} = 2$$

and from Example 5,

$$\operatorname{rank}\begin{pmatrix} 1 & 1 & -1 & 1 \\ 1 & -3 & 1 & 4 \\ 4 & -1 & -1 & 6 \end{pmatrix} = 2.$$

A second characterization of rank should be discussed before returning to our general consideration of systems of linear equations.

Definition Let $A = (a_{ij})$ be an $n \times m$ matrix. The *row space* of A is the subspace of $\mathscr{V}_m$ spanned by the n rows of A. That is,

$$\mathscr{L}\{(a_{11}, a_{12}, \ldots, a_{1m}), \ldots, (a_{n1}, a_{n2}, \ldots, a_{nm})\}$$

is the row space of A.

For example, the row space of the matrix

$$\begin{pmatrix} 2 & 3 \\ 1 & -2 \\ 4 & 8 \end{pmatrix}$$

is $\mathscr{L}\{(2, 3), (1, -2), (4, 8)\} = \mathscr{V}_2$.

The dimension of the row space of a matrix A is often called its "row rank," but it is easily seen that this is just the rank of A as defined above.

Theorem 3.6 If A and B are row-equivalent, then they have the same row space.

Proof It need only be shown that if A is transformed to B with one elementary row operation, then the row spaces are equal. We will consider a type II operation here and leave the others to the reader. Therefore suppose the kth row of B is r times the kth row of A for some nonzero scalar r. If $U_1, \ldots, U_k, \ldots, U_n$ are the rows of A, then $U_1, \ldots, rU_k, \ldots, U_n$ are the rows of B. Now suppose X is in the row space of A, then

$$\begin{aligned} X &= a_1U_1 + \cdots + a_kU_k + \cdots + a_nU_n \qquad \text{for some scalars } a_i \\ &= a_1U_1 + \cdots + (a_k/r)(rU_k) + \cdots + a_nU_n \qquad \text{since } r \neq 0. \end{aligned}$$

Therefore X is in the row space of B and the row space of A is contained in the row space of B. The reverse containment is obtained similarly, proving that the row space is unchanged by an elementary row operation of type II.

Now the row space of a matrix equals the row space of its echelon form, and the rows of a matrix in echelon form are obviously linearly independent. Therefore we have the following theorem.

Theorem 3.7 The rank of a matrix equals the dimension of its row space.

Example 7 Find the dimension of the subspace $\mathscr{S}$ of $\mathscr{V}_4$ given by

$$\mathscr{S} = \mathscr{L}\{(5, 1, -4, 7), (2, 1, -1, 1), (1, -1, -2, 5)\}.$$

The dimension of $\mathscr{S}$, is the rank of the matrix

$$A = \begin{pmatrix} 5 & 1 & -4 & 7 \\ 2 & 1 & -1 & 1 \\ 1 & -1 & -2 & 5 \end{pmatrix}.$$

But the matrix

$$B = \begin{pmatrix} 1 & 0 & -1 & 2 \\ 0 & 1 & 1 & -3 \\ 0 & 0 & 0 & 0 \end{pmatrix}$$

is the echelon form for A. Therefore the dimension of $\mathscr{S}$ is 2.

Notice that in obtaining the dimension of $\mathscr{S}$ we have also obtained a characterization of its elements. For $U \in \mathscr{S}$ if and only if there are scalars a and b such

$$U = a(1, 0, -1, 2) + b(0, 1, 1, -3) = (a, b, -a + b, 2a - 3b).$$

Therefore

$$\mathscr{S} = \{(x, y, z, w) | z = y - x, w = 2x - 3y\}.$$

Example 8 Determine if $\{(2, 2, -4), (-3, -2, 8), (1, 4, 4)\}$ is a basis for $\mathscr{V}_3$.

Consider the matrix A which has the given vectors as rows.

$$A = \begin{pmatrix} 2 & 2 & -4 \\ -3 & -2 & 8 \\ 1 & 4 & 4 \end{pmatrix}.$$

The echelon form for A is

$$\begin{pmatrix} 1 & 0 & -4 \\ 0 & 1 & 2 \\ 0 & 0 & 0 \end{pmatrix}.$$

Therefore the set is not a basis for $\mathscr{V}_3$ because it does not contain three linearly independent vectors.

Problems

1. All the solutions of the system

$$\begin{aligned} x + y + z &= 6 \\ 2x - y + z &= 3 \\ 4x + y + 3z &= 15 \end{aligned}$$

are of the form $x = 3 - 2k$, $y = 3 - k$, $z = 3k$ for some real number k. Determine which of the following systems also have these solutions.

a. $\begin{aligned} x + y + z &= 6 \\ 4x + y + 3z &= 15. \end{aligned}$ b. $\begin{aligned} 3x + 2z &= 9 \\ 3y + z &= 3 \\ x - 2y &= -3. \end{aligned}$ c. $\begin{aligned} x + y - 2z &= 0 \\ 3x + y + z &= 1. \end{aligned}$

2. Suppose the system $A_2X = B_2$ is obtained from $A_1X = B_1$ by adding r times the kth equation to the hth equation. Prove that the two systems are equivalent. That is, prove that an operation of type III yields an equivalent system.

3. Find the echelon form for the following matrices.

a. $\begin{pmatrix} 0 & 2 & 0 \\ 0 & 1 & 3 \\ 1 & 0 & 0 \end{pmatrix}$. b. $\begin{pmatrix} 3 & -1 \\ 4 & 2 \\ 2 & 6 \end{pmatrix}$. c. $\begin{pmatrix} 0 & -3 & 6 & 1 & 6 \\ 0 & -2 & 4 & 1 & 5 \\ 0 & -1 & 2 & 1 & 4 \end{pmatrix}$. d. $\begin{pmatrix} 1 & 4 & 1 & 3 \\ 2 & 8 & 3 & 5 \\ 1 & 4 & 2 & 7 \end{pmatrix}$.

4. Solve the following systems of linear equations by transforming the augmented matrix into echelon form.

a. $\begin{aligned} x + y + z &= 6 \\ x - 2y + 2z &= -2 \\ 3x + 2y - 4z &= 8. \end{aligned}$ b. $\begin{aligned} x + y &= 2 \\ 3x - y &= 2 \\ x - y &= 0. \end{aligned}$ c. $\begin{aligned} x + y + z &= 4 \\ 3x + 2y + z &= 2 \\ x + 2y + 3z &= 3. \end{aligned}$

5. List all possible echelon forms for a 2×2 matrix. Do the same for 2×3 and 3×2 matrices.

6. Characterize the span of the three vectors in Example 8, as was done in Example 7.

7. Use the echelon form to find the row spaces of the following matrices. If the space is not $\mathscr{V}_n$, characterize it as was done in Example 7.

a. $\begin{pmatrix} 2 & 1 \\ 6 & 3 \\ 1 & 2 \end{pmatrix}$. b. $\begin{pmatrix} 2 & 1 & -1 \\ 1 & 2 & 4 \\ 3 & 1 & -3 \end{pmatrix}$. c. $\begin{pmatrix} 3 & 1 & 8 \\ 2 & -1 & -3 \end{pmatrix}$. d. $\begin{pmatrix} 1 & 2 & 0 \\ 4 & 3 & 1 \\ 3 & 0 & 1 \end{pmatrix}$.

e. $\begin{pmatrix} 2 & 1 & 0 & 4 \\ 0 & 1 & 1 & 5 \\ 3 & 0 & 1 & 6 \end{pmatrix}$. f. $\begin{pmatrix} 2 & -1 & 4 & -2 \\ 1 & 1 & 2 & 8 \\ 3 & -1 & 6 & 0 \end{pmatrix}$.

8. Find the rank of each matrix in problem 7.

9. Explain why the rank of an $n \times m$ matrix cannot exceed either n or m.

10. Prove or disprove: If two matrices have the same rank, then they are row-equivalent.

11. Determine if the following sets are linearly independent by finding the rank

of a matrix as in Example 8.

a. $\{(6, 9), (4, 6)\}$. b. $\{(2, -1, 8), (0, 1, -2), (1, -1, 5)\}$
c. $\{(1, 3), (2, 5)\}$. d. $\{(1, 0, 1), (0, 1, 1), (1, 1, 0)\}$.

12. How is the statement "$A \in \mathscr{M}_{n \times m}$" read?

13. Suppose $A \in \mathscr{M}_{n \times m}$ and B is obtained from A with an elementary row operation of type III. Show that A and B have the same row space.

14. Suppose $E = (e_{ij})$ and $F = (f_{ij})$ are $n \times m$ matrices in echelon form. Assume $e_{kh} \neq f_{kh}$, but $e_{ij} = f_{ij}$ for all $i > k$, $1 \leq j \leq m$, and $e_{kj} = f_{kj}$ for all $j < h$. That is, $E \neq F$ and the first entry at which they differ is in the k, h position, counting from the bottom.

    Prove that the two systems of homogeneous equations $EX = \mathbf{0}$ and $FX = \mathbf{0}$ are not equivalent by finding a solution for one system which is not a solution for the other. Why does this prove that a matrix A cannot have both E and F as echelon forms?

§3. Gaussian Elimination

We are now in a position to solve any consistent system of linear equations. But equally important, we can find conditions for the existence and uniqueness of a solution in general.

The solutions of a consistent system of linear equations with augmented matrix in echelon form can easily be obtained directly from the equations. Each equation begins with an unknown that does not appear in any other equation. These unknowns can be expressed in terms of a constant and any unknowns (parameters) that do not appear at the begining of an equation.

Example 1 Given a system with augmented matrix in echelon form

$$\begin{aligned} x_1 + 2x_2 \qquad\qquad\quad - \; x_5 &= 1 \\ x_3 \qquad + 2x_5 &= 0 \\ x_4 + 3x_5 &= 2, \end{aligned}$$

the equations can be rewritten in the form

$$\begin{aligned} x_1 &= 1 - 2x_2 + x_5 \\ x_3 &= -2x_5 \\ x_4 &= 2 - 3x_5. \end{aligned}$$

Therefore, if the unknowns x_2 and x_5 are assigned arbitrary values, these equations determine values for x_1, x_3, and x_4. Thus the solutions for this system involve two parameters and may be written in the form

$$x_1 = 1 - 2s + t, \quad x_2 = s, \quad x_3 = -2t, \quad x_4 = 2 - 3t, \quad x_5 = t$$

$$\text{with } s, t \in R.$$

Transforming a matrix into its echelon form with a sequence of elementary row operations is often called *row reduction* of the matrix, and row reduction of the augmented matrix for a system of linear equations is called *Gaussian elimination* or Gaussian reduction. If the system $AX = B$ is consistent and the rank of A is k, then Gaussian elimination yields a system of k equations in which each of k unknowns has been eliminated from all but one equation. As the preceding example shows, all solutions are easily read directly from such a system. But Gaussian elimination also reveals if a system is inconsistent, for it yields a system that includes the equation $0 = 1$. This situation occurs in Example 3, page 86 and leads to the first general theorem on systems of linear equations.

Theorem 3.8 A system of linear equations is consistent if and only if the rank of its coefficient matrix equals the rank of its augmented matrix.

Proof Elementary row operations do not mix entries from different columns. Therefore, when the augmented matrix A^* of the system $AX = B$ is transformed into a matrix in echelon form with a sequence of elementary row operations, the matrix obtained by deleting the last column is the echelon form for A. Hence the rank of A and the rank of A^* are different if and only if there is a leading 1 in the last column of the echelon form for A^*, that is, if and only if the equation $0 = 1$ is in a system equivalent to the given system $AX = B$. Therefore rank $A \neq$ rank A^* if and only if $AX = B$ is inconsistent.

Example 2 Show that the system

$$\begin{aligned} x + y + z &= 1 \\ x - y - z &= 2 \\ x - 3y - 3z &= 0 \end{aligned}$$

is inconsistent.

The augmented matrix for the system

$$\begin{pmatrix} 1 & 1 & 1 & 1 \\ 1 & -1 & -1 & 2 \\ 1 & -3 & -3 & 0 \end{pmatrix}$$

is row-equivalent to

$$\begin{pmatrix} 1 & 1 & 1 & 1 \\ 0 & -2 & -2 & 1 \\ 0 & 0 & 0 & -1 \end{pmatrix},$$

therefore the rank of A^* is 3. The coefficient matrix A is row equivalent to

$$\begin{pmatrix} 1 & 1 & 1 \\ 0 & -2 & -2 \\ 0 & 0 & 0 \end{pmatrix}.$$

So rank $A = 2 \neq 3 =$ rank A^* and the system is inconsistent. Notice that it was not necessary to obtain the echelon forms to determine that the ranks are not equal.

Now suppose $AX = B$ is a consistent system of n linear equations in m unknowns. Then the rank of A (or A^*) determines the number of parameters needed to express the solution, for an unknown becomes a parameter in the solution when, after Gaussian elimination, it does not begin an equation. Since the rank of A determines the number of equations in the row-reduced system, there will be $m -$ rank A parameters. Thus, in Example 1, there are five unknowns ($m = 5$) and the rank of A is 3, so the solution should have $5 - 3 = 2$ parameters, as is the case.

Example 3 Solve the system

$$x + 2y + z = 3$$
$$2x + 4y - 2z = 2.$$

The augmented matrix A^* for this system has echelon form:

$$\begin{pmatrix} 1 & 2 & 0 & 2 \\ 0 & 0 & 1 & 1 \end{pmatrix},$$

therefore rank $A^* =$ rank $A = 2$ and the system is consistent. Since there are three variables and rank $A = 2$, there is $1 = 3 - 2$ parameter in the solutions. And the solutions are given by $x = 2 - 2t$, $y = t$, $z = 1$, $t \in R$.

The relationship between rank and the number of parameters in a solution leads to the second general theorem.

Theorem 3.9 A consistent system of linear equations has a unique solution if and only if the rank of the coefficient matrix equals the number of unknowns.

Proof Let $AX = B$ be a consistent system of linear equations in m unknowns. There is a unique solution if and only if there are no parameters in the solution. That is, if and only if $m -$ rank $A = 0$ as stated.

Corollary 3.10 A consistent system of linear equations with fewer equations than unknowns has more than one solution.

Proof Let $AX = B$ be a consistent system of n linear equations in m unknowns with $n < m$. Then the rank of A cannot exceed the number of rows, which is n, and the result follows from Theorem 3.9.

Notice that this corollary does not say that a system with fewer equations than unknowns has more than one solution. For example, the system

$$4x - 2y + 6z = 1$$

$$6x - 3y + 9z = 1$$

has no solutions.

Recall that a homogeneous system, $AX = \mathbf{0}$, is always consistent, having the trivial solution $X = \mathbf{0}$. Therefore the problem is to determine when a homogeneous system has nontrivial solutions.

Corollary 3.11 The homogeneous system of linear equations $AX = \mathbf{0}$ has a nontrivial solution if and only if the rank of A is less than the number of unknowns.

Proof This is simply a special case of Theorem 3.9.

Example 4 Determine if the set

$$\{4 + 3t - t^3, 2 + 4t - 5t^3, 2 - t + 4t^3\}$$

is linearly independent in $R[t]$.

Suppose x, y, z are scalars such that

$$x(4 + 3t - t^3) + y(2 + 4t - 5t^3) + z(2 - t + 4t^3) = \mathbf{0}.$$

Collecting coefficients and setting them equal to 0 gives the homogeneous system

$$\begin{aligned} 4x + 2y + 2z &= 0 \\ 3x + 4y - z &= 0 \\ -x - 5y + 4z &= 0, \end{aligned}$$

and the set is linearly independent if this system has only the trivial solution, that is, if the rank of the coefficient matrix equals the number of unknowns. But the coefficient matrix is row-equivalent to

$$\begin{pmatrix} 1 & 0 & 9 \\ 0 & 1 & -1 \\ 0 & 0 & 0 \end{pmatrix},$$

and so has rank 2. Thus the set is linearly dependent.

Example 5 Determine if

$$\{(2, -1, 3, 4), (1, 3, 1, -2), (3, 2, 4, 2)\}$$

is linearly independent in $\mathscr{V}_4$.

This problem can be approached in two ways. First use the fact that the rank of a matrix is the dimension of its row space and consider the matrix

$$\begin{pmatrix} 2 & -1 & 3 & 4 \\ 1 & 3 & 1 & -2 \\ 3 & 2 & 4 & 2 \end{pmatrix}.$$

If the rank of this matrix is 3, then the rows of the matrix and the set are linearly independent.

Second use the above criterion for the existence of nontrivial solutions of homogeneous systems of linear equations. Suppose x, y, z satisfy

$$x(2, -1, 3, 4) + y(1, 3, 1, -2) + z(3, 2, 4, 2) = \mathbf{0},$$

This vector equation yields the homogeneous system

$$\begin{aligned} 2x + y + 3z &= 0 \\ -x + 3y + 2z &= 0 \\ 3x + y + 4z &= 0 \\ 4x - 2y + 2z &= 0 \end{aligned}$$

with coefficient matrix

$$\begin{pmatrix} 2 & 1 & 3 \\ -1 & 3 & 2 \\ 3 & 1 & 4 \\ 4 & -2 & 2 \end{pmatrix},$$

and the set is linearly independent if the rank of this matrix is 3. A little computation shows that both of the above matrices have rank 2, therefore the set is linearly dependent.

Notice that the two matrices in Example 5 are closely related. In fact, the rows of one are the columns of the other.

Definition The *transpose* of the $n \times m$ matrix $A = (a_{ij})$ is the $m \times n$ matrix $A^T = (b_{hk})$, where $b_{hk} = a_{kh}$ for all h and k.

Thus each matrix in Example 5 is the transpose of the other. We will see that, as in Example 5, a matrix and its transpose always have the same rank.

We have seen that, since the dimension of $\mathscr{V}_n$ is n, any set of $n + 1$ vectors in $\mathscr{V}_n$ is linearly dependent. This fact can now be proved without reference to dimension.

Theorem 3.12 A set of $n + 1$ vectors in $\mathscr{V}_n$ is linearly dependent.

Proof Given $\{U_1, \ldots, U_{n+1}\} \subset \mathscr{V}_n$, suppose $x_1, \ldots, x_{n+1}$ satisfy the equation $x_1U_1 + \cdots + x_{n+1}U_{n+1} = \mathbf{0}$. This vector equation yields a system of n (one for each component) homogeneous, linear equations in $n + 1$ unknowns. Since there are fewer equations than unknowns, there is not a unique solution, that is, there exists a nontrivial solution for $x_1, \ldots, x_{n+1}$ and the set is linearly dependent.

Problems

In problems 1 through 10, use Gaussian elimination to determine if the system is consistent and to find all solutions if it is.

1. $\begin{aligned} 2x - 3y &= 12 \\ 3x + 4y &= 1. \end{aligned}$ **2.** $\begin{aligned} 3x - 6y &= 3 \\ 4x - 8y &= 4. \end{aligned}$ **3.** $\begin{aligned} 2x - 6y &= -1 \\ 4x + 9y &= 5. \end{aligned}$

4. $\begin{aligned} -5x + 15y &= 3 \\ 2x - 6y &= 4. \end{aligned}$ **5.** $\begin{aligned} 2x - 3y - 2z &= 2 \\ 6x - 9y - 3z &= -3. \end{aligned}$

6. $\begin{aligned} 2x + y - 6z &= 1 \\ x + y - z &= 2 \\ 3x + 2y - 7z &= 0. \end{aligned}$ **7.** $\begin{aligned} 3x - 6y + 3z &= 9 \\ 2x - 4y + 2z &= 6 \\ 5x - 10y + 5z &= 15. \end{aligned}$ **8.** $\begin{aligned} 3x + 2y + z &= 6 \\ 2x - 4y + 5z &= 3 \\ x + y - z &= 1. \end{aligned}$

9. $\begin{aligned} 4x + 3y - z + w &= -2 \\ 2x + y + z + w &= 2 \\ 5x + 3y + z + w &= -2. \end{aligned}$ **10.** $\begin{aligned} 2x + y - z + 3w &= 0 \\ 5x - 8y + 5z - 3w &= 0 \\ x - 3y + 2z - 2w &= 0 \\ 3x - 2y + z + w &= 0. \end{aligned}$

11. Determine if the following sets of vectors are linearly dependent.

a. $\{2t - t^3, 4 + t^2, t + t^2 + t^3\}$.
b. $\{1 + 2t^2 + 6t^6, t^2 + t^4 + 3t^6, 1 - 2t^4\}$.
c. $\{(2, 4, -3), (4, 0, -5), (-1, 2, 1)\}$.
d. $\{(1, 0, -1, 0), (0, 1, 0, -1), (0, 1, -1, 0), (1, 0, 0, -1)\}$.

12. Find the transpose of

a. $\begin{pmatrix} 2 & 1 \\ 3 & 6 \end{pmatrix}$. b. $\begin{pmatrix} 4 & 2 & 5 \\ 1 & 3 & 7 \end{pmatrix}$. c. $\begin{pmatrix} 2 \\ 5 \\ 6 \end{pmatrix}$. d. $\begin{pmatrix} 5 & 2 & 6 \\ 2 & 5 & 0 \\ 3 & 4 & 2 \end{pmatrix}$. e. $\begin{pmatrix} 3 & 1 & 7 \\ 1 & 0 & 2 \\ 4 & 2 & 0 \\ 1 & 3 & 2 \end{pmatrix}$.

13. Find the rank of each matrix in problem 12 and show that it is equal to the rank of its transpose.

14. Suppose a system of linear equations has more than one solution. Show that it must have an infinite number of solutions.

15. Suppose $\{V_1, \ldots, V_n\}$ is a linearly independent set in $\mathscr{V}_n$. Prove without reference to dimension that this set spans $\mathscr{V}_n$.

§4. Geometric Interpretation

Let $AX = B$ be a system of n linear equations in m unknowns. A solution $X = C$ for this system can be viewed as a point in Cartesian m-space. For although C is an $m \times 1$ matrix, its transpose is essentially an m-tuple of real numbers. Call the set $\{C \in \mathscr{M}_{m\times 1} | AC = B\}$ the *solution set* for $AX = B$. Then the geometric question is: What is the nature of the graph of the solution set in m-space? Of course, if the system is inconsistent, the set is empty and, as we saw in the first section, the solution set is a subspace of $\mathscr{M}_{m\times 1}$ if and only if $B = \mathbf{0}$.

Recall that the graph of a single linear equation in m unknowns is a hyperplane in m-space (a line in the plane if $m = 2$, and a plane in 3-space if $m = 3$). Therefore, a system of n linear equations in m unknowns can be thought of as giving n hyperplanes in m-space, although they need not all be distinct. A solution of the system of equations is then a point in common to all n hyperplanes, and the graph of the solution set is the intersection of the hyperplanes.

Example 1 Determine how the planes with Cartesian equations $4x - 2y + 6z = 1$, $2x + y - 3z = 2$, and $6x - 3y + 9z = 4$ intersect in 3-space.

Gaussian elimination on the system of 3 equations in 3 unknowns yields the inconsistent system

$$\begin{aligned} x &= 0 \\ y - 3z &= 0. \\ 0 &= 1 \end{aligned}$$

Therefore, there is no point common to the three planes. In this case, two of the planes are parallel, with normal direction $(2, -1, 3)$. Since not all three planes are parallel, the two parallel planes intersect the third plane in a pair of parallel lines.

Now suppose the system $AX = B$ is consistent. Then the nature of the graph of the solution set depends on the number of parameters in the solution. If there are no parameters, the system has a unique solution, which is represented by a single point. Thus if two equations in two unknowns have a unique solution, they are represented by two nonparallel lines in the plane intersecting at a point. But in general a solution may involve one or more parameters.

Example 2 Consider the three planes in 3-space given by

$$\begin{aligned} x - 2y + z &= 5 \\ 3x + y - 4z &= 1 \\ x + 5y - 6z &= -9. \end{aligned}$$

Gaussian elimination yields

$$\begin{aligned} x - z &= 1 \\ y - z &= -2. \end{aligned}$$

Therefore the solution involves one parameter and is given by $x = 1 + t$, $y = t - 2$, $z = t$, $t \in R$. The nature of the solution set in 3-space is clear when the solutions are written as

$$(x, y, z) = t(1, 1, 1) + (1, -2, 0) \qquad \text{with} \quad t \in R.$$

So the three planes intersect in the line with direction numbers (1, 1, 1) passing through the point (1, −2, 0).

In the preceding example, one parameter is needed to express all the solutions, and the graph of the solution set is a line. The number of parameters needed to express a solution is geometrically described as the number of *degrees of freedom* in the graph. Thus a line has one degree of freedom. In all of 3-space there are three degrees of freedom, and a single point has no degrees of freedom. In 3-space, this leaves the plane which should have two degrees of freedom. Consider the Cartesian equation of a plane in 3-space, $x + 2y - z = 3$. The solutions of this equation may be written in the form $x = 3 - 2t + s$, $y = t$, $z = s$ with $t, s \in R$. Here the two parameters s and t exhibit the two degrees of freedom. Or the solution could be written as

$$(x, y, z) = (3, 0, 0) + t(-2, 1, 0) + s(1, 0, 1), \qquad t, s \in R.$$

This shows that the plane passes through the point (3, 0, 0) and is parallel to the plane through the origin spanned by (−2, 1, 0) and (1, 0, 1). This idea can be generalized to call the graph of the equation $P = U + tV + sW$, $U, V, W \in \mathscr{V}_m$ with V and W linearly independent, a *plane* in m-space.

Example 3 Consider the intersection of the three hyperplanes in 4-space given by the following system:

$$\begin{aligned} x + y + 2z + 5w &= 5 \\ 3x + y + 8z + 7w &= 9 \\ x - y + 4z - 3w &= -1. \end{aligned}$$

Gaussian elimination yields

$$\begin{aligned} x \quad + 3z + \ w &= 2 \\ y - \ z + 4w &= 3. \end{aligned}$$

Therefore the solution requires two parameters, and the graph, having two degrees of freedom, is a plane. To see this, write the solution as $x = 2 - 3s - t$, $y = 3 + s - 4t$, $z = s$, $w = t$ with $s, t \in R$ or in vector form as

$$(x, y, z, w) = (2, 3, 0, 0) + s(-3, 1, 1, 0) + t(-1, -4, 0, 1), \qquad s, t \in R.$$

This is the plane in 4-space through the point (2, 3, 0, 0) which is parallel to the plane through the origin spanned by $(-3, 1, 1, 0)$ and $(-1, -4, 0, 1)$.

Gaussian elimination provides solutions in parametric form, but Cartesian equations may be obtained by row reduction of a matrix. Sometimes the Cartesian form is more useful. For example, the coefficients in a Cartesian equation for a plane in 3-space provide direction numbers for a normal.

Example 4 Find the normal to the plane in 3-space given by

$$\mathscr{S} = \{(3a + 2b, 2a + 3b, 6a - b) | a, b \in R\}.$$

Although $\mathscr{S}$ is defined using two parameters, it need not be a plane. But

$$(3a + 2b, 2a + 3b, 6a - b) = a(3, 2, 6) + b(2, 3, -1)$$

and (3, 2, 6) and $(2, 3, -1)$ are linearly independent, so there are in fact two degrees of freedom and $\mathscr{S}$ is a plane. Since $\mathscr{S}$ is the row space of

$$\begin{pmatrix} 3 & 2 & 6 \\ 2 & 3 & -1 \end{pmatrix}.$$

the Cartesian equation can be found by row reduction. The matrix

$$\begin{pmatrix} 3 & 2 & 6 \\ 2 & 3 & -1 \end{pmatrix}$$

is row-equivalent to

$$\begin{pmatrix} 1 & 0 & 4 \\ 0 & 1 & -3 \end{pmatrix}.$$

So $\mathscr{S} = \mathscr{L}\{(1, 0, 4), (0, 1, -3)\}$ and $(x, y, z) \in \mathscr{S}$ provided $(x, y, z) = x(1, 0, 4) + y(0, 1, -3)$. Therefore z must equal $4x - 3y$, and the Cartesian equation for $\mathscr{S}$ is $4x - 3y - z = 0$. Thus $(4, -3, -1)$ is a normal direction.

A point P in m-space has m degrees of freedom. If the point is required to satisfy a single linear equation, then there is one constraint on the freedom of P and it has $m - 1$ degrees of freedom, that is, it lies on a hyperplane. If P is to satisfy the consistent system $AX = B$ of n equations in m unknowns, then the number of constraints on P need not be n. Rather it is the rank of the coefficient matrix A. We will say that $AX = B$ has k *independent* equations if the rank of the augmented matrix is k. Thus if P satisfies the consistent system $AX = B$ with k independent equations, then there are k restraints on the freedom of P, or P has $m - k$ degrees of freedom. Recall that the number of parameters required in the solution of a consistent system of equations in m unknowns, is also $m -$ rank A. For example, although three hyperplanes were given in Example 3, they did not consitute three constraints on the points of the graph in 4-space. If they had, there would have been $1 = 4 - 3$ degree of freedom, and the solution would have been a line instead of a plane.

Suppose S is the solution set for a consistent system of linear equations. Then the algebraic character of S depends on whether the system is homogeneous, for S is a subspace only if the system is homogeneous. Otherwise S is the translate of a subspace, see problem 6 below. In contrast, there is no important geometric distinction for the graph of S. For example, a line represents a subspace only if it passes through the origin. But all lines are essentially the same, and any point could have been chosen as the origin. The choice of a particular Cartesian coordinate system is required to obtain our geometric representation of vectors and the corresponding algebraic view of

Euclidean geometry. However, the algebraic view distinguishes one point, the origin, whereas from the geometric perspective, all points are identical. In time it will be necessary to develop ideas that allow us to separate properties which depend on a figure's position from those that do not.

Problems

1. Determine the number of independent equations in the following systems and identify the graph of the solution set.

a. $2x + y = 3$, $x - 2y = 1$, $4x + 7y = 7.$

b. $3x + 9y = 3$, $x + 3y = 1$, $2x + 6y = 2.$

c. $3x - y + 2z = 0$, $x + 4y + z = 0$, $2x - 5y + z = 0.$

d. $4x - 2y + 6z = 0$, $-2x + y - 3z = 0$, $6x - 3y + 9z = 0.$

e. $2x + 4y - 6z + 2w = 2$, $3x + 6y - 9z + 3w = 3.$

f. $x + y = 0$, $y - z = 0$, $x - z = 0.$

2. Determine how the planes, given by the following equations, intersect.

a. $2x + y - z = 0, \quad x + y + 3z = 2.$
b. $2x - y + 3z = 0, \quad 6x - 3y + 9z = 0.$
c. $x + y + z = 0, \quad 2x - y = 0, \quad x + z = 0.$
d. $x + y + z = 2, \quad 2x + y - z = 3, \quad x + 2y + 4z = 3.$

3. Given three distinct planes in 3-space, write their equations in the form $AX = B$. List all possible combinations of the values for the rank of A and the rank of A^*, and describe the geometric condition in each case.

4. Find a Cartesian equation for the following sets by using elementary row operations to simplify a matrix that has the given set as its row space.

a. $\{(2a + 3b, 3a + 5b, 4a + 5b) | a, b \in R\}.$
b. $\{(a, 2a - 3b, 2b - a) | a, b \in R\}.$
c. $\{(a + 4b + 3c, a + b + 2c, 2a - b + 3c) | a, b, c \in R\}.$
d. $\{(a + b, 2a + 3b + c, a + 2c, 3a - 2b + 4c) | a, b, c \in R\}.$
e. $\{(a + b + 2c, -a - b, a - b - 2c, b - a) | a, b, c \in R\}.$

5. Find the intersection of the given hyperplanes, and describe it geometrically.

a. $x + y + z + w = 6$, $2x + y + 4z - 4w = 0$, $3x + 2y + 5z - 5w = 2.$

b. $x + y + 3z + 2w = -2$, $2x + 3y + 8z + 4w = -3$, $x - y - z + 2w = -4.$

c. $x_1 + 2x_2 + 5x_3 - 2x_4 + 4x_5 = 3$, $2x_1 + x_2 + 7x_3 - x_4 + 2x_5 = 3$, $x_1 + 3x_2 + 6x_3 - 5x_4 + 6x_5 = 2.$

6. Suppose $AX = B$ is a consistent system of k independent equations in m unknowns. Let S be the solution set for $AX = B$.

a. Show that if P and Q are any two points in S, then the entire line containing P and Q is in S.

b. Let $\mathscr{S}$ be the solution space for the homogeneous system $AX = \mathbf{0}$. Show that S may be written in the form $S = \{U + V \mid V \in \mathscr{S}\}$ where U is any vector in S. (Geometrically S is a translate of $\mathscr{S}$ and so can be thought of as being parallel to $\mathscr{S}$.)

7. Suppose S and $\mathscr{S}$ are as in problem 6, and $AX = B$ is given by:

$$\begin{aligned} 2x - 3y &= 1 \\ 6x - 9y &= 3. \end{aligned}$$

Show that S and $\mathscr{S}$ are parallel lines, with $\mathscr{S}$ passing through the origin.

8. Suppose S and $\mathscr{S}$ are as in problem 6, and $AX = B$ is given by

$$\begin{aligned} 2x - 4y - 2z &= -2 \\ 3x - 5y + z &= -4 \\ 2x - 2y + 6z &= -4. \end{aligned}$$

a. Find S. Why does part a of problem 6 hold here?
b. Find $\mathscr{S}$.
c. Show that S is parallel to $\mathscr{S}$.

9. Suppose S and $\mathscr{S}$ are as in problem 6, and $AX = B$ is given by

$$\begin{aligned} 2x - 3y - 5z + 4w &= -1 \\ 3x - 2y + w &= 6 \\ 2x + y + 7z - 4w &= 11. \end{aligned}$$

a. Find S and $\mathscr{S}$.
b. Express S as the set $\{U + V \mid V \in \mathscr{S}\}$ for some vector U.
c. Why is the result in part a of problem 6 satisfied in this case?

§5. Determinants: Definition

The determinant can be thought of as a function that assigns a real number to each $n \times n$ matrix. There are several approaches to the definition of this function, and although they differ considerably in their level of abstraction, they all lead to the same properties. For our purposes, we must be able to calculate the value of a determinant and use determinants in mathematical statements. Therefore we will take a simple computational approach to the definition, even though it will then not be possible to prove the central property of determinants.

To motivate the general definition, consider the expression $ad - bc$, for $a, b, c, d \in R$. This expression has occurred several times in this and the previous chapter. For example, we know that the set $\{(a, b), (c, d)\}$ in $\mathscr{V}_2$ is linearly independent if and only if $ad - bc \neq 0$.

Definition The *determinant* of a 2×2 matrix

$$A = \begin{pmatrix} a & b \\ c & d \end{pmatrix}$$

is the number $ad - bc$ and will be denoted by $|A|$, det A, and by

$$\begin{vmatrix} a & b \\ c & d \end{vmatrix}.$$

Thus

$$\begin{vmatrix} 2 & 1 \\ 3 & 5 \end{vmatrix} = 2(5) - 3(1) = 7$$

and

$$\begin{vmatrix} 4 & 8 \\ 3 & 6 \end{vmatrix} = 24 - 24 = 0.$$

Since a 2×2 matrix has rank 2 if and only if its rows are linearly independent, we can make the following statement: For $A \in \mathscr{M}_{2 \times 2}$, rank $A = 2$ if and only if det $A \neq 0$. This can be applied to the system of linear equations

$$ax + by = e$$
$$cx + dy = f.$$

We know that this system has a unique solution if and only if the rank of the coefficient matrix is 2, that is, if and only if

$$\begin{vmatrix} a & b \\ c & d \end{vmatrix} \neq 0.$$

Moreover, if there is a unique solution, it has the form

$$x = \frac{ed - bf}{ad - bc} = \frac{\begin{vmatrix} e & b \\ f & g \end{vmatrix}}{\begin{vmatrix} a & b \\ c & d \end{vmatrix}}, \qquad y = \frac{af - ec}{ad - bc} = \frac{\begin{vmatrix} a & e \\ c & f \end{vmatrix}}{\begin{vmatrix} a & b \\ c & d \end{vmatrix}}.$$

This is an example of Cramer's rule for finding the unique solution of a system of n independent, consistent linear equations in n unknowns. Since there is no pressing need for the general form of this rule, it will be postponed until the proof can be given in matrix form (see Theorem 4.22, page 185). Notice that the particular expression for the solution in terms of determinants is not the only one possible. A number in the form $xw - yz$ can be written as a determinant in many different ways. For example

$$xw - yz = \begin{vmatrix} w & y \\ z & x \end{vmatrix} = \begin{vmatrix} y & -x \\ w & -z \end{vmatrix}.$$

Example 1 Let $U = (x, y)$ and $V = (z, w)$. Then the area of the parallelogram determined by $\mathbf{0}$, U, V and $U + V$ in the plane is the absolute value of

$$\begin{vmatrix} x & y \\ z & w \end{vmatrix}.$$

Let b be the length of the base of the parallelogram and h the height, as in Figure 1. Then if A is the area of the parallelogram, $A = bh$ and $b = |V|$. The height can be found using the angle θ between U and a vector N perpendicular to V. For then

$$h = \|U\| |\cos \theta| = \|U\| \frac{|U \circ N|}{\|U\| \|N\|} = \frac{|U \circ N|}{\|N\|}.$$

Now since $V = (z, w)$ we can let $N = (w, -z)$ and $\|N\| = \sqrt{w^2 + z^2} = \|V\|$.

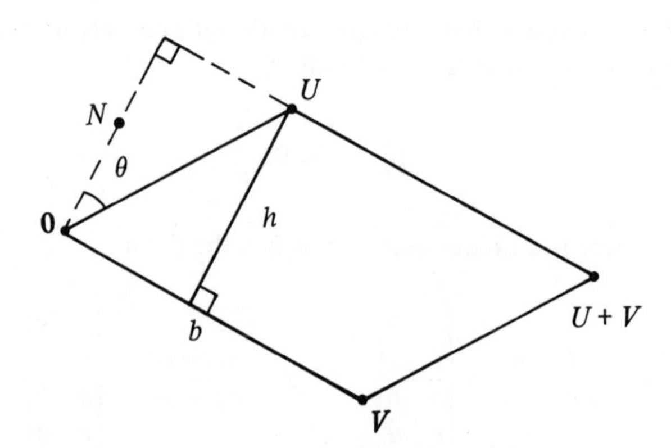

Figure 1

Thus

$$\begin{aligned} A = bh &= \|V\| \frac{|U \circ N|}{\|N\|} = |U \circ N| \\ &= |(x, y) \circ (w, -z)| = |xw - yz| \\ &= \left| \det\begin{pmatrix} x & y \\ z & w \end{pmatrix} \right|. \end{aligned}$$

As a particular case, suppose $U = (2, 5)$ and $V = (7, -3)$. Then

$$\det\begin{pmatrix} 2 & 5 \\ 7 & -3 \end{pmatrix} = -41$$

and the area of the parallelogram with vertices $(0, 0)$, $(2, 5)$, $(7, -3)$, and $(9, 2)$ is $|-41| = 41$.

Although U and V are assumed to be independent in the above example, the calculation is valid provided only that $V \neq \mathbf{0}$. When U and V are linearly dependent, the determinant is zero, but there is also no area.

When it comes to defining the value of the determinant of a 3×3 matrix, there is no expression like $ad - bc$ to work from. Therefore suppose we ask that determinants be defined so that their absolute value is the volume of the parallelepiped in 3-space determined by the origin and three independent vectors. That is, if $A = (a_1, a_2, a_3)$, $B = (b_1, b_2, b_3)$, and $C = (c_1, c_2, c_3)$, then the determinant of the matrix

$$\begin{pmatrix} a_1 & a_2 & a_3 \\ b_1 & b_2 & b_3 \\ c_1 & c_2 & c_3 \end{pmatrix}$$

should be defined so that its value is plus or minus the volume of the parallelepiped with three adjacent sides determined by A, B, and C. By computing the volume V of this parallelepiped, we will see how the determinant of a 3×3 matrix might be defined. Let b be the area of the base determined by B and C, and let h be the height (see Figure 2). Then $V = bh$. If N is a normal to the plane of B and C, then the formula $h = |A \circ N|/\|N\|$ holds in 3-space just as it did in the plane. To find N, we need a Cartesian equation for the plane $\mathscr{L}\{B, C\}$. That is, we must characterize the row space of the matrix

$$\begin{pmatrix} b_1 & b_2 & b_3 \\ c_1 & c_2 & c_3 \end{pmatrix}.$$

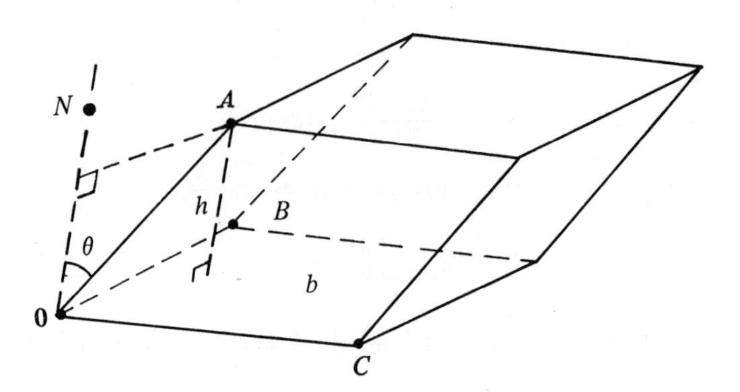

Figure 2

If $b_1c_2 - b_2c_1 \neq 0$, then the echelon form for this matrix is

$$\begin{pmatrix} 1 & 0 & \dfrac{b_3c_2 - b_2c_3}{b_1c_2 - b_2c_1} \\ 0 & 1 & \dfrac{b_1c_3 - b_3c_1}{b_1c_2 - b_2c_1} \end{pmatrix}$$

Thus

$$\mathscr{L}\{B, C\} = \left\{(x, y, z) | z = x\left(\frac{b_3c_2 - b_2c_3}{b_1c_2 - b_2c_1}\right) + y\left(\frac{b_1c_3 - b_3c_1}{b_1c_2 - b_2c_1}\right)\right\}$$

and the equation for the plane $\mathscr{L}\{B, C\}$ can be written as

$$(b_2c_3 - b_3c_2)x - (b_1c_3 - b_3c_1)y + (b_1c_2 - b_2c_1)z = 0.$$

Notice that a particular order has been chosen for the coefficients. Now we can take the vector N to be

$$N = \left(\begin{vmatrix} b_2 & b_3 \\ c_2 & c_3 \end{vmatrix}, -\begin{vmatrix} b_1 & b_3 \\ c_1 & c_3 \end{vmatrix}, \begin{vmatrix} b_1 & b_2 \\ c_1 & c_2 \end{vmatrix}\right).$$

N is called the *cross product* of B and C and is normal to the plane of B and C even if $b_1c_2 - b_2c_1 = 0$. This could be shown directly, but it follows at once from a property of determinants, see problem 2 of Section 6.

The fact that $\|N\| = \|B\|\|C\| \sin \theta$ is proved in Section 3 of the appendix on determinants. Using this equation, the area b equals $\|N\|$, just as in the plane, and

$$V = \|N\| \frac{|A \circ N|}{\|N\|} = |A \circ N|.$$

Therefore the volume is the absolute value of

$$a_1 \begin{vmatrix} b_2 & b_3 \\ c_2 & c_3 \end{vmatrix} + a_2\left(-\begin{vmatrix} b_1 & b_3 \\ c_1 & c_3 \end{vmatrix}\right) + a_3 \begin{vmatrix} b_1 & b_2 \\ c_1 & c_2 \end{vmatrix}.$$

This expression provides the pattern for defining the determinant of 3 × 3 as well as $n \times n$ matrices. As an example, the determinant of the 3 × 3 matrix

$$\begin{pmatrix} 2 & 3 & 4 \\ 1 & 0 & 5 \\ 3 & 2 & 6 \end{pmatrix}$$

should be

$$2\begin{vmatrix} 0 & 5 \\ 2 & 6 \end{vmatrix} + 3\left(-\begin{vmatrix} 1 & 5 \\ 3 & 6 \end{vmatrix}\right) + 4\begin{vmatrix} 1 & 0 \\ 3 & 2 \end{vmatrix} = -20 + 27 + 8 = 15.$$

In order to use the above expression to define the determinant of a 3 × 3 matrix, it is necessary to know how to compute determinants of 2 × 2 matrices. This will be an essential characteristic of our definition. The definition of determinant for $(n + 1) \times (n + 1)$ matrices will use determinants of $n \times n$ matrices. Such a definition is called a definition by induction.

For 1 × 1 matrices, that is (r) with $r \in R$, define the *determinant* to be

$$\det (r) = |(r)| = r.$$

So $|(-3)| = -3$. Now suppose determinant has been defined for all $n \times n$ matrices (n could be 1). Let $A = (a_{ij})$ be an $(n + 1) \times (n + 1)$ matrix. Then for each entry a_{ij} of A, define the *minor* of a_{ij} to be the determinant of the $n \times n$ matrix obtained from A by deleting the ith row and the jth column. The *cofactor* of a_{ij} is defined to be $(-1)^{i+j}$ times the minor of a_{ij} and is denoted by A_{ij}. Then the *determinant* of A, denoted by $|A|$ or $\det A$, is defined by

$$\det A = |A| = a_{11}A_{11} + a_{12}A_{12} + \cdots + a_{1\,n+1}A_{1\,n+1}.$$

The determinant of an $n \times n$ matrix will often be called an $n \times n$ determinant.

Example 2 Let

$$A = \begin{pmatrix} 2 & 1 & 3 \\ -4 & 5 & -1 \\ 0 & 4 & -2 \end{pmatrix}.$$

Then the minor of 4 is the 2×2 determinant

$$\begin{vmatrix} 2 & 3 \\ -4 & -1 \end{vmatrix} = 10.$$

The matrix

$$\begin{pmatrix} 2 & 3 \\ -4 & -1 \end{pmatrix}$$

is obtained by deleting the third row and second column, so the cofactor of 4, A_{32}, is $(-1)^{3+2}10 = -10$.

The minor of 2 in A is

$$\begin{vmatrix} 5 & -1 \\ 4 & -2 \end{vmatrix} = -6$$

and the cofactor of 2 is $A_{11} = (-1)^{1+1}(-6) = -6$.

The determinant of A is given by

$$\begin{aligned} \det A &= a_{11}A_{11} + a_{12}A_{12} + a_{13}A_{13} \\ &= 2(-1)^{1+1}\begin{vmatrix} 5 & -1 \\ 4 & -2 \end{vmatrix} + 1(-1)^{1+2}\begin{vmatrix} -4 & -1 \\ 0 & -2 \end{vmatrix} + 3(-1)^{1+3}\begin{vmatrix} -4 & 5 \\ 0 & 4 \end{vmatrix} \\ &= -12 - 8 - 48 = -68. \end{aligned}$$

Example 3 Show that this definition of determinant agrees with the definition for 2×2 matrices.

$$\begin{aligned} \begin{vmatrix} a_{11} & a_{12} \\ a_{21} & a_{22} \end{vmatrix} &= a_{11}A_{11} + a_{12}A_{12} \\ &= a_{11}(-1)^{1+1}|(a_{22})| + a_{12}(-1)^{1+2}|(a_{21})| \\ &= a_{11}a_{22} - a_{12}a_{21}. \end{aligned}$$

Example 4 Compute the determinant of

$$A = \begin{pmatrix} 3 & 0 & 1 & 0 \\ 1 & 2 & 0 & 0 \\ 0 & 1 & 3 & 0 \\ 0 & 0 & 2 & 1 \end{pmatrix}.$$

$$\begin{aligned} \det A &= 3A_{11} + 0A_{12} + 1A_{13} + 0A_{14} \\ &= 3(-1)^{1+1}\begin{vmatrix} 2 & 0 & 0 \\ 1 & 3 & 0 \\ 0 & 2 & 1 \end{vmatrix} + 1(-1)^{1+3}\begin{vmatrix} 1 & 2 & 0 \\ 0 & 1 & 0 \\ 0 & 0 & 1 \end{vmatrix} \\ &= 3(2)(-1)^{1+1}\begin{vmatrix} 3 & 0 \\ 2 & 1 \end{vmatrix} + (-1)^{1+1}\begin{vmatrix} 1 & 0 \\ 0 & 1 \end{vmatrix} + 2(-1)^{1+2}\begin{vmatrix} 0 & 0 \\ 0 & 1 \end{vmatrix} \\ &= 6(3) + 1 - 2(0) = 19. \end{aligned}$$

If it were not for the presence of several zeros in the above matrix, the evaluation of the 4×4 determinant would involve a considerable amount of arithmatic. A 2×2 determinant involves two products. A 3×3 determinant is the sum of three 2×2 determinants, each multiplied by a real number and therefore involves $3 \cdot 2$ products, each with 3 factors. Similarly, a 4×4 determinant involves $4 \cdot 3 \cdot 2 = 24$ products of four numbers each. In general, to evaluate an $n \times n$ determinant it is necessary to compute $n \cdot (n-1) \cdot \cdots \cdot 3 \cdot 2 = n!$ products of n factors each. Fortunately there are short cuts, so that a determinant need not be evaluated in exactly the way it has been defined. The following theorem is proved in the appendix on determinants, but at this point it is stated without proof.

Theorem 3.13 Let $A = (a_{ij})$ be an $n \times n$ matrix, then for each i, $1 \leq i \leq n$,

$$\det A = a_{i1}A_{i1} + a_{i2}A_{i2} + \cdots + a_{in}A_{in}.$$

This is called the expansion of the determinant along the ith row of A.

And for each j, $1 \leq j \leq n$,

$$\det A = a_{1j}A_{1j} + a_{2j}A_{2j} + \cdots + a_{nj}A_{nj}.$$

This is the expansion of the determinant along the jth column of A.

Example 5 Let

$$A = \begin{pmatrix} 2 & 5 & 1 & 7 \\ 3 & 1 & 0 & -1 \\ 2 & 0 & 0 & 0 \\ 1 & 3 & 0 & 2 \end{pmatrix}.$$

The expansion of det A along the first row, as in the definition, results in four 3×3 determinants. But if det A is expanded along the 3rd column or the 3rd row, then only one 3×3 cofactor need be evaluated. Expansion along the 3rd column of A gives

$$\det A = 1(-1)^{1+3} \begin{vmatrix} 3 & 1 & -1 \\ 2 & 0 & 0 \\ 1 & 3 & 2 \end{vmatrix} = 2(-1)^{2+1} \begin{vmatrix} 1 & -1 \\ 3 & 2 \end{vmatrix} = -2(2+3) = -10.$$

Expansion along the 3rd row of A gives

$$\det A = 2(-1)^{3+1} \begin{vmatrix} 5 & 1 & 7 \\ 1 & 0 & -1 \\ 3 & 0 & 2 \end{vmatrix} = 2(-1)^{1+2} \begin{vmatrix} 1 & -1 \\ 3 & 2 \end{vmatrix} = -2(2+3) = -10.$$

The values are the same as predicted by Theorem 3.13. In the expansions, the first 3×3 minor was expanded along its 2nd row and the second along its 2nd column.

Two corollaries follow easily from Theorem 3.13.

Corollary 3.14 If a row or column of an $n \times n$ matrix A contains only 0's, then det $A = 0$.

Corollary 3.15 If A is an $n \times n$ matrix and B is obtained from A by multiplying a row or column of A by a scalar r, then det $B = r$ det A.

The second corollary agrees with the idea that a determinant measures volume. For if the edges of a parallelogram or parallelepiped determined by one direction are multiplied by r, $r > 0$, then the area or volume is also multiplied by r.

Problems

Evaluate the following determinants:

1. $\begin{vmatrix} 2 & 4 \\ 1 & 3 \end{vmatrix}$. **2.** $|(-5)|$. **3.** $\begin{vmatrix} 2 & 0 & 3 \\ 7 & 2 & 9 \\ 5 & 0 & 8 \end{vmatrix}$. **4.** $\begin{vmatrix} 1 & 5 & 9 \\ 0 & 1 & 2 \\ 0 & 0 & 1 \end{vmatrix}$. **5.** $\begin{vmatrix} 2 & 7 & 0 \\ 3 & 1 & 0 \\ 9 & 2 & 0 \end{vmatrix}$.

6. $\begin{vmatrix} 1 & 2 & 8 & -1 \\ 0 & 0 & 2 & 0 \\ 0 & 3 & 7 & 1 \\ 2 & 1 & 3 & -3 \end{vmatrix}$. **7.** $\begin{vmatrix} 1 & 0 & 0 & 1 \\ 0 & 1 & 1 & 1 \\ 0 & 1 & 0 & 0 \\ 1 & 1 & 1 & 0 \end{vmatrix}$. **8.** $\begin{vmatrix} 1 & 2 & 3 \\ 1 & 2 & 3 \\ 5 & 4 & 1 \end{vmatrix}$.

9. Show that the four points A, B, C, and D are the vertices of a parallelogram with D opposite A and find its area when

a. $A = (0, 0)$, $B = (2, 5)$, $C = (1, -4)$, $D = (3, 1)$.
b. $A = (2, -5)$, $B = (1, 6)$, $C = (4, 7)$, $D = (3, 18)$.
c. $A = (1, 1)$, $B = (2, 4)$, $C = (3, -2)$, $D = (4, 1)$.

10. Find the volume of the parallelepiped in 3-space determined by the vectors

a. $(1, 0, 0)$, $(0, 1, 0)$, $(0, 0, 1)$.
b. $(3, 1, -1)$, $(4, 2, 2)$, $(-1, 2, -3)$.

11. Write out the six terms in the expansion of an arbitrary 3×3 determinant and notice that this sum contains every possible product that can be formed by taking one element from each row and each column. Does this idea generalize to $n \times n$ determinants?

12. Prove Corollary 3.15.

13. Let A be an $n \times n$ matrix. Use induction to prove that $|A^T| = |A|$.

14. For $B, C \in \mathscr{E}_3$, let $B \times C$ denote the cross product of B and C. Compute the following cross products:

a. $(1, 0, 0) \times (0, 1, 0)$.
b. $(0, 1, 0) \times (1, 0, 0)$.
c. $(1, 2, 3) \times (1, 2, 3)$.
d. $(2, 3, 2) \times (-2, 1, 0)$.
e. $(-4, 2, -8) \times (6, -3, 12)$.
f. $(3, 0, 2) \times (-5, 0, 3)$.

15. Suppose the definition of a 3×3 determinant is extended to allow vectors as entries in the first row, then the value of such a determinant is a linear combination of the vectors.

a. Show that if $B = (b_1, b_2, b_3)$ and $C = (c_1, c_2, c_3)$, then

$$B \times C = \begin{vmatrix} E_1 & E_2 & E_3 \\ b_1 & b_2 & b_3 \\ c_1 & c_2 & c_3 \end{vmatrix}$$

where $\{E_1, E_2, E_3\}$ is the standard basis for $\mathscr{E}_3$.

b. Let $A = (a_1, a_2, a_3)$ and show that

$$A \circ (B \times C) = \begin{vmatrix} a_1 & a_2 & a_3 \\ b_1 & b_2 & b_3 \\ c_1 & c_2 & c_3 \end{vmatrix}.$$

$A \circ (B \times C)$ is the *scalar triple product* of A, B, and C and the volume of the parallelepiped determined by A, B, and C in $\mathscr{E}_3$.

16. Use a geometric argument to show that $\{A, B, C\}$ is linearly dependent if $A \circ (B \times C) = 0$.

17. a. Compute $[(0, 1, 2) \times (3, 0, -1)] \times (2, 4, 1)$, and $(0, 1, 2) \times [(3, 0, -1) \times (2, 4, 1)]$.
b. What can be concluded from part a?
c. What can be said about the expression $A \times B \times C$?

§6. Determinants: Properties and Applications

The evaluation of a determinant is greatly simplified if it can be expanded along a row or column with only one nonzero entry. If no such row or column exists, one could be obtained by performing elementary row operations on the matrix. The following theorem shows how such operations affect the value of the determinant.

Theorem 3.16 Suppose A is an $n \times n$ matrix and B is obtained from A by a single elementary row operation, then

1. $|B| = -|A|$ if the elementary row operation is of type I.
2. $|B| = r|A|$ if a row of A is multiplied by a scalar r.
3. $|B| = |A|$ if the elementary row operation is of type III.

Proof The proof of 1 is by induction on the number of rows in A. It is not possible to interchange two rows of a 1×1 matrix, so consider 2×2 matrices. If

$$A = \begin{pmatrix} a & b \\ c & d \end{pmatrix} \quad \text{and} \quad B = \begin{pmatrix} c & d \\ a & b \end{pmatrix},$$

then

$$|B| = cb - da = -(ad - bc) = -|A|.$$

Therefore the property holds for 2×2 matrices. Now assume that 1 holds for all $(n - 1) \times (n - 1)$ matrices with $n \geq 3$. Let A be an $n \times n$ matrix $A = (a_{ij})$ and let $B = (b_{ij})$ be the matrix obtained from A by interchanging two rows. Since $n > 2$, at least one row is not involved in the interchange, say the kth is such a row and expand $|B|$ along this row. Then $|B| = b_{k1}B_{k1} + b_{k2}B_{k2} + \cdots + b_{kn}B_{kn}$. For each index j, the cofactor B_{kj} differs from the cofactor A_{kj} of a_{kj} in A only in that two rows are interchanged. These cofactors are determinants of $(n-1) \times (n-1)$ matrices, so by the induction assumption, $B_{kj} = -A_{kj}$. Since the kth row of B equals the kth row of A, the above expansion for $|B|$ becomes

$$|B| = a_{k1}(-A_{k1}) + a_{k2}(-A_{k2}) + \cdots + a_{kn}(-A_{kn}) = -|A|$$

which is the negative of the expansion of $|A|$ along the kth row of A.

Now 1 holds for all 2×2 matrices, and if it holds for all $(n-1) \times (n-1)$ matrices ($n \geq 3$), then it holds for all $n \times n$ matrices. Therefore 1 is proved by induction.

Part 2 was proved in the previous section (Corollary 3.15); and the induction proof of 3 is left as a problem.

Example 1

$$\begin{vmatrix} -5 & 2 & -3 \\ 4 & -2 & 1 \\ 2 & 3 & 0 \end{vmatrix} = \begin{vmatrix} 7 & -4 & 0 \\ 4 & -2 & 1 \\ 2 & 3 & 0 \end{vmatrix} = 1(-1)^{2+3}\begin{vmatrix} 7 & -4 \\ 2 & 3 \end{vmatrix} = -29.$$

Here the second determinant is obtained from the first by adding three times the second row to the first row.

Elementary row operations can be used to obtain a matrix with zeros in a column but usually not in a row. To do this it is necessary to make changes in the columns of the matrix. *Elementary column operations* are defined just as elementary row operations except they are applied to the columns of a matrix rather than the rows.

Example 2 The following transformation is made with elementary column operations of type III:

$$\begin{pmatrix} 3 & 2 & -1 \\ 2 & 1 & -2 \\ 4 & 3 & -5 \end{pmatrix} \longrightarrow \begin{pmatrix} 3 & 2 & 3 \\ 2 & 1 & 0 \\ 4 & 3 & 1 \end{pmatrix} \longrightarrow \begin{pmatrix} -1 & 2 & 3 \\ 0 & 1 & 0 \\ -2 & 3 & 1 \end{pmatrix}.$$

First 2 times the 2nd column is added to the 3rd and then -2 times the 2nd column is added to the 1st.

Theorem 3.17 Suppose A is an $n \times n$ matrix and B is obtained from A with a single elementary column operation, then

1. $|B| = -|A|$ if the elementary column operation is of type I.
2. $|B| = r|A|$ if a column of A is multiplied by r.
3. $|B| = |A|$ if the elementary column operation is of type III.

Proof Performing an elementary column operation on a matrix has the same effect as performing an elementary row operation on its transpose. Therefore this theorem follows from the corresponding theorem for elementary row operations and the fact that $|A^T| = |A|$ for all $n \times n$ matrices A.

Theorem 3.18 If two rows or two columns of an $n \times n$ matrix A are proportional, then $|A| = 0$.

Proof If A satisfies the hypothesis, then there is an elementary row or column operation of type III which yields a matrix having a row or column of zeros. Thus $|A| = 0$.

It is possible to evaluate any determinant using only elementary row or elementary column operations, that is, without computing any cofactors. To see how this can be done, we will say that the elements a_{ii} of a matrix $A = (a_{ij})$ are on the *main diagonal* of A. And that A is *upper triangular* if all the elements below the main diagonal are zero, that is, if $a_{ij} = 0$ when $i > j$. *Lower triangular* is similarly defined. Then we have the following theorem.

Theorem 3.19 If $A = (a_{ij})$ is an $n \times n$ upper (lower) triangular matrix, then $|A| = a_{11}a_{22} \cdots a_{nn}$.

Proof By induction and left as a problem.

Example 3 Use elementary row operations to evaluate a determinant.

$$\begin{vmatrix} 2 & 4 & -6 \\ 3 & 1 & -4 \\ 2 & 2 & 5 \end{vmatrix} = 2\begin{vmatrix} 1 & 2 & -3 \\ 3 & 1 & -4 \\ 2 & 2 & 5 \end{vmatrix} = 2\begin{vmatrix} 1 & 2 & -3 \\ 0 & -5 & 5 \\ 0 & -2 & 11 \end{vmatrix}$$

$$= -10\begin{vmatrix} 1 & 2 & -3 \\ 0 & 1 & -1 \\ 0 & -2 & 11 \end{vmatrix} = -10\begin{vmatrix} 1 & 2 & -3 \\ 0 & 1 & -1 \\ 0 & 0 & 9 \end{vmatrix}$$

$$= -10 \cdot 1 \cdot 1 \cdot 9 = -90.$$

The technique shown in Example 3 is the same as that used in finding the echelon form of a matrix, except that it is necessary to keep track of the changes in value that arise when elementary row operations of type I and II are used.

Elementary row operations change the value of a determinant by at most a nonzero scalar multiple. Therefore, the determinant of an $n \times n$ matrix A and the determinant of its echelon form can differ at most by a nonzero multiple. Thus if the echelon form for A has a row of zeros, then $|A| = 0$ or:

Theorem 3.20 The rank of an $n \times n$ matrix A is n if and only if $|A| \neq 0$.

Now using determinants, two results from the theory of systems of linear equations can be restated.

Theorem 3.21 1. A system of n linear equations in n unknowns, $AX = B$, has a unique solution if and only if $|A| \neq 0$.

2. A homogeneous system of n linear equations in n unknowns $AX = \mathbf{0}$ has a nontrivial solution if and only if $|A| = 0$.

Example 4 Determine if

$$\{(2, 1, 0, -1), (5, 8, 6, -3), (3, 3, 0, 4), (1, 2, 0, 5)\}$$

is a basis for $\mathscr{V}_4$.

A set of 4 vectors is a basis for $\mathscr{V}_4$ if it is linearly independent, or equivalently if the matrix with the vectors as rows has rank 4, that is, a nonzero determinant. But

$$\begin{vmatrix} 2 & 1 & 0 & -1 \\ 5 & 8 & 6 & -3 \\ 3 & 3 & 0 & 4 \\ 1 & 2 & 0 & 5 \end{vmatrix} = 6(-1)^{2+3}\begin{vmatrix} 2 & 1 & -1 \\ 3 & 3 & 4 \\ 1 & 2 & 5 \end{vmatrix} = -6\begin{vmatrix} 0 & 1 & 0 \\ -3 & 3 & 7 \\ -3 & 2 & 7 \end{vmatrix} = 0.$$

Since two column operations yield a 3×3 matrix with two columns proportional, the determinant is zero and the set fails to be a basis.

Problems

1. Use elementary row and column operations to simplify the evaluation of the following determinants:

a. $\begin{vmatrix} 4 & 5 & 0 \\ 3 & 3 & 1 \\ 3 & -1 & 2 \end{vmatrix}$. b. $\begin{vmatrix} 2 & 1 & 5 \\ 3 & -1 & 2 \\ 4 & 2 & 7 \end{vmatrix}$. c. $\begin{vmatrix} 1 & 1 & 0 & 0 \\ 1 & 1 & 1 & 1 \\ 0 & 1 & 1 & 0 \\ 0 & 0 & 1 & 1 \end{vmatrix}$.

d. $\begin{vmatrix} 1 & 1 & 0 & 1 \\ 0 & 1 & 1 & 0 \\ 1 & 1 & 0 & 0 \\ 1 & 1 & 1 & 1 \end{vmatrix}$. e. $\begin{vmatrix} 2 & 4 & -6 & 4 \\ 2 & 6 & -1 & 7 \\ 6 & 0 & 3 & 6 \\ 1 & 2 & 0 & 2 \end{vmatrix}$. f. $\begin{vmatrix} 0 & 1 & 2 & 3 \\ 1 & 2 & 3 & 0 \\ 2 & 3 & 0 & 1 \\ 3 & 0 & 1 & 2 \end{vmatrix}$.

2. Let $U, V \in \mathscr{E}_3$, and use problem 15 of Section 5 along with the properties of determinants to prove
   a. $U \times V = -V \times U$.
   b. $U \times V$ is perpendicular to both U and V.
   c. If U and V are linearly dependent, then $U \times V = \mathbf{0}$.

3. Use determinants to determine if the following sets are linearly independent:
   a. $\{(4, 2, 3), (1, -6, 1), (2, 3, 1)\}$.
   b. $\{3 + 4i, 2 + i\}$.
   c. $\{t^5 + 2t, t^3 + t + 1, 2t^5 + 5t, t^3 + 1\}$.

4. Is the set $\{(4, 0, 0, 0, 0), (3, -2, 0, 0, 0), (8, 2, 5, 0, 0), (-3, 9, 4, 2, 0), (0, 5, 3, 1, 7)\}$ a basis for $\mathscr{V}_5$?

5. Give induction proofs for the following statements.
   a. An elementary row operation of type III does not affect the value of a determinant.
   b. The determinant of an upper triangular $n \times n$ matrix is the product of the elements on its main diagonal.

6. Suppose $A = (a_{ij})$ is an $n \times n$ matrix with $a_{11} \neq 0$. Show that det $A = (1/(a_{11})^{n-2})$ det B, where $B = (b_{ij})$ is the $(n - 1) \times (n - 1)$ matrix with $2 \leq i, j \leq n$, and $b_{ij} = a_{11}a_{ij} - a_{i1}a_{1j}$. This procedure, called *pivotal condensation*, might be used by a computer to reduce the order of a determinant while introducing only one fraction.

7. Use pivotal condensation to evaluate

   a. $\begin{vmatrix} 3 & 7 & 2 \\ 5 & 1 & 4 \\ 9 & 6 & 8 \end{vmatrix}$. b. $\begin{vmatrix} 5 & 4 & 2 \\ 2 & 7 & 2 \\ 3 & 2 & 3 \end{vmatrix}$. c. $\begin{vmatrix} 2 & 1 & 3 & 2 \\ 1 & 3 & 2 & 1 \\ 2 & 2 & 1 & 1 \\ 3 & 1 & 2 & 1 \end{vmatrix}$.

8. Prove that interchanging two columns in a determinant changes the sign of the determinant. (This does not require induction.)

§7. Alternate Approaches to Rank

The rank of a matrix is defined as the number of nonzero rows in its echelon form, which equals the dimension of its row space. For an $n \times n$

matrix A, the rank is n provided $|A| \neq 0$, and $|A| = 0$ says only that the rank is less than n. However, even if $|A| = 0$ or A is $n \times m$ and the determinant of A is not defined, it is possible to use determinants to find the rank of A.

Definition A $k \times k$ *submatrix* of an $n \times m$ matrix A is a $k \times k$ matrix obtained by deleting $n - k$ rows and $m - k$ columns from A.

The 2×3 matrix

$$\begin{pmatrix} 2 & 1 & 5 \\ 7 & 3 & 8 \end{pmatrix}$$

has three 2×2 submatrices;

$$\begin{pmatrix} 2 & 1 \\ 7 & 3 \end{pmatrix}, \quad \begin{pmatrix} 2 & 5 \\ 7 & 8 \end{pmatrix}, \quad \begin{pmatrix} 1 & 5 \\ 3 & 8 \end{pmatrix}.$$

It has six 1×1 submatrices; (2), (1), (5), etc.

Definition The *determinant rank* of an $n \times m$ matrix A is the largest value of k for which some $k \times k$ submatrix of A has a nonzero determinant.

Example 1 Let

$$A = \begin{pmatrix} 4 & 1 & 3 \\ 2 & 3 & 1 \\ 2 & -2 & 2 \end{pmatrix}.$$

Then

$$\begin{vmatrix} 4 & 1 & 3 \\ 2 & 3 & 1 \\ 2 & -2 & 2 \end{vmatrix} = 0$$

and

$$\begin{vmatrix} 4 & 1 \\ 2 & 3 \end{vmatrix} \neq 0.$$

Therefore the determinant rank of A is 2.

The rank or row rank of A is also 2, for the echelon form for A is

$$\begin{pmatrix} 1 & 0 & 4/5 \\ 0 & 1 & -1/5 \\ 0 & 0 & 0 \end{pmatrix}.$$

Theorem 3.22 For any matrix A, the determinant rank of A equals the rank of A.

Proof Let A be an $n \times m$ matrix with determinant rank s and rank r. We will show first that $s \le r$ and then that $r \le s$.

If the determinant rank of A is s, then A has an $s \times s$ submatrix with nonzero determinant. The s rows of this submatrix are therefore linearly independent. Hence there are s rows of A that are linearly independent, and the dimension of the row space of A is at least s. That is, the rank of A is at least s or $s \le r$.

For the second inequality, the rank of A is r. Therefore the echelon form for A, call it B, has r nonzero rows and r columns containing the leading 1's for these rows. So B has an $r \times r$ submatrix, call it C, with 1's on the main diagonal and 0's elsewhere. Thus $|C| = 1 \ne 0$. Now the r nonzero rows of B span the rows of A, therefore there is a sequence of elementary row operations which transforms these r rows to r rows of A, leaving the remaining $n - r$ rows all zero. This transforms the $r \times r$ submatrix C to an $r \times r$ submatrix of A, and since elementary row operations change the value of a determinant by at most a nonzero multiple, this $r \times r$ submatrix of A has nonzero determinant. That is, the determinant rank of A is at least r or $r \le s$.

Example 2 It may be helpful to look at an example of the procedure used in the second part of the preceding proof.

Let

$$A = \begin{pmatrix} 2 & 4 & 1 & 5 \\ 5 & 10 & 1 & 8 \\ 3 & 6 & 1 & 6 \end{pmatrix}$$

then

$$B = \begin{pmatrix} 1 & 2 & 0 & 1 \\ 0 & 0 & 1 & 3 \\ 0 & 0 & 0 & 0 \end{pmatrix}$$

is the echelon form for A and the rank of A is 2. Now

$$C = \begin{pmatrix} 1 & 0 \\ 0 & 1 \end{pmatrix}$$

is a 2×2 submatrix of B with nonzero determinant. A can be transformed to B with a sequence of elementary row operations; working backward we obtain a sequence that transforms the nonzero rows of B to the first two rows of A (in this case any pair of rows in A could be obtained).

$$\begin{pmatrix} 1 & 2 & 0 & 1 \\ 0 & 0 & 1 & 3 \end{pmatrix} \underset{\text{III}}{\to} \begin{pmatrix} 1 & 2 & 1/2 & 5/2 \\ 0 & 0 & 1 & 3 \end{pmatrix} \underset{\text{II}}{\to} \begin{pmatrix} 1 & 2 & 1/2 & 5/2 \\ 0 & 0 & -3/2 & -9/2 \end{pmatrix}$$

$$\underset{\text{III}}{\to} \begin{pmatrix} 1 & 2 & 1/2 & 5/2 \\ 5 & 10 & 1 & 8 \end{pmatrix} \underset{\text{II}}{\to} \begin{pmatrix} 2 & 4 & 1 & 5 \\ 5 & 10 & 1 & 8 \end{pmatrix}.$$

Performing this sequence of operations on the submatrix C we obtain

$$\begin{pmatrix} 1 & 0 \\ 0 & 1 \end{pmatrix} \underset{\text{III}}{\to} \begin{pmatrix} 1 & 1/2 \\ 0 & 1 \end{pmatrix} \underset{\text{II}}{\to} \begin{pmatrix} 1 & 1/2 \\ 0 & -3/2 \end{pmatrix} \underset{\text{III}}{\to} \begin{pmatrix} 1 & 1/2 \\ 5 & 1 \end{pmatrix} \underset{\text{II}}{\to} \begin{pmatrix} 2 & 1 \\ 5 & 1 \end{pmatrix}.$$

The matrix

$$\begin{pmatrix} 2 & 1 \\ 5 & 1 \end{pmatrix}$$

is a 2×2 submatrix of A, which must have determinant nonzero; in fact, the determinant is -3. Thus the determinant rank of A cannot be less than 2, the rank of A.

Determinant rank might be used to quickly show that the rank of a 3×3 matrix with zero determinant is 2, as in Example 1. But in general, it is not practical to determine the rank of a matrix by finding the determinant rank. For example, if the rank of a 3×5 matrix is 2, it would be necessary to show that all ten, 3×3 submatrices have zero determinant before considering a 2×2 submatrix.

In this chapter, matrices have been used primarily to express systems of linear equations. Thus the rows of a matrix contain the important information, and rank is defined in terms of rows. But in the next chapter, the columns will contain the information needed, so the final approach to rank is in terms of the columns.

Definition The *column space* of an $n \times m$ matrix A is the subspace of $\mathscr{M}_{n\times 1}$ spanned by the m columns of A. The *column rank* of A is the dimension of its column space.

Example 3 Let

$$A = \begin{pmatrix} 2 & 3 & 7 \\ 1 & 5 & 9 \end{pmatrix},$$

then the column space of A is

$$\mathscr{L}\left\{\begin{pmatrix} 2 \\ 1 \end{pmatrix}, \begin{pmatrix} 3 \\ 5 \end{pmatrix}, \begin{pmatrix} 7 \\ 9 \end{pmatrix}\right\} = \mathscr{L}\left\{\begin{pmatrix} 1 \\ 0 \end{pmatrix}, \begin{pmatrix} 0 \\ 1 \end{pmatrix}\right\} = \mathscr{M}_{2\times 1},$$

and the column rank of A is 2. If

$$B = \begin{pmatrix} 2 & 1 & 4 \\ 3 & 0 & 2 \\ -1 & 1 & 2 \end{pmatrix},$$

then

$$\mathscr{L}\left\{\begin{pmatrix} 2 \\ 3 \\ -1 \end{pmatrix}, \begin{pmatrix} 1 \\ 0 \\ 1 \end{pmatrix}, \begin{pmatrix} 4 \\ 2 \\ 2 \end{pmatrix}\right\} = \mathscr{L}\left\{\begin{pmatrix} 1 \\ 0 \\ 1 \end{pmatrix}, \begin{pmatrix} 0 \\ 1 \\ -1 \end{pmatrix}\right\}$$

is the column space of B, and the column rank of B is 2.

The transpose of a matrix turns columns into rows, therefore the column rank of a matrix A is the rank (row rank) of its transpose A^T. Also, transposing a $k \times k$ matrix does not affect its determinant (problem 13, page 115), so the determinant ranks of A^T and A are equal. Therefore,

$$\begin{aligned} \text{column rank } A &= \text{(row) rank } A^T = \text{determinant rank } A^T \\ &= \text{determinant rank } A = \text{(row) rank } A, \end{aligned}$$

proving the following theorem.

Theorem 3.23 The column rank of a matrix equals its rank.

Example 4 Show that no set of $n - 1$ vectors can span the space $\mathscr{V}_n$.

Suppose $\{U_1, \ldots, U_{n-1}\} \subset \mathscr{V}_n$. It is necessary to find a vector $W \in \mathscr{V}_n$ such that the vector equation $x_1U_1 + \cdots + x_{n-1}U_{n-1} = W$ has no solution for $x_1, \ldots, x_{n-1}$. This vector equation yields a system of n linear equations in $n - 1$ unknowns. If the system is denoted by $AX = B$, then A is an $n \times (n - 1)$ matrix, and we are looking for $B = W^T$ so that the rank of A^* is not equal to the rank of A.

Suppose the rank of A is k, then $k \leq n - 1$, and A has k linearly independent columns. But the columns of A are vectors in $\mathscr{M}_{n\times 1}$, which has dimension n. So there exists a column vector B in $\mathscr{M}_{n\times 1}$ which is independent of the columns of A. Let this column vector B determine W by $W = B^T$. Then the rank of the augmented matrix A^* is $k + 1$, and for the vector W, the system is inconsistent. That is

$$W \notin \mathscr{L}\{U_1, \ldots, U_{n-1}\},$$

and the $n - 1$ vectors do not span $\mathscr{V}_n$.

Using elementary column operations, it is possible to define two new relations between matrices following the pattern of row-equivalence. A matrix A is *column-equivalent* to B if B can be obtained from A with a finite sequence of elementary column operations, and A is *equivalent* to B if B can be obtained from A with a finite sequence of elementary row and column operations. The relation of equivalence for matrices will appear later in another context.

Example 5 Show that

$$\begin{pmatrix} 1 & 0 & 3 \\ 2 & 4 & 1 \\ 1 & 4 & -2 \end{pmatrix}$$

is equivalent to

$$\begin{pmatrix} 1 & 0 & 0 \\ 0 & 1 & 0 \\ 0 & 0 & 0 \end{pmatrix}.$$

Using column and row operations, we obtain

$$\begin{pmatrix} 1 & 0 & 3 \\ 2 & 4 & 1 \\ 1 & 4 & -2 \end{pmatrix} \underset{\text{III}}{\xrightarrow{\text{row}}} \begin{pmatrix} 1 & 0 & 3 \\ 0 & 4 & -5 \\ 0 & 4 & -5 \end{pmatrix} \underset{\text{III}}{\xrightarrow{\text{row}}} \begin{pmatrix} 1 & 0 & 3 \\ 0 & 4 & -5 \\ 0 & 0 & 0 \end{pmatrix} \underset{\text{II}}{\xrightarrow{\text{col.}}} \begin{pmatrix} 1 & 0 & 3 \\ 0 & 1 & -5 \\ 0 & 0 & 0 \end{pmatrix} \underset{\text{III}}{\xrightarrow{\text{col.}}} \begin{pmatrix} 1 & 0 & 0 \\ 0 & 1 & 0 \\ 0 & 0 & 0 \end{pmatrix}.$$

Problems

1. Find a matrix equivalent to the given matrix with 1's on the main diagonal and 0's elsewhere.

a. $\begin{pmatrix} 2 & 1 \\ 3 & 6 \\ 1 & 5 \end{pmatrix}$. b. $\begin{pmatrix} 2 & 3 & 1 \\ 2 & -5 & 9 \\ 4 & 2 & 6 \end{pmatrix}$. c. $\begin{pmatrix} 0 \\ 3 \\ 7 \end{pmatrix}$. d. $\begin{pmatrix} 0 & 2 & 3 & 4 & 0 \\ 2 & 4 & 0 & 6 & 2 \\ 3 & 1 & 2 & 1 & 5 \end{pmatrix}$. e. $\begin{pmatrix} 2 & 1 \\ 5 & 9 \end{pmatrix}$

2. Find the column space and thus the rank of the following matrices:

a. $\begin{pmatrix} 2 & 1 \\ 3 & 4 \end{pmatrix}$. b. $\begin{pmatrix} 2 & 1 & 3 \\ -1 & 1 & 0 \\ 1 & 5 & 6 \end{pmatrix}$. c. $\begin{pmatrix} 4 & 0 & 3 \\ 0 & 2 & 1 \\ 0 & 0 & 5 \end{pmatrix}$. d. $\begin{pmatrix} 3 & 1 & 2 & 5 \\ 0 & 2 & 1 & 1 \\ 2 & 0 & 1 & 3 \end{pmatrix}$.

3. Why is it impossible for the column space of A to equal the row space of A, unless A is a 1×1 matrix?

4. The column space of a matrix A is essentially the same as the row space of A^T. Show that even for $n \times n$ matrices, the row space of A need not equal the row space of A^T by considering

a. $A = \begin{pmatrix} 1 & 5 & 2 \\ 2 & 4 & 6 \\ 1 & -1 & 4 \end{pmatrix}$. b. $A = \begin{pmatrix} 2 & 4 & -6 \\ 8 & 8 & -8 \\ 2 & -4 & 10 \end{pmatrix}$.

5. How do you read A^T?

6. Use determinant rank to prove that if $U \times V = \mathbf{0}$ for $U, V \in \mathscr{E}_3$, then U and V are linearly dependent.

7. Use the theory of systems of linear equations to prove that if $\{U_1, \ldots, U_n\}$ span $\mathscr{V}_n$, then $\{U_1, \ldots, U_n\}$ is linearly independent. In your proof, write out the systems used, showing the coefficients and the unknowns.

Review Problems

1. Suppose $AX = B$ is a system of n linear equations in 2 unknowns. Each equation determines a line in the plane. Consider all possible values for the ranks of A and A^* and in each case determine how the n lines intersect.

2. Suppose $AX = \mathbf{0}$ is a system of linear equations obtained in determining if the set $S \subset R_n[t]$ is linearly independent. Find a set S for which this system is obtained if

a. $A = \begin{pmatrix} 4 & 5 \\ 1 & 3 \end{pmatrix}$. b. $A = \begin{pmatrix} 2 & 8 \\ 9 & 3 \\ 1 & 7 \end{pmatrix}$. c. $A = \begin{pmatrix} 3 & 1 & 4 \\ 2 & -1 & 6 \end{pmatrix}$.

3. Suppose $S = \{V_1, \ldots, V_m\} \subset \mathscr{V}_n$ and $m < n$. Use the theory of linear equations and column rank to prove that S does not span $\mathscr{V}_n$.

4. Show that the line in the plane through the points with coordinates (x_1, y_1) and (x_2, y_2) has the equation

$$\begin{vmatrix} x_1 & x_2 & x \\ y_1 & y_2 & y \\ 1 & 1 & 1 \end{vmatrix} = 0.$$

5. Use problem 4 to find an equation for the line through the points with coordinates:

a. (1, 0), (0, 1). b. (3, 4), (2, −5). c. (5, 1), (8, 9).

6. Using problem 4 as a guide, devise a determinant equation for the plane through the three points with coordinates (x_1, y_1, z_1), (x_2, y_2, z_2), and (x_3, y_3, z_3). (What would happen in the equation if the three points were collinear?)

7. Use elementary row operations to find a Cartesian equation for the following planes:

a. $\{(8a + 8b, 4a + 8b, 8a + 11b) | a, b \in R\}$.
b. $\mathscr{L}\{(5, 15, 7), (2, 6, 8), (3, 9, -2)\}$.
c. $\{(2a - 4b + 6c, a + b - 2c, -7a + 2b - c) | a, b, c \in R\}$.
d. $\mathscr{L}\{(2, 3, -2), (-5, -4, 5)\}$.

8. Show that the plane through the points with coordinates (x_1, y_1, z_1), (x_2, y_2, z_2), and (x_3, y_3, z_3) has the equation

$$\begin{vmatrix} x - x_1 & y - y_1 & z - z_1 \\ x_2 - x_1 & y_2 - y_1 & z_2 - z_1 \\ x_3 - x_1 & y_3 - y_1 & z_3 - z_1 \end{vmatrix} = 0.$$

9. Given a system of n linear equations in n unknowns, why is one justified in expecting a unique solution?

10. In the next chapter, we will find that matrix multiplication distributes over matrix addition. That is, $A(U + V) = AU + AV$ provided the products are defined. Use this fact to prove the following:

a. The solution set for a homogeneous system of linear equations is closed under addition (Theorem 3.2, page 80).
b. If $X = C_1$ is a solution for $AX = B$, and $X = C_2$ is a solution for $AX = \mathbf{0}$, then $X = C_1 + C_2$ is a solution for $AX = B$ (problem 6, page 105).

11. Instead of viewing both A and B as being fixed in the system of linear equations $AX = B$, consider only A to be fixed. Then $AX = B$ defines a correspondence from $\mathscr{M}_{m \times 1}$ to $\mathscr{M}_{n \times 1}$.

a. To what vectors do $\begin{pmatrix} 1 \\ 1 \end{pmatrix}$, $\begin{pmatrix} 3 \\ -2 \end{pmatrix}$, and $\begin{pmatrix} 0 \\ 1 \end{pmatrix}$ correspond in $\mathscr{M}_{3 \times 1}$ if

$$A = \begin{pmatrix} 3 & 2 \\ 4 & 1 \\ 3 & 5 \end{pmatrix}?$$

b. To what vectors do $\begin{pmatrix} 0 \\ 0 \\ 0 \end{pmatrix}$, $\begin{pmatrix} -2 \\ 3 \\ -4 \end{pmatrix}$, and $\begin{pmatrix} -1 \\ 1 \\ -1 \end{pmatrix}$ correspond in $\mathcal{M}_{2 \times 1}$ if

$$A = \begin{pmatrix} 1 & 2 & 1 \\ 0 & 4 & 3 \end{pmatrix}?$$

c. Suppose this correspondence satisfies the second condition in the definition for an isomorphism, page 62. What can be concluded about the rank of A, and the relative values of n and m?

4

Linear Transformations

A special type of transformation, the isomorphism, has been touched upon as a natural consequence of introducing coordinates. We now turn to a general study of transformations between vector spaces.

§1. Definitions and Examples

Definition Let $\mathscr{V}$ and $\mathscr{W}$ be vector spaces. If $T(V)$ is a unique vector in $\mathscr{W}$ for each $V \in \mathscr{V}$, then T is a *function, transformation,* or *map* from $\mathscr{V}$, the *domain,* to $\mathscr{W}$, the *codomain.* If $T(V) = W$, then W is the *image* of V under T and V is a *preimage* of W. The set of all images in $\mathscr{W}$, $\{T(V) | V \in \mathscr{V}\}$, is the *image* or *range* of T. The notation $T: \mathscr{V} \to \mathscr{W}$ will be used to denote a transformation from $\mathscr{V}$ to $\mathscr{W}$.

According to this definition, each vector in the domain has one and only one image. However, a vector in the codomain may have many preimages or none at all. For example, consider the map $T: \mathscr{V}_3 \to \mathscr{V}_2$ defined by $T(a, b, c) = (a + b - c, 2c - 2a - 2b)$. For this map, $T(3, 1, 2) = (2, -4)$. Therefore $(2, -4)$ is the image of $(3, 1, 2)$ and $(3, 1, 2)$ is a preimage of $(2, -4)$. But $T(1, 2, 1) = (2, -4)$, so $(1, 2, 1)$ is also a preimage of $(2, -4)$. Can you find any other preimages of $(2, -4)$? On the other hand, $(1, 0)$ has no preimage, for $T(a, b, c) = (1, 0)$ yields the inconsistent system $a + b - c = 1$. $2c - 2a - 2b = 0$. Thus some vectors in the codomain $\mathscr{V}_2$ have many preimages, while others have none.

In the preceding definition, $\mathscr{V}$ and $\mathscr{W}$ could be arbitrary sets and the definition would still be meaningful provided only that the word "vector" were replaced by "element." In fact, elementary calculus is the study of functions from a set of real numbers to the set of real numbers. But a transformation between vector spaces, instead of simply sets of vectors, should be more than simply a rule giving a vector in $\mathscr{W}$ for each vector in $\mathscr{V}$, it should also preserve the vector space structure. Therefore, we begin by restricting our attention to maps that send sets as vector spaces to vector spaces.

Definition Let $\mathscr{V}$ and $\mathscr{W}$ be vector spaces and $T: \mathscr{V} \to \mathscr{W}$. T is a *linear transformation* or a *linear map* if for every $U, V \in \mathscr{V}$ and $r \in R$:

1. $T(U + V) = T(U) + T(V)$; and
2. $T(rU) = rT(U)$.

Thus, under a linear transformation, the image of the sum of vectors is the sum of the images and the image of a scalar multiple is the scalar multiple of the image. We say that a linear transformation "preserves addition and scalar multiplication."

Almost none of the functions studied in calculus are linear maps from the vector space R to R. In fact, $f\colon R \to R$ is linear only if $f(x) = ax$ for some real number a.

The two conditions in the definition of a linear transformation are included in the definition of an isomorphism (see page 62). That is, isomorphisms are linear transformations. But the converse is not true.

Example 1 Define $T\colon \mathscr{V}_3 \to \mathscr{V}_2$ by $T(x, y, z) = (2x + y, 4z)$. Show that T is linear but not an isomorphism.

If $U = (a_1, a_2, a_3)$ and $V = (b_1, b_2, b_3)$, then

$$\begin{aligned} T(U + V) &= T(a_1 + b_1, a_2 + b_2, a_3 + b_3) && \text{Definition of } + \text{ in } \mathscr{V}_3 \\ &= (2[a_1 + b_1] + a_2 + b_2, 4[a_3 + b_3]) && \text{Definition of } T \\ &= (2a_1 + a_2, 4a_3) + (2b_1 + b_2, 4b_3) && \text{Definition of } + \text{ in } \mathscr{V}_2 \\ &= T(a_1, a_2, a_3) + T(b_1, b_2, b_3) && \text{Definition of } T \\ &= T(U) + T(V). \end{aligned}$$

Therefore T preserves addition of vectors. T also preserves scalar multiplication for if $r \in R$, then

$$\begin{aligned} T(rU) &= T(ra_1, ra_2, ra_3) && \text{Definition of scalar multiplication in } \mathscr{V}_3 \\ &= (2ra_1 + ra_2, 4ra_3) && \text{Definition of } T \\ &= r(2a_1, +a_2, 4a_3) && \text{Definition of scalar multiplication in } \mathscr{V}_2 \\ &= rT(a_1, a_2, a_3) && \text{Definition of } T \\ &= rT(U). \end{aligned}$$

Thus T is a linear map. However, T is not an isomorphism for there is not a unique preimage for every vector in the codomain $\mathscr{V}_2$. Given $(a, b) \in \mathscr{V}_2$, there exists a vector $(x, y, z) \in \mathscr{V}_3$ such that $T(x, y, z) = (a, b)$ provided $2x + y = a$ and $4z = b$. Given a and b, this system of two equations in the three unknowns, x, y, z, cannot have a unique solution. Therefore, if a vector (a, b) has one preimage, then it has many. For example, $T(3, 0, 1) = (6, 4) = T(2, 2, 1)$.

Definition A transformation $T: \mathscr{V} \to \mathscr{W}$ is *one to one*, written 1–1, if $T(U) = T(V)$ implies $U = V$.

A map from $\mathscr{V}$ to $\mathscr{W}$ is one to one if each vector in the codomain $\mathscr{W}$ has at most one preimage in the domain $\mathscr{V}$. Thus the map in Example 1 is not 1–1. The property of being one to one is inverse to the property of being a function. For $T: \mathscr{V} \to \mathscr{W}$ is a function provided $T(U) = W$, and $T(U) = X$ implies $W = X$. In fact, the function notation $T(U)$ is used only when U has a unique image.

A map may be linear and one to one and still not be an isomorphism.

Example 2 The map $T: \mathscr{V}_2 \to \mathscr{V}_3$ defined by $T(x, y) = (x, 3x - y, 2x)$ is linear and one to one, but not an isomorphism.

To show that T preserves addition, let $U = (a_1, a_2)$ and $V = (b_1, b_2)$, then

$$\begin{aligned} T(U + V) &= T(a_1 + b_1, a_2 + b_2) \\ &= (a_1 + b_1, 3[a_1 + b_1] - [a_2 + b_2], 2[a_1 + b_1]) \\ &= (a_1, 3a_1 - a_2, 2a_1) + (b_1, 3b_1 - b_2, 2b_1) \\ &= T(a_1, a_2) + T(b_1, b_2) = T(U) + T(V). \end{aligned}$$

The proof that T preserves scalar multiplication is similar, thus T is linear. To see that T is 1–1, suppose $T(U) = T(V)$. Then $(a_1, 3a_1 - a_2, 2a_1) = (b_1, 3b - b_2, 2b_1)$ yields $a_1 = b_1$, $3a_1 - a_2 = 3b_1 - b_2$, and $2a_1 = 2b_1$. These equations have $a_1 = b_1$, $a_2 = b_2$ as the only solution, so $U = V$ and T is 1–1.

However, T is not an isomorphism because some vectors in $\mathscr{V}_3$ have no preimage under T. Given $(a, b, c) \in \mathscr{V}_3$, there exists a vector $(x, y) \in \mathscr{V}_2$ such that $T(x, y) = (a, b, c)$ provided $x = a$, $3x - y = b$, and $2x = c$. For any choice of a, b, c, this is a system of three linear equations in two unknowns, x and y, which need not be consistent. In this case, if $2a \neq c$, there is no solution. For example, (1, 6, 3) has no preimage in $\mathscr{V}_2$ under T. Therefore T does not satisfy the definition of an isomorphism.

Definition A transformation $T: \mathscr{V} \to \mathscr{W}$ is *onto* if for every vector $W \in \mathscr{W}$, there exists a vector $V \in \mathscr{V}$ such that $T(V) = W$.

T maps $\mathscr{V}$ onto the codomain $\mathscr{W}$ if every vector in $\mathscr{W}$ is the image of some vector in $\mathscr{V}$. That is, T is onto if its image or range equals its codomain.

The map in Example 2 is not onto for the image does not contain (1, 6, 3), a vector in the codomain.

The second condition in our definition of an isomorphism requires that an isomorphism be both one to one and onto. It is now possible to give a simple restatement of the definition.

Definition An *isomorphism* from a vector space $\mathscr{V}$ to a vector space $\mathscr{W}$ is a one to one, onto, linear map from $\mathscr{V}$ to $\mathscr{W}$.

An isomorphism is a very special transformation in that isomorphic vector spaces are algebraically identical. But an arbitrary linear transformation may fail to be either one to one or onto.

Example 3 The map $T: \mathscr{V}_2 \to \mathscr{V}_2$ defined by $T(a, b) = (0, 0)$ is linear, but it is neither one to one nor onto.

Example 4 Let

$$T(a_0 + a_1t + \cdots + a_nt^n) = a_1 + 2a_2t + \cdots + na_nt^{n-1}.$$

Then T is a linear map from $R[t]$ to itself which is onto but not 1–1.

The image of a polynomial $P \in R[t]$ under T is the derivative of P with respect to t. One can prove directly that T is linear, but this is also implied by the theorems on differentiation obtained in elementary calculus. T maps $R[t]$ onto $R[t]$; for given any polynomial $b_0 + b_1t + \cdots + b_mt^m \in R[t]$, the polynomial

$$b_0t + \tfrac{1}{2}b_1t^2 + \cdots + \frac{1}{m+1}b_mt^{m+1}$$

is sent to it by T. Thus, every polynomial has a preimage and T is an onto map. But T fails to be one to one. (Why?)

A direct sum decomposition of $\mathscr{V}$ given by $\mathscr{V} = \mathscr{S} \oplus \mathscr{T}$ determines two linear transformations; $P_1: \mathscr{V} \to \mathscr{S}$ and $P_2: \mathscr{V} \to \mathscr{T}$. These *projection maps* are defined by $P_1(V) = U$ and $P_2(V) = W$ where $V = U + W$, $U \in \mathscr{S}$ and $W \in \mathscr{T}$. P_1 and P_2 are well-defined maps because, in a direct sum, U and W are uniquely determined by V. The proof that a projection is linear is left to the reader.

Example 5 Let $\mathscr{S} = \{(x, y, z) | x + y + z = 0\}$ and $\mathscr{T} = \mathscr{L}\{(2, 5, 7)\}$. Then $\mathscr{S} + \mathscr{T}$ is direct and $\mathscr{E}_3 = \mathscr{S} \oplus \mathscr{T}$ expresses $\mathscr{E}_3$ as the direct sum of a plane and a line. Thus there are two projections $P_1 : \mathscr{E}_3 \to \mathscr{S}$ and $P_2 : \mathscr{E}_3 \to \mathscr{T}$. Given $(7, 3, 4) \in \mathscr{E}_3$, one finds that $(7, 3, 4) = (5, -2, -3) + (2, 5, 7)$ with $(5, -2, -3) \in \mathscr{S}$ and $(2, 5, 7) \in \mathscr{T}$. Therefore, $P_1(7, 3, 4) = (5, -2, -3)$ and $P_2(7, 3, 4) = (2, 5, 7)$.

A projection map is much more general than the perpendicular projection found in Euclidean geometry. However, a perpendicular projection can be expressed as a projection map. For example, if $\mathscr{T}$ in Example 5 is replaced by $\mathscr{L}\{(1, 1, 1)\}$, the subspace (line) normal to the plane $\mathscr{S}$, then $P_1(V)$ is at the foot of the perpendicular dropped from V to the plane $\mathscr{S}$. That is, P_1 is the perpendicular projection of $\mathscr{E}_3$ onto $\mathscr{S}$.

A projection map is necessarily onto; for given any vector $U \in \mathscr{S}$, U is in $\mathscr{V}$ and $U = U + \mathbf{0}$, $\mathbf{0} \in \mathscr{T}$. Therefore $P_1(U) = U$, and every vector in $\mathscr{S}$ has a preimage in $\mathscr{V}$. On the other hand, if $\mathscr{T} \neq \{\mathbf{0}\}$, then P_1 fails to be $1 - 1$. This is easily seen, for every vector in $\mathscr{T}$ is sent to the same vector by P_1, namely $\mathbf{0}$. That is, if $W \in \mathscr{T}$, then $W = \mathbf{0} + W$ with $\mathbf{0} \in \mathscr{S}$ and $W \in \mathscr{T}$, so $P_1(W) = \mathbf{0}$.

The linearity conditions have a good geometric interpretation. Generally, a linear map sends parallel lines into parallel lines. To see this, consider the line ℓ with vector equation $P = kU + V$, $k \in R$, for $U, V \in \mathscr{V}_n$, $U \neq \mathbf{0}$. If U and V are viewed as points in Euclidean space, then ℓ is the line through V with direction U. Now suppose $T: \mathscr{V}_n \to \mathscr{V}_n$ is a linear map, then

$$T(P) = T(kU + V) = T(kU) + T(V) = kT(U) + T(V) \qquad \text{for all} \quad k \in R.$$

Therefore if $T(U) \neq \mathbf{0}$, then T sends ℓ into the line through $T(V)$ with direction $T(U)$. Now any line parallel to ℓ has direction U and is sent to a line with direction $T(U)$. What would happen to the parallel line given by $P = kU + W$, $k \in R$, if $T(V) = T(W)$? What would happen to ℓ if $T(U) = \mathbf{0}$?

We can use this geometric interpretation to show that a rotation of the plane about the origin must be linear. Certainly a rotation sends parallel lines into parallel lines, but the converse of the above argument is false. That is, a map may send parallel lines into parallel lines without being linear; consider a translation. However, if $T: \mathscr{E}_2 \to \mathscr{E}_2$ is a rotation about the origin, then $T(\mathbf{0}) = \mathbf{0}$ and T moves geometric figures without distortion. Thus the parallelogram determined by $\mathbf{0}$, U, V and $U + V$ is sent to the parallelogram determined by $\mathbf{0}$, $T(U)$, $T(V)$, and $T(U + V)$. So by the parallelogram rule for addition $T(U + V) = T(U) + T(V)$, and a rotation preserves addition. T also preserves scalar multiplication because a rotation moves a line passing through the origin, $\{rU | r \in R\}$, to a line through the origin without changing

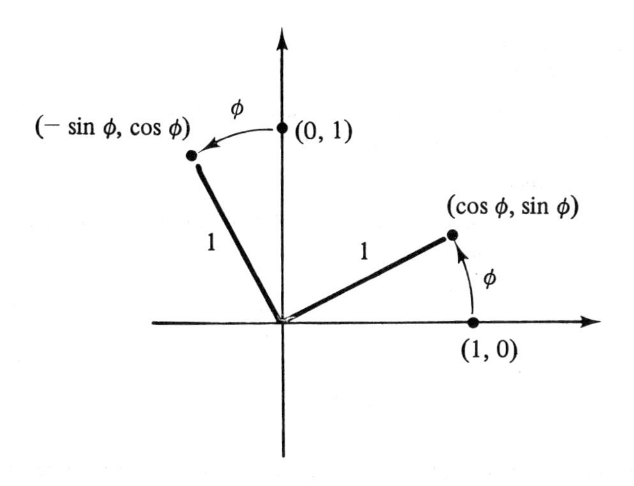

Figure 1

lengths or the relative position of points. Thus a rotation of $\mathscr{E}_2$ about the origin is a linear map, and we can use this fact to obtain a componentwise expression for T. Suppose φ is the angle of rotation with positive angles measured in the counterclockwise direction. Then $T(1, 0) = (\cos \varphi, \sin \varphi)$ and $T(0, 1) = (-\sin \varphi, \cos \varphi)$ as in Figure 1. Now using the linearity of T,

$$\begin{aligned} T(a, b) &= T(a(1, 0) + b(0, 1)) \\ &= aT(1, 0) + bT(0, 1) \\ &= a(\cos \varphi, \sin \varphi) + b(-\sin \varphi, \cos \varphi) \\ &= (a \cos \varphi - b \sin \varphi, a \sin \varphi + b \cos \varphi). \end{aligned}$$

From the geometric point of view, a rotation in $\mathscr{E}_2$ must be 1–1 and onto, but this could also be obtained from the componentwise definition.

Problems

1. Determine which of the following transformations are linear.
   a. $T: \mathscr{V}_2 \to \mathscr{V}_2$; $T(a, b) = (3a - b, b^2)$.
   b. $T: \mathscr{V}_3 \to \mathscr{V}_2$; $T(a, b, c) = (3a + 2c, b)$.
   c. $T: \mathscr{M}_{2\times 2} \to R$; $T\begin{pmatrix} a & b \\ c & d \end{pmatrix} = \det\begin{pmatrix} a & b \\ c & d \end{pmatrix}$.
   d. $T: R_3[t] \to R_3[t]$; $T(a + bt + ct^2) = (a - b)t^2 + 4ct - b$.
   e. $T: \mathscr{V}_3 \to \mathscr{V}_4$; $T(a, b, c) = (a - 5b, 7, a, 0)$.

f. $T: C \to R$; $T(a + bi) = a + b$.
g. $T: C \to C$; $T(a + bi) = i - a$.

2. Suppose $T: \mathscr{V} \to \mathscr{W}$. Show that T is linear if and only if $T(aU + bV) = aT(U) + bT(V)$ for all $a, b \in R$ and all $U, V \in \mathscr{V}$.

3. How do you read "$T: \mathscr{V} \to \mathscr{W}$?"

4. Show that each of the following maps is linear. Find all preimages for the given vector W in the codomain.
a. $T(a, b) = (2a - b, 3b - 4a)$; $W = (-1, 5)$.
b. $T(a, b, c) = (a + b, 3a - c)$; $W = (0, 0)$.
c. $T(a, b) = (a - b, a + b, 2a)$; $W = (2, 5, 3)$.
d. $T(a + bt + ct^2) = (a + b)t + (c - a)t^2$; $W = t$.
e. $T(a + bt + ct^2) = b + 2ct$; $W = t$.
f. $T(a + bt + ct^2) = a + b + (b + c)t + (c - a)t^2$; $W = 3 - 3t^2$.

5. State what it would mean for each of the following maps to be onto.
a. $T: \mathscr{V}_3 \to \mathscr{V}_4$. b. $T: \mathscr{M}_{2 \times 3} \to \mathscr{M}_{2 \times 2}$. c. $T: R_3[t] \to R_2[t]$.

6. Show that each of the following maps is linear. Determine if the maps are one to one or onto.
a. $T: \mathscr{M}_{2 \times 3} \to \mathscr{M}_{2 \times 2}$; $T\begin{pmatrix} a & b & c \\ d & e & f \end{pmatrix} = \begin{pmatrix} a & b - c \\ d - e & f \end{pmatrix}$.
b. $T: C \to C$; $T(a + bi) = 0$.
c. $T: \mathscr{V}_2 \to \mathscr{V}_3$; $T(a, b) = (2a + b, 3b, b - 4a)$.
d. $T: R_3[t] \to R_3[t]$; $T(a + bt + ct^2) = a + b + (b + c)t + (a + c)t^2$.
e. $T: \mathscr{V}_3 \to \mathscr{V}_3$; $T(a, b, c) = (2a + 4b, 2a + 3c, 4b - 3c)$.
f. $T: R[t] \to R[t]$; $T(a_0 + a_1 t + \cdots + a_n t^n)$

$$= a_0 t + \frac{1}{2} a_1 t^2 + \cdots + \frac{1}{n + 1} a_n t^{n+1}.$$

7. Find the following images for the projection maps in Example 5.
a. $P_1(18, 2, 8)$. b. $P_2(18, 2, 8)$. c. $P_1(-1, -1, -12)$.
d. $P_2(0, 0, 14)$. e. $P_1(x, y, -x-y)$. f. $P_2(x, y, -x-y)$.

8. a. Show that a projection map is a linear transformation.
b. Suppose $\mathscr{V} = \mathscr{S} + \mathscr{T}$ but the sum is not direct. Why isn't there a projection map from $\mathscr{V}$ to $\mathscr{S}$?

9. a. What map defines the perpendicular projection of E^2 onto the x axis?
b. What map defines the perpendicular projection of E^2 onto the line with equation $y = 3x$?

10. a. Find the map T that rotates the plane through the angle $\varphi = 60°$. What is $T(1, 0)$?
b. Use the componentwise expression for a rotation to show that any rotation of the plane about the origin is linear.
c. Prove that a rotation about the origin is one to one and onto.

11. A translation in the plane is given by $T(x, y) = (x + h, y + k)$ for some $h, k \in R$. Is a translation linear? 1–1? onto?

12. The reflection of E^2 in the line with equation $y = x$ is given by $T(x, y) = (y, x)$. Is this reflection linear? 1–1? onto?

13. Suppose $T: \mathscr{V} \to \mathscr{W}$ is linear and there exists a nonzero vector $U \in \mathscr{V}$ such that $T(U) = \mathbf{0}$. Prove that T is not one to one.

14. Suppose $T: \mathscr{V} \to \mathscr{W}$ is a linear map, $U_1, \ldots, U_k \in \mathscr{V}$ and $r_1, \ldots, r_k \in R$. Use induction to prove that if $k \geq 2$, then

$$T(r_1U_1 + \cdots + r_kU_k) = r_1T(U_1) + \cdots + r_kT(U_k).$$

15. Suppose $T: \mathscr{V} \to \mathscr{W}$ is linear and $\{V_1, \ldots, V_n\}$ is a basis for $\mathscr{V}$. Prove that the image or range of T is $\mathscr{L}\{T(V_1), \ldots, T(V_n)\}$. That is, the image of T is a subspace of the codomain $\mathscr{W}$.

§2. The Image and Null Spaces

The requirement that a transformation be linear and thus preserve the algebraic structure of a vector space is quite restrictive. In particular, the image of the zero vector under a linear map must be the zero vector. This generally involves two different zero vectors. That is, if $T: \mathscr{V} \to \mathscr{W}$ is linear and $\mathbf{0}_{\mathscr{V}}$, $\mathbf{0}_{\mathscr{W}}$ are the zero vectors in $\mathscr{V}$ and $\mathscr{W}$, respectively, then $T(\mathbf{0}_{\mathscr{V}}) = \mathbf{0}_{\mathscr{W}}$. For, using any vector $V \in \mathscr{V}$, we obtain $T(\mathbf{0}_{\mathscr{V}}) = T(0V) = 0T(V) = \mathbf{0}_{\mathscr{W}}$. The use of different symbols for the zero vectors is helpful here but not necessary. Knowing that T is a map from $\mathscr{V}$ to $\mathscr{W}$ is sufficient to determine what $\mathbf{0}$ means in $T(\mathbf{0}) = \mathbf{0}$.

The requirement that a map preserve the vector space structure also guarantees that the image of a vector space is again a vector space. In general, if T is a map and the set S is contained in its domain, then the set $T[S] = \{T(U) | U \in S\}$ is called the *image* of S. The claim is that if S is a vector space, then so is the image $T[S]$.

Example 1 Define $T: \mathscr{V}_2 \to \mathscr{V}_3$ by

$$T(a, b) = (2a - b, 3b - a, a + b).$$

If $\mathscr{S} = \mathscr{L}\{(1, 2)\}$, then

$$\begin{aligned} T[\mathscr{S}] &= \{T(U) | U \in \mathscr{S}\} = \{T(k, 2k) | k \in R\} \\ &= \{(2k - 2k, 3(2k) - k, k + 2k) | k \in R\} \\ &= \mathscr{L}\{(0, 5, 3)\}. \end{aligned}$$

Note that, in this case, the image of the subspace $\mathscr{S}$ of $\mathscr{V}_2$ is a subspace of $\mathscr{V}_3$.

Theorem 4.1 If $T: \mathscr{V} \to \mathscr{W}$ is linear and $\mathscr{S}$ is a subspace of $\mathscr{V}$, then the image of $\mathscr{S}$, $T[\mathscr{S}]$, is a subspace of $\mathscr{W}$.

Proof It is necessary to show that $T[\mathscr{S}]$ is nonempty and closed under addition and scalar multiplication. Since $\mathscr{S}$ is a subspace of $\mathscr{V}$, $\mathbf{0} \in \mathscr{S}$ and $T(\mathbf{0}) = \mathbf{0}$ because T is linear. Thus $\mathbf{0} \in T[\mathscr{S}]$ and $T[\mathscr{S}]$ is nonempty. To complete the proof it is sufficient to show that if $U, V \in T[\mathscr{S}]$ and $r, s \in R$, then $rU + sV \in T[\mathscr{S}]$. (Why is this sufficient?) But $U, V \in T[\mathscr{S}]$ implies that there exist $W, X \in \mathscr{S}$ such that $T(W) = U$ and $T(X) = V$. And $\mathscr{S}$ is closed under addition and scalar multiplication, so $rW + sX \in \mathscr{S}$. Therefore $T(rW + sX) \in T[\mathscr{S}]$. Now T is linear, so $T(rW + sX) = rT(W) + sT(X) = rU + sV$ completing the proof.

In particular, this theorem shows that the image of T, $T[\mathscr{V}]$, is a subspace of the codomain $\mathscr{W}$. Thus it might be called the *image space* of T. We will denote the image space of T by $\mathscr{I}_T$. Using this notation, $T: \mathscr{V} \to \mathscr{W}$ is onto if and only if $\mathscr{I}_T = \mathscr{W}$.

A map fails to be onto because of the choice of its codomain. If $T: \mathscr{V} \to \mathscr{W}$ is not onto, the codomain might be changed to $\mathscr{I}_T$. Then $T: \mathscr{V} \to \mathscr{I}_T$ is onto; or any transformation maps its domain onto its image. Hence if $T: \mathscr{V} \to \mathscr{W}$ is linear and one to one, then $T: \mathscr{V} \to \mathscr{I}_T$ is an isomorphism.

Example 2 Define $T: \mathscr{V}_2 \to \mathscr{V}_3$ by

$$T(a, b) = (2a + b, b - a, 3a + b).$$

Find the image space of T.

$$\begin{aligned}\mathscr{I}_T &= \{T(a, b) | a, b \in R\} \\ &= \{(2a + b, b - a, 3a + 4b) | a, b \in R\} \\ &= \mathscr{L}\{(2, -1, 3), (1, 1, 4)\} \\ &= \mathscr{L}\{(1, 0, 7/3), (0, 1, 5/3)\} \\ &= \{(x, y, z) | 7x + 5y - 3z = 0\}.\end{aligned}$$

It can be shown that T is linear and 1–1. Thus the fact that $\mathscr{I}_T$ is represented by a plane agrees with the fact that $\mathscr{I}_T$ is isomorphic to $\mathscr{V}_2$.

Example 3 Consider the linear map $T: \mathscr{V}_2 \to \mathscr{V}_2$ given by $T(a, b) = (2a - b, 3b - 6a)$.

T does not map $\mathscr{V}_2$ onto $\mathscr{V}_2$ because

$$\begin{aligned}\mathscr{I}_T &= \{T(a, b)|a, b \in R\} = \{(2a - b, 3b - 6a)|a, b \in R\} \\ &= \mathscr{L}\{(2, -6), (-1, 3)\} = \mathscr{L}\{(1, -3)\} \neq \mathscr{V}_2.\end{aligned}$$

The image of T can be viewed as the line in E^2 with Cartesian equation $y = -3x$. Each point on this line has an entire line of preimages. To see this, consider an arbitrary point $W = (k, -3k)$ in the image. The set of all pre-images for W is

$$\begin{aligned}\{V \in \mathscr{V}_2|T(V) = W\} &= \{(x, y)|T(x, y) = (k, -3k)\} \\ &= \{(x, y)|2x - y = k\}.\end{aligned}$$

Therefore the entire line with equation $y = 2x - k$ is sent by T to $(k, -3k)$. If ℓ_k denotes this line and $P_k = (k, -3k)$, then $T[\ell_k] = P_k$. Every point of the domain is on one of the lines ℓ_k, so the effect of T on the plane is to collapse each line ℓ_k to the point P_k. This is illustrated in Figure 2. The line ℓ_0

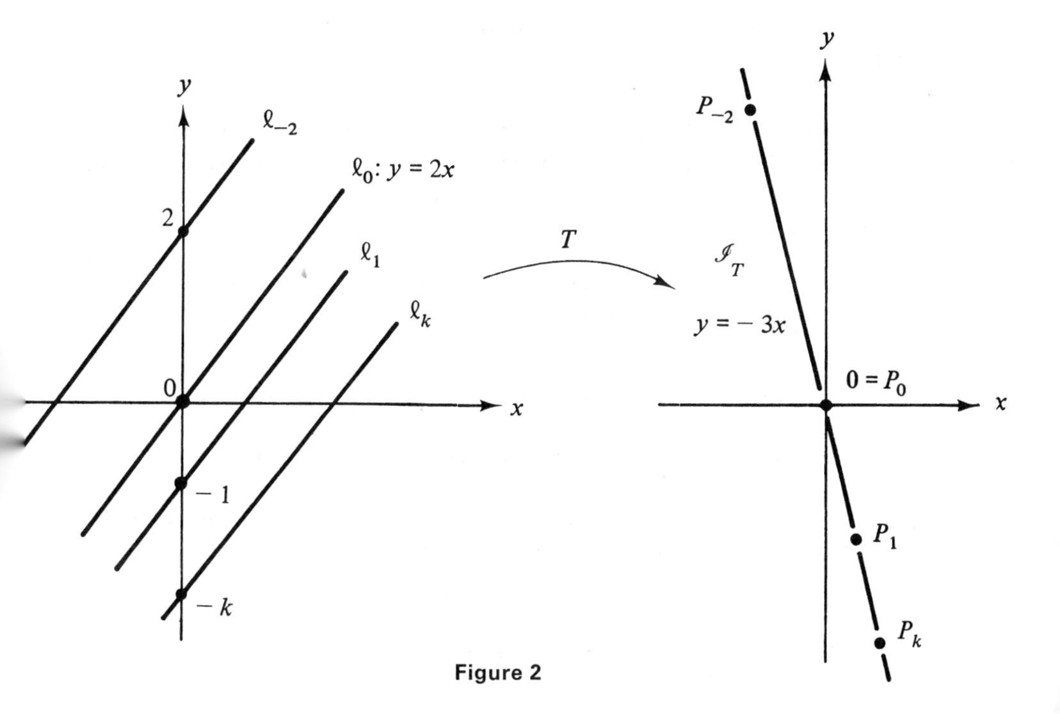

Figure 2

passes through the origin and thus represents a subspace of the domain. This subspace determines much of the nature of T for every line parallel to it is sent to a point. This idea generalizes to all linear maps. That is, knowing what vectors are sent to $\mathbf{0}$, tells much about a linear transformation.

The following theorem is easily proved and left to the reader.

Theorem 4.2 If $T: \mathscr{V} \to \mathscr{W}$ is linear, then $\{V \in \mathscr{V} | T(V) = \mathbf{0}\}$ is a subspace of $\mathscr{V}$.

Definition The *null space* $\mathscr{N}_T$ of a linear map $T: \mathscr{V} \to \mathscr{W}$ is the set of all preimages for $\mathbf{0}_{\mathscr{W}}$. That is, $\mathscr{N}_T = \{V \in \mathscr{V} | T(V) = \mathbf{0}\}$.

The null space is often called the kernel of the map.

In Example 3, the null space of T is represented by the line ℓ_0, and every line parallel to the null space is sent to a single point. Example 5 and problem 13 at the end of this section examine similar situations in 3-space.

As indicated, the null space provides information about a linear map. The simplest result is given by the next theorem.

Theorem 4.3 Suppose $T: \mathscr{V} \to \mathscr{W}$ is linear. T is one to one if and only if $\mathscr{N}_T = \{\mathbf{0}\}$. That is, T is 1–1 if and only if $\mathbf{0}_{\mathscr{V}}$ is the only vector that is mapped to $\mathbf{0}_{\mathscr{W}}$ by T.

Proof $(\Rightarrow)$ Suppose T is 1–1. We must show that $\{\mathbf{0}\} \subset \mathscr{N}_T$ and $\mathscr{N}_T \subset \{\mathbf{0}\}$. The first containment holds because $T(\mathbf{0}) = \mathbf{0}$ for any linear map. So suppose $U \in \mathscr{N}_T$, that is, $T(U) = \mathbf{0}$. Then since T is 1–1, $T(\mathbf{0}) = \mathbf{0} = T(U)$ implies $\mathbf{0} = U$ and $\mathscr{N}_T \subset \{\mathbf{0}\}$. Thus $\mathscr{N}_T = \{\mathbf{0}\}$.

$(\Leftarrow)$ Suppose $\mathscr{N}_T = \{\mathbf{0}\}$. If $T(U) = T(V)$ for some $U, V \in \mathscr{V}$, then $\mathbf{0} = T(U) - T(V) = T(U - V)$. That is, $U - V \in \mathscr{N}_T$ and $U - V = \mathbf{0}$. Therefore $U = V$ and T is 1–1.

Example 4 Determine if the map $T: R_3[t] \to R_3[t]$ is one to one if

$$T(a + bt + ct^2) = a - b + (b - c)t + (c - a)t^2.$$

If $a + bt + ct^2 \in \mathscr{N}_T$, then $T(a + bt + ct^2) = \mathbf{0}$. This yields a system of three homogeneous linear equations in three unknowns:

$$a - b = 0$$

$$b - c = 0$$

$$c - a = 0$$

which has coefficient matrix

$$A = \begin{pmatrix} 1 & -1 & 0 \\ 0 & 1 & -1 \\ -1 & 0 & 1 \end{pmatrix}.$$

The rank of A is 2, therefore there are nontrivial solutions and $\mathcal{N}_T \neq \{\mathbf{0}\}$. So T is not one to one.

A second source of information about a map is provided by the dimension of its null space.

Theorem 4.4 If $T\colon \mathscr{V} \to \mathscr{W}$ is linear and $\mathscr{V}$ is finite dimensional, then

$$\dim \mathscr{V} = \dim \mathcal{N}_T + \dim \mathscr{I}_T.$$

Proof Dimension is the number of vectors in a basis, therefore we must choose bases for the three vector spaces. The best place to start is with $\mathcal{N}_T$, for then a basis for $\mathscr{V}$ can be obtained by extension. Therefore, suppose $\{V_1, \ldots, V_k\}$ is a basis for $\mathcal{N}_T$. Then $\dim \mathcal{N}_T = k$. If $\mathcal{N}_T = \{\mathbf{0}\}$, then $k = 0$ and this basis is empty. Since the dimension of $\mathscr{V}$ is finite, there exists a basis $V_1, \ldots, V_k, V_{k+1}, \ldots, V_n$ for $\mathscr{V}$. Then $\dim \mathscr{V} = n$ and we need a basis for $\mathscr{I}_T$. If $W \in \mathscr{I}_T$, there exists a vector $U \in \mathscr{V}$ such that $T(U) = W$, and there exist scalars $a_1, \ldots, a_n$ such that $U = a_1V_1 + \cdots + a_nV_n$. By problem 14, page 137,

$$\begin{aligned} W = T(U) &= a_1T(V_1) + \cdots + a_kT(V_k) + a_{k+1}T(V_{k+1}) + \cdots + a_nT(V_n) \\ &= a_{k+1}T(V_{k+1}) + \cdots + a_nT(V_n). \end{aligned}$$

Therefore the set $B = \{T(V_{k+1}), \ldots, T(V_n)\}$ spans $\mathscr{I}_T$. To show that B is linearly independent, suppose

$$r_{k+1}T(V_{k+1}) + \cdots + r_nT(V_n) = \mathbf{0}.$$

Then

$$T(r_{k+1}V_{k+1} + \cdots + r_nV_n) = \mathbf{0}$$

and $r_{k+1}V_{k+1} + \cdots + r_nV_n \in \mathcal{N}_T$. Therefore there exist scalars $r_1, \ldots, r_k \in R$ such that

$$r_{k+1}V_{k+1} + \cdots + r_nV_n = r_1V_1 + \cdots + r_kV_k.$$

(Why?) But $\{V_1, \ldots, V_n\}$ is linearly independent as a basis for $\mathscr{V}$, so $r_{k+1} = \cdots = r_n = 0$ and B is a basis for $\mathscr{I}_T$. Now $\dim \mathscr{N}_T + \dim \mathscr{I}_T = k + (n - k) = n = \dim \mathscr{V}$ as claimed.

Definition The dimension of the null space of T, $\mathscr{N}_T$, is the *nullity* of T; and the dimension of the image space $\mathscr{I}_T$ is the *rank* of T.

Thus, Theorem 4.4 states that the nullity of T plus the rank of T equals the dimension of the domain of T.

Example 5 Find the rank and nullity of the map $T: \mathscr{V}_3 \to \mathscr{V}_3$ given by

$$T(a, b, c) = (2a - b + 3c, a - b + c, 3a - 4b + 2c).$$

The vector (x, y, z) is in $\mathscr{N}_T$ if

$$\begin{aligned} 2x - y + 3z &= 0 \\ x - y + z &= 0 \\ 3x - 4y + 2z &= 0. \end{aligned}$$

This homogeneous system is equivalent to $x + 2z = 0$, $y + z = 0$. Therefore

$$\mathscr{N}_T = \{(x, y, z) | x + 2z = 0, y + z = 0\} = \mathscr{L}\{(2, 1, -1)\}.$$

So the nullity of T is 1, and by Theorem 4.4, the rank of T is $2 = 3 - 1 = \dim \mathscr{V}_3 - \dim \mathscr{N}_T$.

If T is thought of as a map sending points in 3-space to points in 3-space, then $\mathscr{N}_T$ is a line ℓ with equation $P = t(2, 1, -1)$, $t \in R$. This line is sent to the origin by T. Moreover, as in Example 3, every line parallel to ℓ is sent to a single point. For a line parallel to ℓ has the vector equation $P = t(2, 1, -1) + V$ for some vector $V \in \mathscr{V}_3$: and

$$T(P) = tT(2, 1, -1) + T(V) = \mathbf{0} + T(V) = T(V).$$

Further since the rank of this map is 2, we know that $\mathscr{I}_T$ is a plane through the origin. In fact,

$$\begin{aligned} \mathscr{I}_T &= \{(2a - b + 3c, a - b + c, 3a - 4b + 2c) | a, b, c \in R\} \\ &= \{(x, y, z) \in \mathscr{V}_3 | x - 5y + z = 0\}. \end{aligned}$$

(Be sure to check this computation.)

Several conclusions follow at once from Theorem 4.4 using the fact that the dimension of $\mathscr{I}_T$ cannot exceed the dimension of the codomain.

Corollary 4.5 Suppose $T: \mathscr{V} \to \mathscr{W}$ is linear and $\mathscr{V}$ is finite dimensional, then

1. $\dim \mathscr{I}_T \leq \dim \mathscr{V}$.
2. T is 1–1 if and only if $\dim \mathscr{I}_T = \dim \mathscr{V}$.
3. T is onto if and only if $\dim \mathscr{I}_T = \dim \mathscr{W}$.
4. If $\dim \mathscr{V} > \dim \mathscr{W}$, then T is not 1–1.
5. If $\dim \mathscr{V} < \dim \mathscr{W}$, then T is not onto.
6. If $\mathscr{V} = \mathscr{W}$, then T is 1–1 if and only if T is onto.

This corollary provides a simple proof that $\mathscr{V}_n$ is not isomorphic to $\mathscr{V}_m$ when $n \neq m$, for any linear map from $\mathscr{V}_n$ to $\mathscr{V}_m$ is either not 1–1 when $n > m$, or it is not onto when $n < m$.

The result of problem 14, page 137, which was used in the proof of Theorem 4.4, is important enough to be stated as a theorem.

Theorem 4.6 Suppose $T: \mathscr{V} \to \mathscr{W}$ is linear and $\{V_1, \ldots, V_n\}$ is a basis for $\mathscr{V}$. If $V = a_1V_1 + \cdots + a_nV_n$, then $T(V) = a_1T(V_1) + \cdots + a_nT(V_n)$.

This states that a linear map is completely determined by its action on a basis. If the images of the basis vectors are known, then the image of any other vector can be found. Again we see how restrictive the linearity requirement is. Compare with the functions studied in calculus for which even a large number of functional values need not determine the function.

The fact that a linear map is determined by its action on a basis means that a linear map can be obtained by arbitrarily assigning images to the vectors in a basis. Suppose $B = \{V_1, \ldots, V_n\}$ is a basis for $\mathscr{V}$ and $W_1, \ldots, W_n$ are any n vectors from a vector space $\mathscr{W}$. Then set $T(V_1) = W_1, \ldots, T(V_n) = W_n$ and define T on the rest of $\mathscr{V}$ by

$$T(V) = a_1W_1 + \cdots + a_nW_n \qquad \text{if} \quad V = a_1V_1 + \cdots + a_nV_n.$$

It is not difficult to show that T is a linear map from $\mathscr{V}$ to $\mathscr{W}$ which sends V_i to W_i, $1 \leq i \leq n$.

Example 6 Given the basis $\{(2, 6), (0, 1)\}$ for $\mathscr{V}_2$, define a linear map $T: \mathscr{V}_2 \to \mathscr{V}_3$ which sends $(2, 6)$ to $(8, 10, 4)$ and $(0, 1)$ to $(-1, 4, 2)$.

Since $(a, b) = \frac{1}{2}a(2, 6) + [b - 3a](0, 1)$, we define T by

$$T(a, b) = \tfrac{1}{2}a(8, 10, 4) + [b - 3a](-1, 4, 2)$$
$$= (7a - b, 4b - 7a, 2b - 4a).$$

Then T is seen to be a linear map from $\mathscr{V}_2$ to $\mathscr{V}_3$ and

$$T(2, 6) = (14 - 6, 24 - 14, 12 - 8) = (8, 10, 4);$$
$$T(0, 1) = (-1, 4, 2)$$

as desired.

We will frequently want to find a linear map that has a certain effect on the vectors of a basis. In such cases we will define the map on the basis and "extend it linearly" to the entire vector space by the foregoing procedure. In general, if $S \subset \mathscr{V}$ and $T: S \to \mathscr{W}$ is given, then to *extend T linearly* to $\mathscr{V}$ is to define T on all of $\mathscr{V}$ in such a way that $T: \mathscr{V} \to \mathscr{W}$ is linear, and for each $U \in S$, $T(U)$ has the given value. However, if S is not a basis for $\mathscr{V}$, then it may not be possible to extent T linearly to $\mathscr{V}$; see problem 10 at the end of this section.

If $T: \mathscr{V} \to \mathscr{W}$ is the extension of $T: S \to \mathscr{W}$, then $T: S \to \mathscr{W}$ is the "restriction of T to S." We will most often wish to restrict a map to a subspace of the domain: If $T: \mathscr{V} \to \mathscr{W}$ and $\mathscr{S}$ is a subspace of $\mathscr{V}$, then the *restriction of T to* $\mathscr{S}$ is the map $T_{\mathscr{S}}: \mathscr{S} \to \mathscr{W}$ defined by $T_{\mathscr{S}}(U) = T(U)$ for all $U \in \mathscr{S}$. Since the domain of the restriction map $T_{\mathscr{S}}$ is a vector space, if T is linear, then so is $T_{\mathscr{S}}$.

Problems

1. Find $\mathscr{I}_T$ and $\mathscr{N}_T$ for the following linear maps:
   a. $T: \mathscr{V}_3 \to \mathscr{V}_2$; $T(a, b, c) = (3a - 2c, b + c)$.
   b. $T: \mathscr{V}_2 \to \mathscr{V}_3$; $T(a, b) = (a - 4b, a + 2b, 2a - b)$.
   c. $T: C \to C$; $T(a + bi) = a + b$.
   d. $T: R_3[t] \to R_3[t]$; $T(a + bt + ct^2) = a + 2c + (b - c)t + (a + 2b)t^2$.

2. Let $T: \mathscr{V}_2 \to \mathscr{V}_2$ be defined by $T(a, b) = (4a - 2b, 3b - 6a)$ and find the images of the following lines under T:
   a. $\{t(1, 5) | t \in R\}$.
   b. $\{t(1, 2) | t \in R\}$.
   c. $\{t(1, 2) + (3, 5) | t \in R\}$.
   d. $\{t(2, 1) + (3, 5) | t \in R\}$.
   e. $\{t(5, 10) + (1, 4) | t \in R\}$.
   f. Under what conditions is one of these lines sent to a single point?

3. Find $T[\mathscr{S}]$ when

a. $T(a, b, c) = (2a + b, 3b, 4b - c)$; $\mathscr{S} = \mathscr{L}\{(-1, 3, 5), (1, 0, 4)\}$.

b. T as in part a and $\mathscr{S} = \{(x, y, z) | y = x + z\}$.

c. $T(a, b, c) = (a + b - c, 2c - b)$; $\mathscr{S} = \mathscr{L}\{(-1, 2, 1)\}$.

d. T as in part c and $\mathscr{S} = \{(x, y, z) | z = 0\}$.

4. Determine if T is 1–1 by examining the null space $\mathscr{N}_T$.

a. $T(a + bt + ct^2) = 2a + b + (a + b + c)t^2$.

b. $T(a + bt + ct^2) = c + (a - b)t + at^2$.

c. $T(a, b, c, d) = (c - 3d, 0, 2a + b, c, 0, c + d)$.

d. $T(a + bt) = a + 2at + (a - b)t^2 + bt^3$.

e. $T(a, b, c) = (a + 2b + c, 3a - b + 2c)$.

5. How could one determine that the maps in parts a and e of problem 4 are not 1–1 by inspection?

6. Suppose $\mathscr{V} = \mathscr{S} \oplus \mathscr{T}$ and $P: \mathscr{V} \to \mathscr{T}$ is a projection map. Find $\mathscr{I}_P$ and $\mathscr{N}_P$.

7. Suppose $T: \mathscr{V} \to \mathscr{W}$ is linear and $S \subset \mathscr{W}$. $T^{-1}[S]$ (read "the complete inverse image of S") is the set of all preimages for vectors in S, that is, $T^{-1}[S] = \{V \in \mathscr{V} | T(V) \in S\}$.

a. What is $T^{-1}[\{\mathbf{0}\}]$?

b. Show that if $\mathscr{S}$ is a subspace of $\mathscr{W}$, then $T^{-1}[\mathscr{S}]$ is a subspace of $\mathscr{V}$.

c. What condition must be satisfied by $T^{-1}[\{W\}]$ for any $W \in \mathscr{W}$ if T is to map $\mathscr{V}$ onto $\mathscr{W}$?

d. What condition must be satisfied by $T^{-1}[\{W\}]$ for any $W \in \mathscr{W}$ if T is to be one to one?

8. $B = \{(4, 3), (2, 2)\}$ is a basis for $\mathscr{V}_2$. Define T on B by $T(4, 3) = (7, 5)$ and $T(2, 2) = (2, 2)$. Extend T linearly to $\mathscr{V}_2$ and find $T(a, b)$.

9. Set $T(4, 1) = (2, 1, -1)$, $T(3, 1) = (4, -2, 1)$, and extend T linearly to a map from $\mathscr{V}_2$ to $\mathscr{V}_3$. What is $T(a, b)$?

10. a. If we set $T(2, 4) = (6, -1, 3)$ and $T(3, 6) = (3, 0, 1)$, can T be extended linearly to a linear transformation from $\mathscr{V}_2$ to $\mathscr{V}_3$?

b. If $T(2, 4) = (2, 0, 6)$ and $T(3, 6) = (3, 0, 9)$, can T be extended to a linear transformation from $\mathscr{V}_2$ to $\mathscr{V}_3$?

11. Show that $T: \mathscr{V}_2 \to \mathscr{V}_2$ is linear if and only if there exist scalars a, b, c, d such that $T(x, y) = (ax + by, cx + dy)$.

12. Prove that if $T: \mathscr{V}_n \to \mathscr{V}_m$ and the ith component of $T(x_1, \ldots, x_n)$ is of the form $a_{i1}x_1 + a_{i2}x_2 + \cdots + a_{in}x_n$ for some scalars $a_{i1}, \ldots, a_{in}$, then T is linear.

13. Define $T: \mathscr{V}_3 \to \mathscr{V}_3$ by $T(a, b, c) = (4a + 2b - 6c, 0, 6a + 3b - 9c)$.

a. Show that $\mathscr{N}_T$ is represented by a plane through the origin.

b. Show that every plane parallel to $\mathscr{N}_T$ is sent by T to a single point.

c. Let $\mathscr{S}$ be the subspace normal to $\mathscr{N}_T$. Show that the restriction map $T_{\mathscr{S}}$ is 1–1 and the image space of $T_{\mathscr{S}}$ is the image space of T.

d. Could $\mathscr{S}$ in part c be replaced by any other subspace?

14. Suppose $\mathscr{V}$ is a finite-dimensional vector space and $T: \mathscr{V} \to \mathscr{W}$ is linear but not 1–1. Find a subspace $\mathscr{S}$ of $\mathscr{V}$ such that $T_{\mathscr{S}}$ is 1–1 and the image of the restriction $T_{\mathscr{S}}$ equals the image of T. Is there a unique choice for $\mathscr{S}$?

15. Suppose $T: \mathscr{V} \to \mathscr{W}$ is linear and $\{U_1, \ldots, U_n\}$ is a basis for $\mathscr{V}$. Show that the vectors $T(U_1), \ldots, T(U_n)$ span the image space $\mathscr{I}_T$.

§3. Algebra of Linear Transformations

Recall the construction of the vector space $\mathscr{F}$, in which vectors are continuous functions from [0, 1] to R. The algebraic structure of $\mathscr{F}$ was defined using addition and multiplication in the codomain of the functions, R. Further, the proof that $\mathscr{F}$ is a vector space relied on the fact that R is itself a vector space. This idea can easily be generalized to other collections of functions with images in a vector space. In particular, the algebraic structure of a vector space can be introduced in the set of all linear transformations from one vector space to another.

A linear transformation preserves the algebraic structure of a vector space and thus belongs to a broad class of functions between algebraic systems which preserve the structure of the system. Such functions are called *homomorphisms*. A map between the field of real numbers and the field of complex numbers would be a homomorphism if it preserved the field structure. Since a linear transformation is an example of a homomorphism, the set of all linear transformations from $\mathscr{V}$ to $\mathscr{W}$ is often denoted $\text{Hom}(\mathscr{V}, \mathscr{W})$.

Definition If $\mathscr{V}$ and $\mathscr{W}$ are vector spaces, then

$$\text{Hom}(\mathscr{V}, \mathscr{W}) = \{T \mid T \text{ is a linear map from } \mathscr{V} \text{ to } \mathscr{W}\}.$$

When $\mathscr{V} = \mathscr{W}$, $\text{Hom}(\mathscr{V}, \mathscr{V})$ will be written as $\text{Hom}(\mathscr{V})$.

Note that the statement $T \in \text{Hom}(\mathscr{V}, \mathscr{W})$ is simply a short way to write "T is a linear transformation from $\mathscr{V}$ to $\mathscr{W}$."

To give $\text{Hom}(\mathscr{V}, \mathscr{W})$ the structure of a vector space, we must define operations of addition and scalar multiplication:

For $S, T \in \text{Hom}(\mathscr{V}, \mathscr{W})$ define $S + T$ by $[S + T](V) = S(V) + T(V)$ for all $V \in \mathscr{V}$. For $T \in \text{Hom}(\mathscr{V}, \mathscr{W})$ and $r \in R$, define rT by $[rT](V) = r[T(V)]$ for all $V \in \mathscr{V}$.

Example 1 Let $S, T \in \operatorname{Hom}(\mathscr{V}_2, \mathscr{V}_3)$ be given by

$$S(a, b) = (a - b, 2a, 3b - a) \quad \text{and} \quad T(a, b) = (b - a, b, a - b).$$

Then $S + T$ is given by

$$[S + T](a, b) = S(a, b) + T(a, b) = (0, 2a + b, 2b)$$

and

$$[rT](a, b) = (rb - ra, rb, ra - rb).$$

Example 2 Let $S, T \in \operatorname{Hom}(R_3[t])$ be defined by

$$S(a + bt + ct^2) = (a + b)t + at^2 \quad \text{and} \quad T(a + bt + ct^2) = (c - a)t^2 - bt.$$

Then

$$[S + T](a + bt + ct^2) = at + ct^2.$$

For example,

$$[S + T](t + 4t^2) = 4t^2.$$

The maps $7S$ and $4T$ have the following effect on the vector $3 - 2t + t^2$:

$$[7S](3 - 2t + t^2) = 7[S(3 - 2t + t^2)] = 7[t + 3t^2] = 7t + 21t^2$$

and

$$[4T](3 - 2t + t^2) = 8t - 8t^2.$$

Theorem 4.7 The set of linear transformations $\operatorname{Hom}(\mathscr{V}, \mathscr{W})$ together with addition and scalar multiplication as defined above is a vector space.

Proof It is necessary to prove that all nine conditions in the definition of a vector space hold, these are listed on page 32.

$\operatorname{Hom}(\mathscr{V}, \mathscr{W})$ is closed under addition if, for any two maps $S, T \in \operatorname{Hom}(\mathscr{V}, \mathscr{W})$, $S + T\colon \mathscr{V} \to \mathscr{W}$ and $S + T$ is linear. $S + T$ is defined for each $V \in \mathscr{V}$ and since $\mathscr{W}$ is closed under addition, the codomain of $S + T$ is

$\mathscr{W}$, therefore $S + T: \mathscr{V} \to \mathscr{W}$. To show that $S + T$ is linear, suppose U, $V \in \mathscr{V}$ and $a, b \in R$, then

$[S + T](aU + bV) = S(aU + bV) + T(aU + bV)$	Definition of + in $\text{Hom}(\mathscr{V}, \mathscr{W})$
$= [a[S(U)] + b[S(V)]] + [a[T(U)] + b[T(V)]]$	$S, T \in \text{Hom}(\mathscr{V}, \mathscr{W})$
$= [a[S(U)] + a[T(U)]] + [b[S(V)] + b[T(V)]]$	Commutativity and associativity of + in $\mathscr{W}$
$= a[S(U) + T(U)] + b[S(V) + T(V)]$	Distributive law in $\mathscr{W}$
$= a[[S + T](U)] + b[[S + T](V)]$	Definition of + in $\text{Hom}(\mathscr{V}, \mathscr{W})$.

Therefore $S + T$ is linear and $S + T \in \text{Hom}(\mathscr{V}, \mathscr{W})$.

To prove that addition is commutative, that is, $S + T = T + S$ for all $S, T \in \text{Hom}(\mathscr{V}, \mathscr{W})$, it is necessary to show that $[S + T](V) = [T + S](V)$ for all $V \in \mathscr{V}$. That is, two maps are equal if and only if they have the same effect on every vector. But

$$[S + T](V) = S(V) + T(V) = T(V) + S(V) = [T + S](V).$$

(At what point in this chain of equalities is the commutivity of addition in $\mathscr{W}$ used?) Therefore addition is commutative in $\text{Hom}(\mathscr{V}, \mathscr{W})$. The proof that addition is associative in $\text{Hom}(\mathscr{V}, \mathscr{W})$ is similar.

Define the zero map $\mathbf{0}: \mathscr{V} \to \mathscr{W}$ by $\mathbf{0}(V) = \mathbf{0}_{\mathscr{W}}$ for all $V \in \mathscr{V}$. The map $\mathbf{0}$ is easily seen to be linear and thus a member of $\text{Hom}(\mathscr{V}, \mathscr{W})$. Further

$$[T + \mathbf{0}](V) = T(V) + \mathbf{0}(V) = T(V) + \mathbf{0} = T(V) \qquad \text{for all} \quad V \in \mathscr{V}.$$

Therefore $T + \mathbf{0} = T$ for any $T \in \text{Hom}(\mathscr{V}, \mathscr{W})$, and $\mathbf{0}$ is the additive identity for $\text{Hom}(\mathscr{V}, \mathscr{W})$. Notice that, in general, there are four different zeros associated with $\text{Hom}(\mathscr{V}, \mathscr{W})$; the number zero, the zero vectors in $\mathscr{V}$ and $\mathscr{W}$, and the zero map in $\text{Hom}(\mathscr{V}, \mathscr{W})$.

For the final property of addition, define $-T$ by $[-T](V) = -[T(V)]$ for all $V \in \mathscr{V}$. It is easily shown that if $T \in \text{Hom}(\mathscr{V}, \mathscr{W})$, then $-T \in \text{Hom}(\mathscr{V}, \mathscr{W})$ and $T + (-T) = \mathbf{0}$. Therefore every element has an additive inverse in $\text{Hom}(\mathscr{V}, \mathscr{W})$.

To complete the proof, it is necessary to prove that all the properties for scalar multiplication hold in $\text{Hom}(\mathscr{V}, \mathscr{W})$. Only one of the distributive laws will be proved here with the other properties left to the reader. To prove that $r(S + T) = rS + rT$ for $r \in R$ and $S, T \in \text{Hom}(\mathscr{V}, \mathscr{W})$, it is necessary to show that $[r(S + T)](V) = [rS + rT](V)$ for any $V \in \mathscr{V}$:

$[r(S + T)](V) = r[(S + T)(V)]$	Definition of scalar multiplication in $\text{Hom}(\mathscr{V}, \mathscr{W})$
$= r[S(V) + T(V)]$	Definition of addition in $\text{Hom}(\mathscr{V}, \mathscr{W})$
$= r[S(V)] + r[T(V)]$	Distributive law in $\mathscr{W}$
$= [rS](V) + [rT](V)$	Definition of scalar multiplication in $\text{Hom}(\mathscr{V}, \mathscr{W})$
$= [rS + rT](V)$	Definition of addition in $\text{Hom}(\mathscr{V}, \mathscr{W})$.

Therefore the maps $r(S + T)$ and $rS + rT$ yield the same image for each vector V and so are equal.

With Theorem 4.7 we have a new class of vector spaces and the term vector applies to a linear transformation. Moreover, any definition or theorem about abstract vector spaces now applies to a space of linear transformations, $\text{Hom}(\mathscr{V}, \mathscr{W})$.

Example 3 Determine if the vectors $T_1, T_2, T_3 \in \text{Hom}(\mathscr{V}_2)$ are linearly independent when $T_1(a, b) = (2a, a - b)$, $T_2(a, b) = (4a + b, a)$, and $T_3(a, b) = (b, 3a)$.

Consider $S = x_1T_1 + x_2T_2 + x_3T_3$, a linear combination of these vectors. The vectors are linearly independent if $S = \mathbf{0}$ implies $x_1 = x_2 = x_3 = 0$. If $S = \mathbf{0}$, then $S(a, b) = \mathbf{0}(a, b) = (0, 0)$ for all (a, b) in $\mathscr{V}_2$. But S is determined by what it does to a basis. Using $\{E_i\}$ we have

$$S(1, 0) = (2x_1 + 4x_2, x_1 + x_2 + 3x_3) = (0, 0)$$

and

$$S(0, 1) = (x_2 + x_3, -x) = (0, 0).$$

These vector equations yield the homogeneous system

$$2x_1 + 4x_2 = 0, \qquad x_1 + x_2 + 3x_3 = 0, \qquad x_2 + x_3 = 0, \qquad -x_1 = 0,$$

which has only the trivial solution. Thus the set $\{T_1, T_2, T_3\}$ is linearly independent in $\text{Hom}(\mathscr{V}_2)$.

Why does Example 3 show that the dimension of the vector space $\text{Hom}(\mathscr{V}_2)$ is at least 3?

Example 4 Consider the three vectors in $\text{Hom}(\mathscr{V}_2, \mathscr{V}_3)$ given by $T_1(a, b) = (a - b, a, 2a - 3b)$, $T_2(a, b) = (3a - b, b, a + b)$, and $T_3(a, b) = (a + b, b - 2a, 7b - 3a)$. Is the set $\{T_1, T_2, T_3\}$ linearly independent?

Suppose $S = x_1T_1 + x_2T_2 + x_3T_3 = \mathbf{0}$. Then $S(1, 0) = S(0, 1) = (0, 0, 0)$ yields two equations in $\mathscr{V}_3$:

$$(x_1 + 3x_2 + x_3, x_1 - 2x_3, 2x_1 + x_2 - 3x_3) = (0, 0, 0)$$

and

$$(-x_1 - x_2 + x_3, x_2 + x_3, -3x_1 + x_2 + 7x_3) = (0, 0, 0).$$

Setting components equal, we obtain a homogeneous system of six linear equations in three unknowns. But the rank of the coefficient matrix is 2, so there are nontrivial solutions. That is, $S = \mathbf{0}$ does not imply that $x_1 = x_2 = x_3 = 0$, and the set $\{T_1, T_2, T_3\}$ is linearly dependent.

If one looks to the last two examples for general ideas, the first point to notice is that again the problem of determining linear dependence or independence leads to a homogeneous system of linear equations. Second, the number of homogeneous equations obtained, 4 and 6, is the product of the dimension of the domain and the dimension of the codomain; $4 = 2 \cdot 2$ and $6 = 2 \cdot 3$. In general, if the maps come from $\text{Hom}(\mathscr{V}_n, \mathscr{V}_m)$, then each of the n vectors in a basis for $\mathscr{V}_n$ yields a vector equation in $\mathscr{V}_m$, and the m components in each vector equation yield m linear equations. Since a system of nm homogeneous linear equations has a nontrivial solution if there are more than nm unknowns, we can conclude that $\dim \text{Hom}(\mathscr{V}_n, \mathscr{V}_m) \leq nm$. We could prove that the dimension of $\text{Hom}(\mathscr{V}_2)$ is 4 by extending the set in Example 3 to a basis. But it will be more informative to construct a standard basis for $\text{Hom}(\mathscr{V}_2)$ from the standard basis for $\mathscr{V}_2$.

Example 5 Show that $\dim \text{Hom}(\mathscr{V}_2) = 4$.

If we follow the pattern set in $\mathscr{V}_n$, $R[t]$, and $\mathscr{M}_{n \times m}$ of finding a simple basis, then we should look for a collection of simple nonzero maps in $\text{Hom}(\mathscr{V}_2)$. Using the standard basis $\{E_i\}$ for $\mathscr{V}_2$, we can obtain a map by choosing images for E_1 and E_2, and then extending linearly. The simplest choice is to send one vector to zero and the other to either E_1 or E_2. This yields four maps $T_{ij}\colon \mathscr{V}_2 \to \mathscr{V}_2$ with $T_{ij}(E_i) = E_j$ and $T_{ij}(E_k) = (0, 0)$ for $k \neq i$; $i, j = 1, 2$. For example, T_{12} sends the first basis vector in $\{E_i\}$ to the second vector in $\{E_i\}$ and it sends the second vector in $\{E_i\}$ to zero. Extending this map linearly to $\mathscr{V}_2$ gives

$$T_{12}(a, b) = aT_{12}(1, 0) + bT_{12}(0, 1) = a(0, 1) + b\mathbf{0} = (0, a).$$

You might check that the other maps are just as simple with $T_{11}(a, b) = (a, 0)$, $T_{21}(a, b) = (b, 0)$, and $T_{22}(a, b) = (0, b)$.

The claim is that $B = \{T_{11}, T_{12}, T_{21}, T_{22}\}$ is a basis for Hom($\mathscr{V}_2$). The proof that B is linearly independent follows the pattern set in Examples 3 and 4. To show that B spans Hom($\mathscr{V}_2$), suppose $T \in$ Hom($\mathscr{V}_2$). Then there exist scalars a_{ij} such that $T(1, 0) = (a_{11}, a_{12})$ and $T(0, 1) = (a_{21}, a_{22})$. (Since T is determined by its action on a basis, these scalars completely determine T.) Now for any vector (x, y):

$$\begin{aligned} T(x, y) &= xT(1, 0) + yT(0, 1) \\ &= (xa_{11} + ya_{21}, xa_{12} + ya_{22}) \\ &= a_{11}(x, 0) + a_{12}(0, x) + a_{21}(y, 0) + a_{22}(0, y) \\ &= a_{11}T_{11}(x, y) + a_{12}T_{12}(x, y) + a_{21}T_{21}(x, y) + a_{22}T_{22}(x, y) \\ &= [a_{11}T_{11} + a_{12}T_{12} + a_{21}T_{21} + a_{22}T_{22}](x, y). \end{aligned}$$

Therefore

$$T = a_{11}T_{11} + a_{12}T_{12} + a_{21}T_{21} + a_{22}T_{22}$$

and B spans Hom($\mathscr{V}_2$). Thus B is a basis for Hom($\mathscr{V}_2$) and the dimension is 4.

In showing that B is a basis for Hom($\mathscr{V}_2$), we have discovered how to find the coordinates of a map with respect to B. In fact, $T\colon (a_{11}, a_{12}, a_{21}, a_{22})_B$. Hence if $T(x, y) = (3x - 4y, 2x)$, then $T(1, 0) = (3, 2)$ and $T(0, 1) = (-4, 0)$, so $T\colon (3, 2, -4, 0)_B$. In the next section we will essentially arrange these coordinates into a matrix to obtain a matrix representation for T.

The technique for constructing a basis for Hom($\mathscr{V}_2$) can easily be generalized.

Theorem 4.8 If dim $\mathscr{V} = n$ and dim $\mathscr{W} = m$, then dim Hom($\mathscr{V}, \mathscr{W}$) = nm.

Proof It is necessary to find a basis for Hom($\mathscr{V}$, $\mathscr{W}$). This means defining nm linear maps from $\mathscr{V}$ to $\mathscr{W}$. Since all we know about $\mathscr{V}$ and $\mathscr{W}$ is their dimensions, the natural place to start is with bases for $\mathscr{V}$ and $\mathscr{W}$.

So suppose $\{V_1, \ldots, V_n\}$ and $\{W_1, \ldots, W_m\}$ are bases for $\mathscr{V}$ and $\mathscr{W}$, respectively. For each i, $1 \le i \le n$, and each j, $1 \le j \le m$, we obtain a map T_{ij} as follows: Let $T_{ij}(V_i) = W_j$, $T_{ij}(V_k) = \mathbf{0}$ if $k \ne i$, and extend T_{ij} to a linear map from $\mathscr{V}$ to $\mathscr{W}$. This gives us nm vectors in Hom($\mathscr{V}$, $\mathscr{W}$), and it is only necessary to show that they are linearly independent and that they span Hom($\mathscr{V}$, $\mathscr{W}$).

Suppose a linear combination of these vectors is the zero map, say

$$S = r_{11}T_{11} + r_{12}T_{12} + \cdots + r_{nm}T_{nm} = \sum_{i=1}^{m}\sum_{j=1}^{n} r_{ij}T_{ij} = \mathbf{0}.$$

Then $S(V_h) = \mathbf{0}(V_h) = \mathbf{0}_{\mathscr{W}}$ for each basis vector V_h. Therefore,

$$\begin{aligned}
\mathbf{0}_{\mathscr{W}} = \mathbf{0}(V_h) &= \left[\sum_{i=1}^{n}\sum_{j=1}^{m} r_{ij}T_{ij}\right](V_h) = \sum_{j=1}^{m}\sum_{i=1}^{n} r_{ij}T_{ij}(V_h) \\
&= \sum_{j=1}^{m}[r_{1j}T_{1j}(V_h) + r_{2j}T_{2j}(V_h) + \cdots + r_{hj}T_{hj}(V_h) + \cdots + r_{nj}T_{nj}(V_h)] \\
&= \sum_{j=1}^{m}[\mathbf{0} + \mathbf{0} + \cdots + \mathbf{0} + r_{hj}W_j + \mathbf{0} + \cdots + \mathbf{0}] \\
&= r_{h1}W_1 + r_{h2}W_2 + \cdots + r_{hm}W_m.
\end{aligned}$$

The W's form a basis for $\mathscr{W}$, so for each integer h, $1 \le h \le n$, $r_{h1} = r_{h2} = \cdots = r_{hm} = 0$ and the maps T_{ij} are linearly independent.

To prove that these maps span Hom($\mathscr{V}$, $\mathscr{W}$), suppose $T \in$ Hom($\mathscr{V}$, $\mathscr{W}$). Since $T(V_i) \in \mathscr{W}$ and $\{W_1, \ldots, W_m\}$ spans $\mathscr{W}$, there exist scalars a_{ij} such that $T(V_i) = a_{i1}W_1 + a_{i2}W_2 + \cdots + a_{im}W_m$, for $1 \le i \le n$. Now T and the linear combination $\sum_{i=1}^{n}\sum_{j=1}^{m} a_{ij}T_{ij}$ have the same effect on the basis $\{V_1, \ldots, V_n\}$, for if $1 \le h \le n$, then

$$\begin{aligned}
\left[\sum_{i=1}^{n}\sum_{j=1}^{m} a_{ij}T_{ij}\right](V_h) &= \sum_{i=1}^{n}\sum_{j=1}^{m} a_{ij}T_{ij}(V_h) \\
&= \sum_{j=1}^{m} a_{hj}W_j \\
&= T(V_h).
\end{aligned}$$

Therefore $T = \sum_{i=1}^{n}\sum_{j=1}^{m} a_{ij}T_{ij}$ and the maps T_{ij} span Hom($\mathscr{V}$, $\mathscr{W}$).

Thus the nm maps T_{ij} form a basis for Hom($\mathscr{V}$, $\mathscr{W}$) and the dimemsion of Hom($\mathscr{V}$, $\mathscr{W}$) is nm.

Example 6 Suppose $B = \{T_{11}, T_{12}, T_{13}, T_{21}, T_{22}, T_{23}\}$ is the basis for $\text{Hom}(\mathscr{V}_2, \mathscr{V}_3)$ obtained using the bases $B_1 = \{(3, 1), (1, 0)\}$ for $\mathscr{V}_2$ and $B_2 = \{(1, 1, 1), (1, 1, 0), (1, 0, 0)\}$ for $\mathscr{V}_3$. Find T_{21}, T_{13} and the coordinates of T with respect to B, given that $T(x, y) = (3x - 2y, 5y, 4y - x)$.

For T_{21}, we know that $T_{21}(3, 1) = (0, 0, 0)$ and $T_{21}(1, 0) = (1, 1, 1)$. Since $(a, b)\colon (b, a - 3b)_{B_1}$, T_{21} is defined by

$$T_{21}(a, b) = b(0, 0, 0) + [a - 3b](1, 1, 1) = (a - 3b, a - 3b, a - 3b).$$

For example, $T_{21}(15, 3) = (6, 6, 6)$. For T_{13},

$$\begin{aligned} T_{13}(a, b) &= bT_{13}(3, 1) + [a - 3b]T_{13}(1, 0) \\ &= b(1, 0, 0) + (0, 0, 0) \\ &= (b, 0, 0). \end{aligned}$$

Therefore $T_{13}(15, 3) = (3, 0, 0)$.

To find the coordinates of T with respect to B, notice that

$$\begin{aligned} T(3, 1) &= (7, 5, 1) \\ &= 1(1, 1, 1) + 4(1, 1, 0) + 2(1, 0, 0) \\ &= 1T_{11}(3, 1) + 4T_{12}(3, 1) + 2T_{13}(3, 1) \\ &= [1T_{11} + 4T_{12} + 2T_{13}](3, 1). \end{aligned}$$

So if $T\colon (a_{11}, a_{12}, a_{13}, a_{21}, a_{22}, a_{23})_B$, then $a_{11} = 1$, $a_{12} = 4$, and $a_{13} = 2$. For the second vector in B_1, $T(1, 0) = (3, 0, -1)\colon (-1, 1, 3)_{B_2}$, therefore the coordinates of T with respect to B are given by $T\colon (1, 4, 2, -1, 1, 3)_B$.

In the process of obtaining the basis $\{T_{ij}\}$ for $\text{Hom}(\mathscr{V}, \mathscr{W})$ we have found how to obtain the coordinates a_{ij} of a map with respect to $\{T_{ij}\}$. The double index notation was natural since two bases were used, but it also strongly suggests the matrix (a_{ij}). In the next section we will see how to associate the transpose of this matrix with the map.

Problems

1. Suppose $T_1, T_2, T_3 \in \text{Hom}(\mathscr{V}_2, \mathscr{V}_3)$ are given by $T_1(a, b) = (3a - b, 2a + 4b, a - 8b)$, $T_2(a, b) = (a + b, b, a - 4b)$, and $T_3(a, b) = (a + 5b, -2a, 3a - 8b)$. Find the following vectors:
   a. $T_1 + T_2$. b. $5T_1$. c. $T_3 - 3T_2$. d. $T_1 - 4T_2 + T_3$.

2. Determine if the following sets of vectors are linearly independent:

a. $\{T_1, T_2, T_3, T_4\} \subset \text{Hom}(\mathscr{V}_2)$; $T_1(a, b) = (a + b, a)$, $T_2(a, b) = (a, a + b)$, $T_3(a, b) = (a + b, 0)$, $T_4(a, b) = (0, b)$.

b. $\{T_1, T_2, T_3\} \subset \text{Hom}(\mathscr{V}_2)$; $T_1(a, b) = (a, b)$, $T_2(a, b) = (b, a)$, $T_3(a, b) = (2a - b, 2b - a)$.

c. $\{T_1, T_2, T_3, T_4\} \subset \text{Hom}(\mathscr{V}_2, \mathscr{V}_3)$; $T_1(a, b) = (a, b, b)$, $T_2(a, b) = (a, a, b)$, $T_3(a, b) = (b, a, a)$, $T_4(a, b) = (b, b, a)$.

d. $\{T_1, T_2, T_3\} \subset \text{Hom}(\mathscr{V}_2, \mathscr{V}_3)$; $T_1(a, b) = (a + b, 2b, b - a)$, $T_2(a, b) = (a - b, 2a, a + b)$, $T_3(a, b) = (a, a + b, b)$.

3. What does the sentence "let $T \in \text{Hom}(\mathscr{V}, \mathscr{W})$" mean?

4. Let $S, T \in \text{Hom}(\mathscr{V}, \mathscr{W})$ and $U, V \in \mathscr{V}$.

a. Identify three different meanings for the plus sign in the equation $[S + T](U + V) = [S + T](U) + [S + T](V)$.

b. What are the meanings of the symbol $\mathbf{0}$ in $\mathbf{0}(U + \mathbf{0}) = \mathbf{0}(U) = \mathbf{0}$?

5. a. Prove that $\text{Hom}(\mathscr{V}, \mathscr{W})$ is closed under scalar multiplication.

b. Prove that $[a + b]T = aT + bT$ for any $T \in \text{Hom}(\mathscr{V}, \mathscr{W})$ and $a, b \in R$.

6. Let $B = \{T_{11}, T_{12}, T_{21}, T_{22}\}$ be the basis constructed for $\text{Hom}(\mathscr{V}_2)$ in Example 5. Find the following images:

a. $T_{12}(3, 5)$. b. $T_{22}(4, -6)$.

c. $[2T_{12} - 3T_{21} + T_{22}](1, 3)$. d. $[5T_{11} + 7T_{12} + 4T_{21}](0, -3)$.

Find the coordinates of the following vectors with respect to B:

e. $T(a, b) = (4a, b - 3a)$. g. $T(a, b) = (4a + b, 0)$.

f. $T(a, b) = (0, 0)$. h. $T(a, b) = (2a + 3b, a - 5b)$.

7. Write out the following sums:

a. $\sum_{i=1}^{3} a_{i2}T_{i2}$. b. $\sum_{j=1}^{2} a_{3j}T_{3j}$. c. $\sum_{i=1}^{3}\sum_{j=1}^{2} a_{ij}T_{ij}$.

8. Write the following sums using the sigma notation:

a. $r_{11}T_{11} + r_{12}T_{12} + r_{21}T_{21} + r_{22}T_{22}$.

b. $a_{41}T_{41} + a_{42}T_{42} + a_{43}T_{43} + a_{44}T_{44}$.

c. $r_1a_{11}T_{11} + r_1a_{12}T_{12} + r_2a_{21}T_{21} + r_2a_{22}T_{22} + r_3a_{31}T_{31} + r_3a_{32}T_{32}$.

9. Suppose T_{ij} are the maps used in Example 6.

a. Find T_{12}. b. Find T_{22}. c. $T_{23}(4, 7) = ?$ d. $T_{11}(5, 9) = ?$

10. If B is the basis for $\text{Hom}(\mathscr{V}_2, \mathscr{V}_3)$ from Example 6, find the coordinates of T with respect to B when

a. $T(x, y) = (x - y, 4x + y, 3y)$. c. $T(x, y) = (x + 2y, x, x - y)$.

b. $T(x, y) = (5x - 3y, 0, 2x + 2y)$. d. $T(x, y) = (3x - 2y, 4y, 2x)$.

11. Use the bases $\{1, t\}$ and $\{1, t, t^2\}$ for $R_2[t]$ and $R_3[t]$ to obtain the basis $B = \{T_{11}, T_{12}, T_{13}, T_{21}, T_{22}, T_{23}\}$ for Hom $(R_2[t], R_3[t])$.

a. What are the maps T_{11}, T_{13}, and T_{21}?

b. What is the vector $2T_{11} + 3T_{21} - 4T_{13}$?

Express the following maps as a linear combination of the T_{ij}'s.

c. $T(a + bt) = at + bt^2$. d. $T(a + bt) = 2b + (a - 4b)t + 5at^2$.

Find the coordinates of the following vectors with respect to B.
e. $T(a + bt) = (a + b)t^2$. g. $T(a + bt) = \mathbf{0}$.
f. $T(a + bt) = 3a - b + (4b - 3a)t + (7a + 5b)t^2$.

12. Use the bases $B_1 = \{(1, 2, 1), (0, 1, -1), (1, 4, 0)\}$ for $\mathscr{V}_3$ and $B_2 = \{(0, 1), (1, 2)\}$ for $\mathscr{V}_2$ to define the maps in the basis $B = \{T_{11}, T_{12}, T_{21}, T_{22}, T_{31}, T_{32}\}$ for $\text{Hom}(\mathscr{V}_3, \mathscr{V}_2)$. Find the following images:
a. $T_{21}(1, 4, 0)$ b. $T_{32}(1, 4, 0)$. c. $T_{22}(0, -5, 5)$.
d. $T_{22}(3, 4, 5)$. e. $T_{12}(1, 0, 0)$.
Find the coordinates of the following maps with respect to B.
f. $T(x, y, z) = (3x - y + z, 2y - x + 4z)$.
g. $T(x, y, z) = (y - 2z, 3x + 4z)$.

13. What are the 4 zeros used in the vector space
a. $\text{Hom}(\mathscr{V}_2, \mathscr{V}_4)$? b. $\text{Hom}(\mathscr{V}_3, R_3[t])$? c. $\text{Hom}(\mathscr{M}_{2\times 3}, \mathscr{M}_{2\times 2})$?

§4. Matrices of a Linear Transformation

We have seen that if $\mathscr{V}$ and $\mathscr{W}$ are finite-dimensional vector spaces, then so is the space of maps $\text{Hom}(\mathscr{V}, \mathscr{W})$. Further each choice of bases for $\mathscr{V}$ and $\mathscr{W}$ yields a basis for $\text{Hom}(\mathscr{V}, \mathscr{W})$. The coordinates of a linear transformation T from $\mathscr{V}$ to $\mathscr{W}$ with respect to such a basis can be arranged as a matrix in several ways. The following example shows which way we should choose and illustrates one of the major uses for such a matrix.

Example 1 Suppose $T\colon R_2[t] \to R_3[t]$ is defined by

$$T(a + bt) = 3a + b + (a - 4b)t + (a + 2b)t^2.$$

An arbitrary image vector has the form $x + yt + zt^2$, and the equation $T(a + bt) = x + yt + zt^2$ yields the system of linear equations:

$$3a + b = x$$
$$a - 4b = y$$
$$a + 2b = z$$

or in matrix form

$$\begin{pmatrix} 3 & 1 \\ 1 & -4 \\ 1 & 2 \end{pmatrix}\begin{pmatrix} a \\ b \end{pmatrix} = \begin{pmatrix} x \\ y \\ z \end{pmatrix}.$$

Therefore the image of any vector $a + bt$ in $R_2[t]$ can be found by matrix multiplication. As an example,

$$\begin{aligned} T(4 - 6t) &= 12 - 6 + (4 + 24)t + (4 - 12)t^2 \\ &= 6 + 28t - 8t^2, \end{aligned}$$

and

$$\begin{pmatrix} 3 & 1 \\ 1 & -4 \\ 1 & 2 \end{pmatrix} \begin{pmatrix} 4 \\ -6 \end{pmatrix} = \begin{pmatrix} 12 - 6 \\ 4 + 24 \\ 4 - 12 \end{pmatrix} = \begin{pmatrix} 6 \\ 28 \\ -8 \end{pmatrix}.$$

Further since

$$\begin{pmatrix} 3 & 1 \\ 1 & -4 \\ 1 & 2 \end{pmatrix} \begin{pmatrix} 1 \\ 3 \end{pmatrix} = \begin{pmatrix} 6 \\ -11 \\ 7 \end{pmatrix},$$

we can conclude that $T(1 + 3t) = 6 - 11t + 7t^2$.

In using the matrix

$$A = \begin{pmatrix} 3 & 1 \\ 1 & -4 \\ 1 & 2 \end{pmatrix}$$

to find images under T, it is clear that the vectors are not being used directly. Instead the columns contain coordinates with respect to the bases $B_1 = \{1, t\}$ and $B_2 = \{1, t, t^2\}$. Thus $4 - 6t\colon (4, -6)_{B_1}$ and $6 + 28t - 8t^2\colon (6, 28, -8)_{B_2}$. In general, the column vector

$$\begin{pmatrix} a \\ b \end{pmatrix}$$

contains the coordinates of $a + bt$ with respect to B_1 and

$$\begin{pmatrix} x \\ y \\ z \end{pmatrix}$$

contains the coordinates of $x + yt + zt^2$ with respect to B_2. Moreover, the matrix A is related to these bases, for $T(1) = 3 + t + t^2$ has coordinates $(3, 1, 1)_{B_2}$ and $T(t) = 1 - 4t + 2t^2$ has coordinates $(1, -4, 2)_{B_2}$. Therefore the two columns in A contain the coordinates with respect to B_2 of $T(1)$ and $T(t)$, the images of the two vectors in B_1.

Using Example 1 as a guide, we associate matrices with a map as follows:

Definition Let $T \in \text{Hom}(\mathscr{V}, \mathscr{W})$ and suppose $B_1 = \{V_1, \ldots, V_n\}$ and $B_2 = \{W_1, \ldots, W_m\}$ are bases for $\mathscr{V}$ and $\mathscr{W}$, respectively. If $T(V_j)\colon (a_{1j}, a_{2j}, \ldots, a_{mj})_{B_2}$ for $1 \leq j \leq n$, then the $m \times n$ matrix (a_{ij}) is the *matrix of T with respect to the bases B_1 and B_2.*

Therefore A is the matrix of T with respect to B_1 and B_2 if the jth column of A contains the coordinates with respect to B_2 for the image of the jth vector in B_1. Since a linear map T is determined by its action on any basis, the matrix for T with respect to B_1 and B_2 contains enough information, in coordinate form, to determine the image of any vector under T.

Example 2 Suppose $T \in \text{Hom}(\mathscr{V}_2, \mathscr{V}_3)$ is given by $T(a, b) = (2a + b, 3a + b, a + 2b)$. Find the matrix A of T with respect to the basis $B_1 = \{(4, 1), (1, 0)\}$ in the domain and the basis $B_2 = \{(1, 1, 1), (1, 1, 0), (1, 0, 0)\}$ in the codomain.

Since

$$T(4, 1) = (9, 13, 6)\colon (6, 7, -4)_{B_2}$$

and

$$T(1, 0) = (2, 3, 1)\colon (1, 2, -1)_{B_2},$$

the matrix A is

$$A = \begin{pmatrix} 6 & 1 \\ 7 & 2 \\ -4 & -1 \end{pmatrix}.$$

Following Example 1, we should expect that if $V\colon (a, b)_{B_1}$ and

$$A\begin{pmatrix} a \\ b \end{pmatrix} = \begin{pmatrix} x \\ y \\ z \end{pmatrix},$$

then $T(V)\colon (x, y, z)_{B_2}$. To check this, choose $V = (4, -6)$ arbitrarily. Then $V = (4, -6)\colon (-6, 28)_{B_1}$ and $T(V) = (2, 5, -8)\colon (-8, 14, -4)_{B_2}$. In this case we do have

$$A\begin{pmatrix} a \\ b \end{pmatrix} = \begin{pmatrix} 6 & 1 \\ 7 & 2 \\ -4 & -1 \end{pmatrix}\begin{pmatrix} -6 \\ 28 \end{pmatrix} = \begin{pmatrix} -8 \\ 14 \\ -4 \end{pmatrix} = \begin{pmatrix} x \\ y \\ z \end{pmatrix}.$$

The proof that coordinates for an image can always be obtained in this way is given by Theorem 4.9.

Example 3 Let T be as in Example 2. Find the matrix A' of T with respect to the standard bases for $\mathscr{V}_2$ and $\mathscr{V}_3$.

$$T(1, 0) = (2, 3, 1)\colon (2, 3, 1)_{\{E_i\}}$$

and

$$T(0, 1) = (1, 1, 2)\colon (1, 1, 2)_{\{E_i\}},$$

therefore

$$A' = \begin{pmatrix} 2 & 1 \\ 3 & 1 \\ 1 & 2 \end{pmatrix}.$$

Examples 2 and 3 show that different bases may give rise to different matrices for a given map. Therefore, one cannot speak of a matrix for a map without reference to a choice of bases. As a further example, consider the two bases $\{(1, 0), (0, 1)\}$ and $\{(3, 1, 1), (1, -4, 2), (5, 1, 2)\}$, the matrix for the map T of Examples 2 and 3 with respect to these two bases is

$$\begin{pmatrix} 1 & 0 \\ 0 & 1 \\ 0 & 0 \end{pmatrix}.$$

Thus by a careful choice of bases, we have obtained a particularly simple matrix for T. We could continue to produce matrices for T in infinite variety by choosing different bases, but not every 3×2 matrix can be a matrix for T. Can you think of a 3×2 matrix which is not the matrix of T with respect to some choice of bases? What about a matrix with a column of zeros?

Theorem 4.9 Let $A = (a_{ij})$ be the matrix of T with respect to B_1 and B_2. If $V\colon (x_1, \ldots, x_n)_{B_1}$ and $T(V)\colon (y_1, \ldots, y_m)_{B_2}$, set $X = (x_1, \ldots, x_n)^T$ and $Y = (y_1, \ldots, y_m)^T$, then $Y = AX$.

Proof Suppose $B_1 = \{V_1, \ldots, V_n\}$ and $B_2 = \{W_1, \ldots, W_m\}$. Then by assumption

$$V = \sum_{j=1}^{n} x_j V_j \quad \text{and} \quad T(V) = \sum_{i=1}^{m} y_i W_i.$$

Further, by the definition of the matrix A,

$$T(V_j) = \sum_{i=1}^{m} a_{ij} W_i.$$

Therefore

$$\begin{aligned} T(V) &= T\left(\sum_{j=1}^{n} x_j V_j\right) = \sum_{j=1}^{n} x_j T(V_j) \\ &= \sum_{j=1}^{n} x_j \left[\sum_{i=1}^{m} a_{ij} W_j\right] = \sum_{i=1}^{m}\left[\sum_{j=1}^{n} a_{ij} x_j\right] W_i \\ &= \left(\sum_{j=1}^{n} a_{1j} x_j\right) W_1 + \left(\sum_{j=1}^{n} a_{2j} x_j\right) W_2 + \cdots + \left(\sum_{j=1}^{n} a_{mj} x_j\right) W_m. \end{aligned}$$

So y_i, the ith coordinate of $T(V)$ with respect to B_2, is $\sum_{j=1}^{n} a_{ij} x_j$. Now

$$\begin{pmatrix} y_1 \\ y_2 \\ \vdots \\ y_m \end{pmatrix} = \begin{pmatrix} \sum_{j=1}^{n} a_{1j} x_j \\ \sum_{j=1}^{n} a_{2j} x_j \\ \vdots \\ \sum_{j=1}^{n} a_{mj} x_j \end{pmatrix} = \begin{pmatrix} a_{11} x_1 + a_{12} x_2 + \cdots + a_{1n} x_n \\ a_{21} x_1 + x_{22} x_2 + \cdots + a_{2n} x_n \\ \vdots \\ a_{m1} x_1 + a_{m2} x_2 + \cdots + a_{mn} x_n \end{pmatrix}$$

$$= \begin{pmatrix} a_{11} & a_{12} & \cdots & a_{1n} \\ a_{21} & a_{22} & \cdots & a_{2n} \\ & & \vdots & \\ a_{m1} & a_{m2} & \cdots & a_{mn} \end{pmatrix} \begin{pmatrix} x_1 \\ x_2 \\ \vdots \\ x_n \end{pmatrix}$$

that is, $Y = AX$.

Loosely, the matrix equation $Y = AX$ states that the product of a matrix A for T times coordinates X for V equals coordinates Y for $T(V)$. Notice the similarity between the matrix equation for a linear transformation, $Y = AX$, and the equation for a linear function from R to R, $y = ax$. We should also observe that the matrix equation $AX = Y$ is in the form of a system of m linear equations in n unknowns, see problem 9 at the end of this section.

Suppose $T \in \text{Hom}(\mathscr{V}_n, \mathscr{V}_m)$ has matrix A with respect to any basis $B = \{V_1, \ldots, V_n\}$ for $\mathscr{V}_n$ and the standard basis for $\mathscr{V}_m$. In this special case, $T(V_j)$ is the transpose of the jth column of A. Then since the rank of T equals the dimension of $\mathscr{I}_T = \mathscr{L}\{T(V_1), \ldots, T(V_n)\}$, the rank of T equals the (column) rank of A.

Example 4 Let $T \in \text{Hom}(\mathscr{V}_4, \mathscr{V}_3)$ be defined by

$$T(a, b, c, d) = (a - b + 2c, 2a - b + d, a - 2c + d).$$

Suppose A is the matrix of T with respect to

$$B = \{(1, 2, 0, 1), (3, 0, 2, 1), (0, -2, 0, 2), (1, 0, -2, 1)\}$$

in the domain and $\{E_i\}$ in the codomain. Show that the rank of A equals the rank of T.

First,

$$\begin{aligned} T(1, 2, 0, 1) &= (-1, 1, 2), \\ T(3, 0, 2, 1) &= (7, 7, 0), \\ T(0, -2, 0, 2) &= (2, 4, 2) \\ T(1, 0, -2, 1) &= (-3, 3, 6). \end{aligned}$$

Therefore

$$A = \begin{pmatrix} -1 & 7 & 2 & -3 \\ 1 & 7 & 4 & 3 \\ 2 & 0 & 2 & 6 \end{pmatrix}.$$

Since A is column-equivalent to

$$A' = \begin{pmatrix} 1 & 0 & 0 & 0 \\ 0 & 1 & 0 & 0 \\ -1 & 1 & 0 & 0 \end{pmatrix},$$

the rank of A is 2. On the other hand, the image space of T is

$$\mathscr{I}_T = \mathscr{L}\{(-1, 1, 2), (7, 7, 7), (2, 4, 2), (-3, 3, 6)\}$$
$$= \mathscr{L}\{(1, 0, -1), (0, 1, 1)\}.$$

So the rank of T is also 2. Notice that $\mathscr{I}_T$ is spanned by the transposed columns of A'. This happens only because A was obtained using the standard basis in the codomain.

In general, if A is a matrix for T, the columns of A do not span $\mathscr{I}_T$, but the ranks of A and T must be equal.

Theorem 4.10 Suppose $T \in \operatorname{Hom}(\mathscr{V}, \mathscr{W})$ has A as its matrix with respect to B_1 and B_2, then rank T = rank A.

Proof If $B_1 = \{V_1, \ldots, V_n\}$, then $\mathscr{I}_T = \mathscr{L}\{T(V_1), \ldots, T(V_n)\}$ and the jth column of A contains the coordinates of $T(V_j)$ with respect to B_2. The columns of A and the spanning vectors of $\mathscr{I}_T$ are related by the isomorphism $L: \mathscr{M}_{m \times 1} \to \mathscr{W}$ where $L((a_1, \ldots, a_m)^T) = W$ if $W: (a_1, \ldots, a_m)_{B_2}$. (The proof that L is an isomorphism is the same as the proof that an m-dimensional vector space is isomorphic to $\mathscr{V}_m$, Theorem 2.14, page 64.) If $A = (a_{ij})$, then $T(V_j):(a_{1j}, \ldots, a_{mj})_{B_2}$. Therefore $L((a_{1j}, \ldots, a_{mj})^T) = T(V_j)$, or L sends the jth column of A to the jth spanning vector of $\mathscr{I}_T$. This means that the restriction of L to the column space of A is a 1–1, onto map from the column space of A to $\mathscr{I}_T$. Therefore dim(column space of A) $= \dim \mathscr{I}_T$ or rank A = rank T.

Example 5 Find the rank of $T: R_3[t] \to R_3[t]$ using a matrix for T, if

$$T(a + bt + ct^2) = 3a + b + 5c + (2a + b + 4c)t + (a + b + 3c)t^2.$$

The matrix of T with respect to the basis $B = \{1, t, t^2\}$ in both the domain and codomain is

$$A = \begin{pmatrix} 3 & 1 & 5 \\ 2 & 1 & 4 \\ 1 & 1 & 3 \end{pmatrix}$$

and

$$A' = \begin{pmatrix} 1 & 0 & 0 \\ 0 & 1 & 0 \\ -1 & 2 & 0 \end{pmatrix}$$

is column-equivalent to A. Hence the rank of A is 2 and rank $T = 2$. Since $\dim \mathscr{I}_T \leq \dim R_3[t]$, T is neither 1–1 nor onto.

The matrix A' can be used to find the image space for T. Since elementary column operations do not change the column space of a matrix, the columns of A' span the column space of A and are coordinates of spanning vectors for $\mathscr{I}_T$. That is, $\mathscr{I}_T = \mathscr{L}\{U, V\}$ where U: $(1, 0, -1)_B$ and V: $(0, 1, 2)_B$. These coordinates give $U = 1 - t^2$ and $V = t + 2t^2$, so $\mathscr{I}_T = \mathscr{L}\{1 - t^2, t + 2t^2\}$. This basis for $\mathscr{I}_T$ is analogous to the special basis obtained for the image space in Example 4.

The matrix A' is not the matrix of T with respect to $\{1, t, t^2\}$ for both the domain and codomain. But bases could be found which would give A' as the matrix for T. In fact, an even simpler matrix can be obtained.

Example 6 Let T and A be as in Example 5. The matrix

$$A'' = \begin{pmatrix} 1 & 0 & 0 \\ 0 & 1 & 0 \\ 0 & 0 & 0 \end{pmatrix}$$

is equivalent to A. That is, A'' can be obtained from A with elementary row and column operations. Find bases B_1 and B_2 for $R_3[t]$ with respect to which A'' is the matrix of T.

Since the third column of A'' has all zeros, the third vector in B_1 must be in $\mathscr{N}_T = \mathscr{L}\{1 + 2t - t^2\}$. Therefore take $1 + 2t + t^2$ as the third vector of B_1. Then the first two vectors in B_1 may be chosen arbitrarily provided the set obtained is linearly independent. Say we take $B_1 = \{t, t^2, 1 + 2t - t^2\}$. Now since $T(t) = 1 + t + t^2$ and $T(t^2) = 5 + 4t + 3t^2$, these must be taken as the first two vectors in B_2. (Why do these two images have to be independent? T is not one to one.) The third vector in B_2 may be any vector in $R_3[t]$ so long as B_2 is linearly independent. So let $B_2 = \{1 + t + t^2, 5 + 4t + 3t^2, 7\}$. Then A'' is the matrix of T with respect to B_1 and B_2.

In order to obtain a simple matrix for a map, such as A'' in Example 6, complicated bases may be needed, while the simple basis $\{1, t, t^2\}$ used in Example 5 produced a 3×3 matrix with no special form for its entries. In some situations a simple matrix is needed while in others the bases should be simple. But in general, it is not possible to have both.

We have been considering various matrices for a particular linear transformation. Suppose instead we fix a choice for the bases B_1 and B_2, and consider all maps in Hom($\mathscr{V}$, $\mathscr{W}$). If dim $\mathscr{V} = n$ and dim $\mathscr{W} = m$, then each map $T \in$ Hom($\mathscr{V}$, $\mathscr{W}$) determines an $m \times n$ matrix A_T which is the matrix of T with respect to B_1 and B_2. Thus there is a map φ defined from Hom($\mathscr{V}$, $\mathscr{W}$) to $\mathscr{M}_{m \times n}$ which sends T to the matrix A_T.

Example 7 Consider the map φ:Hom($\mathscr{V}_2$) $\rightarrow \mathscr{M}_{2\times 2}$ obtained using $B_1 = \{(2, 4), (5, -3)\}$ and $B_2 = \{(1, 0), (3, 1)\}$. Then for $T \in$ Hom($\mathscr{V}_2$), $\varphi(T)$ is the matrix of T with respect to B_1 and B_2.

Suppose $T_1, T_2, T_3, T_4 \in$ Hom($\mathscr{V}_2$) are given by

$$T_1(a, b) = (a, b), \qquad T_2(a, b) = (2a - b, a + b),$$
$$T_3(a, b) = (2a, b), \qquad T_4(a, b) = (-b, a).$$

Then

$$\varphi(T_1) = \begin{pmatrix} -10 & 14 \\ 4 & -3 \end{pmatrix}, \qquad \varphi(T_2) = \begin{pmatrix} -18 & 7 \\ 6 & 2 \end{pmatrix},$$
$$\varphi(T_3) = \begin{pmatrix} -8 & 19 \\ 4 & -3 \end{pmatrix}, \qquad \varphi(T_4) = \begin{pmatrix} -10 & -12 \\ 2 & 5 \end{pmatrix}.$$

Notice that $T_2 = T_3 + T_4$ and

$$\begin{pmatrix} -18 & 7 \\ 6 & 2 \end{pmatrix} = \begin{pmatrix} -8 & 19 \\ 4 & -3 \end{pmatrix} + \begin{pmatrix} -10 & -12 \\ 2 & 5 \end{pmatrix}.$$

That is,

$$\varphi(T_2) = \varphi(T_3 + T_4) = \varphi(T_3) + \varphi(T_4),$$

suggesting that φ preserves addition. Further the map $4T_1$ given by $[4T_1](a, b) = (4a, 4b)$ has

$$\begin{pmatrix} -40 & 56 \\ 16 & -12 \end{pmatrix}$$

as its matrix with respect to B_1 and B_2. So $\varphi(4T_1) = 4[\varphi(T_1)]$, suggesting that φ is a linear transformation.

The map φ is essentially the correspondence that sends maps from $\mathrm{Hom}(\mathscr{V}, \mathscr{W})$ to their coordinates with respect to a basis $\{T_{ij}\}$. The only difference is that instead of writing the coordinates in a row, they are broken up into the columns of a matrix. From this point of view, the correspondance φ should be an isomorphism.

Theorem 4.11 If $\dim \mathscr{V} = n$ and $\dim \mathscr{W} = m$, then $\mathrm{Hom}(\mathscr{V}, \mathscr{W})$ is isomorphic to $\mathscr{M}_{m \times n}$.

Proof Suppose B_1 and B_2 are bases for $\mathscr{V}$ and $\mathscr{W}$, respectively. For each $T \in \mathrm{Hom}(\mathscr{V}, \mathscr{W})$, let $\varphi(T)$ be the matrix of T with respect to B_1 and B_2. It must be shown that φ: $\mathrm{Hom}(\mathscr{V}, \mathscr{W}) \to \mathscr{M}_{m \times n}$ is an isomorphism.

The proof that φ is linear is straightfoward and left to the reader.

To show that φ is 1–1, suppose $T \in \mathscr{N}_\varphi$, then

$$\varphi(T) = \mathbf{0}_{\mathscr{M}_{m \times n}}.$$

That is, the matrix of T with respect to B_1 and B_2 is the zero matrix. Hence the matrix equation for T is $Y = \mathbf{0}X$. Since every image under T has coordinates $Y = \mathbf{0}_{\mathscr{M}_{m \times 1}}$, T is the zero map $\mathbf{0}_{\mathrm{Hom}(\mathscr{V},\mathscr{W})}$. This means that the null space of φ contains only the zero vector, so T is 1–1.

To prove that φ is onto, let $\{T_{ij}\}$ be the basis for $\mathrm{Hom}(\mathscr{V}, \mathscr{W})$ constructed using the bases B_1 and B_2. Then $\varphi(T_{ij})$ is the $m \times n$ matrix with 1 in the ith column and jth row, and zeros everywhere else. The set of all such $m \times n$ matrices spans $\mathscr{M}_{m \times n}$, so φ maps $\mathrm{Hom}(\mathscr{V}, \mathscr{W})$ onto $\mathscr{M}_{m \times n}$. Thus φ is an isomorphism and $\mathrm{Hom}(\mathscr{V}, \mathscr{W}) \cong \mathscr{M}_{n \times m}$.

Theorem 4.11 is an important result for it tells us that the study of linear transformations from one finite-dimensional vector space to another may be carried out in a space of matrices, and conversely. Computations with maps are often handled most easily with matrices; while several theoretical results for matrices are easily derived with linear maps.

Problems

1. Find the matrices of the following maps with respect to the standard bases for the n-tuple spaces.
   a. $T(a, b) = (a, b)$.
   b. $T(a, b) = (0, a)$.
   c. $T(a, b, c) = (0, c, 0)$.
   d. $T(a, b) = (2a - b, 3a + 4b, a + b)$.

2. Use the bases $B_1 = \{1, t\}$ and $B_2 = \{1, t, t^2\}$ to find matrices for the following linear transformations from $\text{Hom}(R_2[t], R_3[t])$.

a. $T(a + bt) = a + bt$. b. $T(a + bt) = (a + b)t^2$.

c. $T(a + bt) = 0$. d. $T(a + bt) = 3a - b + (4b - 3a)t + (7a + 5b)t^2$.

3. Find the matrix of $T \in \text{Hom}(\mathscr{V}_2, \mathscr{V}_3)$ with respect to the bases $\{(7, 3), (5, 2)\}$ and $\{(1, 3, 1), (1, 4, -2), (2, 6, 3)\}$ when

a. $T(a, b) = (0, 0, 0)$.

b. $T(a, b) = (5b - 2a, 20b - 8a, 4a - 10b)$.

c. $T(a, b) = (a - 2b, 0, b)$.

4. Find the matrix of $T(a, b, c) = (a + 3b, b - 2c, a + 6c)$ with respect to B_1 and B_2 when;

a. $B_1 = B_2 = \{E_i\}$.

b. $B_1 = \{(4, 1, -2), (3, -2, 1), (1, 1, 0)\}$, $B_2 = \{E_i\}$.

c. $B_1 = \{(4, 1, -2). (3, -2, 1), (1, 1, 0)\}$,
$B_2 = \{(7, 5, -8), (-3, -4, 9), (1, 2, 3)\}$.

d. $B_1 = \{(2, 1, 0), (4, 0, 1), (-12, 4, 2)\}$,
$B_2 = \{(5, 1, 2), (4, -2, 10), (1, 0, 1)\}$.

e. $B_1 = \{(1, 0, 0), (0, 1, 0), (6, -2, -1)\}$,
$B_2 = \{(1, 0, 1), (3, 1, 0), (8, 4, 2)\}$.

5. Let A be the matrix of $T(a, b, c) = (b - 3a + 2c, 4c - 3a - b, a - c)$ with respect to the standard basis for $\mathscr{V}_3$. Use the matrix formula $Y = AX$ to find:
a. $T(0, 1, 0)$. b. $T(2, -1, 3)$. c. $T(2, 2, 2)$.

6. Let A be the matrix for the map $T(a, b) = (3a - b, b - 4a)$ with respect to the basis $\{(1, 1), (1, 0)\}$ for $\mathscr{V}_2$ (in both the domain and codomain). Use the matrix formula $Y = AX$ to find: a. $T(1, 1)$. b. $T(4, 0)$. c. $T(3, 2)$. d. $T(0, 1)$.

7. Let A be the matrix of $T(a + bt + ct^2) = 2a - c + (3a + b)t + (c - a - 3b)t^2$ with respect to $\{1, t, t^2\}$. Use $Y = AX$ to obtain: a. $T(t)$. b. $T(1 + t^2)$. c. $T(3t - 4)$. d. $T(0)$.

8. Suppose A is the matrix of T with respect to the basis $B_1 = \{V_1, \ldots, V_n\}$ in the domain and $B_2 = \{W_1, \ldots, W_m\}$. What does it mean if:

a. The 3rd column of A has all zeros?

b. The 4th column of A has all zeros except for a 7 in the 2nd row?

c. The 1st column of A contains only two nonzero entries?

d. All nonzero entries of A occur in the first and last rows?

e. $A = (a_{ij})$ and $a_{ij} = 0$ unless $i = j$?

9. Suppose A as the matrix of a map T with respect to B_1 and B_2. View the matrix equation $Y = AX$ as a system of linear equations in the coordinates of vectors. What can be infered about T if:

a. $AX = Y$ is consistent for all Y?

b. $AX = Y$ has a unique solution if consistent?

c. $AX = Y$ is not always consistent?
d. There is at least one parameter in the solution for $AX = Y$ for any Y?

10. The matrix

$$A' = \begin{pmatrix} 1 & 0 \\ 0 & 1 \end{pmatrix}$$

is equivalent to the matrix A in problem 6. Find bases B_1 and B_2 for $\mathscr{V}_2$ with respect to which A' is the matrix for T.

11. The matrix

$$A' = \begin{pmatrix} 1 & 0 & 0 \\ 0 & 1 & 0 \\ 0 & 0 & 0 \end{pmatrix}$$

is equivalent to the matrix A in problem 5. Find bases B_1 and B_2 for $\mathscr{V}_3$ with respect to which T has matrix A'.

12. Let T, A, and A' be as in Example 3. Therefore, A' is column-equivalent to A. Find bases B_1 and B_2 for $\mathscr{V}_4$ and $\mathscr{V}_3$, respectively, so that A' is the matrix of T with respect to B_1 and B_2.

13. Let $B_1 = \{(2, 1, 1), (-1, 3, 1), (1, 0, 2)\}$ and $B_2 = \{(4, 1), (1, 0)\}$. Suppose $\varphi(T)$ is the matrix of T with respect to B_1 and B_2. Define T_1 and T_2 by $T_1(a, b, c) = (3a - b + c, 3c - 4a)$ and $T_2(a, b, c) = (c - b - 2a, a - 2c)$.
a. Find $\varphi(T_1)$, $\varphi(T_2)$, $\varphi(T_1 + T_2)$, and $\varphi(2T_1 + 3T_2)$.
b. Show that $\varphi(T_1 + T_2) = \varphi(T_1) + \varphi(T_2)$.
c. Show that $\varphi(2T_1 + 3T_2) = 2\varphi(T_1) + 3\varphi(T_2)$.

14. Show that the map φ: $\text{Hom}(\mathscr{V}, \mathscr{W}) \to \mathscr{M}_{m \times n}$ is linear.

15. Suppose T, B_1, and B_2 are as in Example 6, page 153.
a. Find the matrix A of T with respect to B_1 and B_2.
b. How does A compare with the coordinates of T with respect to the basis $\{T_{ij}\}$?

16. Suppose $A = (a_{ij})$ is the matrix of $T \in \text{Hom}(\mathscr{V}_3, \mathscr{V}_2)$ with respect to the bases B_1 and B_2, and $B = \{T_{11}, T_{12}, T_{21}, T_{22}, T_{31}, T_{32}\}$ is the basis for $\text{Hom}(\mathscr{V}_3, \mathscr{V}_2)$ constructed using B_1 and B_2. Write the coordinates of T with respect to B in terms of the entries a_{ij} in A.

§5. Composition of Maps

Suppose S and T are maps such that the image of T is contained in the domain of S. Then S and T can be "composed" to obtain a new map. This operation yields a multiplication for some pairs of maps, and using the association between matrices and maps, it leads directly to a general definition for matrix multiplication.

Definition Let $T: \mathscr{V} \to \mathscr{W}$ and $S: \mathscr{W} \to \mathscr{U}$. The *composition* of S and T is the map $S \circ T: \mathscr{V} \to \mathscr{U}$ defined by $[S \circ T](V) = S(T(V))$ for every vector $V \in \mathscr{V}$.

Example 1 Let $T: \mathscr{V}_3 \to \mathscr{V}_2$ and $S: \mathscr{V}_2 \to \mathscr{V}_3$ be given by

$$T(a, b, c) = (a + b, a - c) \quad \text{and} \quad S(x, y) = (x - y, 2x - y, 3y).$$

Then

$$\begin{aligned}[S \circ T](a, b, c) &= S(T(a, b, c)) = S(a + b, a - c) \\ &= ([a + b] - [a - c], 2[a + b] - [a - c], 3[a - c]) \\ &= (b + c, a + 2b + c, 3a - 3c).\end{aligned}$$

The composition $S \circ T$ could be thought of as sending a vector $V \in \mathscr{V}_3$ to $[S \circ T](V)$ by first sending it to $\mathscr{V}_2$ with T and then sending $T(V)$ to $\mathscr{V}_3$ with S. When $V = (1, 2, 0)$ then $T(1, 2, 0) = (3, 1)$, $S(3, 1) = (2, 5, 3)$, and $[S \circ T](1, 2, 0) = (2, 5, 3)$. This is represented pictorially in Figure 3.

The composition $T \circ S$ is also defined for the maps in Example 1, with

$$[T \circ S](x, y) = T(x - y, 2x - y, 3y) = (3x - 2y, x - 4y).$$

Notice that not only are $T \circ S$ and $S \circ T$ not equal, but they have different domains and codomains, for $T \circ S: \mathscr{V}_2 \to \mathscr{V}_2$ and $S \circ T: \mathscr{V}_3 \to \mathscr{V}_3$.

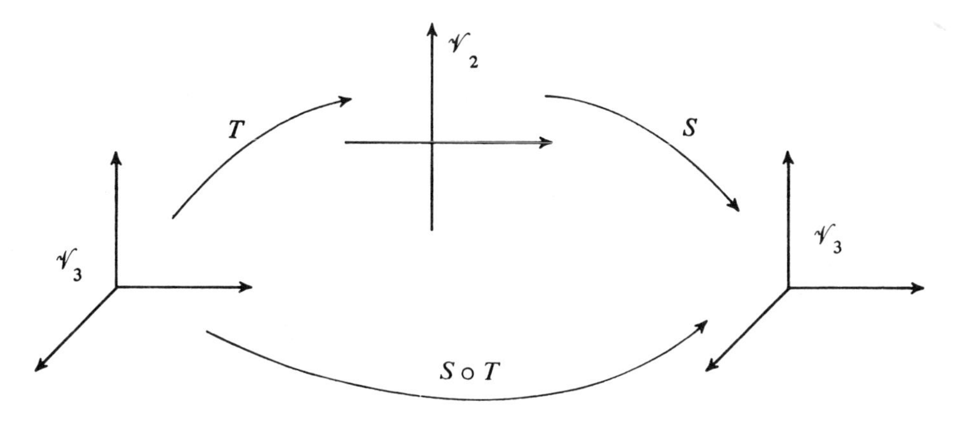

Figure 3

Example 2 Let $T\colon \mathscr{V}_3 \to \mathscr{V}_3$ and $S\colon \mathscr{V}_2 \to \mathscr{V}_3$ be defined by

$$T(a, b, c) = (a - c, 0, b + c) \quad \text{and} \quad S(x, y) = (y, -y, x + y).$$

Then $S \circ T$ is undefined because the codomain of T is $\mathscr{V}_3$ while the domain of S is $\mathscr{V}_2$. However, $T \circ S$ is defined, with $[T \circ S](x, y) = (-x, 0, x)$. Thus it may be possible to compose two maps in only one way.

Composed maps are encountered frequently in calculus. The chain rule for differentiation is introduced in order to differentiate composed functions. For example, the function $y = \sqrt{\sin x}$ is the composition of $z = f(x)$ and $y = g(z)$ where $f(x) = \sin x$ and $g(z) = \sqrt{z}$. For then,

$$y = [g \circ f](x) = g(f(x)) = g(\sin x) = \sqrt{\sin x}.$$

Composition may be viewed as a multiplication of maps. This is in contrast to scalar multiplication, which is only the multiplication of a number times a map. The most obvious characteristic of this multiplication is that it is not defined for all pairs of maps. But when composition is defined, it satisfies some of the properties associated with multiplication of real numbers.

Theorem 4.12

1. The composition of linear maps is linear.
2. Composition of maps is associative; if $T_1 \circ [T_2 \circ T_3]$ and $[T_1 \circ T_2] \circ T_3$ are defined, then

$$T_1 \circ [T_2 \circ T_3] = [T_1 \circ T_2] \circ T_3.$$

3. For linear maps, composition distributes over addition; if $T_1 \circ [T_2 + T_3]$ and $T_1 \circ T_2 + T_1 \circ T_3$ are defined, then

$$T_1 \circ [T_2 + T_3] = T_1 \circ T_2 + T_1 \circ T_3.$$

Similarly

$$[T_1 + T_2] \circ T_3 = T_1 \circ T_3 + T_2 \circ T_3,$$

provided the maps may be composed and added.

Proof 1 and 3 are left to the reader. For 2, suppose V is in the domain of T_3, then

$$(T_1 \circ [T_2 \circ T_3])(V) = T_1([T_2 \circ T_3](V)) = T_1(T_2(T_3(V)))$$
$$= [T_1 \circ T_2](T_3(V)) = ([T_1 \circ T_2] \circ T_3)(V).$$

Since the maps $T_1 \circ [T_2 \circ T_3]$ and $[T_1 \circ T_2] \circ T_3$ have the same effect on V, they are equal, and composition of maps is associative.

If S and T are linear maps from a vector space to itself, the product $S \circ T$ is always defined. Further $S \circ T$: $\mathscr{V} \rightarrow \mathscr{V}$ is linear, so Hom($\mathscr{V}$) is closed under composition. Thus composition defines a multiplication on the space of linear transformations Hom($\mathscr{V}$). This is a special situation for up to this point we have considered only one other multiplicative operation within a vector space, namely, the cross product in $\mathscr{E}_3$.

Consider the algebraic system consisting of Hom($\mathscr{V}$) together with addition and multiplication given by composition, while disregarding scalar multiplication. This algebraic system is not a vector space. It is rather a set of elements, maps, on which two operations, addition and multiplication, are defined, and as such it is similar to the real number system. If we check through the basic properties of the real numbers listed on page 2, we find that addition of maps satisfies all the properties listed for addition in R. Of the properties listed for multiplication in R, we have seen, in Theorem 4.12, that Hom($\mathscr{V}$) is closed under multiplication and that the associative law for multiplication holds in Hom($\mathscr{V}$). Moreover, the distributive laws hold in Hom($\mathscr{V}$). An algebraic system satisfying just these properties of addition and multiplication is called a *ring*. Thus the set of linear maps, Hom($\mathscr{V}$), together with addition and multiplication (composition) form a ring.

It is easy to see that Hom($\mathscr{V}$) does not satisfy all the field properties listed for R.

Example 3 Let $S, T \in$ Hom($\mathscr{V}_2$) be given by

$$S(a, b) = (2a - b, a + 3b) \quad \text{and} \quad T(a, b) = (a + b, 2b - a).$$

Show that $S \circ T \neq T \circ S$.

We have

$$[S \circ T](a, b) = S(T(a, b)) = S(a + b, 2b - a) = (3a, 7b - 2a),$$

and

$$[T \circ S](a, b) = T(S(a, b)) = T(2a - b, a + 3b) = (3a + 2b, 7b).$$

Therefore $S \circ T \neq T \circ S$ and multiplication is not commutative in $\text{Hom}(\mathscr{V}_2)$. You might check to see that the same result is obtained with almost any pair of maps from $\text{Hom}(\mathscr{V}_2)$.

A ring of maps does share one additional property with the real numbers; it has a multiplicative identity. We define the *identity map* $I: \mathscr{V} \to \mathscr{V}$ by $I(V) = V$ for all $V \in \mathscr{V}$. It is then immediate that $I \in \text{Hom}(\mathscr{V})$ and $I \circ T = T \circ I = T$ for any map $T \in \text{Hom}(\mathscr{V})$. Since a multiplicative identity exists, the ring $\text{Hom}(\mathscr{V})$ is called a *ring with identity*.

Since there is an identity for composition, we can look for multiplicative inverses.

Definition Given $T: \mathscr{V} \to \mathscr{W}$. Suppose $S: \mathscr{W} \to \mathscr{V}$ satisfies $S \circ T = I_{\mathscr{V}}$ and $T \circ S = I_{\mathscr{W}}$. Then S is the *inverse* of T. T is said to be *invertible*, and its inverse is denoted by T^{-1}.

Example 4 Suppose $T: \mathscr{V}_2 \to \mathscr{V}_2$ is given by $T(a, b) = (2a + 3b, 3a + 4b)$. Show that T is invertible by finding T^{-1}.

Write $T(a, b) = (x, y)$. If a map S exists such that $S \circ T = I$, then $S(x, y) = (a, b)$. Therefore, S may be obtained by solving

$$2a + 3b = x$$

$$3a + 4b = y$$

for a and b in terms of x and y. These equations have the unique solution

$$a = -4x + 3y$$

$$b = 3x - 2y.$$

Therefore S should be defined by

$$S(x, y) = (-4x + 3y, 3x - 2y).$$

Using this definition,

$$\begin{aligned}[S \circ T](a, b) &= S(T(a, b)) = S(2a + 3b, 3a + 4b) \\ &= (-4(2a + 3b) + 3(3a + 4b), 3(2a + 3b) - 2(3a + 4b)) \\ &= (a, b).\end{aligned}$$

That is, $S \circ T = I$ as desired. Since composition is not commutative, it is necessary to check that $T \circ S = I$ also, in order to show that S is T^{-1}.

Example 5 Define $T \in \text{Hom}(R_3[t], \mathscr{V}_3)$ by

$$T(a + bt + ct^2) = (a + b + c, 2a + b + 2c, a + b).$$

Given that T is invertible, find T^{-1}.

If

$$T(a + bt + ct^2) = (x, y, z),$$

then

$$T^{-1}(x, y, z) = a + bt + ct^2.$$

That is, a, b, and c must be expressed in terms of x, y, and z. The system

$$a + b + c = x, \qquad 2a + b + 2c = y, \qquad a + b = z,$$

has the solution

$$a = -2x + y + z, \qquad b = 2x - y, \qquad c = x - z$$

for any x, y, and z. Therefore, T^{-1} should be defined by

$$T^{-1}(x, y, z) = -2x + y + z + (2x - y)t + (x - z)t^2.$$

It must now be shown that $T^{-1} \circ T = I_{R_3[t]}$ and $T \circ T^{-1} = I_{\mathscr{V}_3}$. This is left to the reader.

In the real number system, every nonzero number has a multiplicative inverse. However, a similar statement cannot be made for the nonzero maps in $\text{Hom}(\mathscr{V}_2)$.

Example 6 Suppose $T \in \text{Hom}(\mathscr{V}_2)$ is defined by $T(a, b) = (a, 0)$. Show that even through $T \neq \mathbf{0}$, T has no inverse.

Assume a map $S \in \text{Hom}(\mathscr{V}_2)$ exists such that $S \circ T = I$. Then $S(T(V)) = V$ for all $V \in \mathscr{V}_2$. But $\mathscr{N}_T \neq \{\mathbf{0}\}$; for example, $T(0, 1) = (0, 0) = T(0, 2)$. Therefore $S(0, 0)$ must be $(0, 1)$ on the one hand and $(0, 2)$ on the other. Since this violates the definition of a map, S cannot exist. In this case, S fails to exist because T is not one to one.

Suppose we had asked that $T \circ S = I$. Then $T(S(W)) = W$ for all $W \in \mathscr{V}_2$. But now $\mathscr{I}_T \neq \mathscr{V}_2$; in particular, $(0, 1) \notin \mathscr{I}_T$. So if W is taken to be $(0, 1)$, then W has no preimage under T. Therefore no matter how $S(0, 1)$ is defined, $T(S(0, 1)) \neq (0, 1)$, and this map S cannot exist either. Here S fails to exist because T is not onto.

Before generalizing the implications of this example, in Theorem 4.14, a few facts concerning inverses should be stated. They are not difficult to prove, so their proofs are left to the reader.

Theorem 4.13

1. If T is invertible, then T has a unique inverse.
2. If T is linear and invertible, then T^{-1} is linear and invertible with $(T^{-1})^{-1} = T$.
3. If S and T are invertible and $S \circ T$ is defined, then $S \circ T$ is invertible with $(S \circ T)^{-1} = T^{-1} \circ S^{-1}$.

Notice that the inverse of a product is the product of the inverses in the reverse order. This change in order is necessary because composition is not commutative. In fact, even if the composition $(S^{-1} \circ T^{-1}) \circ (S \circ T)$ is defined, it need not be the identity map.

Theorem 4.14 A linear map $T: \mathscr{V} \to \mathscr{W}$ is invertible if an only if it is one to one and onto.

Proof ($\Rightarrow$) Assume T^{-1} exists. To see that T is 1–1, suppose $V \in \mathscr{N}_T$. Then since T^{-1} is linear, $T^{-1}(T(V)) = T^{-1}(\mathbf{0}) = \mathbf{0}$. But

$$T^{-1}(T(V)) = [T^{-1} \circ T](V) = I(V) = V.$$

Therefore $V = \mathbf{0}$ and $\mathscr{N}_T = \{\mathbf{0}\}$, so T is 1–1.

To see that T is onto, suppose $W \in \mathscr{W}$. Since T^{-1} is defined on $\mathscr{W}$, there is a vector $U \in \mathscr{V}$, such that $T^{-1}(W) = U$. But then

$$T(U) = T(T^{-1}(W)) = [T \circ T^{-1}](W) = I(W) = W.$$

Thus $\mathscr{W} \subset \mathscr{I}_T$ and T is onto.

($\Leftarrow$) Assume T is 1–1 and onto. For each $W \in \mathscr{W}$, there exists a vector $V \in \mathscr{V}$, such that $T(V) = W$, because T maps $\mathscr{V}$ onto $\mathscr{W}$. Further V is uniquely determined by W since T is 1–1. Therefore we may define $S: \mathscr{W} \to \mathscr{V}$ by $S(W) = V$ where $T(V) = W$. Now if $X \in \mathscr{V}$, $[S \circ T](X) = S(T(X)) = X$

and if $Y \in \mathscr{W}$, then $[T \circ S](Y) = T(S(Y)) = Y$. Thus S is T^{-1} and T is invertible.

Theorem 4.14 shows that a linear map is invertible if and only if it is an isomorphism. However, the essential requirement is that the map be one to one, for if $T: \mathscr{V} \to \mathscr{W}$ is one to one but not onto, then $T: \mathscr{V} \to \mathscr{I}_T$ is both one to one and onto. So an invertible map may be obtained without changing how T is defined on $\mathscr{V}$. On the other hand, if T is not 1–1, then there is no such simple change which will correct the defect. Since being 1–1 is essential to inverting a map, a linear transformation is said to be *nonsingular* if it is one to one; otherwise it is *singular*. A singular map cannot be inverted even if its codomain is redefined, for it collapses a subspace of dimension greater than zero to the single vector $\mathbf{0}$.

The property of being invertible is not equivalent to being nonsingular or 1–1. However, in the improtant case of a map from a finite-dimensional vector space to itself, the two conditions are identical.

Theorem 4.15 If $\mathscr{V}$ is finite dimensional and $T \in \text{Hom}(\mathscr{V})$, then the following conditions are equivalent:

1. T is invertible.
2. T is 1–1. (Nullity $T = 0$ or $\mathscr{N}_T = \{\mathbf{0}\}$.)
3. T is onto. (Rank $T = \dim \mathscr{V}$.)
4. If A is a matrix for T with respect to some choice of bases, then $\det A \neq 0$.
5. T is nonsingular.

Proof These follow at once from Theorem 4.14 together with the two facts:

$$\dim \mathscr{V} = \text{rank } T + \text{nullity } T \qquad \text{and} \qquad \text{rank } T = \text{rank } A.$$

We can see how the conditions in Theorem 4.15 are related by considering an arbitrary map $T \in \text{Hom}(\mathscr{V}_2)$. From problem 11, page 145, we know that $T(a, b) = (ra + sb, ta + ub)$ for some scalars r, s, t, u. If we write $T(a, b) = (x, y)$, then T^{-1} exists provided the equations

$$ra + sb = x$$

$$ta + ub = y$$

can be solved for a and b given any values for x and y. (This was done in Example 5.) From our work with systems of linear equations, we know that such a solution exists if and only if the coefficient matrix

$$A = \begin{pmatrix} r & s \\ t & u \end{pmatrix}$$

has rank 2. But A is the matrix of T with respect to the standard basis for $\mathscr{V}_2$. So T^{-1} exists if and only if det $A \neq 0$ or rank $T = 2$. Further, the rank of A is 2 if and only if the equations have a unique solution for every choice of x and y. That is, if and only if T is 1–1 and onto.

Returning to the ring of maps Hom($\mathscr{V}$), Examples 3 and 5 can be generalized to show that if the dimension of $\mathscr{V}$ is at least 2, then multiplication in Hom($\mathscr{V}$) is not commutative and nonzero elements need not have multiplicative inverses. So the ring Hom($\mathscr{V}$) differs from the field of real numbers in two important respects. Suppose we briefly consider the set of nonsingular maps in Hom($\mathscr{V}$) together with the single operation of composition. This algebraic system is similar to the set of all nonzero real numbers together with the single operation of multiplication. The only difference is that composition of maps is not commutative. That is, for the set of all nonsingular maps in Hom($\mathscr{V}$) together with composition (1) the set is closed under composition; (2) composition is associative; (3) there is an identity element; and (4) every element has an inverse under composition. Such an algebraic system with one operation, satisfying these four properties is called a *group*. The nonzero real numbers together with multiplication also form a group. However, since multiplication of numbers is commutative, this system is called a *commutative group*. Still another commutative group is formed by the set of all real numbers together with the single operation of addition.

A general study of groups, rings, and fields is usually called modern algebra. We will not need to make a general study of such systems, but in time we will have occasion to examine general fields, rings of polynomials, and groups of permutations.

Problems

1. Find $S \circ T$ and/or $T \circ S$ when

a. $T(a, b) = (2a - b, 3b)$, $S(x, y) = (y - x, 4x)$.

b. $T(a, b, c) = (a - b + c, a + c)$, $S(x, y) = (x, x - y, y)$.

c. $T(a, b) = (a, 3b + a, 2a - 4b, b)$, $S(a, b, c) = (2a, b)$.

d. $T(a + bt) = 2a - bt^2$, $S(a + bt + ct^2) = c + bt^2 - 3at^3$.

2. For $T \in$ Hom($\mathscr{V}$), T^n is the map obtained by composing T with itself n times. If $\mathscr{V} = \mathscr{V}_2$ and $T(a, b) = (2a, b - a)$, find

a. T^2. b. T^3. c. T^n. d. $T^2 - 2T$. e. $T + 3I$.

3. Show that $T^2 = I$ when $T(a, b) = (a, 3a - b)$.

4. Is T^2 meaningful for any linear map T?

5. Suppose $T \in \text{Hom}(\mathscr{V}, \mathscr{W})$ and $S \in \text{Hom}(\mathscr{W}, \mathscr{U})$. Prove that $S \circ T \in \text{Hom}(\mathscr{V}, \mathscr{U})$. That is, prove that the composition of linear maps is linear.

6. Suppose $T_1(a, b, c) = (a, b + c, a - c)$, $T_2(a, b) = (2a - b, a, b - a)$ and $T_3(a, b) = (0, 3a - b, 2b)$. Show that the maps $T_1 \circ [T_2 + T_3]$ and $T_1 \circ T_2 + T_1 \circ T_3$ are equal.

7. a. Prove that the distributive law in problem 6 holds for any three linear maps for which the expressions are defined.
   b. Why are there two distributive laws given for the ring of maps $\text{Hom}(\mathscr{V})$ when only one was stated for the field of real numbers R?

8. Find all maps $S \in \text{Hom}(\mathscr{V}_2)$ such that $S \circ T = T \circ S$ when
   a. $T(a, b) = (a, 0)$.
   b. $T(a, b) = (2a, b)$.
   c. $T(a, b) = (2a - b, 3a)$.
   d. $T(a, b) = (a + 2b, 3b - 2a)$.

9. Suppose $T \in \text{Hom}(\mathscr{V})$ and let $\mathscr{S} = \{S \in \text{Hom}(\mathscr{V}) | S \circ T = T \circ S\}$.
   a. Prove that $\mathscr{S}$ is a subspace of $\text{Hom}(\mathscr{V})$.
   b. Show that if $\dim \mathscr{V} \geq 1$, then $\dim \mathscr{S} \geq 1$ also.

10. Determine if T is invertible and if so find its inverse.
   a. $T(a, b, c) = (a - 2b + c, 2a + b - c, b - 3a)$.
   b. $T(a, b) = (a - 2b, b - a)$.
   c. $T(a, b, c) = a + c + bt + (-a - b)t^2$.
   d. $T(a + bt) = (2a + b, 3a + b)$.
   e. $T(a + bt) = a + 3at + bt^2$.
   f. $T(a, b, c) = (a - 2c, 2a + b, b + 3c)$.

11. Suppose $T: \mathscr{V} \to \mathscr{W}$ is linear and invertible.
   a. Prove that T has only one inverse.
   b. Prove that T^{-1} is linear.

12. Show by example that the set of all nonsingular maps in $\text{Hom}(\mathscr{V}_2)$ together with $\mathbf{0}$ is not closed under addition. Therefore the set of nonsingular maps together with addition and composition does not form a ring.

13. Find an example of a linear map from a vector space to itself which is
   a. one to one but not onto.
   b. onto but not one to one.

14. Suppose the composition $S \circ T$ is one to one and onto.
   a. Prove that T is one to one.
   b. Prove that S is onto.
   c. Show by example that T need not be onto and S need not be 1–1.

15. Suppose $\mathscr{V}$ is a finite-dimensional vector space and $T \in \text{Hom}(\mathscr{V})$. Prove that if there exists a map $S \in \text{Hom}(\mathscr{V})$ such that $S \circ T = I$, then $T \circ S = I$. That is, if $S \circ T = I$ (or $T \circ S = I$), then T is nonsingular and $S = T^{-1}$.

16. Suppose $S(a, b) = (4a + b, 3a + b)$ and $T(a, b) = (3a + b, 5a + 2b)$.
   a. Find $S \circ T$.
   b. Find S^{-1}, T^{-1}, and $(S \circ T)^{-1}$.
   c. Find $T^{-1} \circ S^{-1}$ and check that it equals $(S \circ T)^{-1}$.
   d. Show that $S^{-1} \circ T^{-1} \circ S \circ T \neq I$. That is, $(S \circ T)^{-1} \neq S^{-1} \circ T^{-1}$.

17. Prove that if S and T are invertible and $S \circ T$ is defined, then $(S \circ T)^{-1} = T^{-1} \circ S^{-1}$.

18. We know from analytic geometry that if T_α and T_β are rotations of the plane through the angles α and β, respectively, then $T_\alpha \circ T_\beta = T_{\alpha+\beta}$. Show that the equation $T_\alpha \circ T_\beta(1, 0) = T_{\alpha+\beta}(1, 0)$ yields the addition formulas for the sine and cosine functions.

§6. Matrix Multiplication

The multiplication of linear transformations, given by composition in conjunction with the correspondence between maps and matrices, yields a general definition for matrix multiplication. The product of matrices should be defined so that the isomorphisms between spaces of maps and spaces of matrices preserve multiplication as well as addition and scalar multiplication. That is, if S and T are maps with matrices A and B with respect to some choice of bases, then $S \circ T$ should have the matrix product AB as its matrix with respect to the same choice of bases. An example will lead us to the proper definition.

Example 1 Let $S, T \in \text{Hom}(\mathscr{V}_2)$ be given by

$$S(x, y) = (a_{11}x + a_{12}y, a_{21}x + a_{22}y),$$
$$T(x, y) = (b_{11}x + b_{12}y, b_{21}x + b_{22}y),$$

then

$$A_S = \begin{pmatrix} a_{11} & a_{12} \\ a_{21} & a_{22} \end{pmatrix} \quad \text{and} \quad A_T = \begin{pmatrix} b_{11} & b_{12} \\ b_{21} & b_{22} \end{pmatrix}$$

are the matrices of S and T with respect to the standard basis for $\mathscr{V}_2$. And since

$$S \circ T(x, y) = ([a_{11}b_{11} + a_{12}b_{21}]x + [a_{11}b_{12} + a_{12}b_{22}]y,$$
$$[a_{21}b_{11} + a_{22}b_{21}]x + [a_{21}b_{12} + a_{22}b_{22}]y),$$

the matrix of $S \circ T$ with respect to $\{E_i\}$ is

$$A_{S \circ T} = \begin{pmatrix} a_{11}b_{11} + a_{12}b_{21} & a_{11}b_{12} + a_{12}b_{22} \\ a_{21}b_{11} + a_{22}b_{21} & a_{21}b_{12} + a_{22}b_{22} \end{pmatrix}$$

Therefore the product of A_S and A_T should be defined so that

$$\begin{pmatrix} a_{11} & a_{12} \\ a_{21} & a_{22} \end{pmatrix}\begin{pmatrix} b_{11} & b_{12} \\ b_{21} & b_{22} \end{pmatrix} = \begin{pmatrix} a_{11}b_{11} + a_{12}b_{21} & a_{11}b_{12} + a_{12}b_{22} \\ a_{21}b_{11} + a_{22}b_{21} & a_{21}b_{12} + a_{22}b_{22} \end{pmatrix}.$$

Notice that the product should be defined with the same row times column rule used to represent systems of linear equations in matrix form. For the first column of $A_{S \circ T}$ is the matrix product

$$\begin{pmatrix} a_{11} & a_{12} \\ a_{21} & a_{22} \end{pmatrix}\begin{pmatrix} b_{11} \\ b_{21} \end{pmatrix}$$

and the second column is

$$\begin{pmatrix} a_{11} & a_{12} \\ a_{21} & a_{22} \end{pmatrix}\begin{pmatrix} b_{12} \\ b_{22} \end{pmatrix}.$$

In other words, the element in the ith row and jth column of $A_{S \circ T}$ is the sum of the products of the elements in the ith row of A_S times the elements in the jth column of A_T. This "row by column" rule will be used to define multiplication in general. The summation notation is very useful for this. If we set $A_{S \circ T} = (c_{ij})$ above, then $c_{ij} = a_{i1}b_{1j} + a_{i2}b_{2j}$ for each i and j, therefore

$$c_{ij} = \sum_{k=1}^{2} a_{ik}b_{kj}.$$

A sum of this form will be characteristic of an entry in a matrix product. Notice that in the sum $\sum_{i=1}^{m} a_{ik}b_{kj}$, the elements a_{ik} come from the ith row of the first matrix and the elements b_{kj} come from the jth column of the second.

Definition Let $A = (a_{ij})$ be be an $n \times m$ matrix and $B = (b_{ij})$ an $m \times p$ matrix. The *product* AB of A and B is the $n \times p$ matrix (c_{ij}) where

$$c_{ij} = \sum_{k=1}^{m} a_{ik}b_{kj} = a_{i1}b_{1j} + a_{i2}b_{2j} + \cdots + a_{im}b_{mj},$$

$$1 \le i \le n, \qquad 1 \le j \le p.$$

This definition yields $A_S A_T = A_{S \circ T}$ when A_S, A_T, and $A_{S \circ T}$ are as in Example 1. The definition also agrees with our previous definition of matrix product when B is $m \times 1$. Not all matrices can be multiplied, the product AB is defined only if the number of columns in A equals the number of rows in B. What does this correspond to in the composition of maps?

Example 2 Let

$$A = (a_{ij}) = \begin{pmatrix} 2 & 4 \\ 1 & 3 \\ 2 & -1 \end{pmatrix}$$

and

$$B = (b_{ij}) = \begin{pmatrix} 1 & 2 & -1 & 0 \\ -1 & 1 & 2 & -2 \end{pmatrix}.$$

Since A is 3×2 and B is 2×4, AB is defined but BA is not. The product $AB = (c_{ij})$ is a 3×4 matrix with

$$c_{ij} = \sum_{k=1}^{2} a_{ik} b_{kj}.$$

Thus

$$c_{11} = \sum_{k=1}^{2} a_{1k} b_{k1} = 2 \cdot 1 + 4(-1) = -2$$

and

$$c_{34} = \sum_{k=1}^{2} a_{3k} b_{k4} = 2 \cdot 0 + (-1) \cdot (-2) = 2.$$

Computing all the entries c_{ij} gives

$$AB = \begin{pmatrix} -2 & 8 & 6 & -8 \\ -2 & 5 & 5 & -6 \\ 3 & 3 & -4 & 2 \end{pmatrix}.$$

Example 3 Some other examples of matrix products are:

$$\begin{pmatrix} 3 & -1 \\ 2 & -4 \end{pmatrix} \begin{pmatrix} 2 & 2 \\ 1 & 3 \end{pmatrix} = \begin{pmatrix} 5 & 3 \\ 0 & -8 \end{pmatrix}, \quad \begin{pmatrix} 4 \\ 2 \end{pmatrix} (1 \quad 3 \quad 0) = \begin{pmatrix} 4 & 12 & 0 \\ 2 & 6 & 0 \end{pmatrix},$$

$$\begin{pmatrix} 4 & 1 & -2 \\ 1 & 3 & 5 \end{pmatrix}\begin{pmatrix} 2 & -3 \\ -2 & 0 \\ 3 & 1 \end{pmatrix} = \begin{pmatrix} 0 & -14 \\ 11 & 2 \end{pmatrix},$$

$$\begin{pmatrix} 2 & -3 \\ -2 & 0 \\ 3 & 1 \end{pmatrix}\begin{pmatrix} 4 & 1 & -2 \\ 1 & 3 & 5 \end{pmatrix} = \begin{pmatrix} 5 & -7 & -19 \\ -8 & -2 & 4 \\ 13 & 6 & -1 \end{pmatrix}.$$

The complicated row by column definition for matrix multiplication is chosen to correspond to composition of linear transformations. Once this correspondence is established in the next theorem, all the properties relating to composition, such as associativity or invertibility, can be carried over to matrix multiplication.

Theorem 4.16 Let $T\colon \mathscr{V} \to \mathscr{W}$ and $S\colon \mathscr{W} \to \mathscr{U}$ be linear with $\dim \mathscr{V} = p$, $\dim \mathscr{W} = m$, and $\dim \mathscr{U} = n$. Suppose B_1, B_2, B_3 are bases of $\mathscr{V}$, $\mathscr{W}$, $\mathscr{U}$, respectively, and that A_T is the $m \times p$ matrix of T with respect to B_1 and B_2, A_S is the $n \times m$ matrix of S with respect to B_2 and B_3, and $A_{S \circ T}$ is the $n \times p$ matrix of $S \circ T$ with respect to B_1 and B_3.

Then $A_S A_T = A_{S \circ T}$.

Proof The proof is of necessity a little involved, but it only uses the definitions for a matrix of a map and matrix multiplication.

Let

$$B_1 = \{V_1, \ldots, V_p\}, \qquad B_2 = \{W_1, \ldots, W_m\}, \qquad B_3 = \{U_1, \ldots, U_n\}.$$

If $A_S = (a_{hk})$, then

$$S(W_k) = \sum_{h=1}^{n} a_{hk} U_h \qquad \text{for} \quad 1 \le k \le m.$$

If $A_T = (b_{kj})$, then

$$T(V_j) = \sum_{k=1}^{m} b_{kj} W_k \qquad \text{for} \quad 1 \le j \le p.$$

And if $A_{S \circ T} = (c_{hj})$, then

$$S \circ T(V_j) = \sum_{h=1}^{n} c_{hj} U_h \qquad \text{for} \quad 1 \le j \le p.$$

The proof consists in finding a second expression for $S \circ T(V_j)$.

$$S \circ T(V_j) = S(T(V_j)) = S\left(\sum_{k=1}^{m} b_{kj} W_k\right) = \sum_{k=1}^{m} b_{kj} S(W_k)$$

$$= \sum_{k=1}^{m} b_{kj}\left[\sum_{h=1}^{n} a_{hk} U_h\right] = \sum_{h=1}^{n}\left[\sum_{k=1}^{m} a_{hk} b_{kj}\right] U_h.$$

The last equality is obtained by rearranging the nm terms in the previous sum. Now the two expressions for $S \circ T(V_j)$ give

$$\sum_{h=1}^{n} c_{hj} U_h = \sum_{h=1}^{n} \sum_{k=1}^{m} a_{hk} b_{kj} U_h.$$

But the U's are linearly independent, so $c_{hj} = \sum_{k=1}^{m} a_{hk} b_{kj}$ for each h and j. That is, $(a_{hk})(b_{kj}) = (c_{hj})$ or $A_S A_T = A_{S \circ T}$

If this proof appears mysterious, try showing that it works for the spaces and maps in Example 1, or you might consider the case $\mathscr{V} = \mathscr{V}_2$, $\mathscr{W} = \mathscr{V}_3$, and $\mathscr{U} = \mathscr{V}_2$.

For all positive integers n and m, let φ: $\text{Hom}(\mathscr{V}_m, \mathscr{V}_n) \to \mathscr{M}_{n \times m}$ with $\varphi(T)$ the matrix of T with respect to the standard bases. For any particular n and m we know that φ is an isomorphism, and with the above theorem, φ preserves multiplication where it is defined. That is, if $T \in \text{Hom}(V_p, \mathscr{V}_m)$ and $S \in \text{Hom}(\mathscr{V}_m, \mathscr{V}_n)$, then $\varphi(S \circ T) = \varphi(S)\varphi(T)$. Thus φ can be used to carry the properties for composition of maps over to multiplication of matrices.

Theorem 4.17 Matrix multiplication is associative and distributive over addition.

That is,

$$A(BC) = (AB)C, \qquad A(B + C) = AB + AC, \qquad (A + B)C = AC + BC$$

hold for all matrices A, B, and C for which the expressions are defined.

Proof Consider one of the distributive laws. Suppose

$$A \in \mathscr{M}_{n \times m}, \qquad B,\ C \in \mathscr{M}_{m \times p}.$$

Then there exist maps $T \in \text{Hom}(\mathscr{V}_m, \mathscr{V}_n)$ and S, $U \in \text{Hom}(\mathscr{V}_p, \mathscr{V}_m)$ such that $\varphi(T) = A$, $\varphi(S) = B$, and $\varphi(U) = C$. Using these maps,

$$\begin{aligned} A(B + C) &= \varphi(T)[\varphi(S) + \varphi(U)] \\ &= \varphi(T)\varphi(S + U) && \varphi \text{ preserves addition} \\ &= \varphi(T \circ [S + U]) && \varphi \text{ preserves multiplication} \\ &= \varphi(T \circ S + T \circ U) && \text{Distributive law for maps} \\ &= \varphi(T \circ S) + \varphi(T \circ U) && \varphi \text{ preserves addition} \\ &= \varphi(T)\varphi(S) + \varphi(T)\varphi(U) && \varphi \text{ preserves multiplication} \\ &= AB + AC. \end{aligned}$$

The map φ carries the algebraic properties of $\text{Hom}(\mathscr{V}_n)$ to $\mathscr{M}_{n \times n}$, making $\mathscr{M}_{n \times n}$ a ring with identity. Thus φ: $\text{Hom}(\mathscr{V}_n) \rightarrow \mathscr{M}_{n \times n}$ is a (ring) homomorphism. The identity in $\mathscr{M}_{n \times n}$ is $\varphi(I)$, the $n \times n$ matrix with 1's on the main diagonal and 0's elsewhere. This matrix is denoted by I_n and is called the *$n \times n$ identity matrix*. If A is an $n \times n$ matrix, then $AI_n = I_nA = A$. If B is an $n \times m$ matrix with $n \neq m$, then $I_nB = B$ and the product of B times I_n is not defined, but $BI_m = B$.

Definition Let A be an $n \times n$ matrix and suppose there exists an $n \times n$ matrix B such that $AB = BA = I_n$. Then B is the *inverse* of A, written A^{-1}, and A is said to be *nonsingular*. Otherwise A is *singular*.

Theorem 4.18 Let $T \in \text{Hom}(\mathscr{V}_n)$ and $A \in \mathscr{M}_{n \times n}$ be the matrix of T with respect to the standard basis for $\mathscr{V}_n$, i.e., $A = \varphi(T)$. Then A is nonsingular if and only if T is nonsingular.

Corollary 4.19 An $n \times n$ matrix A is nonsingular if an only if $|A| \neq 0$.

These two facts are easily proved and left as problems.

Example 4 Show that

$$A = \begin{pmatrix} 2 & 1 \\ 5 & 3 \end{pmatrix}$$

is nonsingular and find A^{-1}.

We have that $|A| = 1$, so A is nonsingular and there exist scalars a, b, c, d such that

$$\begin{pmatrix} 2 & 1 \\ 5 & 3 \end{pmatrix}\begin{pmatrix} a & b \\ c & d \end{pmatrix} = I_2 = \begin{pmatrix} 1 & 0 \\ 0 & 1 \end{pmatrix}.$$

This matrix equation yields four linear equations which have the unique solution $a = 3$, $b = -1$, $c = -5$, and $d = 2$. One can verify that if

$$B = \begin{pmatrix} 3 & -1 \\ -5 & 2 \end{pmatrix},$$

then $AB = BA = I_2$ and $B = A^{-1}$. This is not a good way to find the inverse of any matrix. In particular, we will see that the inverse of a 2×2 matrix A can be written down directly from A.

The definition of A^{-1} requires that $AB = I_n$ and $BA = I_n$, but it is not necessary to check both products. Problem 15 on page 175 yields the following result for matrices.

Theorem 4.20 Let $A, B \in \mathcal{M}_{n \times n}$. If $AB = I_n$ or $BA = I_n$, then A is nonsingular and $B = A^{-1}$.

The amount of computation needed to invert an $n \times n$ nonsingular matrix increases an n increases, and although there is generally no quick way to find an inverse, there are two general techniques that have several useful consequences. The first technique results from the following fact.

Theorem 4.21 Suppose $A = (a_{ij})$ is an $n \times n$ matrix with A_{ij} the cofactor of a_{ij}. If $h \neq k$, then

$$\sum_{j=1}^{n} a_{hj}A_{kj} = 0 \qquad \text{and} \qquad \sum_{i=1}^{n} a_{ih}A_{ik} = 0.$$

(The first equation shows that the sum of the elements from the hth row times the cofactors from the kth row is zero, if $h \neq k$. Recall that if $h = k$, then $\sum_{j=1}^{n} a_{hj}A_{hj} = \det A$.)

Proof The proof consists in showing that $\sum_{j=1}^{n} a_{hj}A_{kj}$ is the expansion of a determinant with two equal rows.

Consider the matrix $B = (b_{ij})$ obtained from A by replacing its kth row with its hth row. (This is not an elementary row operation.) Then $b_{ij} = a_{ij}$ when $i \neq k$ and $b_{kj} = a_{hj}$, for all j. Since B and A differ only in their kth rows, the cofactors of corresponding elements from their kth rows are equal. That is, $B_{kj} = A_{kj}$, for $1 \leq j \leq n$. Therefore, expanding $|B|$ along the kth row yields,

$$|B| = \sum_{j=1}^{n} b_{kj}B_{kj} = \sum_{j=1}^{n} a_{hj}A_{kj}.$$

This gives the first equation, for B has two equal rows, and $|B| = 0$.

The second equality contains the corresponding result for columns and it may be obtained similarly.

Using Theorem 4.21, we see that the sum of any row in A times a row of cofactors is either 0 or $|A|$. Therefore using a matrix with the rows of cofactors written as columns yields the following:

$$\begin{pmatrix} a_{11} & a_{12} & \cdots & a_{1n} \\ a_{21} & a_{22} & \cdots & a_{2n} \\ & & \vdots & \\ a_{n1} & a_{n2} & \cdots & a_{nn} \end{pmatrix} \begin{pmatrix} A_{11} & A_{21} & \cdots & A_{n1} \\ A_{12} & A_{22} & \cdots & A_{n2} \\ & & \vdots & \\ A_{1n} & A_{2n} & \cdots & A_{nn} \end{pmatrix}$$

$$= \begin{pmatrix} |A| & 0 & \cdots & 0 \\ 0 & |A| & \cdots & 0 \\ & & \vdots & \\ 0 & 0 & \cdots & |A| \end{pmatrix} = |A|I_n.$$

So when $|A| \neq 0$, the matrix of cofactors can be used to obtain A^{-1}.

Definition Let $A = (a_{ij})$ be an $n \times n$ matrix. The *adjoint (matrix)* of A, denoted A^{adj}, is the transpose of the cofactor matrix (A_{ij}). That is, $A^{\text{adj}} = (A_{ij})^T$ and the element in the ith row and jth column of A^{adj} is the cofactor of the element in the jth row and ith column of A.

Now the above matrix equation can be rewritten as $AA^{\text{adj}} = |A|I_n$. Therefore if $|A| \neq 0$, we can divide by $|A|$ to obtain

$$A\left[\frac{1}{|A|}A^{\text{adj}}\right] = I_n \quad \text{or} \quad A^{-1} = \frac{1}{|A|}A^{\text{adj}}.$$

This is the adjoint formula for finding the inverse of a matrix.

Example 5. Consider the matrix

$$A = \begin{pmatrix} 2 & 1 \\ 5 & 3 \end{pmatrix}$$

from Example 4. Since $|A| = 1$,

$$A^{-1} = A^{\text{adj}} = \begin{pmatrix} 3 & -1 \\ -5 & 2 \end{pmatrix}.$$

In general, for 2×2 matrices

$$\begin{pmatrix} a & b \\ c & d \end{pmatrix}^{\text{adj}} = \begin{pmatrix} d & -b \\ -c & a \end{pmatrix},$$

so if

$$\begin{pmatrix} a & b \\ c & d \end{pmatrix}$$

is nonsingular, then

$$\begin{pmatrix} a & b \\ c & d \end{pmatrix}^{-1} = \frac{1}{ad - bc}\begin{pmatrix} d & -b \\ -c & a \end{pmatrix}.$$

As an example,

$$\begin{pmatrix} 2 & 3 \\ -2 & 7 \end{pmatrix}^{-1} = \frac{1}{20}\begin{pmatrix} 7 & -3 \\ 2 & 2 \end{pmatrix} = \begin{pmatrix} 7/20 & -3/20 \\ 1/10 & 1/10 \end{pmatrix}.$$

Example 6 Find the inverse of the matrix

$$A = \begin{pmatrix} 2 & 1 & 3 \\ 2 & 4 & 0 \\ 1 & 2 & 1 \end{pmatrix}.$$

We have

$$A^{-1} = \frac{1}{|A|} A^{\text{adj}} = \frac{1}{6}\begin{pmatrix} 4 & 5 & -12 \\ -2 & -1 & 6 \\ 0 & -3 & 6 \end{pmatrix} = \begin{pmatrix} 2/3 & 5/6 & -2 \\ -1/3 & -1/6 & 1 \\ 0 & -1/2 & 1 \end{pmatrix}.$$

The adjoint formula for the inverse of a matrix can be applied to certain systems of linear equations. Suppose $AX = B$ is a consistent system of n independent linear equations in n unknowns. Then $|A| \neq 0$ and A^{-1} exists. Therefore, we can multiply the equation $AX = B$ by A^{-1} to obtain $X = A^{-1}B$. This is the unique solution for

$$A(A^{-1}B) = (AA^{-1})B = I_nB = B.$$

The solution $X = A^{-1}B$ is the matrix form of Cramer's rule.

Theorem 4.22 (Cramer's rule) Let $AX = B$ be a consistent system of n independent linear equations in n unknowns with $A = (a_{ij})$, $X = (x_j)$, and $B = (b_j)$. Then the unique solution is given by

$$x_j = \frac{\begin{vmatrix} a_{11} & \cdots & a_{1,j-1} & b_1 & a_{1,j+1} & \cdots & a_{1n} \\ a_{21} & \cdots & a_{2,j-1} & b_2 & a_{2,j+1} & \cdots & a_{2n} \\ & \vdots & & & & \vdots & \\ a_{n1} & \cdots & a_{n,j-1} & b_n & a_{n,j+1} & \cdots & a_{nn} \end{vmatrix}}{|A|} \qquad \text{for} \quad 1 \leq j \leq n.$$

That is, x_j is the ratio of two determinants with the determinant in the numerator obtained from $|A|$ by replacing the jth column with the column of constant terms, B.

Proof The proof simply consists in showing that the jth entry in the matrix solution $X = A^{-1}B$ has the required form.

$$\begin{pmatrix} x_1 \\ x_2 \\ \vdots \\ x_n \end{pmatrix} = X = A^{-1}B = \frac{A^{\text{adj}}B}{|A|} = \frac{1}{|A|}\begin{pmatrix} A_{11} & A_{21} & \cdots & A_{n1} \\ A_{12} & A_{22} & \cdots & A_{n2} \\ & & \vdots & \\ A_{1n} & A_{2n} & \cdots & A_{nn} \end{pmatrix}\begin{pmatrix} b_1 \\ b_2 \\ \vdots \\ b_n \end{pmatrix}.$$

Therefore

$$x_j = \frac{1}{|A|}[A_{1j}b_1 + A_{2j}b_2 + \cdots + A_{nj}b_n]$$

for each j, but $b_1A_{1j} + b_2A_{2j} + \cdots + b_nA_{nj}$ is the expansion, along the jth column, of the determinant of the matrix obtained from A by replacing the jth column with B. This is the desired expression for x_j.

Example 7 Use Cramer's rule to solve the following system for x, y, and z:

$$2x + y - z = 3$$
$$x + 3y + 2z = 1$$
$$x + 2y + 2z = 2.$$

The determinant of the coefficient matrix is 5, therefore the equations are independent and Cramer's rule can be applied:

$$x = \frac{\begin{vmatrix} 3 & 1 & -1 \\ 1 & 3 & 2 \\ 2 & 2 & 2 \end{vmatrix}}{5}, \quad y = \frac{\begin{vmatrix} 2 & 3 & -1 \\ 1 & 1 & 2 \\ 1 & 2 & 2 \end{vmatrix}}{5}, \quad z = \frac{\begin{vmatrix} 2 & 1 & 3 \\ 1 & 3 & 1 \\ 1 & 2 & 2 \end{vmatrix}}{5}.$$

This gives $x = 12/5$, $y = -1$, and $z = 4/5$.

Problems

1. Compute the following matrix products:

a. $\begin{pmatrix} 2 & -1 \\ 3 & 5 \end{pmatrix}\begin{pmatrix} 1 & 4 \\ 3 & 0 \end{pmatrix}$. b. $(2 \quad 1 \quad 3)\begin{pmatrix} 4 \\ -5 \\ 1 \end{pmatrix}$. c. $\begin{pmatrix} 4 & -1 \\ -2 & 3 \\ 1 & 2 \end{pmatrix}\begin{pmatrix} 2 & 1 \\ 5 & 1 \end{pmatrix}$.

d. $\begin{pmatrix} 7 \\ -2 \\ 3 \end{pmatrix}(4 \quad 2 \quad -3)$. e. $\begin{pmatrix} 6 & 9 \\ 4 & 6 \end{pmatrix}\begin{pmatrix} 3 & -6 \\ -2 & 4 \end{pmatrix}$. f. $\begin{pmatrix} 2 & 1 & 5 \\ 0 & 4 & 2 \\ 1 & 1 & 0 \end{pmatrix}\begin{pmatrix} -1 & 4 & 3 \\ 6 & -3 & 0 \\ 3 & 2 & 2 \end{pmatrix}$.

2. Let $S(x, y) = (2x - 3y, x + 2y, y - 3x)$, $T(x, y, z) = (x + y + z, 2x - y + 3z)$ and suppose A_S, A_T are the matrices of S and T with respect to the standard bases for $\mathscr{V}_2$ and $\mathscr{V}_3$. Show that the product $A_T A_S$ is the matrix of $T \circ S$ with respect to the standard bases.

3. Use the adjoint formula to find the inverse of:

a. $\begin{pmatrix} 2 & 4 \\ 1 & 3 \end{pmatrix}$. b. $\begin{pmatrix} 4 & -1 \\ 3 & 2 \end{pmatrix}$. c. $\begin{pmatrix} 2 & 4 \\ 4 & 6 \end{pmatrix}$. d. $\begin{pmatrix} 3 & 2 & 0 \\ 2 & 1 & 1 \\ 3 & 0 & 2 \end{pmatrix}$.

e. $\begin{pmatrix} 1 & 0 & 0 \\ 0 & 0 & 1 \\ 0 & 1 & 0 \end{pmatrix}$. f. $\begin{pmatrix} 1 & 2 & 0 \\ 0 & 1 & 4 \\ 3 & 0 & 1 \end{pmatrix}$.

4. a. Prove that the product of two $n \times n$ nonsingular matrices is nonsingular.

b. Suppose $A_1, A_2, \ldots, A_k$ are $n \times n$ nonsingular matrices. Show that $A_1 A_2 \cdots A_k$ is nonsingular and

$$(A_1 A_2 \cdots A_k)^{-1} = A_k^{-1} A_{k-1}^{-1} \cdots A_2^{-1} A_1^{-1}.$$

5. a. Prove Theorem 4.18.
b. Prove Corollary 4.19.
c. Prove Theorem 4.20.

6. Use Cramer's rule to solve the following systems of equations:
a. $2x - 3y = 5$
$x + 2y = 4.$
b. $3x - 5y - 2z = 1$
$x + y - z = 4$
$-x + 2y + z = 5.$
c. $3a + 2c = 5$
$4b + 5d = 1$
$a - 2c + d = 0$
$4b + 8c = 5.$

7. $B = \{(1, -2, -2), (1, -1, 1), (2, -3, 0)\}$ is a basis for $\mathscr{V}_3$. If the vector (a, b, c) has coordinates x, y, z with respect to B, then they satisfy the following system of linear equations:

$$x + y + 2z = a, \qquad -2x - y - 3z = b, \qquad -2x + y = c.$$

Use the matrix form of Cramer's rule to find the coordinates of a. $(1, 0, 0)$, b. $(1, 0, 3)$, c. $(1, 1, 8)$, d. $(4, -6, 0)$, with respect to the basis B.

8. Let $B = \{(4, 1, 4), (-3, 5, 2), (3, 2, 4)\}$ and use the matrix form of Cramer's rule to find the coordinates with respect to B of a. $(4, -1, 2)$. b. $(-3, 5, 2)$. c. $(0, 1, 0)$. d. $(0, 0, 2)$.

9. Let $T(a, b) = (2a - 3b, 3b - 3a)$. If A is the matrix of T with respect to the standard basis for $\mathscr{V}_2$, then A^{-1} is the matrix of T^{-1} with respect to the standard basis. Find A and A^{-1} and use A^{-1} to find a. $T^{-1}(0, 1)$. b. $T^{-1}(-5, 6)$. c. $T^{-1}(1, 3)$. d. $T^{-1}(x, y)$.

10. Let $T(a, b, c) = (2a + b + 3c, 2a + 2b + c, 4a + 2b + 4c)$ and use the same technique as in problem 9 to find a. $T^{-1}(0, 0, 8)$. b. $T^{-1}(9, 1, 12)$. c. $T^{-1}(x, y, z)$.

§7. Elementary Matrices

The second useful technique for finding the inverse of a matrix is based on the fact that every nonsingular $n \times n$ matrix is row-equivalent to I_n. That is, the $n \times n$ identity matrix is the echelon form for an $n \times n$ matrix of rank n. It is now possible to represent the elementary row operations that transform a matrix to its echelon form with multiplication by special matrices called "elementary matrices." Multiplication on the left by an elementary matrix will give an elementary row operation, while multiplication on the right yields an elementary column operation. There are three basic types of ele- elementary matrices, corresponding to the three types of elementary operations, and each is obtained from I_n with an elementary row operation.

An $n \times n$ *elementary matrix of type I*, $E_{h,k}$, is obtained by interchanging the hth and kth rows of I_n. If A is an $n \times m$ matrix, then multiplying A on the left by $E_{h,k}$ interchanges the hth and kth rows of A.

Example 1

$$E_{2,3}\begin{pmatrix} 2 & 1 \\ 4 & 5 \\ 7 & 8 \end{pmatrix} = \begin{pmatrix} 1 & 0 & 0 \\ 0 & 0 & 1 \\ 0 & 1 & 0 \end{pmatrix}\begin{pmatrix} 2 & 1 \\ 4 & 5 \\ 7 & 8 \end{pmatrix} = \begin{pmatrix} 2 & 1 \\ 7 & 8 \\ 4 & 5 \end{pmatrix}.$$

$$E_{1,3}\begin{pmatrix} 9 & 2 & 5 & 0 \\ 0 & 7 & 1 & 6 \\ 1 & 4 & 2 & 9 \\ 2 & 1 & 4 & 4 \end{pmatrix} = \begin{pmatrix} 0 & 0 & 1 & 0 \\ 0 & 1 & 0 & 0 \\ 1 & 0 & 0 & 0 \\ 0 & 0 & 0 & 1 \end{pmatrix}\begin{pmatrix} 9 & 2 & 5 & 0 \\ 0 & 7 & 1 & 6 \\ 1 & 4 & 2 & 9 \\ 2 & 1 & 4 & 4 \end{pmatrix} = \begin{pmatrix} 1 & 4 & 2 & 9 \\ 0 & 7 & 1 & 6 \\ 9 & 2 & 5 & 0 \\ 2 & 1 & 4 & 4 \end{pmatrix}.$$

An $n \times n$ *elementary matrix of type II*, $E_h(r)$, is obtained from I_n by multiplying the hth row by r, a nonzero scalar. Multiplication of a matrix on the left by an elementary matrix of type II multiplies a row of the matrix by a nonzero scalar.

Example 2

$$E_1(1/2)\begin{pmatrix} 2 & 4 & 8 \\ 4 & 5 & 3 \end{pmatrix} = \begin{pmatrix} 1/2 & 0 \\ 0 & 1 \end{pmatrix}\begin{pmatrix} 2 & 4 & 8 \\ 4 & 5 & 3 \end{pmatrix} = \begin{pmatrix} 1 & 2 & 4 \\ 4 & 5 & 3 \end{pmatrix}.$$

$$E_2(-3)\begin{pmatrix} 3 & 2 & 4 \\ 1 & -2 & 1 \\ 4 & 5 & 8 \end{pmatrix} = \begin{pmatrix} 1 & 0 & 0 \\ 0 & -3 & 0 \\ 0 & 0 & 1 \end{pmatrix}\begin{pmatrix} 3 & 2 & 4 \\ 1 & -2 & 1 \\ 4 & 5 & 8 \end{pmatrix} = \begin{pmatrix} 3 & 2 & 4 \\ -3 & 6 & -3 \\ 4 & 5 & 8 \end{pmatrix}.$$

An $n \times n$ *elementary matrix of type III*, $E_{h,k}(r)$, is obtained from I_n by adding r times the hth row to the kth row. Multiplication of a matrix on the left by an elementary matrix of type III performs an elementary row operation of type III.

Example 3

$$E_{1,2}(-2)\begin{pmatrix} 2 & 3 & 1 \\ 4 & 2 & 5 \end{pmatrix} = \begin{pmatrix} 1 & 0 \\ -2 & 1 \end{pmatrix}\begin{pmatrix} 2 & 3 & 1 \\ 4 & 2 & 5 \end{pmatrix} = \begin{pmatrix} 2 & 3 & 1 \\ 0 & -4 & 3 \end{pmatrix}.$$

$$E_{3,1}(-4)\begin{pmatrix} 4 & 8 & 6 & 2 \\ 0 & 2 & 5 & 1 \\ 1 & 3 & 1 & 0 \end{pmatrix} = \begin{pmatrix} 1 & 0 & -4 \\ 0 & 1 & 0 \\ 0 & 0 & 1 \end{pmatrix}\begin{pmatrix} 4 & 8 & 6 & 2 \\ 0 & 2 & 5 & 1 \\ 1 & 3 & 1 & 0 \end{pmatrix} = \begin{pmatrix} 0 & -4 & 2 & 2 \\ 0 & 2 & 5 & 1 \\ 1 & 3 & 1 & 0 \end{pmatrix}.$$

Theorem 4.23 Elementary matrices are nonsingular, and the inverse of an elementary matrix is an elementary matrix of the same type.

Proof One need only check that

$$E_{h,k}^{-1} = E_{h,k}, \qquad E_h(r)^{-1} = E_h(1/r), \qquad E_{h,k}(r)^{-1} = E_{h,k}(-r).$$

The notations used above for the various elementary matrices are useful in defining the three types, but it is generally not necessary to be so explicit in representing an elementary matrix. For example, suppose A is an $n \times n$ nonsingular matrix. Then A is row-equivalent to I_n and there is a sequence of elementary row operations that transforms A to I_n. This can be expressed as $E_p \cdots E_2 E_1 A = I_n$ where $E_1, E_2, \ldots, E_p$ are the elementary matrices that perform the required elementary row operations. In this case, it is neither important nor useful to specify exactly which elementary matrices are being used. This example also provides a method for obtaining the inverse of A. For if $E_p \cdots E_2 E_1 A = I_n$, then the product $E_p \cdots E_2 E_1$ is the inverse of A. Therefore

$$A^{-1} = E_p \cdots E_2 E_1 = E_p \cdots E_2 E_1 I_n.$$

That is, the inverse of A can be obtained from I_n by performing the same sequence of elementary row operations on I_n that must be applied to A in order to transform A to I_n. In practice it is not necessary to keep track of the elementary matrices $E_1, \ldots, E_p$, for the elementary row operations can be performed simultaneously on A and I_n.

Example 4 Use elementary row operations to find the inverse of

$$A = \begin{pmatrix} 2 & 1 \\ 5 & 3 \end{pmatrix},$$

the matrix of Examples 4 and 5 in the previous section.

Place A and I_2 together and transform A to I_2 with elementary row operations.

$$\begin{pmatrix} 2 & 1 & 1 & 0 \\ 5 & 3 & 0 & 1 \end{pmatrix} \xrightarrow[\text{II}]{} \begin{pmatrix} 1 & 1/2 & 1/2 & 0 \\ 5 & 3 & 0 & 1 \end{pmatrix} \xrightarrow[\text{III}]{} \begin{pmatrix} 1 & 1/2 & 1/2 & 0 \\ 0 & 1/2 & -5/2 & 1 \end{pmatrix}$$

$$\xrightarrow[\text{II}]{} \begin{pmatrix} 1 & 1/2 & 1/2 & 0 \\ 0 & 1 & -5 & 2 \end{pmatrix} \xrightarrow[\text{III}]{} \begin{pmatrix} 1 & 0 & 3 & -1 \\ 0 & 1 & -5 & 2 \end{pmatrix}.$$

Therefore

$$\begin{pmatrix} 2 & 1 \\ 5 & 3 \end{pmatrix}^{-1} = \begin{pmatrix} 3 & -1 \\ -5 & 2 \end{pmatrix}.$$

This agrees with our previous results. Comparing the three ways in which A^{-1} has been found should make it clear that the adjoint formula provides the simplest way to obtain the inverse of a 2×2 matrix.

For larger matrices the use of elementary row operations can be simpler than the adjoint method. However, it is much easier to locate and correct errors in the computation of the adjoint matrix than in a sequence of elementary row operations.

Example 5 Use elementary row operations to find

$$\begin{pmatrix} 0 & 1 & 3 \\ 1 & 2 & 0 \\ 2 & 4 & 2 \end{pmatrix}^{-1}.$$

We have

$$\begin{pmatrix} 0 & 1 & 3 & 1 & 0 & 0 \\ 1 & 2 & 0 & 0 & 1 & 0 \\ 2 & 4 & 2 & 0 & 0 & 1 \end{pmatrix} \xrightarrow[\mathrm{I}]{} \begin{pmatrix} 1 & 2 & 0 & 0 & 1 & 0 \\ 0 & 1 & 3 & 1 & 0 & 0 \\ 2 & 4 & 2 & 0 & 0 & 1 \end{pmatrix} \xrightarrow[\mathrm{III}]{} \begin{pmatrix} 1 & 2 & 0 & 0 & 1 & 0 \\ 0 & 1 & 3 & 1 & 0 & 0 \\ 0 & 0 & 2 & 0 & -2 & 1 \end{pmatrix}$$

$$\xrightarrow[\mathrm{II}]{} \begin{pmatrix} 1 & 2 & 0 & 0 & 1 & 0 \\ 0 & 1 & 3 & 1 & 0 & 0 \\ 0 & 0 & 1 & 0 & -1 & 1/2 \end{pmatrix} \xrightarrow[\mathrm{III}]{} \begin{pmatrix} 1 & 2 & 0 & 0 & 1 & 0 \\ 0 & 1 & 0 & 1 & 3 & -3/2 \\ 0 & 0 & 1 & 0 & -1 & 1/2 \end{pmatrix}$$

$$\xrightarrow[\mathrm{III}]{} \begin{pmatrix} 1 & 0 & 0 & -2 & -5 & 3 \\ 0 & 1 & 0 & 1 & 3 & -3/2 \\ 0 & 0 & 1 & 0 & -1 & 1/2 \end{pmatrix}.$$

So

$$\begin{pmatrix} 0 & 1 & 3 \\ 1 & 2 & 0 \\ 2 & 4 & 2 \end{pmatrix}^{-1} = \begin{pmatrix} -2 & -5 & 3 \\ 1 & 3 & -3/2 \\ 0 & -1 & 1/2 \end{pmatrix}.$$

This technique for finding an inverse is based on the fact that if A is nonsingular, then $A^{-1} = E_p \cdots E_2 E_1$ for some sequence of elementary matrices. This expression can be rewritten in the form $A = (E_p \cdots E_2E_1)^{-1}$, and using problem 4 on page 186 we have $A = E_1^{-1}E_2^{-1} \cdots E_p^{-1}$. Since the inverse of an elementary matrix is itself an elementary matrix, we have proved the following.

Theorem 4.24 Every nonsingular matrix can be expressed as a product of elementary matrices.

The importance of this theorem lies in the existence of such a representation. However, it is possible to find an explicit representation for a nonsingular matrix A by keeping track of the elementary matrices used to transform A to I_n. This also provides a good exercise in the use of elementary matrices.

Example 6 Express

$$\begin{pmatrix} 0 & 3 \\ 2 & 8 \end{pmatrix}$$

as a product of elementary matrices.

$$\begin{pmatrix} 0 & 3 \\ 2 & 8 \end{pmatrix} \xrightarrow[\begin{pmatrix} 0 & 1 \\ 1 & 0 \end{pmatrix}]{} \begin{pmatrix} 2 & 8 \\ 0 & 3 \end{pmatrix} \xrightarrow[\begin{pmatrix} 1/2 & 0 \\ 0 & 1 \end{pmatrix}]{} \begin{pmatrix} 1 & 4 \\ 0 & 3 \end{pmatrix} \xrightarrow[\begin{pmatrix} 1 & 0 \\ 0 & 1/3 \end{pmatrix}]{} \begin{pmatrix} 1 & 4 \\ 0 & 1 \end{pmatrix} \xrightarrow[\begin{pmatrix} 1 & -4 \\ 0 & 1 \end{pmatrix}]{} \begin{pmatrix} 1 & 0 \\ 0 & 1 \end{pmatrix}.$$

Therefore

$$\begin{aligned} \begin{pmatrix} 0 & 3 \\ 2 & 8 \end{pmatrix} &= \begin{pmatrix} 0 & 1 \\ 1 & 0 \end{pmatrix}^{-1} \begin{pmatrix} 1/2 & 0 \\ 0 & 1 \end{pmatrix}^{-1} \begin{pmatrix} 1 & 0 \\ 0 & 1/3 \end{pmatrix}^{-1} \begin{pmatrix} 1 & -4 \\ 0 & 1 \end{pmatrix}^{-1} \\ &= \begin{pmatrix} 0 & 1 \\ 1 & 0 \end{pmatrix} \begin{pmatrix} 2 & 0 \\ 0 & 1 \end{pmatrix} \begin{pmatrix} 1 & 0 \\ 0 & 3 \end{pmatrix} \begin{pmatrix} 1 & 4 \\ 0 & 1 \end{pmatrix}. \end{aligned}$$

Is this expression unique?

If a matrix A is multiplied on the right by an elementary matrix, the result is equivalent to performing an elementary column operation. Multiplying A on the right by $E_{h,k}$ interchanges the hth and kth columns of A, multiplication by $E_h(r)$ on the right multiplies the hth column of A by r, and multiplication by $E_{h,k}(r)$ on the right adds r times the kth column to the hth column (note the change in order).

Example 7 Let

$$A = \begin{pmatrix} 0 & 1 & 0 & 5 \\ 5 & 2 & 2 & 6 \\ 2 & 3 & 5 & 9 \end{pmatrix},$$

then

$$AE_{1,2} = \begin{pmatrix} 0 & 1 & 0 & 5 \\ 5 & 2 & 2 & 6 \\ 2 & 3 & 5 & 9 \end{pmatrix} \begin{pmatrix} 0 & 1 & 0 & 0 \\ 1 & 0 & 0 & 0 \\ 0 & 0 & 1 & 0 \\ 0 & 0 & 0 & 1 \end{pmatrix} = \begin{pmatrix} 1 & 0 & 0 & 5 \\ 2 & 5 & 2 & 6 \\ 3 & 2 & 5 & 9 \end{pmatrix}.$$

$$AE_3(1/2) = \begin{pmatrix} 0 & 1 & 0 & 5 \\ 5 & 2 & 2 & 6 \\ 2 & 3 & 5 & 9 \end{pmatrix} \begin{pmatrix} 1 & 0 & 0 & 0 \\ 0 & 1 & 0 & 0 \\ 0 & 0 & 1/2 & 0 \\ 0 & 0 & 0 & 1 \end{pmatrix} = \begin{pmatrix} 0 & 1 & 0 & 5 \\ 5 & 2 & 1 & 6 \\ 2 & 3 & 5/2 & 9 \end{pmatrix}.$$

$$AE_{4,2}(-5) = \begin{pmatrix} 0 & 1 & 0 & 5 \\ 5 & 2 & 2 & 6 \\ 2 & 3 & 5 & 9 \end{pmatrix} \begin{pmatrix} 1 & 0 & 0 & 0 \\ 0 & 1 & 0 & -5 \\ 0 & 0 & 1 & 0 \\ 0 & 0 & 0 & 1 \end{pmatrix} = \begin{pmatrix} 0 & 1 & 0 & 0 \\ 5 & 2 & 2 & -4 \\ 2 & 3 & 5 & -6 \end{pmatrix}.$$

The fact that a nonsingular matrix can be expressed as a product of elemenatry matrices is used to prove that the determinant of a product is the product of the determinants. But first we need the following special case.

Lemma 4.25 If A is an $n \times n$ matrix and E is an $n \times n$ elementary matrix, then $|AE| = |A||E| = |EA|$.

Proof First since an elementary matrix is obtained from I_n with a single elementary row operation, the properties of determinants give

$$|E_{h,k}| = -|I_n| = -1, \qquad |E_h(r)| = r|I_n| = r, \qquad |E_{h,k}(r)| = |I_n| = 1.$$

Then AE and EA are obtained from A with a single column and row operation so

$$|AE_{h,k}| = -|A| = |E_{h,k}A|,$$
$$|AE_h(r)| = r|A| = |E_h(r)A|,$$
$$|AE_{h,k}(r)| = |A| = |E_{h,k}(r)A|.$$

Combining these two sets of equalities proves that $|AE| = |A|\,|E| = |EA|$.

Theorem 4.26 If A and B are $n \times n$ matrices, then $|AB| = |A|\,|B|$.

Proof *Case* 1. Suppose B is nonsingular. Then $B = E_1E_2\cdots E_q$ for some sequence of elementary matrices $E_1, \ldots, E_q$, and the proof is by induction on q.

When $q = 1$, $B = E_1$ and $|AB| = |A|\,|B|$ by the lemma.

Assume $|AB| = |A|\,|B|$ for any matrix B which is the product of $q - 1$ elementary matrices, $q \geq 2$. Then

$$
\begin{aligned}
|AB| = |AE_1\cdots E_{q-1}E_q| &= |(AE_1\cdots E_{q-1})E_q| \\
&= |AE_1\cdots E_{q-1}|\,|E_q| && \text{By the lemma} \\
&= |A|\,|E_1\cdots E_{q-1}|\,|E_q| && \text{By assumption} \\
&= |A|\,|(E_1\cdots E_{q-1})E_q| && \text{By the lemma} \\
&= |A|\,|B|.
\end{aligned}
$$

Case 2. Suppose B is singular. Let S and T be the maps in Hom($\mathscr{V}_n$) which have A and B as their matrices with respect to the standard basis. B is singular, therefore T is singular which implies that $S \circ T$ is singular. [If $T(U) = T(V)$, then $S \circ T(U) = S \circ T(V)$.] Therefore AB, the matrix of $S \circ T$ with respect to the standard basis, is singular. Thus $|AB| = 0$, but B is singular, so $|B| = 0$, and we have $|AB| = |A|\,|B|$.

Since every $n \times n$ matrix is either singular or nonsingular, the proof is complete.

Corollary 4.27 If A is nonsingular, then $|A^{-1}| = 1/|A|$.

Corollary 4.28 If A and B are $n \times n$ matrices such that the rank of AB is n, then A and B each have rank n.

The proofs for these corollaries are left to the reader.

The second corollary is a special case of the following theorem.

Theorem 4.29 If A and B are matrices for which AB is defined, then rank $AB \leq$ rank A and rank $AB \leq$ rank B.

Rather than prove this theorem directly, it is easier to prove the corresponding result for linear transformations.

Theorem 4.30 Let S and T be linear maps defined on finite-dimensional vector spaces for which $S \circ T$ is defined, then rank $S \circ T \leq$ rank T and rank $S \circ T \leq$ rank S.

Proof Suppose $\{U_1, \ldots, U_n\}$ is a basis for the domain of T, and rank $T = k$. Then dim $\mathscr{L}\{T(U_1), \ldots, T(U_n)\} = k$, and since a linear map cannot increase dimension, rank $S \circ T =$ dim $\mathscr{L}\{S \circ T(U_1), \ldots, S \circ T(U_n)\}$ cannot exceed k. That is, rank $S \circ T \leq$ rank T. On the other hand, $\{S \circ T(U_1), \ldots, S \circ T(U_n)\}$ spans $\mathscr{I}_{S \circ T}$ and is a subset of $\mathscr{I}_S$. Therefore dim $\mathscr{I}_{S \circ T} \leq$ dim $\mathscr{I}_S$ or rank $S \circ T \leq$ rank S.

Now Theorem 4.29 for matrices follows from Theorem 4.30 using the isomorphism between linear transformations and matrices.

Problems

1. Let $A = \begin{pmatrix} 2 & 4 \\ 5 & 1 \end{pmatrix}$, $B = \begin{pmatrix} 5 & 2 \\ 1 & 2 \\ 3 & 8 \end{pmatrix}$, and $C = \begin{pmatrix} 1 & 2 & 3 & 2 \\ 2 & 1 & 0 & 3 \\ 0 & 4 & 2 & 8 \end{pmatrix}$. Find the elementary matrix that performs each of the following operations and use it to carry out the operation:

a. Add $-5/2$ times the first row of A to the second row.
b. Interchange the first and second rows of B.
c. Divide the first row of A by 2.
d. Interchange the second and third rows of C.
e. Subtract twice the first row of C from the second row.
f. Add -3 times the second row of B to the third row.
g. Divide the third row of C by 4.

2. a. Find the elementary matrices E_i used to perform the following sequence of elementary row operations:

$$\begin{pmatrix} 2 & 6 \\ 5 & 9 \end{pmatrix} \xrightarrow[E_1]{} \begin{pmatrix} 1 & 3 \\ 5 & 9 \end{pmatrix} \xrightarrow[E_2]{} \begin{pmatrix} 1 & 3 \\ 0 & -6 \end{pmatrix} \xrightarrow[E_3]{} \begin{pmatrix} 1 & 3 \\ 0 & 1 \end{pmatrix} \xrightarrow[E_4]{} \begin{pmatrix} 1 & 0 \\ 0 & 1 \end{pmatrix}.$$

b. Compute the product $E_4E_3E_2E_1$ and check that it is $\begin{pmatrix} 2 & 6 \\ 5 & 9 \end{pmatrix}^{-1}$.
c. Find the inverses of E_1, E_2, E_3, and E_4.
d. Compute the product $E_1^{-1}E_2^{-1}E_3^{-1}E_4^{-1}$.

3. Write the following matrices as products of elementary matrices:

a. $\begin{pmatrix} 0 & 3 \\ 5 & -10 \end{pmatrix}$. b. $\begin{pmatrix} 1 & 4 \\ 3 & 8 \end{pmatrix}$. c. $\begin{pmatrix} 2 & 4 & 6 \\ 0 & 1 & 1 \\ 3 & 7 & 9 \end{pmatrix}$. d. $\begin{pmatrix} 2 & 1 & 2 \\ 0 & 1 & 2 \\ 1 & 0 & 1 \end{pmatrix}$.

4. Use elementary row operations to find the inverse of:

a. $\begin{pmatrix} 3 & 1 & 4 \\ 1 & 0 & 2 \\ 2 & 5 & 0 \end{pmatrix}$. b. $\begin{pmatrix} 3 & 2 \\ 5 & 4 \end{pmatrix}$. c. $\begin{pmatrix} 0 & 1 & 0 & 4 \\ 5 & 0 & 8 & 1 \\ 2 & 3 & 4 & 4 \\ 3 & 0 & 6 & 0 \end{pmatrix}$. d. $\begin{pmatrix} 0 & 3 & 5 \\ 2 & 3 & 0 \\ 2 & 5 & 3 \end{pmatrix}$.

5. We know that if $A, B \in \mathscr{M}_{n \times n}$ and $AB = I_n$, then $BA = I_n$. What is wrong with the following "proof" of this fact?
Since $AB = I_n$, $(AB)A = I_nA = A = AI_n$. Therefore $A(BA) - AI_n = \mathbf{0}$ or $A(BA - I_n) = \mathbf{0}$. But $A \neq \mathbf{0}$ since $AB = I_n \neq \mathbf{0}$, therefore $BA - I_n = \mathbf{0}$ and $BA = I_n$.

6. Let $A, B \in \mathscr{M}_{n \times n}$. Prove that if $AB = \mathbf{0}$ and A is nonsingular, then $B = \mathbf{0}$.

7. a. Prove that if A is row-equivalent to B, then there exists a nonsingular matrix Q such that $B = QA$.
   b. Show that Q is unique if A is $n \times n$ and nonsingular.
   c. Show by example that Q need not be unique if A is $n \times n$ and singular.

8. Find a nonsingular matrix Q such that $B = QA$ when

a. $A = \begin{pmatrix} 2 & 4 & 8 \\ 1 & 3 & 2 \end{pmatrix}$, $B = \begin{pmatrix} 1 & 0 & 8 \\ 0 & 1 & -2 \end{pmatrix}$. b. $A = \begin{pmatrix} 1 & 4 \\ 1 & 5 \\ 3 & 2 \end{pmatrix}$, $B = \begin{pmatrix} 1 & 0 \\ 0 & 1 \\ 0 & 0 \end{pmatrix}$.

c. $A = \begin{pmatrix} 1 & 2 & 4 \\ 3 & 5 & 1 \\ 4 & 6 & -6 \end{pmatrix}$, $B = \begin{pmatrix} 1 & 0 & -18 \\ 0 & 1 & 11 \\ 0 & 0 & 0 \end{pmatrix}$.

d. $A = \begin{pmatrix} 2 & 3 \\ 1 & 5 \end{pmatrix}$, $B = \begin{pmatrix} 1 & 0 \\ 0 & 1 \end{pmatrix}$.

9. Let $A = \begin{pmatrix} 1 & 2 & 0 \\ 2 & 4 & 1 \\ 4 & 8 & 0 \end{pmatrix}$ and $B = \begin{pmatrix} 0 & 1 & -3 \\ 1 & 2 & -4 \end{pmatrix}$. Find the elementary matrix which performs the following elementary column operations and perform the operation by multiplication: a. Divide the second column of A by 2. b. Subtract twice the first column of A from the second. c. Interchange the first and second columns of B. d. Multiply the second column of B by 3 and add it to the third column. e. Interchange the second and third columns of A.

10. Suppose $B = AP$, A, B, and P matrices with P nonsingular. Prove that A is column-equivalent to B.

11. Find a nonsingular matrix P such that $B = AP$ when

a. $A = \begin{pmatrix} 3 & 4 & 2 \\ 6 & 9 & 8 \end{pmatrix}$, $B = \begin{pmatrix} 1 & 0 & 0 \\ 0 & 1 & 0 \end{pmatrix}$. b. $A = \begin{pmatrix} 2 & 5 \\ 1 & 3 \\ 4 & 7 \end{pmatrix}$, $B = \begin{pmatrix} 1 & 0 \\ 0 & 1 \\ 5 & -6 \end{pmatrix}$.

12. Find nonsingular matrices Q and P such that $B = QAP$ when

a. A as in 8a, $B = \begin{pmatrix} 1 & 0 & 0 \\ 0 & 1 & 0 \end{pmatrix}$. b. A as in 8c, $B = \begin{pmatrix} 1 & 0 & 0 \\ 0 & 1 & 0 \\ 0 & 0 & 0 \end{pmatrix}$.

c. A as in 11b, $B = \begin{pmatrix} 1 & 0 \\ 0 & 1 \\ 0 & 0 \end{pmatrix}$.

13. Prove that if A is nonsingular, then $|A^{-1}| = 1/|A|$.

14. Prove that rank $AB \leq$ rank A using the corresponding result for linear transformations.

15. Use rank to prove that if $A, B \in \mathcal{M}_{n \times n}$ and $AB = I_n$, then A is nonsingular.

Review Problems

1. Let $\begin{pmatrix} 2 & 3 \\ 1 & 4 \\ 5 & 1 \end{pmatrix}$ be the matrix of $T \in \text{Hom}(\mathscr{V}_2, \mathscr{V}_3)$ with respect to the bases $\{(0, 1), (1, 3)\}$ and $\{(-2, 1, 0), (3, 1, -4), (0, -2, 3)\}$ and find: a. $T(2, 5)$. b. $T(1, 3)$. c. $T(a, b)$.

2. Find the linear map $T: \mathscr{V}_3 \to \mathscr{V}_3$ which has I_n as its matrix with respect to B_1 and B_2 if:

a. $B_1 = B_2$.

b. $B_1 = \{(1, 0, 1), (1, 1, 0), (0, 0, 1)\}$, $B_2 = \{(2, -3, 5), (0, 4, 2), (-4, 3, 1)\}$.

3. Prove the following properties of isomorphism for any vector spaces $\mathscr{V}$, $\mathscr{W}$, and $\mathscr{U}$:

a. $\mathscr{V} \cong \mathscr{V}$.

b. If $\mathscr{V} \cong \mathscr{W}$, then $\mathscr{W} \cong \mathscr{V}$.

c. If $\mathscr{V} \cong \mathscr{W}$ and $\mathscr{W} \cong \mathscr{U}$, then $\mathscr{V} \cong \mathscr{U}$.

4. Define $T \in \text{Hom}(\mathscr{V}_2)$ by $T(a, b) = (2a - 3b, b - a)$.

a. Find the matrix A of T with respect to the standard basis.

b. Prove that there do not exist bases for $\mathscr{V}_2$ with respect to which $\begin{pmatrix} 1 & 0 \\ 0 & 0 \end{pmatrix}$ is the matrix for T.

c. Find bases for $\mathscr{V}_2$ with respect to which $\begin{pmatrix} 4 & -6 \\ -3 & 1 \end{pmatrix}$ is the matrix for T.

5. Suppose $\mathscr{V}$ and $\mathscr{W}$ are finite-dimensional vector spaces. Prove that $\mathscr{V} \cong \mathscr{W}$ if and only if $\dim \mathscr{V} = \dim \mathscr{W}$.

6. Let $T(a, b, c) = (2a - 3c, b + 4c, 8a + 3b)$.

a. Express $\mathscr{N}_T$ as the span of independent vectors.

b. Find bases B_1 and B_2 for $\mathscr{V}_3$ with respect to which $\begin{pmatrix} 2 & 0 & 0 \\ -1 & 1 & 0 \\ 3 & 0 & 0 \end{pmatrix}$ is the matrix of T.

7. Let $T(a, b, c, d) = (a + 3b + d, 2b - c + 2d, 2a + 3c - 4d, 2a + 4b + c)$. Find bases for $\mathscr{V}_4$ with respect to which A is the matrix of T when

a. $A = \begin{pmatrix} 1 & 0 & 0 & 0 \\ 0 & 1 & 0 & 0 \\ 0 & 0 & 0 & 0 \\ 0 & 0 & 0 & 0 \end{pmatrix}$. b. $A = \begin{pmatrix} 1 & 0 & 0 & 0 \\ 0 & 1 & 0 & 0 \\ 1 & 0 & 0 & 0 \\ 0 & 1 & 0 & 0 \end{pmatrix}$.

8. Let A be an $n \times n$ matrix such that $A^2 = A$, prove that either $A = I_n$ or A is singular. Find a 2×2 matrix A with all nonzero entries such that $A^2 = A$.

9. Suppose $T(a, b, c) = (2a + 3b + c, 6a + 9b + 3c, 5a - 3b + c)$.

a. Show that the image space $\mathscr{I}_T$ is a plane. Use elementary row operations on a matrix with $\mathscr{I}_T$ as row space to find a Cartesian equation for $\mathscr{I}_T$.

b. Use Gaussian elimination to show that the null space $\mathscr{N}_T$ is a line.

c. Show that any line parallel to $\mathscr{N}_T$ is sent to a single point by T.

d. Find bases for $\mathscr{V}_3$ with respect to which $\begin{pmatrix} 1 & 0 & 0 \\ 0 & 1 & 0 \\ 0 & 0 & 0 \end{pmatrix}$ is the matrix for T.

e. Find bases for $\mathscr{V}_3$ with respect to which $\begin{pmatrix} 1 & 0 & 1 \\ 0 & 1/3 & 0 \\ 0 & 0 & 0 \end{pmatrix}$ is the matrix for T.

10. Suppose $T \in \text{Hom}(\mathscr{V})$. Prove that $\mathscr{N}_T \cap \mathscr{I}_T = \{\mathbf{0}\}$ if and only if $\mathscr{N}_{T^2} = \mathscr{N}_T$.

11. Define $T: \mathscr{V}_3 \to \mathscr{V}_3$ by $T(a, b, c) = (a + b, b + c, a - c)$.

a. Show that T^{-1} does not exist.

b. Find $T^{-1}[\mathscr{S}]$ if $\mathscr{S} = \{(x, y, z) | x - y + z = 0\}$ and show that $T[T^{-1}[\mathscr{S}]] \neq \mathscr{S}$. (The complete inverse image of a set is defined in problem 7, page 145.)

c. Find $T^{-1}[\mathscr{T}]$ if $\mathscr{T} = \{(x, y, z) | 3x - 4y + z = 0\}$.

d. Find $T^{-1}[\{(x, y, z) | x - y - z = 0\}]$.

12. For which values of k does T fail to be an isomorphism?

a. $T \in \text{Hom}(\mathscr{V}_2)$; $T(a, b) = (3a + b, ka - 4b)$.

b. $T \in \text{Hom}(\mathscr{V}_2)$; $T(a, b) = (2a - kb. ka + b)$.

c. $T \in \text{Hom}(\mathscr{V}_3)$; $T(a, b, c) = (ka + 3c, 2a - kb, 3c - kb)$.

13. The definition of determinant was motivated by showing that it gives area in the plane and should give volume in 3-space. Relate this geometric idea with the fact that $S \in \text{Hom}(\mathscr{E}_2)$ or $S \in \text{Hom}(\mathscr{E}_3)$ is singular if and only if any matrix for S has zero determinant.

14. For the given matrix A, show that some matrix in the sequence $I_n, A, A^2, A^3, \ldots$ is a linear combination of those preceding it.

a. $A = \begin{pmatrix} 1 & 2 \\ 0 & 3 \end{pmatrix}$. b. $A = \begin{pmatrix} 1 & 2 \\ 3 & 6 \end{pmatrix}$. c. $A = \begin{pmatrix} 1 & -1 & 1 \\ 1 & -1 & 1 \\ 0 & 0 & 2 \end{pmatrix}$.

d. $A = \begin{pmatrix} 2 & -1 & 0 \\ 0 & 3 & 1 \\ 1 & 1 & -1 \end{pmatrix}$. e. $A = \begin{pmatrix} 2 & -1 & 0 \\ 0 & -1 & 2 \\ 0 & 0 & 3 \end{pmatrix}$. f. $A = \begin{pmatrix} 0 & 0 & 0 & 0 \\ 1 & 0 & 0 & 0 \\ 0 & 1 & 0 & 0 \\ 0 & 0 & 1 & 0 \end{pmatrix}$.

15. Suppose $A \in \mathscr{M}_{n \times n}$. Show that there exists an integer $k \leq n^2$ and scalars $a_0, a_1, \ldots, a_k$ such that

$$a_0 I_n + a_1 A + a_2 A^2 + \ldots + a_k A^k = \mathbf{0}.$$

16. a. Suppose $T \in \text{Hom}(\mathscr{V})$ and the dimension of $\mathscr{V}$ is n. Restate problem 15 in terms of T.

b. Find two different proofs for this fact.

17. Suppose $\mathscr{V}$ is a three-dimensional vector space, not necessarily $\mathscr{V}_3$, and $\mathscr{W}$ is a vector space of infinite dimension. Show that there is a one to one, linear map from $\mathscr{V}$ to $\mathscr{W}$.

18. Associate with each ordered pair of complex numbers (z, w) the 2×2 matrix $\begin{pmatrix} z & w \\ -\bar{w} & \bar{z} \end{pmatrix}$, where the bar denotes the complex conjugate (defined on page 277). Use this association to define the "quaternionic multiplication"

$$(z, w)(u, v) = (zu - w\bar{v}, zv + w\bar{u}).$$

a. Show that the definition of this product follows from a matrix product.

b. Use Theorem 4.17 to show that this multiplication is associative.

c. Show by example that quaternionic multiplication is not commutative.

19. a. Suppose A is $n \times n$ and B is $n \times m$. Show that if A is nonsingular, then rank $AB =$ rank B.

b. Suppose C is $n \times m$ and D is $m \times m$. Show that if D is nonsingular, then rank $CD =$ rank C.

c. Is it true that if rank $E \leq$ rank F, then rank $EF \geq$ rank E?

5

Change of Basis

In this chapter we will restrict our attention to finite-dimensional vector spaces. Therefore, given a basis, each vector can be associated with an ordered n-tuple of coordinates. Since coordinates do not depend on the vector alone, they reflect in part the choice of basis. The situation is similar to the measure of slope in analytic geometry, where the slope of a line depends on the position of the coordinate axes. In order to determine the influence of a choice of basis, we must examine how different choices affect the coordinates of vectors and the matrices of maps.

§1. Change in Coordinates

Let B be a basis for the vector space $\mathscr{V}$, with $\dim \mathscr{V} = n$. Then each vector $V \in \mathscr{V}$ has coordinates with respect to B, say $V\colon (x_1, \ldots, x_n)_B$. If a new basis B' is chosen for $\mathscr{V}$, and $V\colon (x_1', \ldots, x_n')_{B'}$, then the problem is to determine how the new coordinates x_i' are related to the old coordinates x_i. Suppose $B = \{U_1, \ldots, U_n\}$ and $B' = \{U_1', \ldots, U_n'\}$, then

$$x_1U_1 + \cdots + x_nU_n = V = x_1'U_1' + \cdots + x_n'U_n'.$$

So a relationship between the two sets of coordinates can be obtained by expressing the vectors from one basis in terms of the other basis. Therefore, suppose that for each j, $1 \le j \le n$,

$$U_j' = p_{1j}U_1 + \cdots + p_{nj}U_n = \sum_{i=1}^{n} p_{ij}U_i.$$

That is, $U_j'\colon (p_{1j}, \ldots, p_{nj})_B$. The $n \times n$ matrix $P = (p_{ij})$ will be called the *transition matrix* from the basis B to the basis B'. Notice that the jth column of the transition matrix P contains the coordinates of U_j' with respect to B. The scalars p_{ij} could be indexed in other ways, but this notation will conform with the notation for matrices of a linear transformation.

Now the equation

$$x_1U_1 + \cdots + x_nU_n = x_1'U_1' + \cdots + x_n'U_n'$$

becomes

$$x_1U_1 + \cdots + x_nU_n = x_1'\left(\sum_{i=1}^{n} p_{i1}U_i\right) + \cdots + x_n'\left(\sum_{i=1}^{n} p_{in}U_i\right),$$

and since $\{U_1, \ldots, U_n\}$ is linearly independent, the coefficient of U_i on the left equals the coefficient of U_i on the right. For example, the terms involving U_1 on the right are $x_1'p_{11}U_1 + x_2'p_{12}U_1 + \cdots + x_n'p_{1n}U_1$, and this sum equals $(\sum_{j=1}^n p_{1j}x_j')U_1$, so that the equation becomes

$$x_1U_1 + \cdots + x_nU_n = \left(\sum_{j=1}^n p_{1j}x_j'\right)U_1 + \left(\sum_{j=1}^n p_{2j}x_j'\right)U_2 + \cdots + \left(\sum_{j=1}^n p_{nj}x_j'\right)U_n.$$

Thus

$$x_1 = \sum_{j=1}^n p_{1j}x_j', \qquad x_2 = \sum_{j=1}^n p_{2j}x_j', \qquad \ldots, \qquad x_n = \sum_{j=1}^n p_{nj}x_j'.$$

The general relationship between the new and old coordinates,

$$x_i = \sum_{j=1}^n p_{ij}x_j'$$

should suggest the matrix product

$$\begin{pmatrix} x_1 \\ x_2 \\ \vdots \\ x_n \end{pmatrix} = \begin{pmatrix} p_{11} & p_{12} & \cdots & p_{1n} \\ p_{21} & p_{22} & \cdots & p_{2n} \\ & & \vdots & \\ p_{n1} & p_{n2} & \cdots & p_{nn} \end{pmatrix} \begin{pmatrix} x_1' \\ x_2' \\ \vdots \\ x_n' \end{pmatrix}. \tag{1}$$

If the column of coordinates for V with respect to B is denoted by X and the column of coordinates for V with respect to B' by X', then the matrix equation (1) becomes $X = PX'$. That is, the product of the transition matrix P from B to B' times the coordinates of a vector V with respect to B' yields the coordinates of V with respect to B.

Example 1 $B = \{(2, 1), (1, 0)\}$ and $B' = \{(0, 1), (1, 4)\}$ are bases for $\mathscr{V}_2$. By inspection $(0, 1)$: $(1, -2)_B$ and $(1, 4)$: $(4, -7)_B$, therefore the transition matrix from B to B' is

$$P = \begin{pmatrix} 1 & 4 \\ -2 & -7 \end{pmatrix}.$$

To check Eq. (1) for this change in coordinates, consider a particular vector from $\mathscr{V}_2$, say $(3, -5)$. Since $(3, -5)\colon (-5, 13)_B$ and $(3, -5)\colon (-17, 3)_{B'}$,

$$X = \begin{pmatrix} -5 \\ 13 \end{pmatrix} \quad \text{and} \quad X' = \begin{pmatrix} -17 \\ 3 \end{pmatrix}.$$

Therefore

$$PX' = \begin{pmatrix} 1 & 4 \\ -2 & -7 \end{pmatrix}\begin{pmatrix} -17 \\ 3 \end{pmatrix} = \begin{pmatrix} -5 \\ 13 \end{pmatrix} = X$$

as desired.

It might be helpful to work through the derivation of the formula $X = PX'$ for a particular example, see problem 4 at the end of this section.

Example 2 Suppose the problem is to find the coordinates of several vectors in $\mathscr{V}_2$ with respect to the basis $B = \{(3, 4), (2, 2)\}$. Rather than obtain the coordinates of each vector separately, the transition matrix from B to $\{E_i\}$ can be used to obtain coordinates with respect to B from the coordinates with respect to $\{E_i\}$, i.e., the components. In this case the standard basis is viewed as the new basis and the coordinates of E_i with respect to B are found to be $(1, 0)\colon (-1, 2)_B$ and $(0, 1)\colon (1, -3/2)_B$. Therefore the coordinates of an arbitrary vector (a, b) with respect to B are given by the product

$$\begin{pmatrix} -1 & 1 \\ 2 & -3/2 \end{pmatrix}\begin{pmatrix} a \\ b \end{pmatrix}.$$

Example 3 Find the transition matrix P from the basis $B = \{t + 1, 2t^2, t - 1\}$ to $B' = \{4t^2 - 6t, 2t^2 - 2, 4t\}$ for the space $R_3[t]$.

A little computation shows that $4t^2 - 6t\colon(-3, 2, -3)_B$, $2t^2 - 2\colon(-1, 1, 1)_B$, and $4t\colon (2, 0, 2)_B$. Therefore

$$P = \begin{pmatrix} -3 & -1 & 2 \\ 2 & 1 & 0 \\ -3 & 1 & 2 \end{pmatrix}.$$

A change of basis can be viewed either as a change in the names (coordinates) for vectors or as a transformation. The first view was used in deriving the formula $X = PX'$ and can be illustrated geometrically with the space $\mathscr{V}_2$. Suppose B is the standard basis for $\mathscr{V}_2$ and $B' = \{(3, 1), (1, 2)\}$.

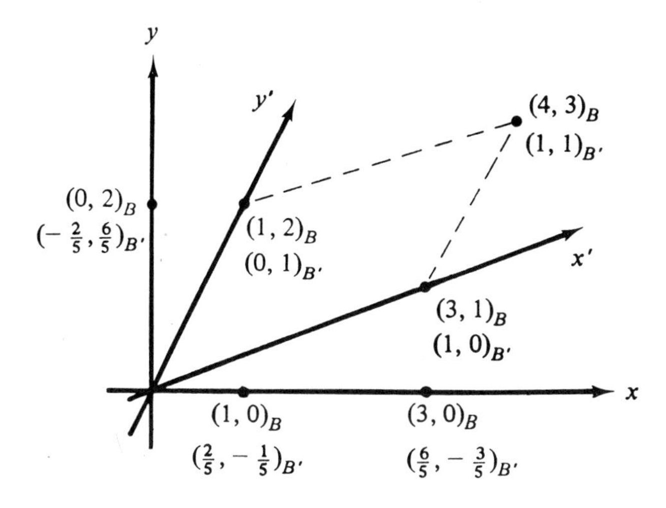

Figure 1

Then using our interpretation of vectors as points in the Cartesian plane E^2, the two sets of coordinates give two names for each point in the plane. This is shown in Figure 1 where the two sets of coordinate names are listed for a few points. The vectors represented by the points are not indicated, but their components are the same as the coordinates with respect to the standard basis B. The familiar x and y axes of the Cartesian coordinate system represent the subspaces spanned by (1, 0) and (0, 1), respectively. But the two lines representing the span of each vector in B' can also be viewed as coordinate axes, the x' axis representing $\mathscr{L}\{(3, 1)\}$ and the y' axis representing $\mathscr{L}\{(1, 2)\}$. These two axes define a non-Cartesian coordinate system in E^2 corresponding to B'. It is not difficult to plot points in such a coordinate system using the parallelogram rule for addition of vectors. For example, if A is the point with coordinates $(1/2, -1)$ in the $x'y'$ system, then it is the point representing the vector $\frac{1}{2}(3, 1) + (-1)(1, 2)$. This point is plotted in Figure 2 where the ordered pairs denote vectors, not coordinates. This $x'y'$ coordinate system looks strange because of our familarity with Euclidean geometry, but most of the bases we have considered for $\mathscr{V}_2$ would yield similar non-Cartesian coordinate systems in E^2.

The second way to view a change of basis is as the linear transformation which sends the old basis to the new basis. With $B = \{U_1, \ldots, U_n\}$ and $B' = \{U_1', \ldots, U_n'\}$, there is a linear transformation from $\mathscr{V}$ to itself which sends U_i to U_i' for each i. This map is defined by

$$T(x_1 U_1 + \cdots + x_n U_n) = x_1 U_1' + \cdots + x_n U_n'.$$

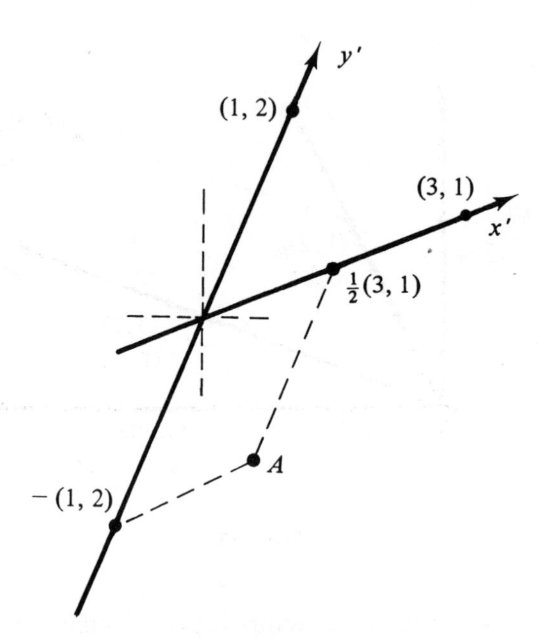

Figure 2

Therefore T sends the vector having coordinates $x_1, \ldots, x_n$ with respect to B to the vector having the same coordinates with respect to B'. Consider the above example in $\mathscr{V}_2$. In this case, the map T is given by

$$T(a, b) = T(a(1, 0) + b(0, 1)) = a(3, 1) + b(1, 2) = (3a + b, a + 2b).$$

Since T sends $\mathscr{V}_2$ to itself, it can be represented as a map from the plane to the plane, as in Figure 3. T sends the x axis to the x' axis and the y axis to the y' axis, so the xy coordinate system is used in the domain of T and the $x'y'$ system in the codomain. When viewed as a map, this change of basis for $\mathscr{V}_2$ is again seen to be too general for Euclidean geometry, for not only isn't the Cartesian coordinate system sent to a Cartesian system, but the unit square with vertices (0, 0), (1, 0), (0, 1), (1, 1) is sent to the parallelogram with vertices (0, 0), (3, 1), (1, 2), (4, 3). Therefore the map T, arising from this change in basis, distorts length, angle, and area. This does not imply that this change in basis is geometrically useless, for it provides an acceptable coordinate change in both affine geometry and projective geometry.

Now to bring the two views for a change of basis together, suppose $B = \{U_1, \ldots, U_n\}$ and $B' = \{U_1', \ldots, U_n'\}$ with P the transition matrix from B to B' and T the linear transformation given by $T(U_i) = U_i'$. Then

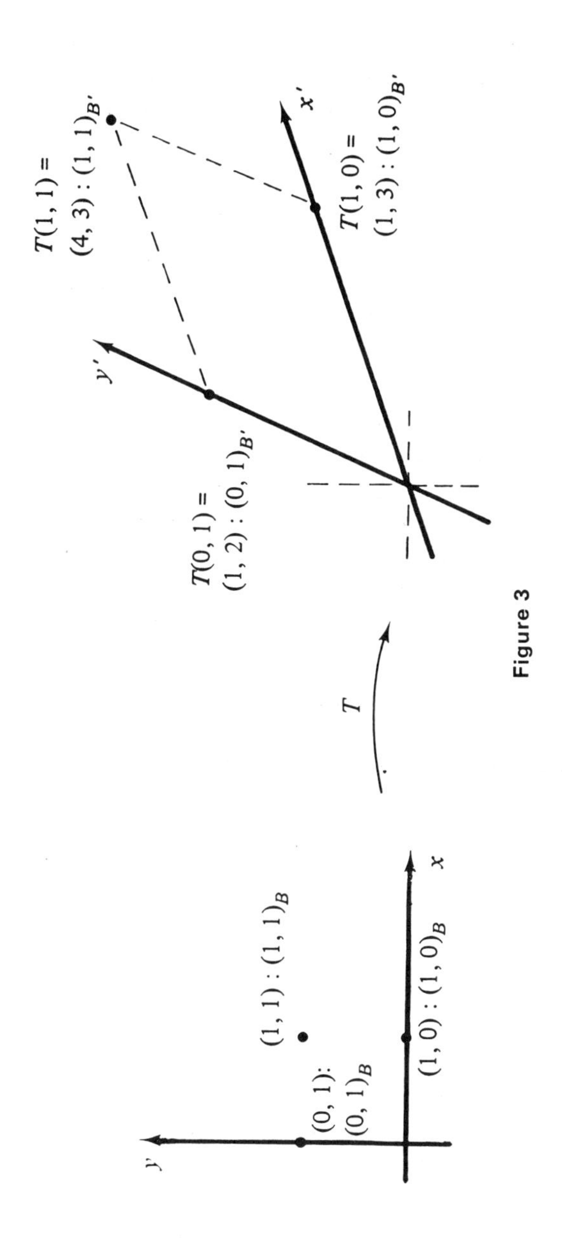

Figure 3

P is the matrix of T with respect to the basis B in both the domain and codomain. This is immediate for if A is the matrix of T with respect to B, then the jth column of A contains the coordinates of $T(U_j)$ with respect to B. But $T(U_j) = U_j'$ and the coordinates of U_j' with respect to B are in the jth column of P, so $A = P$. This shows that the transition matrix P is nonsingular, for T is nonsingular in that it sends a basis to a basis.

Theorem 5.1 If P is the transition matrix from B to B', then

1. P is nonsingular.
2. $X' = P^{-1}X$. That is, coordinates of vectors with respect to B' are obtained from coordinates with respect to B upon multiplication by P^{-1}.
3. P^{-1} is the transition matrix from B' to B.

Proof 1 has been obtained above. 2 is obtained from the equation $PX' = X$ on multiplication by the matrix P^{-1}. For 3, suppose Q is the transition matrix from B' to B. Then the jth column of Q contains the coordinates of U_j (the jth basis vector in B) with respect to B'. By 2 these coordinates are given by $P^{-1}X$ where X contains the coordinates of U_j with respect to B. But $X = (0, \ldots, 0, 1, 0, \ldots, 0)^T = E_j^T$. Therefore the product $P^{-1}X = P^{-1}E_j^T$ is the jth column of P^{-1}. That is, $Q = P^{-1}$ as claimed.

Example 4 Find the transition matrix P from $B = \{(3, -2), (4, -2)\}$ to the standard basis for $\mathscr{V}_2$.

The transition matrix from $\{E_i\}$ to B is

$$\begin{pmatrix} 3 & 4 \\ -2 & -2 \end{pmatrix}.$$

By Theorem 5.1, this is P^{-1}, therefore

$$P = \frac{1}{2}\begin{pmatrix} -2 & -4 \\ 2 & 3 \end{pmatrix} = \begin{pmatrix} -1 & -2 \\ 1 & 3/2 \end{pmatrix}.$$

Theorem 5.2 Suppose $\dim \mathscr{V} = n$, B is a basis for $\mathscr{V}$, and $S = \{U_1, \ldots, U_n\} \subset \mathscr{V}$. Let A be the $n \times n$ matrix containing the coordinates of U_j with respect to B in its jth column, $1 \leq j \leq n$. Then S is a basis for $\mathscr{V}$ if and only if A is nonsingular.

Proof ($\Rightarrow$) If S is a basis, A is the transition matrix from B to S and therefore nonsingular.

(⇐) Let $B = \{V_1, \ldots, V_n\}$ and let T be the map in Hom($\mathscr{V}$) that has A as its matrix with respect to B. (Why does such a map exist?) Then A nonsingular implies that T is nonsingular. Thus $T(V_1) = U_1, \ldots, T(V_n) = U_n$ are linearly independent. Since the dimension of $\mathscr{V}$ is n, S is a basis for $\mathscr{V}$.

Problems

1. Find the transition matrix from B to B' when
  a. $B = \{(2, 3), (0, 1)\}$, $B' = \{(6, 4), (4, 8)\}$.
  b. $B = \{(5, 1), (1, 2)\}$, $B' = \{(1, 0), (0, 1)\}$.
  c. $B = \{(1, 1, 1), (1, 1, 0), (1, 0, 0)\}$,
  $B' = \{(2, 0, 3), (-1, 4, 1), (3, 2, 5)\}$.
  d. $B = \{t, 1, t^2\}$, $B' = \{3 + 2t + t^2, t^2 - 4, 2 + t\}$.
  e. $B = \{t, t^2 + 1, t^2 + t\}$, $B' = \{t^2 + 3t - 3, 4t^2 + t + 2, t^2 - 2t + 1\}$.
  f. $B = \{T_1, T_2, T_3\}$, $B' = \{S_1, S_2, S_3\}$, with $T_1(a, b, c) = 2a$, $T_2(a, b, c) = a - b$, $T_3(a, b, c) = c$, $S_1(a, b, c) = 3a - b + 4c$, $S_2(a, b, c) = 4b + 5c$, $S_3(a, b, c) = a + b + c$. [That is, B and B' are bases for Hom($\mathscr{V}_3$).]

2. Use a transition matrix and coordinates with respect to the standard basis to find the coordinates of the given vectors with respect to the basis $B = \{(3, 1), (5, 2)\}$.
  a. $(4, 3)$. b. $(1, 3)$. c. $(-3, -2)$. d. (a, b).

3. Use a transition matrix and coordinates with respect to $\{E_i\}$ in $\mathscr{V}_3$ to find the coordinates of the given vectors with respect to the basis $B = \{(2, -2, 1), (3, 0, 2), (3, 5, 3)\}$.
  a. $(1, 1, 1)$. b. $(2, 3, 2)$. c. $(7, 1, 5)$.

4. Work through the derivation of the formula $X = PX'$ for the following examples:
  a. $\mathscr{V} = \mathscr{V}_2$, $B = \{(2, 1), (1, 0)\}$, $B' = \{(0, 1), (1, 4)\}$.
  b. $\mathscr{V} = R_3[t]$, $B = \{1, t, t^2\}$, $B' = \{1 + t^2, 1 + t, t + t^2\}$.

5. a. Prove that $(AB)^T = B^TA^T$ and $(A^{-1})^T = (A^T)^{-1}$.
  b. Show that the relationships between coordinates with respect to two different bases can be written with the coordinates as row vectors to give

$$(x_1, \ldots, x_n) = (x_1', \ldots, x_n')P^T$$

and

$$(x_1', \ldots, x_n') = (x_1, \ldots, x_n)(P^{-1})^T.$$

6. For each basis B' below, make two sketches for the change from the standard basis (B) for $\mathscr{V}_2$ to B', representing the change first as a renaming of points and second as a transformation from the plane to the plane. How do the transformations affect the geometry of the plane?
  a. $B' = \{(1, 1), (4, -1)\}$. b. $B' = \{(1, 0), (2, 1)\}$.
  c. $B' = \{(2, 0), (0, 2)\}$. d. $B' = \{(1/\sqrt{2}, 1/\sqrt{2}), (-1/\sqrt{2}, 1/\sqrt{2})\}$.
  e. $B' = \{(-1, 0), (0, 1)\}$. f. $B' = \{(-1, 0), (0, -1)\}$.

7. Suppose B_1 and B_2 are bases for a vector space $\mathscr{V}$ and A is the matrix of the identity map I: $\mathscr{V} \to \mathscr{V}$ with respect to B_1 in the domain and B_2 in the codomain. Show that A is the transition matrix from B_2 to B_1.

8. Prove that a transition matrix is nonsingular without reference to the linear map derived from the change in basis.

§2. Change in the Matrix of a Map

If T is a linear transformation from $\mathscr{V}$ to $\mathscr{W}$, then each choice of bases for $\mathscr{V}$ and $\mathscr{W}$ yields a matrix for T. In order to determine how two such matrices are related, it is only necessary to combine the relationships involved. Suppose A is the matrix of T with respect to B_1 in $\mathscr{V}$ and B_2 in $\mathscr{W}$ and A' is the matrix of T with respect to B_1' and B_2'. Then A' should be related to A by transition matrices. Therefore let P be the transition matrix from the basis B_1 to B_1' for $\mathscr{V}$ and let Q be the transition matrix from the basis B_2 to B_2' for $\mathscr{W}$. Now the matrices of a map and the transition matrices are used with coordinates, so for $U \in \mathscr{V}$ let X and X' be the coordinates of U with respect to B_1 and B_1', respectively. Then $T(U) \in \mathscr{W}$, so let Y and Y' be the coordinates of $T(U)$ with respect to B_2 and B_2', respectively. These coordinates are related as follows:

$$X = PX', \qquad Y = QY', \qquad Y = AX, \qquad Y' = A'X'.$$

These equations yield $QY' = Y = AX = A(PX')$ or $Y' = Q^{-1}(A(PX'))$. But $Y' = (Q^{-1}AP)X'$ states that $Q^{-1}AP$ is the matrix of T with respect to B_1' and B_2' or $A' = Q^{-1}AP$.

Example 1 Let $T \in \text{Hom}(\mathscr{V}_3, \mathscr{V}_2)$ be given by $T(a, b, c) = (3a - c, 4b + 2c)$. Let

$$B_1 = \{(1, 1, 1), (1, 1, 0), (1, 0, 0)\}, \qquad B_2 = \{(0, 1), (1, -2)\},$$

$$B_1' = \{(2, -1, 3), (1, 0, 2), (0, 1, 1)\}, \qquad B_2' = \{(1, 1), (2, 1)\}.$$

Then the matrix of T with respect to B_1 and B_2 is

$$A = \begin{pmatrix} 10 & 10 & 6 \\ 2 & 3 & 3 \end{pmatrix},$$

and

$$A' = \begin{pmatrix} 1 & 7 & 13 \\ 1 & -3 & -7 \end{pmatrix}$$

is the matrix of T with respect to B_1' and B_2'. The transition matrix from B_1 to B_1' is

$$P = \begin{pmatrix} 3 & 2 & 1 \\ -4 & -2 & 0 \\ 3 & 1 & -1 \end{pmatrix}$$

and the transition matrix from B_2 to B_2' is

$$Q = \begin{pmatrix} 3 & 5 \\ 1 & 2 \end{pmatrix},$$

with

$$Q^{-1} = \begin{pmatrix} 2 & -5 \\ -1 & 3 \end{pmatrix}.$$

Now the formula $Q^{-1}AP = A'$ can be verified for this example:

$$\begin{aligned} Q^{-1}AP &= \begin{pmatrix} 2 & -5 \\ -1 & 3 \end{pmatrix}\begin{pmatrix} 10 & 10 & 6 \\ 2 & 3 & 3 \end{pmatrix}\begin{pmatrix} 3 & 2 & 1 \\ -4 & -2 & 0 \\ 3 & 1 & -1 \end{pmatrix} \\ &= \begin{pmatrix} 2 & -5 \\ -1 & 3 \end{pmatrix}\begin{pmatrix} 8 & 6 & 4 \\ 3 & 1 & -1 \end{pmatrix} = \begin{pmatrix} 1 & 7 & 13 \\ 1 & -1 & -7 \end{pmatrix} = A'. \end{aligned}$$

Theorem 5.3 *A and A' are matrices of some linear map from $\mathscr{V}$ to $\mathscr{W}$ if an only if there exist nonsingular matrices P and Q such that $A' = Q^{-1}AP$.*

Proof ($\Rightarrow$) This has been obtained above.

($\Leftarrow$) If A and A' are $n \times m$ matrices, then $\dim \mathscr{V} = m$ and $\dim \mathscr{W} = n$. Choose a basis $B_1 = \{V_1, \ldots, V_m\}$ for $\mathscr{V}$ and a basis $B_2 = \{W_1, \ldots, W_n\}$ for $\mathscr{W}$. Then there exists a map $T \in \mathrm{Hom}(\mathscr{V}, \mathscr{W})$ which has A as its matrix with respect to B_1 and B_2. We now use P and Q to construct new bases for

$\mathscr{V}$ and $\mathscr{W}$. If $P = (p_{ij})$, then there exist vectors $V'_1, \ldots, V'_m$ such that

$$V'_1: (p_{11}, p_{21}, \ldots, p_{m1})_{B_1}, \ldots, V'_m: (P_{1m}, p_{2m}, \ldots, p_{mm})_{B_1}.$$

Let $B'_1 = \{V'_1, \ldots, V'_m\}$. Then P is the matrix of coordinates of the vectors from B'_1 with respect to the basis B_1 and P is nonsingular, so by Theorem 5.2, B'_1 is a basis for $\mathscr{V}$. Further the transition matrix from B_1 to B'_1 is P. We define $B'_2 = \{W'_1, \ldots, W'_n\}$ similarly with W'_i: $(q_{1i}, q_{2i}, \ldots, q_{ni})_{B_2}$ for $1 \leq i \leq n$, where $Q = (q_{ij})$. Then B'_2 is a basis for $\mathscr{W}$ and the transition matrix from B_2 to B'_2 is Q. Thus, by the first part of this theorem, the matrix of T with respect to B'_1 and B'_2 is $Q^{-1}AP$, completing the proof.

The matrix equation $A' = Q^{-1}AP$ expresses the relationship between two matrices for a map. Examination of this relation would reveal properties shared by all possible matrices for a given map. If such an examination encompassed all possible choices of bases, then Q^{-1} would represent an arbitrary nonsingular matrix and could be denoted more simply by Q.

Definition Two $n \times m$ matrices A and B are *equivalent* if there exist nonsingular matrices P and Q such that $B = QAP$.

Recall that in Chapter 3 we called two matrices equivalent if one could be obtained from the other using elementary row and column operations. It should be clear that these two definitions are equivalent, as stated in problem 3 at the end of this section.

Using the term equivalent, Theorem 5.3 can be restated as follows: Two matrices are equivalent if and only if they are the matrices for some map with respect to two choices of bases in the domain and codomain. In Chapter 4 several examples and problems produced different matrices for the same map and hence equivalent matrices. Examples 2 through 6 on pages 157 to 162 contain several pairs of equivalent matrices.

Example 2 Show that the 3×3 matrix

$$A = \begin{pmatrix} 3 & 1 & 0 \\ 0 & -2 & 4 \\ 1 & 0 & 3 \end{pmatrix}$$

is equivalent to the identity matrix I_3.

Consider the map $T \in \text{Hom}(\mathscr{V}_3)$ which has A as its matrix with respect to the standard basis in both the domain and codomain. Then T is given by

$$T(a, b, c) = (3a + b, 4c - 2b, a + 3c).$$

Since $|A| \neq 0$, T is nonsingular and the set

$$\begin{aligned} B &= \{T(1, 0, 0), T(0, 1, 0), T(0, 0, 1)\} \\ &= \{(3, 0, 1), (1, -2, 0), (0, 4, 3)\} \end{aligned}$$

is a basis for $\mathscr{V}_3$. Now the matrix of T with respect to the standard basis in the domain and B in the codomain is I_3, therefore A is equivalent to I_3. (How might this example be generalized to a theorem?)

Equivalence of matrices is obtained by considering a general transformation and arbitrary choices of bases. A very important relationship is obtained in the special case of a map from a vector space to itself if the matrices are obtained using the same basis in both the domain and codomain. To see how this relation should be defined, suppose $T \in \text{Hom}(\mathscr{V})$ has the matrix A with respect to the basis B and the matrix A' with respect to B'. Let P be the transition matrix from B to B'. Then with the same change of basis in both the domain and the codomain, the relation $A' = Q^{-1}AP$ becomes $A' = P^{-1}AP$.

Definition Two $n \times n$ matrices A and B are *similar* if there exists a nonsingular matrix P such that $B = P^{-1}AP$.

Example 3 Suppose $T \in \text{Hom}(\mathscr{V}_3)$ is defined by

$$T(a, b, c) = (2a + 4c, 3b, 4a + 8c).$$

Let B be the standard basis and let $B' = \{(1, 0, 2), (0, 2, 0), (2, 1, -1)\}$. Show that the matrix of T with respect to B is similar to the matrix of T with respect to B'.

A little computation shows that the matrix of T with respect to B is

$$A = \begin{pmatrix} 2 & 0 & 4 \\ 0 & 3 & 0 \\ 4 & 0 & 8 \end{pmatrix}$$

and that

$$A' = \begin{pmatrix} 10 & 0 & 0 \\ 0 & 3 & 0 \\ 0 & 0 & 0 \end{pmatrix}$$

is the matrix of T with respect to B'. The transition matrix from B to B' is

$$P = \begin{pmatrix} 1 & 0 & 2 \\ 0 & 2 & 0 \\ 2 & 0 & -1 \end{pmatrix}$$

and

$$P^{-1} = \begin{pmatrix} 1/5 & 0 & 2/5 \\ 0 & 1/2 & 0 \\ 2/5 & 0 & -1/5 \end{pmatrix}.$$

Therefore we have

$$P^{-1}AP = P^{-1}\begin{pmatrix} 2 & 0 & 4 \\ 0 & 3 & 0 \\ 4 & 0 & 8 \end{pmatrix}\begin{pmatrix} 1 & 0 & 2 \\ 0 & 2 & 0 \\ 2 & 0 & -1 \end{pmatrix} = \begin{pmatrix} 1/5 & 0 & 2/5 \\ 0 & 1/2 & 0 \\ 2/5 & 0 & -1/5 \end{pmatrix}\begin{pmatrix} 10 & 0 & 0 \\ 0 & 6 & 0 \\ 20 & 0 & 0 \end{pmatrix}$$

$$= \begin{pmatrix} 10 & 0 & 0 \\ 0 & 3 & 0 \\ 0 & 0 & 0 \end{pmatrix} = A'.$$

Theorem 5.4 Two $n \times n$ matrices are similar if and only if they are matrices for some linear transformation from a vector space to itself using the same basis in the domain and codomain.

Proof Half of this theorem has been obtained above and the other half follows as in the proof of Theorem 5.3.

Similarity is a special case of the relation equivalence, or similar matrices are also equivalent. Therefore, since the matrices A and A' of Example 3 are similar, they are equivalent. However, equivalent matrices need not be similar. For example the equivalent matrices A and I_3 of Example 2 are not similar; $P^{-1}I_3P = I_3 \neq A$ for all nonsingular matrices P. We will find that similarity is a much more complex relation than equivalence; and the study of similarity will lead to many useful concepts. Before begining such a study, it would be well to consider the general class of relations, called equivalence relations, to which both similarity and equivalence belong.

Problems

1. Let $T(a, b) = (2a - b, 3a - 4b, a + b)$, $B_1 = \{(1, 0), (0, 1)\}$, $B_1' = \{(2, 3), (1, 2)\}$, $B_2 = \{(1, 0, 0), (0, 1, 0), (0, 0, 1)\}$, and $B_2' = \{(1, 1, 1), (1, 0, 0), (1, 0, 1)\}$. Find the matrices A, A', P, Q, and show that $A' = Q^{-1}AP$.
2. Let $T(a, b) = (4a - b, 3a + 2b)$, $B = \{(1, 1), (-1, 0)\}$, and $B' = \{(0, 1), (-1, 3)\}$. Find the matrices A, A', P and show that $A' = P^{-1}AP$.
3. Show that A is equivalent to B if and only if B can be obtained from A by elementary row and column operations.
4. a. Show that $\begin{pmatrix} 2 & 1 \\ 3 & 4 \end{pmatrix}$ is equivalent to I_2.

   b. What condition would insure that a 2×2 matrix is equivalent to the identity matrix I_2?
5. Show that $\begin{pmatrix} 1 & 1 & 1 \\ -3 & 1 & 5 \\ 2 & 1 & 0 \end{pmatrix}$ is equivalent to $\begin{pmatrix} 1 & 0 & 0 \\ 0 & 1 & 0 \\ 0 & 0 & 0 \end{pmatrix}$ by

   a. using a linear transformation.

   b. using the result of problem 3.
6. Suppose $T \in \operatorname{Hom}(\mathscr{V})$ has A as its matrix with respect to B. Prove that if $A' = P^{-1}AP$ for some nonsingular matrix P, then A' is the matrix of T with respect to some basis for $\mathscr{V}$.
7. Find all $n \times n$ matrices which are similar to I_n.
8. In what ways do the relations of equivalence and similarity differ?
9. a. Show that $\begin{pmatrix} 0 & 1 \\ 1 & 0 \end{pmatrix}$ is similar to $\begin{pmatrix} 1 & 0 \\ 0 & -1 \end{pmatrix}$.

   b. Show that $\begin{pmatrix} 0 & 1 \\ -1 & 0 \end{pmatrix}$ is not similar to any matrix of the form $\begin{pmatrix} x & 0 \\ 0 & y \end{pmatrix}$, where x and y are real numbers.
10. Let $A = \begin{pmatrix} 12 & -4 \\ 12 & -2 \end{pmatrix}$, $B = \begin{pmatrix} 4 & 0 \\ 0 & 6 \end{pmatrix}$, and $P = \begin{pmatrix} 2 & 2 \\ 4 & 3 \end{pmatrix}$.

    a. Show that $B = P^{-1}AP$.

    b. Let P_1 and P_2 denote the two columns of P. Show that $AP_1 = 4P_1$ and $AP_2 = 6P_2$.
11. Suppose

$$P^{-1}AP = \begin{pmatrix} r_1 & 0 & \cdots & 0 \\ 0 & r_2 & \cdots & 0 \\ & & \vdots & \\ 0 & 0 & \cdots & r_n \end{pmatrix}.$$

Let P_j denote the jth column of P and prove that for each column, $AP_j = r_jP_j$, $1 \leq j \leq n$. (Notice that this generalizes the result obtained in problem 10.)

12. Let A' be the matrix $P^{-1}AP$ of problem 11, with r's on the diagonal and 0's elsewhere. If A' is the matrix of $T \in \text{Hom}(\mathscr{V})$ with respect to the basis $B' = \{U_1', \ldots, U_n'\}$, then what are the images of the vectors in B'?

§3. Equivalence Relations

An equivalence relation identifies elements that share certain properties. It is much like the relation of equality except that equivalent objects need not be identical. The congruence relation between triangles in plane geometry provides a familiar example. Two congruent triangles are equal in all respects except for their position in the plane. Two such triangles cannot be called equal, yet the relationship between them satisfies three major properties of the equality relation. First, every triangle is congruent to itself. Second, congruence does not depend on the order in which two triangles are considered. And third, if one triangle is congruent to a second and the second is congruent to a third, then the first is congruent to the third. We will use these three properties, shared by equality and congruence of triangles, to characterize an equivalence relation. In order to define such a relation, we need a notation for an arbitrary relation. Therefore suppose S is a set with $a, b \in S$. Let the symbol $\sim$ (tilde) denote a relation defined for elements of S; read $a \sim b$ as "a is related to b." Essentially a relation is a collection of ordered pairs of elements from S written as $a \sim b$. For example, the order relation for real numbers is expressed in this way: $2 < 5$.

Definition A relation $\sim$ defined on a set S is an *equivalence relation* on S, if for all $a, b, c \in S$

1. $a \sim a$ for all $a \in S$ $\qquad \sim$ is *reflexive*.
2. If $a \sim b$, then $b \sim a$ $\qquad \sim$ is *symmetric*.
3. If $a \sim b$ and $b \sim c$, then $a \sim c$ $\qquad \sim$ is *transitive*.

This definition is quite general, for nothing is said about the set S; its elements need not even by vectors—for the congruence relation above they are triangles. And further, the three properties give no hint as to how a particular equivalence relation might be defined.

It should be clear that the relation of equality is an equivalence relation on any set. At the other extreme is the relation defined by $a \sim b$ for all $a, b \in S$

which relates everything to everything else. This easily satisfies the definition of an equivalence relation, but it is not very interesting.

Example 1 Similarity is an equivalence relation on the set of all $n \times n$ matrices. That is, if $S = \mathscr{M}_{n \times n}$ and $A \sim B$ means "A is similar to B" for $A, B \in \mathscr{M}_{n \times n}$, then $\sim$ is an equivalence relation on S.

To justify this, it is necessary to verify that the three conditions in the definition of an equivalence relation hold for similarity.

Reflexive: For any $n \times n$ matrix A, $A = I_n^{-1}AI_n$, therefore A is similar to A and similarity is reflexive.

Symmetric: Suppose A is similar to B. Then there exists a nonsingular matrix P such that $B = P^{-1}AP$. Therefore $PBP^{-1} = PP^{-1}APP^{-1} = A$, or $A = Q^{-1}BQ$ with $Q = P^{-1}$. Q is nonsingular since it is the inverse of a nonsingular matrix, so B is similar to A and similarity is symmetric.

Transitive: Suppose A is similar to B and B is similar to C. Then $B = P^{-1}AP$ and $C = Q^{-1}BQ$, with P and Q nonsingular. So $C = Q^{-1}BQ = Q^{-1}(P^{-1}AP)Q = (PQ)^{-1}A(PQ)$. Since the product of nonsingular matrices is nonsingular, A is similar to C and similarity is transitive.

This completes the proof that similarity is an equivalence relation.

Each equivalence relation identifies elements that share certain properties. For similarity, the collection of all matrices similar to a given matrix would contain all possible matrices for some linear map using the same basis in the domain and the codomain. Such a collection of equivalent elements is called an equivalence class.

Definition Suppose $\sim$ is an equivalence relation defined on the set S. For any element a in S, the set of all elements equivalent to a is called the *equivalence class of a* and will be denoted by $[a]$. Therefore $[a] = \{b \in S | a \sim b\}$.

Example 2 What is the quivalence class of

$$\begin{pmatrix} 2 & 1 \\ 3 & 5 \end{pmatrix}$$

for the equivalence relation of similarity?

Using the definition, one answer is

$$\left[\begin{pmatrix} 2 & 1 \\ 3 & 5 \end{pmatrix}\right] = \left\{P^{-1}\begin{pmatrix} 2 & 1 \\ 3 & 5 \end{pmatrix}P \middle| P \in \mathscr{M}_{2 \times 2}, \text{ and } P \text{ is nonsingular}\right\}.$$

Therefore

$$\begin{pmatrix} 2 & 1 \\ 3 & 5 \end{pmatrix} \in \left[\begin{pmatrix} 2 & 1 \\ 3 & 5 \end{pmatrix}\right]$$

using $P = I_2$, and

$$\begin{pmatrix} 4 & 1 \\ 5 & 3 \end{pmatrix} \in \left[\begin{pmatrix} 2 & 1 \\ 3 & 5 \end{pmatrix}\right]$$

using

$$P = \begin{pmatrix} 2 & -1 \\ -1 & 1 \end{pmatrix}.$$

A second answer uses the map T defined by $T(a, b) = (2a + b, 3a + 5b)$. The matrix of T with respect to the standard basis is

$$\begin{pmatrix} 2 & 1 \\ 3 & 5 \end{pmatrix}.$$

Therefore,

$$\left[\begin{pmatrix} 2 & 1 \\ 3 & 5 \end{pmatrix}\right] = \{A \in \mathcal{M}_{2\times 2} | A \text{ is the matrix of } T \text{ with respect to some basis for } \mathscr{V}_2\}.$$

Using the basis $B = \{(2, -1), (1, 0)\}$,

$$T(2, -1) = (3, 1)\colon (-1, 5)_B,$$
$$T(1, 0) = (2, 3)\colon (-3, 8)_B,$$

and

$$\begin{pmatrix} -1 & -3 \\ 5 & 8 \end{pmatrix} \in \left[\begin{pmatrix} 2 & 1 \\ 3 & 5 \end{pmatrix}\right].$$

Notice that neither of these two answers provides a simple way to determine if say $\begin{pmatrix} 1 & -3 \\ 2 & 1 \end{pmatrix}$ or $\begin{pmatrix} 6 & 1 \\ -1 & 1 \end{pmatrix}$ is in $\left[\begin{pmatrix} 2 & 1 \\ 3 & 5 \end{pmatrix}\right]$.

Not all equivalence classes of 2×2 matrices under similarity are so complicated. For example,

$$\left[\begin{pmatrix} 0 & 0 \\ 0 & 0 \end{pmatrix}\right] = \left\{\begin{pmatrix} 0 & 0 \\ 0 & 0 \end{pmatrix}\right\} \quad \text{and} \quad [I_2] = \{I_2\}.$$

Example 3 Isomorphism defines an equivalence relation on a set of vector spaces. Suppose S is the set of all finite-dimensional vector spaces and let $\mathscr{V} \sim \mathscr{W}$ for $\mathscr{V}, \mathscr{W} \in S$ mean "$\mathscr{V}$ is isomorphic to $\mathscr{W}$." Isomorphism is reflexive since $I_{\mathscr{V}}: \mathscr{V} \to \mathscr{V}$ is an isomorphism and it is symmetric for if $T: \mathscr{V} \to \mathscr{W}$ is an isomorphism, then $T^{-1}: \mathscr{W} \to \mathscr{V}$ is also an isomorphism. Finally the relation is transitive for it is not difficult to show that if $T_1: \mathscr{V} \to \mathscr{W}$ and $T_2: \mathscr{W} \to \mathscr{U}$ are isomorphisms, then $T_2 \circ T_1: \mathscr{V} \to \mathscr{U}$ is also an isomorphism.

What is the nature of an equivalence class for this relation? Since $\mathscr{V}$ is isomorphic to $\mathscr{W}$ if and only if $\dim \mathscr{V} = \dim \mathscr{W}$, there is one equivalence class for each dimension. With $\mathscr{V}_0$ denoting the zero vector space, the equivalence classes can easily be listed, $[\mathscr{V}_0], [\mathscr{V}_1], [\mathscr{V}_2], \ldots, [\mathscr{V}_n], \ldots$, and every finite-dimensional vector space is in one of these equivalence classes; $R_3[t] \in [\mathscr{V}_3]$, $\mathscr{M}_{2\times 2} \in [\mathscr{V}_4]$, $\text{Hom}(\mathscr{V}_2, \mathscr{V}_3) \in [\mathscr{V}_6]$, and so on.

Since an equivalence class in S contains equivalent elements, the character of a class and the differences between classes provide information about both the equivalence relation and the elements of S. The following theorem gives the basic properties of equivalence classes.

Theorem 5.5 Suppose $\sim$ is an equivalence relation on the set S, then

1. $a \in [a]$ for every element a in S.
2. $a \sim b$ if and only if $[a] = [b]$.
3. For any two equivalence classes $[a]$ and $[b]$, either $[a] = [b]$ or $[a]$ and $[b]$ have no elements in common.

Proof 1 and 3 are left as problems. For 2 there are two parts.

($\Rightarrow$) Suppose $a \sim b$. To prove that $[a] = [b]$ it is necessary to show that $[a] \subset [b]$ and $[b] \subset [a]$. Therefore suppose $x \in [a]$, that is, $a \sim x$. By the symmetry of $\sim$, $b \sim a$, and by transitivity, $b \sim a$ and $a \sim x$ imply that $b \sim x$. Therefore $x \in [b]$ and $[a] \subset [b]$. Similarly $[b] \subset [a]$. Therefore if $a \sim b$, then $[a] = [b]$.

($\Leftarrow$) Suppose $[a] = [b]$. By 1, $a \in [a]$, so $a \in [b]$ or $b \sim a$. Now by symmetry $a \sim b$, as desired.

This theorem shows that an equivalence relation defined on a set S divides S into a collection of equivalence classes in such a way that each element of S is in one and only one of the classes. Such a multually exclusive and totally exhaustive division is called a *partition* of S. For the equivalence relation, elements within an equivalence class are essentially identical, and the important differences are between the equivalence classes.

Due to notation, the element a appears to play a special role in the equivalence class $[a]$. However, the preceding theorem shows that if $c \in [a]$, then the class $[a]$ could also be written as $[c]$. It might be better to say that the element *a represents* the equivalence class $[a]$. The equivalence class

$$\left[\begin{pmatrix} 2 & 1 \\ 3 & 5 \end{pmatrix}\right]$$

could also be represented by

$$\begin{pmatrix} -1 & -3 \\ 5 & 8 \end{pmatrix}$$

and written as

$$\left[\begin{pmatrix} -1 & -3 \\ 5 & 8 \end{pmatrix}\right].$$

Although any element in an equivalence class can represent that class, it is often possible to pick a "best" or "simplest" representative. Thus the equivalence classes of vector spaces under isomorphism were denoted $[\mathscr{V}_n]$ using the n-tuple spaces as representatives. When one special element is chosen to represent each equivalence class in S, the choices are called *canonical forms* for the elements of S. Selecting one such element from each equivalence class may also be called a cross section of the equivalence classes. Choosing a canonical form to represent each equivalence class is not done arbitrarily, rather the element chosen should exemplify the properties shared by the elements of the class. If $a \in S$, then the canonical form in the class $[a]$ is called the *canonical form for a*.

The relation of equivalence for $n \times m$ matrices provides a good example of a choice of canonical forms. The proof that equivalence is an equivalence relation on the set $\mathscr{M}_{n \times m}$ follows the pattern for similarity in Example 1. To choose a set of canonical forms, we pick exactly one matrix from each equivalence class. A way to do this has been hinted at in several problems and examples, and follows from the fact that equivalence depends only on rank.

Theorem 5.6 If A is an $n \times m$ matrix of rank k, then A is equivalent to the $n \times m$ matrix with only k nonzero entries appearing as 1's on the main diagonal. Denote this matrix by

$$\begin{pmatrix} I_k & \mathbf{0} \\ \mathbf{0} & \mathbf{0} \end{pmatrix}$$

where the zeros represent blocks of zeros in the matrix.

Proof Suppose A is the matrix of $T \in \operatorname{Hom}(\mathscr{V}, \mathscr{W})$ with respect to some choice of bases. Then it need only be shown that there are bases for $\mathscr{V}$ and $\mathscr{W}$ with respect to which $\begin{pmatrix} I_k & \mathbf{0} \\ \mathbf{0} & \mathbf{0} \end{pmatrix}$ is the matrix of T. Since

$$\dim \mathscr{V} = \dim \mathscr{I}_T + \dim \mathscr{N}_T,$$

a basis $B_1 = \{U_1, \ldots, U_k, U_{k+1}, \ldots, U_m\}$ can be obtained for $\mathscr{V}$, so that $\{T(U_1), \ldots, T(U_k)\}$ is a basis for $\mathscr{I}_T$ and $\{U_{k+1}, \ldots, U_m\}$ is a basis for $\mathscr{N}_T$. (If this is not clear, refer to the proof of Theorem 4.4 on page 141.) $\mathscr{I}_T$ is a subspace of $\mathscr{W}$, so there is a basis for $\mathscr{W}$ of the form $B_2 = \{T(U_1), \ldots, T(U_k), W_{k+1}, \ldots, W_n\}$. Now the matrix of T with respect to B_1 and B_2 contains 0's everywhere except for 1's in the first k entries of its main diagonal. That is, the matrix of T with respect to B_1 and B_2 is $\begin{pmatrix} I_k & \mathbf{0} \\ \mathbf{0} & \mathbf{0} \end{pmatrix}$. Since A and $\begin{pmatrix} I_k & \mathbf{0} \\ \mathbf{0} & \mathbf{0} \end{pmatrix}$ are matrices for T with respect to different bases, they are equivalent.

Theorem 5.6 shows that each equivalence class of matrices under equivalence contains a matrix in the form $\begin{pmatrix} I_k & \mathbf{0} \\ \mathbf{0} & \mathbf{0} \end{pmatrix}$. That each equivalence class contains exactly one matrix in this form follows from the fact that rank QAP = rank A, if Q and P are nonsingular (problem 19, page 198). Therefore matrices of this form can be taken as canonical forms for matrices under equivalence. For example, the matrix

$$\begin{pmatrix} 2 & 1 & 3 & 1 \\ 1 & 0 & 4 & 2 \\ 3 & 1 & 7 & 3 \end{pmatrix}$$

has rank 2, therefore

$$\begin{pmatrix} 1 & 0 & 0 & 0 \\ 0 & 1 & 0 & 0 \\ 0 & 0 & 0 & 0 \end{pmatrix}$$

would be taken as its canonical form. The matrix

$$\begin{pmatrix} 4 & 1 \\ 2 & 3 \\ 5 & 1 \end{pmatrix}$$

also has rank 2, so

$$\begin{pmatrix} 1 & 0 \\ 0 & 1 \\ 0 & 0 \end{pmatrix}$$

is its canonical form.

It terms of linear maps, Theorem 5.6 states that, given any map T, there is some choice of bases for which the matrix of T has the form $\begin{pmatrix} I_k & \mathbf{0} \\ \mathbf{0} & \mathbf{0} \end{pmatrix}$, where k is the rank of T.

The selection of matrices in the form $\begin{pmatrix} I_k & \mathbf{0} \\ \mathbf{0} & \mathbf{0} \end{pmatrix}$ for canonical forms under equivalence reflects the fact that two $n \times m$ matrices are equivalent if and only if they have the same rank. That is, the rank of a matrix is not affected by equivalence, and this single property can be used to determine if one $n \times m$ matrix is equivalent to another.

Definition Suppose $\sim$ is an equivalence relation on S. A property is an *invariant* or *invariant under* $\sim$ provided that when $a \in S$ satisfies the property and $a \sim b$, then b also satisfies the property.

We have seen that not only is rank invariant under equivalence of matrices, but the rank of a particular matrix completely determines which equivalence class it belongs in. Dimension plays a similar role for the equivalence relation defined on finite-dimensional vector spaces by isomorphism. Dimension is invariant under isomorphism, and the equivalence class to which a vector space belongs is completely determined by its dimension. In fact, the invariance of dimension was used to choose the n-tuple spaces $\mathscr{V}_n$ as canonical forms for vector spaces under isomorphism. Rank and dimension are said to be "complete sets of invariants" for equivalence and isomorphism, respectively. In general, a *complete set of invariants* for an equivalence relation is a list of properties that can be used to determine if two elements are equivalent. Often more than one invariant must be checked to determine if two elements are equivalent. For example, it is not possible to determine if two triangles are congruent by simply checking a single property such as area or the length of a side.

Two invariants are easily obtained for similarity. First since similar matrices are equivalent, rank is invariant under similarity. A second invariant is given by the determinant.

Theorem 5.7 If A and B are similar matrices, then $|A| = |B|$.

Proof There exists a nonsingular matrix P such that $B = P^{-1}AP$. Therefore

$$|B| = |P^{-1}AP| = |P^{-1}|\,|A|\,|P| = (1/|P|)\,|A|\,|P| = |A|.$$

It is not difficult to show that rank and determinant do not constitute a complete set of invariants for matrices under similarity. Consider the matrices

$$\begin{pmatrix} 1 & 0 \\ 0 & 1 \end{pmatrix} \quad \text{and} \quad \begin{pmatrix} 0 & 1 \\ -1 & 0 \end{pmatrix}.$$

Each has rank 2 and determinant 1, but they are obviously not similar. $[P^{-1}I_2P = I_2$ for any nonsingular matrix $P.]$

Problems

1. Which of the properties reflexive, symmetric, transitive hold and which fail to hold for the following relations?
   a. The order relation $\leq$ for real numbers.
   b. The relation $\sim$ defined on real numbers by $x \sim y$ if $x^2 + y^2 = 4$.
   c. The relation $\sim$ defined on $S = \{1, 3\}$ by $1 \sim 3$, $1 \sim 1$, $3 \sim 1$ and $3 \sim 3$.
   d. The relation $\sim$ defined on $S = R$ by $1 \sim 3$, $1 \sim 1$, $3 \sim 1$, $3 \sim 3$.
   e. The relation $\sim$ defined on $S = R$ by $a \sim b$ if $ab \geq 0$.
2. What is wrong with the following "proof" that requiring an equivalence relation to be reflexive is superfulous?
   If $a \sim b$, then by symmetry $b \sim a$, and by transitivity, $a \sim b$ and $b \sim a$ implies $a \sim a$. Therefore if $\sim$ is symmetric and transitive, $a \sim a$ for all $a \in S$.
3. Define the relation $\sim$ on the set of integers Z as follows: For $a, b \in Z$, $a \sim b$ if $a - b = 5c$ for some $c \in Z$. That is, $a \sim b$ if $a - b$ is divisable by 5. For example, $20 \sim 40$, $13 \sim -2$, but $2 \nsim 8$.
   a. Prove that $\sim$ is an equivalence relation on Z.
   b. How many equivalence classes does Z have under $\sim$ and what does each look like?

c. What would be a "good" choice for canonical elements under $\sim$?
d. What property of integers is invariant under this relation?

4. Define $\sim$ on R by $a \sim b$ if $a - b$ is rational.
a. Prove that $\sim$ is an equivalence relation on R.
b. Describe the equivalence classes [1], [2/3], [$\sqrt{2}$], and [π].
c. How many equivalence classes are there for this relation?

5. Let S be the set of all directed line segments in the plane E^2. That is, an element $\overrightarrow{PQ} \in S$ is the arrow from the point P to the point Q. Define $\sim$ on S by, $\overrightarrow{PQ} \sim \overrightarrow{XY}$ if $\overrightarrow{PQ}$ and $\overrightarrow{XY}$ have the same direction and length.
a. Show that $\sim$ is an equivalence relation on S.
b. How might an equivalence class be drawn in E^2?
c. Choose a set of canonical forms for the equivalence classes.
d. How might the equivalence classes be related to the geometric use of vectors?

6. Let $T: \mathscr{V}_2 \rightarrow \mathscr{V}_2$ be given by $T(a, b) = (2a - b, 3b - 6a)$. Define $\sim$ on $\mathscr{V}_2$ by $U \sim V$ if $T(U) = T(V)$.
a. Prove that $\sim$ is an equivalence relation on $\mathscr{V}_2$.
b. Give a geometric description of the equivalence classes.

7. Define $T: \mathscr{V}_3 \rightarrow \mathscr{V}_2$ by $T(a, b, c) = (b - 2a + 4c, 4a - 2b - 8c)$ and for $U, V \in \mathscr{V}_3$, define $U \sim V$ by $T(U) = T(V)$. Then $\sim$ is an equivalence relation on $\mathscr{V}_3$ as in problem 6.
a. What is our usual designation for [**0**]?
b. Describe the equivalence classes for $\sim$ geometrically.

8. A matrix A is row-equivalent to B if $B = QA$ for some nonsingular matrix Q (problem 7a, page 000.)
a. Show that row-equivalence is an equivalence relation on $\mathscr{M}_{n \times m}$.
b. Find an invariant under row-equivalence.
c. Choose canonical forms for matrices under row-equivalence.
d. How many equivalence classes does $\mathscr{M}_{2 \times 2}$ have under row-equivalence?

9. Let S be the set of all triangles in the plane and let $\sim$ be the congruence relation between triangles. List some properties of triangles which are invariant under this equivalence relation.

10. If $\begin{pmatrix} I_k & \mathbf{0} \\ \mathbf{0} & \mathbf{0} \end{pmatrix}$ is the canonical form for an $n \times m$ matrix under equivalence, then what are the sizes of the three blocks of zeros?

§4. Similarity

We have seen that every matrix has a very simple canonical form under equivalence, and that rank can be used to determine when two matrices are equivalent. In contrast, we have not mentioned a canonical form for similarity nor have we found a complete set of invariants that could be used to determine when any two matrices are similar. With the results for equivalence in mind, one might hope that every $n \times n$ matrix is similar to a matrix with all entries off the main diagonal equal to zero. Although the problem of obtaining a canonical form for similarity is not so easily resolved, there are many situations in which the matrices under consideration are similar to such matrices.

Definition An $n \times n$ matrix $A = (a_{ij})$ is *diagonal* if $a_{ij} = 0$ when $i \neq j$. An $n \times n$ matrix is *diagonalizable* if it is similar to a diagonal matrix.

Since a diagonal matrix $A = (a_{ij})$ has zero entries off the main diagonal, it is often written as diag $(a_{11}, a_{22}, \ldots, a_{nn})$. Thus

$$\text{diag}(2, 5, 3) = \begin{pmatrix} 2 & 0 & 0 \\ 0 & 5 & 0 \\ 0 & 0 & 3 \end{pmatrix}.$$

The matrix

$$A = \begin{pmatrix} 2 & 0 & 4 \\ 0 & 3 & 0 \\ 4 & 0 & 8 \end{pmatrix}$$

of Example 3, page 211, was found to be diagonalizable, for A is similar to diag(10, 3, 0).

Our immediate goal will be to determine when a matrix is diagonalizable. That not every $n \times n$ matrix is similar to a diagonal matrix can be seen by considering the matrix

$$\begin{pmatrix} 0 & 0 \\ 1 & 0 \end{pmatrix}.$$

This matrix is similar to matrices of the form

$$\begin{pmatrix} a & b \\ c & d \end{pmatrix}^{-1} \begin{pmatrix} 0 & 0 \\ 1 & 0 \end{pmatrix} \begin{pmatrix} a & b \\ c & d \end{pmatrix} = \frac{1}{ad - bc} \begin{pmatrix} -ab & -b^2 \\ a^2 & ab \end{pmatrix},$$

where a, b, c, and d satisfy $ad - bc \neq 0$. Such a matrix is diagonal only if $a^2 = b^2 = 0$. But then $ad - bc = 0$, so $\begin{pmatrix} 0 & 0 \\ 1 & 0 \end{pmatrix}$ is not similar to a diagonal matrix.

What can be concluded from the fact that an $n \times n$ matrix A is similar to the diagonal matrix $D = \text{diag}(\lambda_1, \lambda_2, \ldots, \lambda_n)$? One way to find out would be to consider the map $T \in \text{Hom}(\mathscr{V}_n)$ having matrix A with respect to some basis B and matrix D with respect to B'. Then the transition matrix P from B to B' satisfies $P^{-1}AP = D$. Since D is a diagonal matrix, the action of T on the basis B' is particularly simple. For if $B' = \{U_1', \ldots, U_n'\}$, then $T(U_1') = \lambda_1 U_1', \ldots, T(U_n') = \lambda_n U_n'$. If the equations $T(U_j') = \lambda_j U_j'$ are written using the matrix A, they become $AP_1 = \lambda_1 P_1, \ldots, AP_n = \lambda_n P_n$ where $P_1, \ldots, P_n$ are the coordinates of $U_1', \ldots, U_n'$ with respect to B. But by definition, the jth column of P contains the coordinates of U_j' with respect to B. This relationship, $AP_j = \lambda_j P_j$, between A, the columns of P, and the diagonal entries of D gives a necessary and sufficient condition for an $n \times n$ matrix to be diagonalizable.

Theorem 5.8 An $n \times n$ matrix A is diagonalizable if and only if there exist n linearly independent column vectors $P_1, \ldots, P_n$ such that $AP_1 = \lambda_1 P_1, \ldots, AP_n = \lambda_n P_n$ for some scalars $\lambda_1, \ldots, \lambda_n$. Further, if P is the $n \times n$ matrix with P_j as its jth column, then $P^{-1}AP = \text{diag}(\lambda_1, \lambda_2, \ldots, \lambda_n)$.

Proof ($\Rightarrow$) This direction was obtained above.

($\Leftarrow$) Suppose $AP_j = \lambda_j P_j$ for $1 \leq j \leq n$, and P_j is the jth column of the $n \times n$ matrix P. Since the n columns of P are linearly independent by assumption, P is nonsingular, and we must show that $P^{-1}AP$ is a diagonal matrix.

Write P as $(P_1, \ldots, P_n)$. Then since matrix multiplication is row by column, the product AP can be written as $(AP_1, \ldots, AP_n)$. Therefore, using the hypothesis, $AP = (\lambda_1 P_1, \ldots, \lambda_n P_n)$. That is, AP is obtained from P by multiplying each column of P by a scalar, or AP is obtained from P by n elementary column operations of the second type, and the product of the n elementary matrices that perform these operations is $\text{diag}(\lambda_1, \ldots, \lambda_n)$. Therefore

$$AP = P\,\text{diag}(\lambda_1, \ldots, \lambda_n) \qquad \text{or} \qquad P^{-1}AP = \text{diag}(\lambda_1, \ldots, \lambda_n).$$

Definition Suppose A is an $n \times n$ matrix and V is a nonzero column vector such that $AV = \lambda V$ for some scalar λ. Then λ is a *characteristic value* of A and V is a *characteristic vector* for A corresponding to λ.

Several other terms are used for characteristic value and characteristic vector. Most common is eigenvalue and eigenvector, but the adjectives proper, latent, and principal can also be found.

Example 1 Given that 3 and 1 are the characteristic values for

$$A = \begin{pmatrix} -5 & 12 \\ -4 & 9 \end{pmatrix},$$

find all characteristic vectors for A and a matrix P such that $P^{-1}AP$ is diagonal.

A vector

$$V = \begin{pmatrix} a \\ b \end{pmatrix}$$

is a characteristic vector for A corresponding to 3, if $AV = 3V$ or

$$\begin{pmatrix} -5 & 12 \\ -4 & 9 \end{pmatrix}\begin{pmatrix} a \\ b \end{pmatrix} = 3\begin{pmatrix} a \\ b \end{pmatrix} = \begin{pmatrix} 3a \\ 3b \end{pmatrix}.$$

The equations $-8a + 12b = 0$ and $-4a + 6b = 0$ are dependent, yielding the condition $2a = 3b$. Therefore, every vector of the form $(3k, 2k)^T$ with $k \neq 0$ is a characteristic vector for the characteristic value 3. A characteristic vector V corresponding to 1 satisfies $AV = V$ or

$$\begin{pmatrix} -5 & 12 \\ -4 & 9 \end{pmatrix}\begin{pmatrix} a \\ b \end{pmatrix} = \begin{pmatrix} a \\ b \end{pmatrix}.$$

This system is equivalent to $a - 2b = 0$, so all vectors of the form $(2k, k)^T$ with $k \neq 0$ are characteristic vectors of A corresponding to 1.

A is a 2×2 matrix and the vectors $(3, 2)^T$ and $(2, 1)^T$ are two linearly independent characteristic vectors for A. Therefore, by Theorem, 5.8, A is diagonalizable, and if

$$P = \begin{pmatrix} 3 & 2 \\ 2 & 1 \end{pmatrix},$$

then

$$P^{-1}AP = \text{diag}(3, 1) = \begin{pmatrix} 3 & 0 \\ 0 & 1 \end{pmatrix}.$$

Since there are many characteristic vectors for A, there are many matrices that diagonalize A. For example, $(10, 5)^T$ and $(6, 4)^T$ are linearly independent characteristic vectors corresponding to 1 and 3, respectively. So if

$$Q = \begin{pmatrix} 10 & 6 \\ 5 & 4 \end{pmatrix},$$

then

$$Q^{-1}AQ = \begin{pmatrix} 1 & 0 \\ 0 & 3 \end{pmatrix}.$$

Many characteristic vectors were obtained for each characteristic value in Example 1. In general, if λ is a characteristic value for an $n \times n$ matrix A, then there exists a nonzero vector V such that $AV = \lambda V$. This condition can be rewritten as follows:

$$\mathbf{0} = AV - \lambda V = AV - \lambda(I_n V) = (A - \lambda I_n)V.$$

(Why must the identity matrix be introduced?) Therefore, if λ is a characteristic value for A, then the homogeneous system of linear equations $(A - \lambda I_n)X = \mathbf{0}$ has a nontrivial solution, namely, $X = V \neq \mathbf{0}$. Since a homogeneous system with one nontrivial solution has an infinite number of solutions, we see why there are many characteristic vectors for each value. Conversely, given λ such that $(A - \lambda I_n)X = \mathbf{0}$ has a nontrivial solution $X = W \neq \mathbf{0}$, then $(A - \lambda I_n)W = \mathbf{0}$ and $AW = \lambda W$. Therefore, λ is a characteristic value for A if and only if the homogeneous system $(A - \lambda I_n)X = \mathbf{0}$ has a nontrivial solution. But $(A - \lambda I_n)X = \mathbf{0}$ is a system of n equations in n unknowns, therefore it has a nontrivial solution if and only if the determinant of the coefficient matrix $A - \lambda I_n$ is zero. This proves:

Theorem 5.9 Let A be an $n \times n$ matrix and $\lambda \in R$. λ is a characteristic value of A if and only if $\det(A - \lambda I_n) = 0$.

Example 2 For the matrix A of Example 1,

$$\det(A - \lambda I_2) = \left| \begin{pmatrix} -5 & 12 \\ -4 & 9 \end{pmatrix} - \lambda \begin{pmatrix} 1 & 0 \\ 0 & 1 \end{pmatrix} \right| = \begin{vmatrix} -5 - \lambda & 12 \\ -4 & 9 - \lambda \end{vmatrix}$$

$$= \lambda^2 - 4\lambda + 3.$$

The roots of the equation $\lambda^2 - 4\lambda + 3 = 0$ are 3 and 1, the characteristic values given for A in Example 1.

Since the expansion of a determinant contains all possible products taking one element from each row and each column, the determinant of the matrix $A - \lambda I_n$ is an nth degree polynomial in λ.

Definition For an $n \times n$ matrix A, $\det(A - \lambda I_n)$ is the *characteristic polynomial* of A, $\det(A - \lambda I_n) = 0$ is the *characteristic equation* of A, and the roots of the equation $\det(A - \lambda I_n) = 0$ are the *characteristic roots* of A.

The roots of a polynomial equation need not be real numbers. Yet in our vector spaces, the scalars must be real numbers. Therefore, a characteristic root of a matrix need not be a characteristic value. Such a situation occurs with the matrix

$$A = \begin{pmatrix} 0 & 1 \\ -1 & 0 \end{pmatrix},$$

for the characteristic equation for A is $\lambda^2 + 1 = 0$. In the next chapter, complex vector spaces are defined and the equation $AV = \lambda V$ with $\lambda \in C$ can be considered.

Example 3 Determine if

$$A = \begin{pmatrix} 3 & 6 & 6 \\ 0 & 2 & 0 \\ -3 & -12 & -6 \end{pmatrix}$$

is diagonalizable.

The characteristic equation, $\det(A - \lambda I_3) = 0$, for A is

$$\begin{vmatrix} 3-\lambda & 6 & 6 \\ 0 & 2-\lambda & 0 \\ -3 & -12 & -6-\lambda \end{vmatrix} = 0 \qquad \text{or} \qquad (2-\lambda)(\lambda^2 + 3\lambda) = 0,$$

obtained by expanding along the second row. The characteristic roots 2, -3, and 0 are real, so they are characteristic values for A. A is diagonalizable if it has three linearly independent characteristic vectors. Solving the system of linear equations $AV = \lambda V$ for each characteristic value λ yields the characteristic vectors $(12, -5, 3)^T$, $(1, 0, -1)^T$, and $(2, 0, -1)^T$ for 2, -3, and 0, respectively. For example, when $\lambda = -3$, the system of linear equations

$AV = 3V$ is

$$\begin{pmatrix} 3 & 6 & 6 \\ 0 & 2 & 0 \\ -3 & -12 & -6 \end{pmatrix}\begin{pmatrix} a \\ b \\ c \end{pmatrix} = -3\begin{pmatrix} a \\ b \\ c \end{pmatrix} \quad \text{or} \quad \begin{pmatrix} 6 & 6 & 6 \\ 0 & 5 & 0 \\ -3 & -12 & -3 \end{pmatrix}\begin{pmatrix} a \\ b \\ c \end{pmatrix} = \begin{pmatrix} 0 \\ 0 \\ 0 \end{pmatrix}.$$

This system is equivalent to $a + c = 0$ and $b = 0$, therefore $(k, 0, -k)^T$ is a characteristic vector for each $k \neq 0$. The three characteristic vectors above are easily seen to be independent, so A is diagonalizable. In fact, if

$$P = \begin{pmatrix} 12 & 1 & 2 \\ -5 & 0 & 0 \\ 3 & -1 & -1 \end{pmatrix},$$

then

$$P^{-1}AP = \begin{pmatrix} 2 & 0 & 0 \\ 0 & -3 & 0 \\ 0 & 0 & 0 \end{pmatrix}.$$

As the next theorem shows, the linear independence of the three characteristic vectors in the preceding example follows from the fact that the three characteristic values are distinct.

Theorem 5.10 Distinct characteristic values correspond to linearly independent characteristic vectors, or if $\lambda_1 \neq \lambda_2$ and V_1 and V_2 are characteristic vectors for λ_1 and λ_2, respectively, then V_1 and V_2 are linearly independent.

Proof Suppose $aV_1 + bV_2 = \mathbf{0}$ for $a, b \in R$. It must be shown that $a = b = 0$ using the fact that $AV_1 = \lambda_1 V_1$ and $AV_2 = \lambda_2 V_2$. These three equations give

$$\mathbf{0} = A\mathbf{0} = A(aV_1 + bV_2) = aAV_1 + bAV_2 = a\lambda_1 V_1 + b\lambda_2 V_2.$$

Now assume $b \neq 0$ and work for a contradiction. If $b \neq 0$, then $V_2 = (-a/b)V_1$, so

$$\mathbf{0} = a\lambda_1 V_1 + b\lambda_2(-a/b)V_1 = a(\lambda_1 - \lambda_2)V_1.$$

We have $V_1 \neq \mathbf{0}$ since V_1 is a characteristic vector, and $\lambda_1 \neq \lambda_2$ by assumption, therefore $a = 0$. But then $\mathbf{0} = aV_1 + bV_2 = bV_2$ implies $b = 0$, for V_2 is also a characteristic vector. This contradicts the assumption $b \neq 0$, so $b = 0$. Now $\mathbf{0} = aV_1 + bV_2 = aV_1$ implies $a = 0$ and the vectors V_1 and V_2 are linearly independent.

Corollary 5.11 If $\lambda_1, \ldots, \lambda_k$ are distinct characteristic values for a matrix A and $V_1, \ldots, V_k$ are corresponding characteristic vectors, then $V_1, \ldots, V_k$ are linearly independent.

Proof By induction.

Corollary 5.12 An $n \times n$ matrix is diagonalizable if it has n distinct (real) characteristic values.

The converse of Corollary 5.12 is false. For example, I_n is obviously diagonalizable but all n of its characteristic values are the same; the solutions of $(1 - \lambda)^n = 0$. As a less trivial example consider the following.

Example 4 Determine if

$$A = \begin{pmatrix} 3 & 2 & -1 \\ -2 & -2 & 2 \\ 3 & 6 & -1 \end{pmatrix}$$

is diagonalizable. The characteristic equation of A, $\lambda^3 - 12\lambda + 16 = 0$, has roots 2, 2, and -4. Since the roots are not distinct, Corollary 5.12 does not apply, and it must be determined if A has three independent characteristic vectors. It is not difficult to show that only one independent characteristic vector corresponds to -4; therefore A is diagonalizable if there are two linearly independent characteristic vectors corresponding to 2. Such vectors must satisfy the equation

$$\begin{pmatrix} 3 & 2 & -1 \\ -2 & -2 & 2 \\ 3 & 6 & -1 \end{pmatrix}\begin{pmatrix} a \\ b \\ c \end{pmatrix} = 2\begin{pmatrix} a \\ b \\ c \end{pmatrix}.$$

This system is equivalent to the single equation $a + 2b - c = 0$. Since the solutions for a single equation in three unknowns involve two parameters, there are two independent characteristic vectors for the characteristic value 2. Thus A is diagonalizable.

Example 5 Determine if

$$A = \begin{pmatrix} 3 & -2 & 0 \\ -1 & 3 & -1 \\ -5 & 7 & -1 \end{pmatrix}$$

is diagonalizable. The characteristic equation for A is $\lambda^3 - 5\lambda^2 + 8\lambda - 4 = 0$ and it has roots 2, 2, and 1. Again there is only one independent characteristic vector corresponding to 1, and it is necessary to determine if A has two linearly independent characteristic vectors for 2. But $(a, b, c)^T$ is a characteristic vector for 2 if $a - 2b = 0$ and $a - b + c = 0$. Since the solutions for this system depend on only one parameter, there do not exist two linearly independent characteristic vectors for 2, and A is not diagonalizable.

There are many situations in which it is very useful to replace a matrix by a diagonal matrix. A geometric example of this is considered in Chapter 7, but there are several nongeometric applications. Some of these are surveyed in Section 6 of Chapter 8, which briefly considers applications of the Jordan canonical form. This section could be read at this point since a Jordan form for a diagonalizable matrix A is any diagonal matrix similar to A.

Problems

1. Find a matrix P and a diagonal matrix D such that $P^{-1}AP = D$ when A is

   a. $\begin{pmatrix} 3 & 1 \\ -5 & -3 \end{pmatrix}$. b. $\begin{pmatrix} -4 & -3 \\ 8 & 7 \end{pmatrix}$. c. $\begin{pmatrix} 2 & 0 & 6 \\ 3 & 3 & 2 \\ 1 & 0 & 3 \end{pmatrix}$. d. $\begin{pmatrix} 3 & 2 & -1 \\ -2 & -2 & 1 \\ 3 & 6 & -3 \end{pmatrix}$.

2. Determine if the following matrices are diagonalizable.

   a. $\begin{pmatrix} 1 & -4 \\ 2 & 5 \end{pmatrix}$. b. $\begin{pmatrix} 2 & 1 & 2 \\ 3 & 5 & 0 \\ 0 & 0 & 6 \end{pmatrix}$. c. $\begin{pmatrix} 5 & 0 & -3 & 1 \\ 8 & 1 & -6 & 2 \\ 4 & 0 & -2 & 1 \\ 0 & 0 & 0 & 1 \end{pmatrix}$. d. $\begin{pmatrix} 1 & 0 & 1 \\ 3 & 2 & 4 \\ 1 & 0 & 1 \end{pmatrix}$.

3. Find all characteristic vectors for the matrix in Example 4 and a matrix P that diagonalizes A.
4. Find all characteristic vectors for the matrix in Example 5.
5. Why is the zero vector excluded from being a characteristic vector?
6. Suppose A is an $n \times n$ matrix with all n roots of its characteristic equation equal to r, is A then similar to diag $(r, r, \ldots, r)$?
7. Why are there many characteristic vectors for each characteristic value of a matrix?

8. With the answer to problem 7 in mind, find at least two characteristic values for each of the following matrices *by inspection*.

a. $\begin{pmatrix} 2 & 3 & 0 \\ 0 & 4 & 1 \\ 0 & 0 & 3 \end{pmatrix}$. b. $\begin{pmatrix} 2 & 2 & 0 \\ 4 & 4 & 7 \\ 0 & 0 & 5 \end{pmatrix}$. c. $\begin{pmatrix} 1 & 8 & 3 \\ 0 & 7 & 0 \\ 3 & 8 & 1 \end{pmatrix}$.

9. a. A 2×2 matrix can fail to have any (real) characteristic values. Can there be a 3×3 matrix with no (real) characteristic values?
b. Find a 3×3 matrix with only one independent characteristic vector.

10. a. What are the characteristic values of a diagonal matrix?
b. What does this say of the characteristic values of a diagonalizable matrix?

11. Suppose A is the matrix of $T \in \text{Hom}(\mathscr{V})$ with respect to a basis B.
a. What can be said of T if 0 is a characteristic value for A?
b. What subspace of $\mathscr{V}$ corresponds to the set of characteristic vectors for A corresponding to the characteristic value 0?
c. If $A = \text{diag}(a_1, a_2, \ldots, a_n)$ and $B = \{U_1, \ldots, U_n\}$, then what do the vectors $T(U_1), \ldots, T(U_n)$ equal?

§5. Invariants for Similarity

A complete set of invariants for matrices under similarity would be a list of matrix properties that could be checked to determine if two matrices are similar. Such a list could also be used to choose a set of canonical forms for matrices under similarity. The results of the last section do not yield such a complete list of invariants, but they provide a begining.

The characteristic values of a diagonal matrix D are the entries on its main diagonal. If A is similar to D, then the characteristic values of A are on the main diagonal of D. Therefore, the characteristic values and the characteristic polynomial of a diagonalizable matrix are invariant under similarity. This result is easily extended to all $n \times n$ matrices.

Theorem 5.13 Similar matrices have the same characteristic polynomial.

Proof Suppose A and B are similar $n \times n$ matrices with $B = P^{-1}AP$. Then the characteristic polynomial of B is

$$|B - \lambda I_n| = |P^{-1}AP - \lambda I_n| = |P^{-1}AP - \lambda(P^{-1}I_nP)| = |P^{-1}(A - \lambda I_n)P|$$
$$= |P^{-1}|\,|A - \lambda I_n|\,|P| = |A - \lambda I_n|,$$

which is the characteristic polynomial of A.

Since the characteristic polynomial is invariant, its n coefficients provide n numerical invarients for an $n \times n$ matrix A under similarity.

Example 1 Consider the matrix

$$A = \begin{pmatrix} 2 & 0 & 0 \\ 2 & 3 & -3 \\ -1 & 6 & -3 \end{pmatrix}.$$

The characteristic polynomial of A is

$$18 - 9\lambda + 2\lambda^2 - \lambda^3 = (2 - \lambda)(9 + \lambda^2).$$

So the numbers 18, -9, and 2 are invariants for A. That is, if B is similar to A, then the characteristic polynomial of B will have these coefficients. Notice that these three numerical invariants are real while two of the three characteristic roots are complex. These three numerical invariants can be related to both A and the characteristic roots of A. In this case, 18 is the determinant of A and the product of the three characteristic roots; and 2 is the sum of the main diagonal entries of A, as well as the sum of the three characteristic roots.

Definition If $A = (a_{ij})$ is an $n \times n$ matrix, then the *trace* of A, denoted tr A, is $a_{11} + a_{22} + \cdots + a_{nn}$.

The determinant and the trace of A are two of the coefficients in the characteristic polynomial. In general, if $\lambda_1, \ldots, \lambda_n$ are the n characteristic roots of an $n \times n$ matrix A, then the characteristic polynomial for A may be written in the form

$$\begin{aligned} |A - \lambda I_n| &= (\lambda_1 - \lambda)(\lambda_2 - \lambda)\cdots(\lambda_n - \lambda) \\ &= s_n - s_{n-1}\lambda + s_{n-2}\lambda^2 - \cdots + (-1)^{n-1}s_1\lambda^{n-1} + (-1)^n\lambda^n. \end{aligned}$$

Neglecting the minus signs, the coefficients $s_n, s_{n-1}, \ldots, s_1$ give n numerical invariants for A. The defining equation shows that each of these n invariants is a sum of products of characteristic roots. In fact, it is not difficult to see that s_k is the sum of all possible products containing k of the characteristic roots $\lambda_1, \ldots, \lambda_n$. The value of s_k does not depend on the order in which the roots are written down, thus s_k is called the "kth symmetric function" of the n characteristic roots. The functions s_n and s_1 are simplest with

$$s_n = \lambda_1\lambda_2\cdots\lambda_n$$

and

$$s_1 = \lambda_1 + \lambda_2 + \cdots + \lambda_n.$$

In terms of the entries in A, $s_n = \det A$ since it is the constant term in the characteristic polynomial. And an examination of how the terms in λ^{n-1} are obtained in the expansion of $\det(A - \lambda I_n)$ shows that $s_1 = \operatorname{tr} A$. s_{n-1} also has a fairly simple expression in terms of the elements of A. (See problem 3 at the end of this section.)

We now have $n + 1$ numerical invariants for an $n \times n$ matrix under similarity, n from the characteristic polynomial and the rank. But this is not a complete set of invariants. This can be seen by considering

$$\begin{pmatrix} 3 & 0 \\ 0 & 3 \end{pmatrix} \quad \text{and} \quad \begin{pmatrix} 3 & 0 \\ 1 & 3 \end{pmatrix}.$$

These two matrices cannot be similar, $P^{-1}(3I_2)P = 3I_2$ for any P, yet they both have rank 2, trace 6, and determinant 9. The search for additional invariants is more easily conducted with linear transformations than with their matrices. Since similarity arose from the consideration of different matrices for the same map, all of the above results for matrices extend directly to maps.

Definition A scalar λ is a *characteristic value* for $T \in \operatorname{Hom}(\mathscr{V})$, if $T(V) = \lambda V$ for some nonzero vector $V \in \mathscr{V}$. A nonzero vector $V \in \mathscr{V}$ such that $T(V) = \lambda V$ is a *characteristic vector* for T corresponding to the characteristic value λ.

Since matrices of a map with respect to different bases are similar, and it has been shown that the characteristic polynomial is invariant under similarity, the following definition can be made.

Definition Let $T \in \operatorname{Hom}(\mathscr{V})$ and suppose A is the matrix for T with respect to some basis for $\mathscr{V}$. The *characteristic polynomial* and the *characteristic equation* for T are the characteristic polynomial and the characteristic equation for A, respectively. The *characteristic* roots of T are the roots of its characteristic polynomial.

Theorem 5.9 gives the following result for maps:

Theorem 5.14 Let $T \in \operatorname{Hom}(\mathscr{V})$. λ is a characteristic value for T if and only if λ is a real characteristic root for T.

Maps and matrices correspond via a basis. If A is the matrix of T with respect to the basis B, then characteristic vectors for A are the coordinates with respect to B of characteristic vectors for T.

Example 2 Find the characteristic values and characteristic vectors of the map $T \in \text{Hom}(R_3[t])$ defined by

$$T(a + bt + ct^2) = 2a - c + 3bt + (2c - a)t^2.$$

The matrix of T with respect to the basis $B = \{1, t, t^2\}$ is

$$A = \begin{pmatrix} 2 & 0 & -1 \\ 0 & 3 & 0 \\ -1 & 0 & 2 \end{pmatrix},$$

which gives $(3 - \lambda)^2(1 - \lambda)$ as the characteristic polynomial of T. Therefore 3, 3, and 1 are the characteristic values of T. A characteristic vector V for T corresponding to 3 has coordinates $(x, y, z)_B$ which satisfy

$$\begin{pmatrix} 2 & 0 & -1 \\ 0 & 3 & 0 \\ -1 & 0 & 2 \end{pmatrix}\begin{pmatrix} x \\ y \\ z \end{pmatrix} = 3\begin{pmatrix} x \\ y \\ z \end{pmatrix}.$$

This yields $x = -z$ and y is arbitrary, so $(h, k, -h)_B$ are coordinates of a characteristic vector for T provided $h^2 + k^2 \neq 0$. Therefore every vector of the form $h + kt - ht^2$, $h^2 + k^2 \neq 0$, is a characteristic vector for T corresponding to 3.

The characteristic vectors for 1 could be found without using the matrix A, for $a + bt + ct^2$ is a characteristic vector for 1 if

$$T(a + bt + ct^2) = 1(a + bt + ct^2).$$

That is,

$$2a - c + 3bt + (2c - a)t^2 = a + bt + ct^2,$$

which yields $a = c$, $b = 0$. So the vectors $k + kt^2$, $k \neq 0$, are characteristic vectors of T corresponding to 1.

The matrix A above is diagonalizable, and therefore we should also call the map T diagonalizable.

Definition $T \in \text{Hom}(\mathscr{V})$ is *diagonalizable* if the characteristic vectors of T span $\mathscr{V}$.

We will see that a diagonalizable map has a particularly simple action on a vector space.

Suppose λ is a characteristic value for the map $T \in \text{Hom}(\mathscr{V})$, and let $\mathscr{S}(\lambda) = \{V \in \mathscr{V} | T(V) = \lambda V\}$. That is, $\mathscr{S}(\lambda)$ is the set of all characteristic vectors for λ plus the zero vector. It is not difficult to show that $\mathscr{S}(\lambda)$ is a subspace of $\mathscr{V}$. The map T operates on $\mathscr{S}(\lambda)$ simply as scalar multiplication, so T sends $\mathscr{S}(\lambda)$ into itself.

Definition Let $T \in \text{Hom}(\mathscr{V})$. A subspace $\mathscr{S}$ of $\mathscr{V}$ is an *invariant subspace* of T if $T[\mathscr{S}] \subset \mathscr{S}$.

An invariant subspace need not be a space of characteristic vectors for example, $\mathscr{V}$ is an invariant subspace for any map in $\text{Hom}(\mathscr{V})$.

The simple nature of a diagonalizable map may be stated as follows: If $T \in \text{Hom}(\mathscr{V})$ is diagonalizable, then $\mathscr{V}$ can be expressed as the direct sum of invariant subspaces of characteristic vectors. If there are more than two distinct characteristic values, then this requires an extension of our definition of direct sum.

Example 3 For the map $T \in \text{Hom}(R_3[t])$ of Example 2, the invariant subspaces of characteristic vectors are

$$\mathscr{S}(3) = \mathscr{L}\{1 - t^2, t\} \qquad \text{and} \qquad \mathscr{S}(1) = \mathscr{L}\{1 + t^2\}.$$

T is diagonalizable, so the sum $\mathscr{S}(3) + \mathscr{S}(1)$ is $R_3[t]$ and $\mathscr{S}(3) \cap \mathscr{S}(1) = \{\mathbf{0}\}$. That is, $R_3[t] = \mathscr{S}(3) \oplus \mathscr{S}(1)$. Thus given any polynomial $P \in R_3[t]$, there exist unique polynomials $U \in \mathscr{S}(3)$ and $V \in \mathscr{S}(1)$, such that $P = U + V$. Then the action of T on P is given simply by $T(P) = 3U + 1V$.

Example 4 Let $T \in \text{Hom}(\mathscr{V}_3)$ have the matrix

$$A = \begin{pmatrix} 3 & 2 & -1 \\ -2 & -2 & 2 \\ 3 & 6 & -1 \end{pmatrix}$$

with respect to the standard basis. A is the diagonalizable matrix considered in Example 4, page 229, with characteristic values 2, 2, -4. Since T is diagonalizable, $\mathscr{V}_3 = \mathscr{S}(2) \oplus \mathscr{S}(-4)$, where

$$\mathscr{S}(2) = \mathscr{L}\{(0, 1, 2), (1, 0, 1)\} \qquad \text{and} \qquad \mathscr{S}(-4) = \mathscr{L}\{(1, -2, 3)\}$$

are the invariant subspaces of characteristic vectors. Geometrically, as a map from 3-space to 3-space, T sends the plane $\mathscr{S}(2)$ into itself on multiplication by 2 and the line $\mathscr{S}(-4)$ into itself on multiplication by -4.

Even if a map is not diagonalizable, it acts simply on a subspace if it has a characteristic value. Thus every map from $\mathscr{V}_3$ to $\mathscr{V}_3$ must operate on some subspace by scalar multiplication, for every third degree polynomial equation has at least one real root. However, if a map on $\mathscr{V}$ is not diagonalizable, then there will not be enough invariant subspaces of characteristic vectors to generate $\mathscr{V}$.

By now you may suspect that finding a complete set of invariants for similarity is not a simple task. Since this is the case and much can be done without them, the problem of obtaining such a set and the canonical forms associated with it will be postponed. However, it is possible to give a general idea of how one might proceed. We have said that a diagonalizable map on $\mathscr{V}$ determines a direct sum decomposition of $\mathscr{V}$ into invariant subspaces of characteristic vectors. The idea is to show that an arbitrary map on $\mathscr{V}$ determines an essentially unique direct sum decomposition of $\mathscr{V}$ into invariant subspaces. There are usually many different invariant subspaces for any one map; so the problem is to find a process which selects subspaces in a well defined way. It is possible to construct an invariant subspace by starting with any nonzero vector $U \in \mathscr{V}$ and its images under T. To see how, let

$$T^m(U) = T(T^{m-1}(U)) \qquad \text{for} \quad m = 2, 3, 4, \ldots.$$

If the dimension of $\mathscr{V}$ is n, then there exists an integer k, $k \leq n$, such that the vectors $U, T(U), T^2(U), \ldots, T^{k-1}(U)$ are linearly independent and the vectors $U, T(U), T^2(U), \ldots, T^{k-1}(U), T^k(U)$ are linearly dependent. (Why?) Now since $T^k(U) \in \mathscr{L}\{U, T(U), \ldots, T^{k-1}(U)\}$, the subspace $\mathscr{S}(U, T) = \mathscr{L}\{U, T(U), \ldots, T^{k-1}(U)\}$ is an invariant subspace of T. Although different vectors need not produce different invariant subspaces, this procedure can yield invariant subspaces in infinite variety.

Example 5 Let $T \in \operatorname{Hom}(\mathscr{V}_3)$ have the matrix

$$A = \begin{pmatrix} 3 & -2 & 0 \\ -1 & 3 & -1 \\ -5 & 7 & -1 \end{pmatrix}$$

with respect to the standard basis. (A is the nondiagonalizable matrix considered in Example 5, page 230.) Suppose we take $U = (0, 0, 1)$, then

U, $T(U) = (0, -1, -1)$, and $T^2(U) = (2, -2, -6)$ are linearly independent. Therefore $\mathscr{S}(U, T)$, the invariant subspace generated by U, is all of $\mathscr{V}_3$; and $B' = \{U, T(U), T^2(U)\}$ is a basis for $\mathscr{V}_3$. Since $T^3(U) = 5T^2(U) - 8T(U) + 4U$, the matrix of T with respect to B' is

$$A' = \begin{pmatrix} 0 & 0 & 4 \\ 1 & 0 & -8 \\ 0 & 1 & 5 \end{pmatrix}.$$

Compare the characteristic equation for A, $\lambda^3 - 5\lambda^2 + 8\lambda - 4 = 0$, with the relation $T^3(U) - 5T^2(U) + 8T(U) - 4U = \mathbf{0}$ obtained above. This suggests that A' might be chosen as the canonical representative for the equivalence class of A under similarity. In fact, you will find that almost any choice for U leads to the same matrix. Of course, not every vector and its images can generate $\mathscr{V}_3$ since T has two independent characteristic vectors.

The matrix A' above is simple and easily obtained using the concept of an invariant subspace. However, there is another matrix similar to A, which in some sense is simpler than A' in that it is closer to being diagonal and has the characteristic values on the main diagonal. (See problem 11 below.) Example 5 and problem 11 do not provide a general procedure for obtaining a simple matrix similar to any given matrix. But they do indicate that the characteristic polynomial and its factors are somehow involved. A complete solution to the problem of finding invariants and canonical forms for matrices under similarity is obtained in Chapter 8.

Problems

1. Find the characteristic polynomial for each of the following maps.
   a. $T(a, b) = (3a + b, a - 2b)$.
   b. $T(a, b, c) = (a + 3b - 2c, 2a + b, 4b - 4c)$.
   c. $T(a + bt) = 2a + b + (3a - 4b)t$.
   d. $T(a + bt + ct^2) = a + b - 3c + (2a - c)t^2$.
2. Use the determinant and trace to obtain the characteristic polynomial for
   a. $\begin{pmatrix} 2 & 4 \\ 1 & 3 \end{pmatrix}$. b. $\begin{pmatrix} 0 & 1 \\ -1 & 0 \end{pmatrix}$. c. $\begin{pmatrix} 1 & 0 \\ 0 & -1 \end{pmatrix}$. d. $\begin{pmatrix} 2 & 5 \\ 5 & 1 \end{pmatrix}$.
3. a. Find a general expression for the coefficient s_{n-1} in the characteristic polynomial of an $n \times n$ matrix $A = (a_{ij})$. Consider s_2 for a 3×3 matrix first.
   b. Use the determinant (s_3), s_2, and the trace (s_1) to obtain the characteristic polynomials of
   i. $\begin{pmatrix} 3 & 2 & 1 \\ 0 & -2 & 1 \\ 1 & 0 & 1 \end{pmatrix}$. ii. $\begin{pmatrix} 1 & 2 & -3 \\ 1 & 5 & 2 \\ 2 & 4 & -6 \end{pmatrix}$. iii. $\begin{pmatrix} 1 & 0 & 0 \\ 0 & 1 & 0 \\ 0 & 1 & 1 \end{pmatrix}$.

4. Show by example that the rank of an $n \times n$ matrix A is independent of the characteristic polynomial. That is, the rank and the n values $s_1, \ldots, s_n$ give $n + 1$ independent invariants for A under similarity.

5. Let $T(a, b, c) = (a + b, -a/2 + b + c/2, c - b)$.
  a. Find a line ℓ, through the origin, that is an invariant subspace of T.
  b. Show that the plane through the origin with normal ℓ is an invariant subspace of T.
  c. Use part b to conclude that a proper, invariant subspace of T need not be a subspace of characteristic vectors.

6. Without using matrices or linear equations, show that if T has λ as a characteristic value, then it has many characteristic vectors corresponding to λ.

7. Suppose λ is a characteristic value for $T \in \text{Hom}(\mathscr{V})$.
  a. Prove that $\mathscr{S}(\lambda) = \{V \in \mathscr{V} \mid T(V) = \lambda V\}$ is a subspace of $\mathscr{V}$.
  b. What is our usual notation for $\mathscr{S}(\lambda)$ when $\lambda = 0$?

8. Determine if the following maps are diagonalizable.
  a. $T(a, b, c) = (b, a, c)$.
  b. $T(a + bt) = 4a + 4bt$.
  c. $T(a + bt + ct^2) = a + c + (a + 4b)t + 4ct^2$.
  d. $T(a, b) = (2a - b, a + 3b)$.
  e. $T(a, b, c) = (2a, a - 2b, b + 3c)$.

9. Let $T(a, b, c) = (2a + c, a + 3b - c, a + 2c)$.
  a. Show that T is diagonalizable.
  b. Find invariant subspaces of characteristic vectors for T and show that $\mathscr{V}_3$ is the direct sum of these spaces.

10. Suppose $T \in \text{Hom}(\mathscr{V}_3)$ has 1 as a characteristic value and $\mathscr{S}(1)$ is a line. Suppose further that $\mathscr{T}$ is an invariant subspace for T such that $\mathscr{V}_3 = \mathscr{S}(1) \oplus \mathscr{T}$. Show that every plane π parallel to $\mathscr{T}$ is sent into itself by T, i.e., $T[\pi] \subset \pi$.

11. Let T be the map in Example 5, page 236.
  a. Show that $\{U, T(U), T^2(U)\}$ is a basis for $\mathscr{V}_3$ when $U = (1, 1, 0)$.
  b. Find the matrix of T with respect to the basis in part a.
  c. Find the invariant subspace for T generated by $U = (0, 1, 2)$ and its images.
  d. Find the matrix of T with respect to the basis $\{(0, 1, 2), T(0, 1, 2), (1, 1, 1)\}$.

Review Problems

1. Let $T \in \text{Hom}(\mathscr{V}_3, \mathscr{V}_2)$ be defined by $T(a, b, c) = (2a - b + 4c, a + 2b - 3c)$, and suppose A is the matrix of T with respect to the standard bases.
  a. Find the canonical form A' for A under equivalence.
  b. Find basis B_1' and B_2' for $\mathscr{V}_3$ and $\mathscr{V}_2$ such that A' is the matrix of T with respect to B_1' and B_2'.

c. Find the transition matrix P from $\{E_i\}$ to B_1' and Q from $\{E_i\}$ to B_2'.
d. Show that $Q^{-1}AP = A'$.

2. a. Let S be the set of all systems of linear equations in m unknowns. Prove that equivalence of systems of linear equations defines an equivalence relation on S.
b. What would be a good choice of canonical forms for this relation?

3. Let $S = \mathscr{M}_{n \times n}$ and define the relation $\sim$ by $A \sim B$ if $|A| = |B|$.
a. Show that $\sim$ is an equivalence relation on S.
b. How is this relation related to equivalence and similarity of matrices?
c. Choose a set of canonical forms for $\sim$, when $n = 3$.

4. Show that every 2×2 matrix of the form $\begin{pmatrix} a & b \\ b & c \end{pmatrix}$ is diagonalizable.

5. Determine which rotations in the plane have invariant subspaces of characteristic vectors. That is, for which values of φ does $T(a, b) = (a \cos \varphi - b \sin \varphi, a \sin \varphi + b \cos \varphi)$ have such invariant subspaces?

6. Which of the following matrices are diagonalizable?

a. $\begin{pmatrix} 0 & 0 & 0 \\ 1 & 1 & 0 \\ 0 & 2 & 0 \end{pmatrix}$. b. $\begin{pmatrix} 0 & 0 & 0 \\ 1 & 1 & 0 \\ 0 & 2 & 2 \end{pmatrix}$. c. $\begin{pmatrix} 0 & 0 & 0 \\ 1 & 1 & 0 \\ 2 & 0 & 1 \end{pmatrix}$. d. $\begin{pmatrix} 0 & 0 & 1 \\ 0 & 0 & 1 \\ 1 & 1 & 0 \end{pmatrix}$.

7. Let $\mathscr{S}$, $\mathscr{T}$, and $\mathscr{U}$ be subspaces of a vector space $\mathscr{V}$. Devise definitions for the sum $\mathscr{S} + \mathscr{T} + \mathscr{U}$ and the direct sum $\mathscr{S} \oplus \mathscr{T} \oplus \mathscr{U}$ which generalize the definitions for the terms for two subspaces.

8. Let $T \in \text{Hom}(\mathscr{V}_4)$ be given by

$$T(a, b, c, d) = (a + d, -3b, 2c - 3a + 3d, 3a - d).$$

a. Show that T is diagonalizable.
b. Find the invariant subspaces of characteristic vectors $\mathscr{S}(2)$, $\mathscr{S}(-2)$, and $\mathscr{S}(-3)$.
c. Show that $\mathscr{V}_4 = \mathscr{S}(2) \oplus \mathscr{S}(-2) \oplus \mathscr{S}(-3)$. (See problem 7.)

9. Define $T \in \text{Hom}(\mathscr{V}_3)$ by

$$T(a, b, c) = (4a + 2b - 4c, 2a - 4c, 2a + 2b - 2c).$$

a. Find the characteristic polynomial of T.
b. Show that T is not diagonalizable, even though the characteristic equation has three distinct characteristic roots.
c. Let $U = (1, 0, 0)$ and show that $B = \{U, T(U), T^2(U)\}$ is a basis for $\mathscr{V}_3$; or that $\mathscr{S}(U, T) = \mathscr{V}_3$.
d. Find the matrix of T with respect to the basis B of part c.

10. Prove that if A has two linearly independent characteristic vectors corresponding to the characteristic value $\lambda = r$, then $(\lambda - r)^2$ is a factor of the characteristic polynomial of A. [A need not be diagonalizable.]

11. Give an example of a map T for which $\mathscr{S}(U, T) \neq \mathscr{S}(V, T)$ whenever U and V are linearly independent.

12. If $A = \begin{pmatrix} 6 & 8 \\ -5 & -8 \end{pmatrix}$, $A' = \begin{pmatrix} -4 & 0 \\ 0 & 2 \end{pmatrix}$, and $P = \begin{pmatrix} -4 & -2 \\ 5 & 1 \end{pmatrix}$, then $P^{-1}AP = A'$. Suppose A is the matrix of $T \in \text{Hom}(R_2[t])$ with respect to the basis $B = \{t, 1 - 5t\}$.

a. Find $T(a + bt)$.

b. Use the matrix P to obtain a basis B' with respect to which A' is the matrix of T.

Inner Product Spaces

The vector space $\mathscr{E}_2$ was introduced in Chapter 1 to provide an algebraic basis for the geometric properties of length and angle. However, the definition of an abstract vector space was molded on $\mathscr{V}_2$ and did not include the additional algebraic structure of $\mathscr{E}_2$. $\mathscr{E}_2$ is an example of an inner product space and we now turn to a general study of such spaces. This study will culminate in showing that a certain type of matrix, called symmetric, is always diagonalizable. Inner product spaces also have applications in areas outside of geometry, for example, in the study of the function spaces encountered in the field of functional analysis.

§1. Definition and Examples

Recall that the space $\mathscr{E}_2$ was obtained from $\mathscr{V}_2$ by adding the dot product to the algebraic structure of $\mathscr{V}_2$. To generalize this idea we will require that a similar operation, defined on an abstract vector space, satisfy the basic properties of a dot product listed in Theorem 1.2 on page 19.

Definition Let $\mathscr{V}$ be a real vector space and let $\langle\ ,\ \rangle$ be a function that assigns real numbers to pairs of vectors, that is, $\langle U, V\rangle \in R$ for each $U, V \in \mathscr{V}$. $\langle\ ,\ \rangle$ is an *inner product* if the following conditions hold for all $U, V, W \in \mathscr{V}$ and $r, s \in R$:

1. $\langle U, V\rangle = \langle V, U\rangle$ — $\langle\ ,\ \rangle$ is *symmetric.*
2. $\langle rU + sV, W\rangle = r\langle U, W\rangle + s\langle V, W\rangle$.
3. If $U \neq \mathbf{0}$, then $\langle U, U\rangle > 0$; $\langle \mathbf{0}, \mathbf{0}\rangle = 0$ — $\langle\ ,\ \rangle$ is *positive definite.*

The vector space $\mathscr{V}$ together with an inner product $\langle\ ,\ \rangle$ is an *inner product space* and will be denoted $(\mathscr{V}, \langle\ ,\ \rangle)$ or simply $\mathscr{E}$.

Since the inner product of two vectors is a scalar, an inner product is sometimes called a *scalar product*, and finite-dimensional inner product spaces are often called *Euclidean spaces.*

An inner product is a function of two variables, the vectors U and V. If the second variable is held constant, say $V = V_0$, then a real-valued function T is obtained, $T(U) = \langle U, V_0\rangle$. The second condition above requires that T be linear, thus it is said that an inner product is linear in the first variable. But using the symmetry of an inner product,

$$\begin{aligned}\langle U, rV + sW\rangle = \langle rV + sW, U\rangle &= r\langle V, U\rangle + s\langle W, U\rangle \\ &= r\langle U, V\rangle + s\langle U, W\rangle.\end{aligned}$$

Therefore an inner product is linear in each variable and so is said to be *bilinear*.

Example 1 The dot product in $\mathscr{E}_n$ is an inner product, and $\mathscr{E}_n$ or $(\mathscr{V}_n, \circ)$ is an inner product space.

Recall that if $U = (a_1, \ldots, a_n)$ and $V = (b_1, \ldots, b_n)$, then

$$U \circ V = a_1b_1 + \cdots + a_nb_n = \sum_{i=1}^{n} a_ib_i.$$

The dot product is symmetric because multiplication in R is commutative. To check that the dot product is linear in the first variable, suppose $U = (a_1, \ldots, a_n)$, $V = (b_1, \ldots, b_n)$, and $W = (c_1, \ldots, c_n)$. For $r, s \in R$, $rU + sV = (ra_1 + sb_1, \ldots, ra_n + sb_n)$, therefore

$$(rU + sV) \circ W = \sum_{i=1}^{n} (ra_i + sb_i)c_i = r\sum_{i=1}^{n} a_ic_i + s\sum_{i=1}^{n} b_ic_i$$
$$= rU \circ W + sV \circ W.$$

The verification that the dot product is positive definite is left to the reader.

Example 2 Let $B = \{(1, 3), (4, 2)\}$. For this basis, define $\langle\ ,\ \rangle_B$ on $\mathscr{V}_2$ by

$$\langle U, V\rangle_B = a_1b_1 + a_2b_2,$$

where $U\colon (a_1, a_2)_B$ and $V\colon (b_1, b_2)_B$. Then $\langle\ ,\ \rangle_B$ is an inner product on $\mathscr{V}_2$.

We will justify that $\langle\ ,\ \rangle_B$ is positive definite here. Suppose $U\colon (a_1, a_2)_B$, then $U \neq \mathbf{0}$ if and only if the coordinates satisfy $(a_1)^2 + (a_2)^2 > 0$, that is, $U \neq \mathbf{0}$ if and only if $\langle U, U\rangle_B > 0$.

The inner product space $(\mathscr{V}_2, \langle\ ,\ \rangle_B)$ is not the same space as $\mathscr{E}_2 = (\mathscr{V}_2, \circ)$. For example, $\langle(1, 3), (1, 3)\rangle_B = 1$, since $(1, 3)\colon (1, 0)_B$ but $(1, 3) \circ (1, 3) = 10$. As another example, $(-2, 4)\colon (2, -1)_B$ and $(1, -7)\colon (-3, 1)_B$, therefore

$$\langle(-2, 4), (1, -7)\rangle_B = 2(-3) + (-1)1 = -7.$$

Obviously every basis for $\mathscr{V}_2$ gives rise to an inner product on $\mathscr{V}_2$, and there are many ways to make $\mathscr{V}_2$ into an inner product space. However, not all such inner products are different; $\langle\ ,\ \rangle_{\{(1,0),(0,1)\}}$ and $\langle\ ,\ \rangle_{\{(0,1),(1,0)\}}$ both have the same effect as the dot product on vectors from $\mathscr{V}_2$. Are there any other bases B for $\mathscr{V}_2$ for which $\langle\ ,\ \rangle_B$ is the dot product?

The use of a basis to define an inner product can easily be applied to any finite-dimensional vector space. A polynomial space provides a good example.

Example 3 Choose $B = \{1, t, t^2\}$ as a basis for $R_3[t]$, and define the function $\langle \ , \ \rangle_B$ by

$$\langle a_1 + a_2 t + a_3 t^2, b_1 + b_2 t + b_3 t^2 \rangle_B = a_1 b_1 + a_2 b_2 + a_3 b_3.$$

It is easily shown that $\langle \ , \ \rangle_B$ satisfies the conditions for an inner product, therefore $R_3[t]$ together with $\langle \ , \ \rangle_B$ yields an inner product space of polynomials. In this space,

$$\langle t, t \rangle_B = 1, \qquad \langle t, t^2 \rangle_B = 0, \qquad \langle 3 + 2t, 4t + t^2 \rangle_B = 8,$$

and so on.

An inner product space need not be finite dimensional. The integral can be used to provide an inner product for functions.

Example 4 Consider the vector space $\mathscr{F}$ of all continuous real-valued functions with domain [0, 1]. For any two functions $f, g \in \mathscr{F}$, define

$$\langle f, g \rangle = \int_0^1 f(x)g(x)\, dx.$$

Thus

$$\langle x, x \rangle = \int_0^1 x^2\, dx = 1/3,$$

$$\langle 6x, \sqrt{1 - x^2} \rangle = \int_0^1 6x\sqrt{1 - x^2}\, dx = 2,$$

$$\langle \sin \pi x, \cos \pi x \rangle = \int_0^1 \sin \pi x \cos \pi x\, dx = \tfrac{1}{2} \int_0^1 \sin 2\pi x\, dx = 0.$$

This function is an inner product on $\mathscr{F}$ if the three conditions in the definition hold. Symmetry is immediate and bilinearity follows easily from the basic properties of the integral. However, the fact that $\langle \ , \ \rangle$ is positive definite is not obvious. It must be shown that $\langle f, f \rangle = \int_0^1 [f(x)]^2\, dx > 0$ for any nonzero function f. But to say $f \neq \mathbf{0}$ only means there is a point p at which $f(p) \neq 0$, and without using continuity it is not possible to prove that $\langle f, f \rangle > 0$. Suppose $f(p) \neq 0$ and $f^2(p) = k > 0$. Then by continuity, there is a positive number δ such that $f^2(x) > k/2$ for all values of x such that $p - \delta < x < p + \delta$. For simplicity, assume $0 < p - \delta$ and $p + \delta < 1$, then

$$\langle f, f\rangle = \int_0^1 f^2(x)dx = \int_0^{p-\delta} f^2(x)dx + \int_{p-\delta}^{p+\delta} f^2(x)dx + \int_{p+\delta}^1 f^2(x)dx$$
$$\geq \int_{p-\delta}^{p+\delta} f^2(x)dx > \int_{p-\delta}^{p+\delta} \tfrac{1}{2}k\,dx = \delta k > 0.$$

Thus $\langle\ ,\ \rangle$ is positive definite and defines an inner product on $\mathscr{F}$.

Example 5 Since polynomials can be thought of as continuous functions, the integral can be used to define various inner products on a polynomial space. For $P, Q \in R_3[t]$, set

$$\langle P(t), Q(t)\rangle = \int_0^1 P(t)Q(t)\,dt,$$

then $(R_3[t], \langle\ ,\ \rangle)$ is a different inner product space from the one constructed in Example 3. Computing the inner product for any of the three pairs of polynomials considered in Example 3 shows that this is another space, for

$$\langle t, t\rangle = \int_0^1 t^2\,dt = 1/3,$$
$$\langle t, t^2\rangle = \int_0^1 t^3\,dt = 1/4,$$
$$\langle 3 + 2t, 4t + t^2\rangle = 61/6.$$

Example 6 The function $\langle\ ,\ \rangle$ defined on $\mathscr{V}_2$ by

$$\langle (a, b), (c, d)\rangle = 3ac - ad - bc + 2bd$$

is an inner product.

$\langle\ ,\ \rangle$ is easily seen to be symmetric. For bilinearity, let $r, s \in R$, then

$$\begin{aligned}
&\langle r(a_1, b_1) + s(a_2, b_2), (c, d)\rangle \\
&\quad = \langle (ra_1 + sa_2, rb_1 + sb_2), (c, d)\rangle \\
&\quad = 3(ra_1 + sa_2)c - (ra_1 + sa_2)d - (rb_1 + sb_2)c + 2(rb_1 + sb_2)d \\
&\quad = r(3a_1c - a_1d - b_1c + 2b_1d) + s(3a_2c - a_2d - b_2c + 2b_2d) \\
&\quad = r\langle (a_1, b_1), (c, d)\rangle + s\langle (a_2, b_2), (c, d)\rangle.
\end{aligned}$$

Thus $\langle\ ,\ \rangle$ is linear in the first variable, and since $\langle\ ,\ \rangle$ is symmetric, it is bilinear. Finally, if $(a, b) \neq (0, 0)$, then

$$\begin{aligned}\langle (a, b), (a, b)\rangle &= 3a^2 - 2ab + 2b^2 \\ &= 2a^2 + (a - b)^2 + b^2 > 0\end{aligned}$$

and the function is positive definite. The inner product space obtained from $\mathscr{V}_2$ using this inner product is neither $\mathscr{E}_2$ nor the space obtained in Example 2, for $\langle (1, 0), (0, 1)\rangle = 3$ and $\langle (1, 3), (1, 3)\rangle = 13$.

Problems

1. Evaluate the following dot products:
a. $(2, 1)\circ(3, -4)$. b. $(4, 1, 2)\circ(2, -10, 1)$.
c. $(-3, 2)\circ(-3, 2)$. d. $(0, 1, 3, 1)\circ(2, 5, -3, 0)$.

2. Let $B = \{(0, 1), (1, 3)\}$ be a basis for $\mathscr{V}_2$ and define $\langle\ ,\ \rangle_B$ by $\langle U, V\rangle_B = a_1a_2 + b_1b_2$ where $V\colon (a_1, b_1)_B$ and $U\colon (a_2, b_2)_B$.
a. Evaluate i. $\langle (0, 1), (0, 1)\rangle_B$. ii. $\langle (1, 0), (1, 0)\rangle_B$.
iii. $\langle (1, 3), (1, 3)\rangle_B$. iv. $\langle (1, 3), (0, 1)\rangle_B$.
v. $\langle (2, 4), (1, 0)\rangle_B$. vi. $\langle (-2, 0), (1, 1)\rangle_B$.
b. Show that $\langle\ ,\ \rangle_B$ is an inner product.
c. Show that if $V\colon (a, b)_B$, then $\langle V, (0, 1)\rangle_B = a$ and $\langle V, (1, 3)\rangle_B = b$.

3. Suppose $\mathscr{V}$ has $B = \{U_1, \ldots, U_n\}$ as a basis and $\langle\ ,\ \rangle_B$ is defined by $\langle V, W\rangle_B = \sum_{i=1}^{n} a_ib_i$ where $V\colon (a_1, \ldots, a_n)_B$ and $W\colon (b_1, \ldots, b_n)_B$.
a. Prove that $\langle\ ,\ \rangle_B$ is an inner product for $\mathscr{V}$.
b. Show that $V = \langle V, U_1\rangle_B U_1 + \cdots + \langle V, U_n\rangle_B U_n$.

4. Let $B = \{1, 1 + t, 1 + t + t^2\}$ and define $\langle\ ,\ \rangle_B$ on $R_3[t]$ as in problem 3. Compute the following inner products:
a. $\langle 1 + t, 1 + t\rangle_B$. b. $\langle 1, 1 + t + t^2\rangle_B$.
c. $\langle 2t - t^2, 4 + t - 3t^2\rangle_B$. d. $\langle 3 - t^2, 2 + 4t + 6t^2\rangle_B$.

5. Let $(\mathscr{V}, \langle\ ,\ \rangle)$ be an inner product space. Prove that $\langle \mathbf{0}, V\rangle = 0$ for every vector $V \in \mathscr{V}$.

6. For $U, V \in \mathscr{V}_2$, let (U^T, V^T) be the 2×2 matrix with U^T and V^T as columns. Define $b(U, V) = \det(U^T, V^T)$.
a. Find $b((2, 3), (1, 4))$, $b((2, 3), (2, 3))$, and $b((1, 4), (2, 3))$.
b. Show that b is a bilinear function.
c. Show that b is neither symmetric nor positive definite.

7. Define $b(f, g) = f(1)g(1)$ for $f, g \in \mathscr{F}$.
a. Find $b(2x, 4x - 3)$ and $b(x^2, \sin x)$.
b. Show that b is bilinear.
c. Does b define an inner product on $\mathscr{F}$?

8. Let $f((a, b), (c, d)) = (a \quad b)\begin{pmatrix} 4 & 2 \\ 2 & 3 \end{pmatrix}\begin{pmatrix} c \\ d \end{pmatrix}$.

a. Find $f((1, -2), (1, -2))$ and $f((5, -3), (-1, 14))$.
b. Show that f is bilinear.
c. Does f define an inner product on $\mathscr{V}_2$?

9. Suppose $\langle\ ,\ \rangle$ is an inner product defined on $\mathscr{V}_2$. Let $\langle(1, 0), (1, 0)\rangle = x$, $\langle(1, 0), (0, 1)\rangle = y$, and $\langle(0, 1), (0, 1)\rangle = z$. Show that

$$\begin{aligned}\langle(a, b), (c, d)\rangle &= xac + yad + ybc + zbd \\ &= (a \quad b)\begin{pmatrix} x & y \\ y & z \end{pmatrix}\begin{pmatrix} c \\ d \end{pmatrix}.\end{aligned}$$

10. Find the matrix $\begin{pmatrix} x & y \\ y & z \end{pmatrix}$ of problem 9 for

a. The dot product.
b. The inner product in Example 2, page 243.
c. The inner product in Example 6, page 245.

11. Suppose x, y, z are arbitrary real numbers, not all zero. Does the bilinear function

$$f((a, b), (c, d)) = (a \quad b)\begin{pmatrix} x & y \\ y & z \end{pmatrix}\begin{pmatrix} c \\ d \end{pmatrix}$$

define an inner product on $\mathscr{V}_2$?

§2. The Norm and Orthogonality

When representing a vector U from $\mathscr{E}_2$ geometrically by an arrow, the length of the arrow is $\sqrt{U \circ U}$. This idea of length can easily be extended to any inner product space, although there will not always be such a clear geometric interpretation.

Definition Let $(\mathscr{V}, \langle\ ,\ \rangle)$ be an inner product space and $U \in \mathscr{V}$, then the *norm* of U is $\|U\| = \sqrt{\langle U, U\rangle}$. U is a *unit vector* if $\|U\| = 1$.

Consider the function space $\mathscr{F}$ with the integral inner product. In this space, the norm of a function f is $(\int_0^1 f^2(x)dx)^{\frac{1}{2}}$. Thus $\|x^3\| = (\int_0^1 x^6 dx)^{\frac{1}{2}} = 1/\sqrt{7}$ and $\|e^x\| = \sqrt{\frac{1}{2}(e^2 - 1)}$. Since the norm of x^3 is $1/\sqrt{7}$, the norm of $\sqrt{7}x^3$ is 1, and $\sqrt{7}x^3$ is a unit vector in the inner product space $(\mathscr{F}, \langle\ ,\ \rangle)$.

In general, if $U \neq \mathbf{0}$, then $(1/\|U\|)U$ is a unit vector.

A basic relation between the inner product of two vectors and their norms is given by the Schwartz Inequality.

Theorem 6.1 (Schwartz Inequality) Suppose U and V are vectors from an inner product space, then $\langle U, V\rangle^2 \le \|U\|^2\|V\|^2$ or $|\langle U, V\rangle| \le \|U\|\,\|V\|$.

Proof Consider the vector $xU + V$ for any $x \in R$. Since $\langle\ ,\ \rangle$ is positive definite, $0 \le \langle xU + V, xU + V\rangle$. Using the bilinearity and symmetry of $\langle\ ,\ \rangle$, we have

$$0 \le x^2\langle U, U\rangle + 2x\langle U, V\rangle + \langle V, V\rangle = \|U\|^2x^2 + 2\langle U, V\rangle x + \|V\|^2,$$

for all real values of x. This means that the parabola with equation $y = \|U\|^2x^2 + 2\langle U, V\rangle x + \|V\|^2$ is entirely above or tangent to the x axis, so there is at most one real value for x that yields $y = 0$ and the discriminant $(2\langle U, V\rangle)^2 - 4\|U\|^2\|V\|^2$ must be nonpositive. But $(2\langle U, V\rangle)^2 - 4\|U\|^2\|V\|^2 \le 0$ is the desired relation.

Corollary 6.2 (Triangle Inequality) If U and V are elements of an inner product space, then $\|U + V\| \le \|U\| + \|V\|$.

Proof

$$\begin{aligned}\|U + V\|^2 &= \langle U + V, U + V\rangle = \|U\|^2 + 2\langle U, V\rangle + \|V\|^2 \\ &\le \|U\|^2 + 2\|U\|\,\|V\| + \|V\|^2 = (\|U\| + \|V\|)^2.\end{aligned}$$

If the vectors U, V, $U + V$ from $\mathscr{E}_2$ are represented by the sides of a triangle in the plane, then the triangle inequality states that the length of one side of the triangle cannot exceed the sum of the lengths of the other two sides.

The Schwartz inequality justifies generalizing the angle measure in $\mathscr{E}_2$ to any inner product space, for if U and V are nonzero vectors, then $\langle U, V\rangle^2 \le \|U\|^2\|V\|^2$ implies that

$$-1 \le \frac{\langle U, V\rangle}{\|U\|\,\|V\|} \le 1.$$

Therefore, there exists exactly one angle θ such that $0 \le \theta \le \pi$ and

$$\cos\theta = \frac{\langle U, V\rangle}{\|U\|\,\|V\|}.$$

Notice that any function with $[-1, 1]$ in its range could have been chosen, but the geometric significance in $\mathscr{E}_2$ requires that the cosine function be used.

Definition If U and V are nonzero vectors in an inner product space, then the *angle* between U and V is

$$\cos^{-1}\left(\frac{\langle U, V\rangle}{\|U\|\,\|V\|}\right).$$

With this definition it is possible to speak of the angle between two polynomials or two functions.

If the definition of angle is extended to the zero vector by making the angle between it and any other vector zero, then the relation $\langle U, V\rangle = \|U\|\,\|V\| \cos\theta$ holds for all vectors in an inner product space. For the space $\mathscr{E}_2$, the definition of angle was motivated by the law of cosines in the Cartesian plane. Now in an arbitrary inner product space the law of cosines is a consequence of the definition of angle.

Theorem 6.3 (Law of Cosines) If U and V are vectors in an inner product space and θ is the angle between them, then

$$\|U - V\|^2 = \|U\|^2 + \|V\|^2 - 2\|U\|\,\|V\| \cos\theta.$$

Proof This follows at once from

$$\|U - V\|^2 = \langle U - V, U - V\rangle = \langle U, U\rangle - 2\langle U, V\rangle + \langle V, V\rangle.$$

The angle between two vectors from an arbitrary inner product space need not be interpreted geometrically. Therefore it will often not be of interst to simply compute angles between vectors. However, the angle of $90°$ or $\pi/2$ plays a central role in all inner product spaces, much as it does in Euclidean geometry.

Definition Let $(\mathscr{V}, \langle\ ,\ \rangle)$ be an inner product space. $U, V \in \mathscr{V}$ are *orthogonal* if $\langle U, V\rangle = 0$.

The angle between nonzero orthogonal vectors is $\pi/2$, therefore orthogonal vectors in $\mathscr{E}_2$ or $\mathscr{E}_n$ might also be called perpendicular. In $\mathscr{F}$, with the integral inner product used in Section 1, the functions x and $3x - 2$

are orthogonal, as are the functions $\sin \pi x$ and $\cos \pi x$. As another example, consider the inner product space $(\mathscr{V}, \langle\ ,\ \rangle_B)$ where the inner product $\langle\ ,\ \rangle_B$ is defined using coordinates with respect to a basis B. In this space any two distinct vectors from B are orthogonal. Thus (2, 1) and (1, 0) are orthogonal in the space $(\mathscr{V}_2, \langle\ ,\ \rangle_B)$ with $B = \{(2, 1), (1, 0)\}$.

The definition of orthogonal implies that the zero vector is orthogonal to every vector including itself. Can an inner product space have a nonzero vector that is orthogonal to itself?

The Pythagorean theorem can be stated in any inner product space and is proved by simply replacing V with $-V$ in the law of cosines.

Theorem 6.4 (Pythagorean Theorem) If U and V are orthogonal vectors, then $\|U + V\|^2 = \|U\|^2 + \|V\|^2$.

Since an inner product space has more algebraic structure than a vector space, one might expect restrictions on the choice of a basis. To see what might be required, consider the inner product space $\mathscr{E}_2$. Since $\mathscr{E}_2$ is essentially the Cartesian plane viewed as an inner product space, a basis $\{U, V\}$ for $\mathscr{E}_2$ should yield a Cartesian coordinate system. That is, the coordinate axes $\mathscr{L}\{U\}$ and $\mathscr{L}\{V\}$ should be perpendicular. Further, the points rU and rV should be r units from the origin. This requires that U and V be orthogonal and unit vectors.

Definition A basis $\{V_1, \ldots, V_n\}$ for a vector space $\mathscr{V}$ is an *orthonormal basis* for the inner product space $(\mathscr{V}, \langle\ ,\ \rangle)$ if $\langle V_i, V_j\rangle = 0$ when $i \neq j$ and $\langle V_i, V_i\rangle = 1$, for $1 \leq i, j \leq n$.

That is, an orthonormal basis is a basis consisting of unit vectors that are mutually orthogonal. The standard basis $\{E_i\}$ is an orthonormal basis for $\mathscr{E}_n$, but there are many other orthonormal bases for $\mathscr{E}_n$.

There is a useful notation for expressing the conditions in the above definition. The *Kronecker delta* is defined by

$$\delta_{ij} = \begin{cases} 0 & \text{if } i \neq j \\ 1 & \text{if } i = j. \end{cases}$$

Using the Kronecker delta, $\{V_1, \ldots, V_n\}$ is an orthonormal basis for the inner product space $(\mathscr{V}, \langle\ ,\ \rangle)$ if and only if $\langle V_i, V_j\rangle = \delta_{ij}$. The Kronecker delta is useful any time a quantity is zero when two indexes are unequal. For example, the $n \times n$ identity matrix may be written as $I_n = (\delta_{ij})$.

Example 1 $\{(1/\sqrt{2}, 1/\sqrt{2}), (1/\sqrt{2}, -1/\sqrt{2})\}$ is an orthonormal basis for $\mathscr{E}_2$ and $\{(1/\sqrt{2}, 0, 1/\sqrt{2}), (1/\sqrt{3}, 1/\sqrt{3}, -1/\sqrt{3}), (-1/\sqrt{6}, 2/\sqrt{6}, 1/\sqrt{6})\}$ is an orthonormal basis for $\mathscr{E}_3$.

Example 2 An orthonormal basis for $\mathscr{E}_2$ can be written in one of the following two forms for some angle φ:

$$\{(\cos \varphi, \sin \varphi), (-\sin \varphi, \cos \varphi)\} \quad \text{or} \quad \{(\cos \varphi, \sin \varphi), (\sin \varphi, -\cos \varphi)\}.$$

To verify this, suppose $\{U, V\}$ is an orthonormal basis for $\mathscr{E}_2$. If $U = (a, b)$, then $1 = \|U\|^2 = a^2 + b^2$ implies that $|a| \leq 1$. Therefore there exists an angle θ such that $a = \cos \theta$. Then $b^2 = 1 - \cos^2\theta$ implies $b = \pm\sin \theta$. Since $\pm\sin \theta = \sin \pm\theta$ and $\cos \theta = \cos \pm\theta$, there is an angle φ such that $U = (\cos \varphi, \sin \varphi)$. Similarly there exists an angle α such that $V = (\cos \alpha, \sin \alpha)$. Now

$$0 = U \circ V = \cos \varphi \cos \alpha + \sin \varphi \sin \alpha = \cos (\varphi - \alpha),$$

so φ and α differ by an odd multiple of $\pi/2$ and $\cos \alpha = (-\sin \varphi)(\pm 1)$ and $\sin \alpha = (\cos \varphi)(\pm 1)$. Therefore $U \circ V = 0$ implies that $V = \pm(-\sin \varphi, \cos \varphi)$.

Consider the change of basis from the standard basis for $\mathscr{E}_2$ to the first form for an orthonormal basis in Example 2. If $T\colon \mathscr{E}_2 \to \mathscr{E}_2$ is the linear transformation obtained from this change, then

$$T(1, 0) = (\cos \varphi, \sin \varphi) \quad \text{and} \quad T(0, 1) = (-\sin \varphi, \cos \varphi).$$

Therefore T is given by

$$T(x, y) = (x \cos \varphi - y \sin \varphi, x \sin \varphi + y \cos \varphi).$$

That is, this change in basis is equivalent to a rotation through the angle φ. We will see, in Example 2, page 262, that the change from the standard basis to the second form for an orthonormal basis is equivalent to a reflection in a line passing through the origin.

Example 3 Let $B = \{(2, 5), (1, 4)\}$ and let $\langle\ ,\ \rangle_B$ be the inner product for $\mathscr{V}_2$ defined using coordinates with respect to B. Then

$$\langle(2, 5), (2, 5)\rangle_B = 1 = \langle(1, 4), (1, 4)\rangle_B$$

and

$$\langle(2, 5), (1, 4)\rangle_B = 0.$$

Therefore B is an orthonormal basis for the inner product space $(\mathscr{V}_2, \langle\ ,\ \rangle_B)$. Notice that the standard basis for $\mathscr{V}_2$ is not an orthonormal basis for this inner product space. For example, the norm of (1, 0) is $\sqrt{41}/3$, so (1, 0) is not a unit vector in this space.

The result of this example can be extended to an arbitrary finite-diminisional inner product space.

Theorem 6.5 B is an orthonormal basis for $(\mathscr{V}, \langle\ ,\ \rangle)$ if and only if $\langle\ ,\ \rangle$ equals $\langle\ ,\ \rangle_B$. That is, if $U\colon (a_1, \ldots, a_n)_B$ and $V\colon (b_1, \ldots, b_n)_B$, then B is orthonormal if and only if $\langle U, V\rangle = a_1b_1 + \cdots + a_nb_n$.

This theorem is easily proved. It shows that the orthonormal bases are precisely those bases for which the inner product can be computed with coordinates much as the dot product was defined in $\mathscr{E}_2$.

The coordinates of a vector with respect to an orthonormal basis are easily expressed in terms of the inner product. This follows at once from Theorem 6.5.

Corollary 6.6 If $B = \{V_1, \ldots, V_n\}$ is an orthonormal basis for $(\mathscr{V}, \langle\ ,\ \rangle)$, then the ith coordinate of U with respect to B is $\langle U, V_i\rangle$. That is, $U\colon (\langle U, V_1\rangle, \ldots, \langle U, V_n\rangle)_B$.

In some circumstances it is either not possible or not necessary to obtain an orthonormal basis for an inner product space. In such a case one of the following types of sets is often useful.

Definition Let S be a subset of nonzero vectors from an inner product space $(\mathscr{V}, \langle\ ,\ \rangle)$. S is an *orthogonal set* if $\langle U, V\rangle = 0$ for all $U, V \in S$ with $U \neq V$. An orthogonal set S is an *orthonormal set* if $\|U\| = 1$ for all $U \in S$.

An orthonormal basis is obviously both an orthonormal set and an orthogonal set. On the other hand, an orthonormal subset of an inner product space need not be a basis for the space. For example, the set $\{(1, 0, 0), (0, 1, 0)\}$ is orthonormal in $\mathscr{E}_3$ but it is not a basis for $\mathscr{E}_3$. An orthonormal set can be obtained from an orthogonal set by simply dividing each vector by its norm.

Thus the orthogonal subset $\{(1, 2, 1), (1, -1, 1)\}$ of $\mathscr{E}_3$ yields the orthonormal set $\{(1/\sqrt{6}, 2/\sqrt{6}, 1/\sqrt{6}), (1/\sqrt{3}, -1/\sqrt{3}, 1/\sqrt{3})\}$.

Theorem 6.7 An orthogonal set is linearly independent.

Proof Suppose S is an orthogonal subset of the inner product space $(\mathscr{V}, \langle\ ,\ \rangle)$. S is linearly independent if no finite subset is linearly dependent; note that $\mathscr{V}$ need not be finite dimensional.

Therefore, suppose $a_1V_1 + \cdots + a_kV_k = \mathbf{0}$ for some k vectors from S. Then

$$\begin{aligned} 0 = \langle \mathbf{0}, V_i \rangle &= \langle a_1V_1 + \cdots + a_iV_i + \cdots + a_kV_k, V_i \rangle \\ &= a_1\langle V_1, V_1 \rangle + \cdots + a_i\langle V_i, V_i \rangle + \cdots + a_k\langle V_k, V_i \rangle \\ &= a_i\langle V_i, V_i \rangle. \end{aligned}$$

Since $V_i \neq \mathbf{0}$, $a_i = 0$ for each i, $1 \leq i \leq k$, and S is linearly independent.

Example 4 An orthonormal set of functions from $\mathscr{F}$ with the integral inner product can be constructed using the fact that

$$\int_0^1 \sin 2n\pi x\, dx = \int_0^1 \cos 2n\pi x\, dx = 0$$

for any positive integer n.

For example,

$$\begin{aligned} \langle \sin 2n\pi x, \cos 2m\pi x \rangle &= \int_0^1 \sin 2n\pi x \cos 2m\pi x\, dx \\ &= \int_0^1 \sin 2(n + m)\pi x\, dx + \int_0^1 \sin 2(n - m)\pi x\, dx \\ &= 0. \end{aligned}$$

Therefore the functions $\sin 2n\pi x$ and $\cos 2m\pi x$ are orthogonal for all positive integers n and m. Further, since

$$\int_0^1 \sin^2 2n\pi x\, dx = \tfrac{1}{2}\int_0^1 1 - \cos 4n\pi x\, dx = \tfrac{1}{2},$$

the norm of $\sin 2n\pi x$ is $1/\sqrt{2}$ and $\sqrt{2} \sin 2n\pi x$ is a unit vector. Proceeding in this manner, it can be shown that

$$\{1, \sqrt{2}\sin 2\pi x, \sqrt{2}\cos 2\pi x, \ldots, \sqrt{2}\sin 2n\pi x, \sqrt{2}\cos 2n\pi x, \ldots\}$$

is an orthonormal subset of $\mathscr{F}$ with the integral inner product.

By Theroem 6.7, this set is linearly independent. But since it can be shown that it does not span, it is not a basis for $\mathscr{F}$. However, using the theory of infinite series, it can be shown that for every function f in $\mathscr{F}$, if $0 < x < 1$, then

$$f(x) = a_0 + 2\sum_{n=1}^{\infty}(a_n \cos 2n\pi x + b_n \sin 2n\pi x)$$

where

$$a_0 = \langle f, 1\rangle, \qquad a_n = \langle f(x), \cos 2n\pi x\rangle, \qquad b_n = \langle f(x), \sin 2n\pi x\rangle.$$

This expression for $f(x)$ is called the *Fourier series* of f.

Problems

1. Use the integral inner product for $\mathscr{F}$ to find the norm of
   a. x. b. $2x + 1$. c. $\cos 8\pi x$.
2. Verify that the Schwartz inequality holds for
   a. $U = (1, 3)$, $V = (2, 1)$ in $\mathscr{E}_2$.
   b. $U = x$, $V = \sqrt{1 - x^2}$ in $\mathscr{F}$.
3. Show that the Schwartz inequality for vectors in $\mathscr{E}_n$ yields the inequality
$$\left(\sum_{i=1}^{n} a_i b_i\right)^2 \le \left(\sum_{i=1}^{n} (a_i)^2\right)\left(\sum_{i=1}^{n} (b_i)^2\right)$$
   for all real numbers a_i, b_i.
4. Prove that $\langle U, V\rangle^2 = \|U\|^2\|V\|^2$ if and only if U and V are linearly dependent.
5. Find two orthonormal bases for $\mathscr{E}_2$ containing the vector $(3/5, 4/5)$.
6. Show that each of the following sets is orthonormal in $\mathscr{E}_3$ and for each set find an orthonormal basis for $\mathscr{E}_3$ containing the two vectors.
   a. $\{(1/\sqrt{2}, 0, -1/\sqrt{2}), (0, 1, 0)\}$.
   b. $\{(-1/3, 2/3, -2/3), (0, 1/\sqrt{2}, 1/\sqrt{2})\}$.
   c. $\{(2/7, 3/7, 6/7), (6/7, 2/7, -3/7)\}$.
7. Let S be a nonempty subset of an inner product space $(\mathscr{V}, \langle\,,\rangle)$. Show that the set of all vectors orthogonal to every vector in S is a subspace of $\mathscr{V}$.
8. $B = \{(4, 1), (2, 3)\}$ is a basis for $\mathscr{V}_2$.
   a. Find an inner product defined on $\mathscr{V}_2$ for which B is an orthonormal basis.

b. What is the norm of $(2, -7)$ in the inner product space obtained in a?

9. Let $S = \{(1, 0, 2, 1), (2, 3, -1, 0), (-3, 2, 0, 3)\}$.
a. Show that S is an orthogonal subset of $\mathscr{E}_4$.
b. Obtain an orthonormal set from S.

10. Following Example 4:
a. Show that if $n \neq m$, then $\sin 2n\pi x$ is orthogonal to $\sin 2m\pi x$.
b. Find the Fourier series expansion of $f(x) = x$.

§3. The Gram-Schmidt Process

We have seen that orthonormal bases are the most useful bases for a given inner product space. However, an inner product space often arises with a basis that is not orthonormal. In such a case it is usually helpful to first obtain an orthonormal basis from the given basis. The Gram-Schmidt process provides a procedure for doing this. The existence of such a procedure will also show that every finite-dimensional inner product space has an orthonormal basis.

For an indication of how this procedure should be defined, suppose $\mathscr{S} = \mathscr{L}\{U_1, U_2\}$ is a 2-dimensional subspace of an inner product space. Then the problem is to find an orthonormal basis $\{V_1, V_2\}$ for $\mathscr{S}$. Setting $V_1 = U_1/\|U_1\|$ yields a unit vector. To obtain V_2, consider $W_2 = U_2 + xV_1$. For any $x \in R$, $\mathscr{L}\{V_1, W_2\} = \mathscr{L}\{U_1, U_2\}$. Therefore it is necessary to find a value for x such that $\langle W_2, V_1\rangle = 0$. Then V_2 can be set equal to $W_2/\|W_2\|$. (How do we know that W_2 cannot be $\mathbf{0}$?) W_2 is orthogonal to V_1 if

$$0 = \langle U_2 + xV_1, V_1\rangle = \langle U_2, V_1\rangle + x\langle V_1, V_1\rangle = \langle U_2, V_1\rangle + x.$$

Therefore when $x = -\langle U_2, V_1\rangle$, $\langle W_2, V_1\rangle = 0$.

Example 1 Suppose $\mathscr{S} = \mathscr{L}\{U_1, U_2\}$ is the subspace of $\mathscr{E}_3$ spanned by $U_1 = (2, 1, -2)$ and $U_2 = (0, 1, 2)$. Using the above technique,

$$V_1 = U_1/\|U_1\| = (2/3, 1/3, -2/3)$$

and W_2 is given by

$$\begin{aligned} W_2 &= U_2 - (U_2 \circ V_1)V_1 \\ &= (0, 1, 2) - [(0, 1, 2) \circ (2/3, 1/3, -2/3)](2/3, 1/3, -2/3) \\ &= (2/3, 4/3, 4/3). \end{aligned}$$

Setting $V_2 = W_2/\|W_2\| = (1/3, 2/3, 2/3)$ yields

$$\{V_1, V_2\} = \{(2/3, 1/3, -2/3), (1/3, 2/3, 2/3)\},$$

an orthonormal basis for $\mathscr{S}$ constructed from U_1 and U_2.

There is a good geometric illustration which shows why $W_2 = U_2 - \langle U_2, V_1\rangle V_1$ is orthogonal to V_1. Since V_1 is a unit vector,

$$\langle U_2, V_1\rangle V_1 = (\|U_2\|\cos\theta)V_1.$$

Since $\cos\theta$ is the ratio of the side adjacent to the angle θ, and the hypotenuse, this vector is represented by the point P at the foot of the perpendicular dropped from U_2 to the line determined by $\mathbf{0}$ and V_1, see Figure 1. Thus $W_2 = U_2 - \langle U_2, V_1\rangle V_1$ can be represented by the arrow from P to U_2. That is, W_2 and V_1 are represented by perpendicular arrows. The vector $\langle U_2, V_1\rangle V_1$ is called the "orthogonal projection" of U_2 on V_1. Figure 1 shows that subtracting the orthogonal projection of U_2 on V_1 from U_2 yields a vector orthogonal to V_1. In general, the orthogonal projection of one vector on another is defined as in $\mathscr{E}_2$ with V_1 replaced by $U_1/\|U_1\|$.

Definition If U_1, U_2 are vectors from an inner product space with $U_1 \neq \mathbf{0}$, then the *orthogonal projection* of U_2 on U_1 is the vector

$$\left(\frac{\langle U_2, U_1\rangle}{\langle U_1, U_1\rangle}\right)U_1.$$

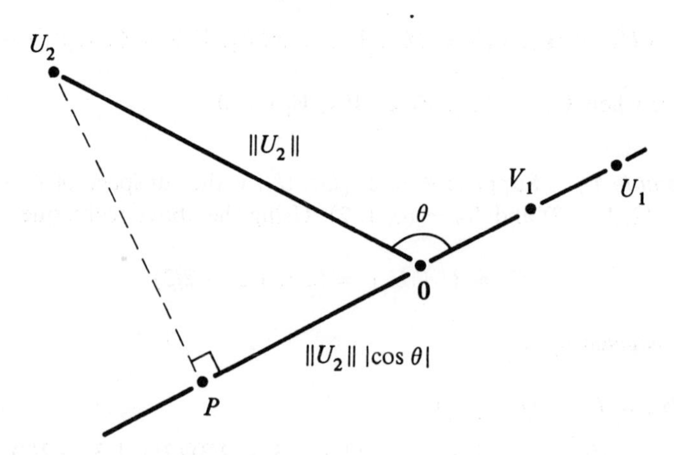

Figure 1

Suppose now that it is desired to find an orthonormal basis for the 3-dimensional space $\mathscr{L}\{U_1, U_2, U_3\}$. Using the above procedure, we can obtain V_1, and V_2 such that $\{V_1, V_2\}$ is an orthonormal set and $\mathscr{L}\{V_1, V_2\} = \mathscr{L}\{U_1, U_2\}$. To obtain a vector W_3 from U_3 orthogonal to V_1 and V_2, one might consider subtracting the orthogonal projections of U_3 on V_1 and V_2 from U_3. Therefore let

$$W_3 = U_3 - \langle U_3, V_1\rangle V_1 - \langle U_3, V_2\rangle V_2.$$

Then

$$\begin{aligned}\langle W_3, V_1\rangle &= \langle U_3, V_1\rangle - \langle U_3, V_1\rangle\langle V_1, V_1\rangle - \langle U_3, V_2\rangle\langle V_2, V_1\rangle \\ &= \langle U_3, V_1\rangle - \langle U_3, V_1\rangle = 0.\end{aligned}$$

So W_3 is orthogonal to V_1. Similarly W_3 is orthogonal to V_2, and setting $V_3 = W_3/\|W_3\|$ yields an orthonormal basis $\{V_1, V_2, V_3\}$ for $\mathscr{L}\{U_1, U_2, U_3\}$. This construction should indicate how the following theorem is proved by induction.

Theorem 6.8 (Gram-Schmidt Process) Let $\mathscr{E}$ be a finite-dimensional inner product space with basis $\{U_1, \ldots, U_n\}$. Then $\mathscr{E}$ has an orthonormal basis $\{V_1, \ldots, V_n\}$ defined inductively by $V_1 = U_1/\|U_1\|$ and for each k, $1 < k \leq n$, $V_k = W_k/\|W_k\|$ where

$$W_k = U_k - \langle U_k, V_1\rangle V_1 - \cdots - \langle U_k, V_{k-1}\rangle V_{k-1}.$$

Further $\mathscr{L}\{V_1, \ldots, V_k\} = \mathscr{L}\{U_1, \ldots, U_k\}$.

Example 2 Use the Gram-Schmidt process to obtain an orthonormal basis for

$$\mathscr{L}\{(1, 0, 1, 0), (3, 1, -1, 0), (8, -7, 0, 3)\} \subset \mathscr{E}_4.$$

Set $V_1 = (1/\sqrt{2}, 0, 1/\sqrt{2}, 0)$. Then

$$\begin{aligned}W_2 &= (3, 1, -1, 0) - [(3, 1, -1, 0) \circ (1/\sqrt{2}, 0, 1/\sqrt{2}, 0)](1/\sqrt{2}, 0, 1/\sqrt{2}, 0) \\ &= (2, 1, -2, 0),\end{aligned}$$

and $V_2 = W_2/\|W_2\| = (2/3, 1/3, -2/3, 0)$ is a unit vector orthogonal to V_1. Finally subtracting the orthogonal projections of $(8, -7, 0, 3)$ on V_1

and V_2 from $(8, -7, 0, 3)$ gives

$$\begin{aligned} W_3 &= (8, -7, 0, 3) - [(8, -7, 0, 3) \circ (1/\sqrt{2}, 0, 1/\sqrt{2}, 0)](1/\sqrt{2}, 0, 1/\sqrt{2}, 0) \\ &\quad - [(8, -7, 0, 3) \circ (2/3, 1/3, -2/3, 0)](2/3, 1/3, -2/3, 0) \\ &= (8, -7, 0, 3) - (4, 0, 4, 0) - (2, 1, -2, 0) = (2, -8, -2, 3). \end{aligned}$$

So $V_3 = W_3/\|W_3\| = (2/9, -8/9, -2/9, 1/3)$ and the orthonormal basis is

$$\{(1/\sqrt{2}, 0, 1/\sqrt{2}, 0), (2/3, 1/3, -2/3, 0), (2/9, -8/9, -2/9, 1/3)\}.$$

In practice it is often easier to obtain an orthogonal basis such as $\{U_1, W_2, \ldots, W_n\}$ and then divide each vector by its length to get an orthonormal basis. This postpones the introduction of radicals and even allows for the elimination of fractions by scalar multiplication. That is, if $(1/2, 0, 1/3, 0)$ is orthogonal to a set of vectors, then so is $(3, 0, 2, 0)$. In Example 2 the second and third orthogonal vectors are given by

$$\begin{aligned} &(3, 1, -1, 0) - \frac{(3, 1, -1, 0) \circ (1, 0, 1, 0)}{(1, 0, 1, 0) \circ (1, 0, 1, 0)}(1, 0, 1, 0) \\ &\qquad = (3, 1, -1, 0) - (1, 0, 1, 0) = (2, 1, -2, 0) \end{aligned}$$

and

$$\begin{aligned} &(8, -7, 0, 3) - \frac{(8, -7, 0, 3) \circ (1, 0, 1, 0)}{(1, 0, 1, 0) \circ (1, 0, 1, 0)}(1, 0, 1, 0) \\ &\qquad - \frac{(8, -7, 0, 3) \circ (2, 1, -2, 0)}{(2, 1, -2, 0) \circ (2, 1, -2, 0)}(2, 1, -2, 0) \\ &\qquad = (8, -7, 0, 3) - (4, 0, 4, 0) - (2, 1, -2, 0) \\ &\qquad = (2, -8, -2, 3). \end{aligned}$$

Now the desired orthonormal basis is obtained from the orthogonal set $\{(1, 0, 1, 0), (2, 1, -2, 0), (2, -8, -2, 3)\}$ when each vector is divided by its length.

Corollary 6.9

1. Every finite-dimensional inner product space has an orthonormal basis.
2. An orthonormal subset of a finite-dimensional inner product space $\mathscr{E}$ can be extended to an orthonormal basis for $\mathscr{E}$.

Proof These follow at once from the corresponding results for vector spaces, using the Gram-Schmidt process.

Example 3 Suppose $\mathscr{E}$ is the inner product space obtained by defining $\langle\ ,\ \rangle$ on $\mathscr{V}_2$ by

$$\langle(a, b), (c, d)\rangle = 3ac - 2ad - 2bc + 5bd.$$

Find an orthonormal basis for $\mathscr{E}$.

An orthonormal basis can be constructed from the standard basis for $\mathscr{V}_2$. A vector orthogonal to (1, 0) is given by

$$(0, 1) - \frac{\langle(0, 1), (1, 0)\rangle}{\langle(1, 0), (1, 0)\rangle}(1, 0) = (0, 1) - (-2/3, 0) = (2/3, 1).$$

Therefore $\{(1, 0), (2, 3)\}$ is an orthogonal basis for $\mathscr{E}$, and dividing each vector by its length gives the orthonormal basis $\{(1/\sqrt{3}, 0), (2/\sqrt{33}, 3/\sqrt{33})\}$ for $\mathscr{E}$.

Definition Let $\mathscr{S}$ be a subspace of an inner product space $\mathscr{E}$. The set

$$\mathscr{S}^\perp = \{V \in \mathscr{E} | \langle U, V\rangle = 0 \text{ for all } U \in \mathscr{S}\}$$

is the *orthogonal complement* of $\mathscr{S}$. The notation $\mathscr{S}^\perp$ is often read "$\mathscr{S}$ perp."

Example 4 If $\mathscr{S}$ is the line $\mathscr{L}\{(1, 3, -2)\}$ in $\mathscr{E}_3$, then what is the orthogonal complement of $\mathscr{S}$?

Suppose $(x, y, z) \in \mathscr{S}^\perp$, then $(x, y, z) \circ (1, 3, -2) = 0$ or $x + 3y - 2z = 0$. Conversely, if a, b, c satisfy the equation $a + 3b - 2c = 0$, then $(a, b, c) \circ (1, 3, -2) = 0$ and $(a, b, c) \in \mathscr{S}^\perp$. Therefore $\mathscr{S}^\perp$ is the plane with Cartesian equation $x + 3y - 2z = 0$. $(1, 3, -2)$ are direction numbers normal to this plane, so $\mathscr{S}^\perp$ is perpendicular to the line $\mathscr{S}$ in the geometric sense.

As a consequence of problem 7, page 254, the orthogonal complement of a subspace is also a subspace. So if $\mathscr{S}$ is a subspace of an inner product space $\mathscr{E}$, then the subspace $\mathscr{S} + \mathscr{S}^\perp$ exists. This sum must be direct for if $U \in \mathscr{S}$ and $U \in \mathscr{S}^\perp$, then $\langle U, U\rangle = 0$ and $U = \mathbf{0}$ (Why?) If $\mathscr{E}$ is finite dimensional, then Corollary 6.9 can be used to show that $\mathscr{S} \oplus \mathscr{S}^\perp$ is a direct sum decomposition of $\mathscr{E}$.

Theorem 6.10 If $\mathscr{S}$ is a subspace of a finite-dimensional inner product space $\mathscr{E}$, then $\mathscr{E} = \mathscr{S} \oplus \mathscr{S}^\perp$.

Proof As a subspace of $\mathscr{E}$, $\mathscr{S}$ is an inner product space and therefore has an orthonormal basis $\{V_1, \ldots, V_k\}$. This basis can be extended to an orthonormal basis $\{V_1, \ldots, V_k, V_{k+1}, \ldots, V_n\}$ for $\mathscr{E}$. The proof will be complete if it can be shown that the vectors $V_{k+1}, \ldots, V_n \in \mathscr{S}^\perp$. (Why is this?) Therefore consider V_j with $j > k$. If $U \in \mathscr{S}$, then there exist scalars $a_1, \ldots, a_k$ such that $U = a_1V_1 + \cdots + a_kV_k$, so

$$\langle U, V_j\rangle = a_1\langle V_1, V_j\rangle + \cdots + a_k\langle V_k, V_j\rangle = 0.$$

Therefore $V_j \in \mathscr{S}^\perp$ for $k < j \leq$ n and $\mathscr{E} = \mathscr{S} \oplus \mathscr{S}^\perp$.

Problems

1. Use the Gram-Schmidt process to obtain an orthonormal basis for the spaces spanned by the following subsets of $\mathscr{E}_n$.
   a. $\{(2, 0, 0), (3, 0, 5)\}$.
   b. $\{(2, 3), (1, 2)\}$.
   c. $\{(1, -1, 1), (2, 1, 5)\}$.
   d. $\{(2, 2, 2, 2), (3, 2, 0, 3), (0, -2, 0, 6)\}$.
2. Construct an orthonormal basis from $\{1, t, t^2\}$ for the inner product space
$$(R_3[t], \langle\ ,\ \rangle) \text{ where } \langle P, Q\rangle = \int_0^1 P(t)Q(t)\,dt.$$
3. Let $\mathscr{E} = (\mathscr{V}_2, \langle\ ,\ \rangle)$ where the inner product is given by
$$\langle(a, b), (c, d)\rangle = 4ac - ad - bc + 2bd.$$
   Use the Gram-Schmidt process to find an orthonormal basis for $\mathscr{E}$
   a. starting with the standard basis for $\mathscr{V}_2$.
   b. starting with the basis $\{(2, 2), (-3, 7)\}$.
4. Define an inner product on $\mathscr{V}_3$ by
$$\langle(a_1, a_2, a_3), (b_1, b_2, b_3)\rangle$$
$$= a_1b_1 - 2a_1b_2 - 2a_2b_1 + 5a_2b_2 + a_2b_3 + a_3b_2 + 4a_3b_3.$$
   Use the Gram-Schmidt process to construct an orthonormal basis for the inner product space $(\mathscr{V}_3, \langle\ ,\ \rangle)$ from the standard basis for $\mathscr{V}_3$.
5. Carefully write an induction proof for the Gram-Schmidt process.
6. Find $\mathscr{S}^\perp$ for the following subsets of $\mathscr{E}_n$:
   a. $\mathscr{S} = \mathscr{L}\{(2, 1, 4), (1, 3, 1)\}$.
   b. $\mathscr{S} = \mathscr{L}\{(1, 3)\}$.
   c. $\mathscr{S} = \mathscr{L}\{(2, -1, 3), (-4, 2, -6)\}$.
   d. $\mathscr{S} = \mathscr{L}\{(1, 0, -2)\}$.
   e. $\mathscr{S} = \mathscr{L}\{(2, 0, 1, 0), (1, 3, 0, 2), (0, 2, 2, 0)\}$.

7. Let T be a linear map from $\mathscr{E}_3$ to $\mathscr{E}_3$ such that $T(U) \circ T(V) = U \circ V$ for all $U, V \in \mathscr{E}_3$. Suppose a subspace $\mathscr{S}$ is invariant under T, and prove that $\mathscr{S}^\perp$ is also invariant.

8. a. Verify that $T(a, b, c) = (\frac{3}{5}a - \frac{4}{5}c, b, \frac{4}{5}a + \frac{3}{5}c)$ satisfies the hypotheses of problem 7.
   b. Find a line $\mathscr{S}$ and the plane $\mathscr{S}^\perp$ that are invariant under this T.

9. Verify that $\mathscr{E}_2 = \mathscr{S} \oplus \mathscr{S}^\perp$ when $\mathscr{S} = \mathscr{L}\{(1, 4)\}$. What is the geometric interpretation of this direct sum decomposition of the plane?

10. Verify that $\mathscr{E}_4 = \mathscr{S} \oplus \mathscr{S}^\perp$ when $\mathscr{S}$ is the plane spanned by $(1, 0, 2, 4)$ and $(2, 0, -1, 0)$.

§4. Orthogonal Transformations

What type of map sends an inner product space into an inner product space? Since an inner product space is a vector space together with an inner product, the map should preserve both the vector space structure and the inner product. That is, it must be linear and the inner product of the images of two vectors must equal the inner product of the two vectors.

Definition Let $(\mathscr{V}, \langle\ ,\ \rangle_{\mathscr{V}})$ and $(\mathscr{W}, \langle\ ,\ \rangle_{\mathscr{W}})$ be inner product spaces and $T\colon \mathscr{V} \to \mathscr{W}$ a linear map. T is *orthogonal* if for all $U, V \in \mathscr{V}$,

$$\langle T(U), T(V)\rangle_{\mathscr{W}} = \langle U, V\rangle_{\mathscr{V}}.$$

Since the norm of a vector is defined using the inner product, it is preserved by an orthogonal map. Thus the image of a nonzero vector is also nonzero, implying that every orthogonal map is nonsingular.

Example 1 The rotation about the origin in $\mathscr{E}_2$ through the angle φ is an orthogonal transformation of $\mathscr{E}_2$ to itself.

It is not difficult to show that $T(U) \circ T(V) = U \circ V$ for all $U, V \in \mathscr{E}_2$ when

$$T(a, b) = (a \cos \varphi - b \sin \varphi, a \sin \varphi + b \cos \varphi).$$

Use the fact that $\sin^2 \varphi + \cos^2 \varphi = 1$.

The distance between two points A and B in the plane (or $\mathscr{E}_n$) is the norm of the difference $A - B$. If $T\colon \mathscr{E}_2 \to \mathscr{E}_2$ is orthogonal, then

$$\|T(A) - T(B)\| = \|T(A - B)\| = \|A - B\|.$$

That is, an orthogonal map of $\mathscr{E}_2$ to itself preserves the Euclidean distance between points.

Example 2 The reflection of $\mathscr{E}_2$ in a line through the origin is an orthogonal transformation.

Suppose T is the reflection of $\mathscr{E}_2$ in the line ℓ with vector equation $P = tU$, $t \in R$, where U is a unit vector. For any $V \in \mathscr{E}_2$, let $W = (V \circ U)U$; then W is the orthogonal projection of V on U. Since W is the midpoint of the line between V and $T(V)$, $W = \frac{1}{2}(V + T(V))$, see Figure 2. Therefore $T(V) = 2W - V = 2(V \circ U)U - V$.

The reflection T is linear for

$$\begin{aligned} T(r_1V_1 + r_2V_2) &= 2[(r_1V_1 + r_2V_2) \circ U]U - (r_1V_1 + r_2V_2) \\ &= r_1 2(V_1 \circ U)U - r_1V_1 + r_2 2(V_2 \circ U)U - r_2V_2 \\ &= r_1T(V_1) + r_2T(V_2), \end{aligned}$$

and T preserves the dot product for

$$\begin{aligned} T(V_1) \circ T(V_2) &= [2(V_1 \circ U)U - V_1] \circ [2(V_2 \circ U)U - V_2] \\ &= 4(V_1 \circ U)(V_2 \circ U) - 2(V_1 \circ U)(U \circ V_2) \\ &\quad - 2(V_2 \circ U)(V_1 \circ U) + V_1 \circ V_2 \\ &= V_1 \circ V_2. \end{aligned}$$

Thus T is an orthogonal transformation. If $U = (\cos \alpha, \sin \alpha)$, as in Figure 2, and $V = (a, b)$, then a short computation shows that the reflection T is

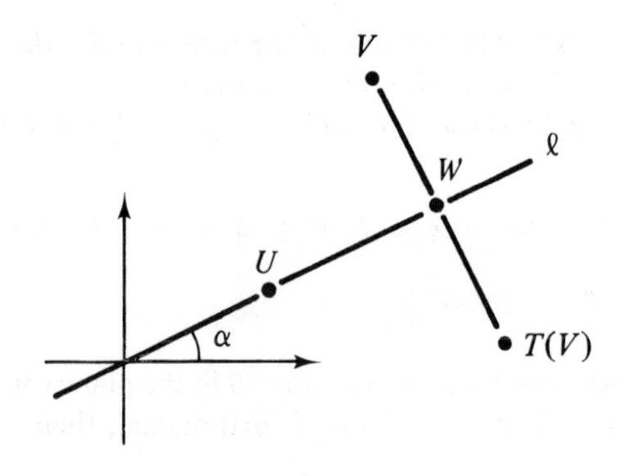

Figure 2

given by

$$T(a, b) = (a \cos 2\alpha + b \sin 2\alpha, a \sin 2\alpha - b \cos 2\alpha).$$

Notice that T transforms the standard basis for $\mathscr{E}_2$ to $\{(\cos \varphi, \sin \varphi), (\sin \varphi, -\cos \varphi)\}$ when $\varphi = 2\alpha$. This is the second form for an orthonormal basis for $\mathscr{E}_2$ obtained in Example 2, page 251.

We will find in Theorem 6.13 that the only orthogonal transformations of $\mathscr{E}_2$ to itself are rotations about the origin and reflections in lines through the origin.

The dot product was defined in $\mathscr{E}_2$ to introduce the geometric properties of length and angle. Since an orthogonal map from $\mathscr{E}_2$ to itself preserves the dot product, it preserves these geometric properties. However, the converse is not true. That is, there are transformations of the plane into itself that preserve length and angle but are not linear. A translation given by $T(V) = V + P$, P a nonzero, fixed vector, is such a transformation.

Theorem 6.11 Let $T \in \text{Hom}(\mathscr{V}, \mathscr{W})$ and suppose $B = \{V_1, \ldots, V_n\}$ is an orthonormal basis for $(\mathscr{V}, \langle\ ,\ \rangle_{\mathscr{V}})$. Then T is orthogonal if and only if $\{T(V_1), \ldots, T(V_n)\}$ is an orthonormal set in $(\mathscr{W}, \langle\ ,\ \rangle_{\mathscr{W}})$.

Proof ($\Rightarrow$) If T is orthogonal, then

$$\langle T(V_i), T(V_j)\rangle_{\mathscr{W}} = \langle V_i, V_j\rangle_{\mathscr{V}} = \delta_{ij}.$$

($\Leftarrow$) Suppose $\langle T(V_i), T(V_j)\rangle_{\mathscr{W}} = \delta_{ij}$ and let $U, V \in \mathscr{V}$. If $U = \sum_{i=1}^n a_i V_i$ and $V = \sum_{j=1}^n b_j V_j$, then

$$\begin{aligned}\langle T(U), T(V)\rangle_{\mathscr{W}} &= \left\langle \sum_{i=1}^n a_i T(V_i), \sum_{j=1}^n b_j T(V_j)\right\rangle_{\mathscr{W}} \\ &= \sum_{i=1}^n a_i \sum_{j=1}^n b_j \langle T(V_i), T(V_j)\rangle_{\mathscr{W}} \\ &= \sum_{i=1}^n \sum_{j=1}^n a_i b_j \delta_{ij} = \sum_{i=1}^n a_i b_i = \langle U, V\rangle_B = \langle U, V\rangle_{\mathscr{V}}.\end{aligned}$$

Definition Two inner product spaces are *isomorphic* if there exists an orthogonal map from one onto the other.

Theorem 6.12 Any two n-dimensional inner product spaces are isomorphic.

Proof Suppose $\mathscr{E}$ and $\mathscr{E}'$ are n-dimensional inner product spaces. Let $B = \{V_1, \ldots, V_n\}$ and $B' = \{V_1', \ldots, V_n'\}$ be orthonormal bases for $\mathscr{E}$ and $\mathscr{E}'$, respectively. Set $T(V_i) = V_i'$, $1 \leq i \leq n$, and extend T linearly to all of $\mathscr{E}$. Then by Theorem 6.11, T is orthogonal. Since T is clearly onto, the spaces are isomorphic.

Thus any n-dimensional inner product space is algebraically indistinguishable from $\mathscr{E}_n$. This result corresponds to the fact that every n-dimensional vector space is isomorphic to the space of n-tuples, $\mathscr{V}_n$. It of course does not say that $\mathscr{E}_n$ is the only n-dimensional inner product space.

Suppose $T: \mathscr{E} \to \mathscr{E}$ is an orthogonal map and $B = \{V_1, \ldots, V_n\}$ is an orthonormal basis for $\mathscr{E}$. What is the nature of the matrix A for T with respect to B? If $A = (a_{ij})$ then $T(V_k)\colon (a_{1k}, \ldots, a_{nk})_B$. The fact that $\langle\ ,\ \rangle = \langle\ ,\ \rangle_B$ and $\langle T(V_i), T(V_j)\rangle = \delta_{ij}$ implies $a_{1i}a_{1j} + \cdots + a_{ni}a_{nj} = \delta_{ij}$. But this sum is the entry from the ith row and jth column of the matrix product $A^T A$. That is, $A^T A = (\delta_{ij}) = I_n$, or the inverse of A is its transpose.

Definition A nonsingular matrix A is *orthogonal* if $A^{-1} = A^T$.

Notice that the condition $A^{-1} = A^T$ implies that the n rows (or the n transposed columns) of A form an orthonormal basis for $\mathscr{E}_n$, and conversely. Using this idea, the orthogonal maps from $\mathscr{E}_2$ to itself can easily be characterized.

Theorem 6.13 If $T: \mathscr{E}_2 \to \mathscr{E}_2$ is orthogonal, then T is either a rotation or a reflection.

Proof If A is the matrix of T with respect to the standard basis, then A is orthogonal. Thus if

$$A = \begin{pmatrix} a & c \\ b & d \end{pmatrix},$$

then $B = \{(a, b), (c, d)\}$ is an orthonormal basis for $\mathscr{E}_2$. Now Example 2, page 251, states that there exists an angle φ such that

$$B = \{(\cos\varphi, \sin\varphi), \pm(-\sin\varphi, \cos\varphi)\}.$$

Taking the plus sign for B gives

$$A = \begin{pmatrix} \cos\varphi & -\sin\varphi \\ \sin\varphi & \cos\varphi \end{pmatrix}.$$

Since A is the matrix of T with respect to the standard basis,

$$T(x, y) = (x\cos\varphi - y\sin\varphi, x\sin\varphi + y\cos\varphi).$$

That is, T is the rotation about the origin, through the angle φ.

Using the minus sign for B gives

$$A = \begin{pmatrix} \cos\varphi & \sin\varphi \\ \sin\varphi & -\cos\varphi \end{pmatrix},$$

and T is given by

$$T(x, y) = (x\cos\varphi + y\sin\varphi, x\sin\varphi - y\cos\varphi).$$

That is, T is the reflection in the line through the origin with direction numbers $(\cos\varphi/2, \sin\varphi/2)$.

Example 3 Determine the nature of the orthogonal map from $\mathscr{E}_2$ to itself given by

$$T(x, y) = (-\tfrac{4}{5}x + \tfrac{3}{5}y, \tfrac{3}{5}x + \tfrac{4}{5}y).$$

The matrix of T with respect to the standard basis is

$$\begin{pmatrix} -4/5 & 3/5 \\ 3/5 & 4/5 \end{pmatrix}.$$

With $\cos\varphi = -4/5$ and $\sin\varphi = 3/5$, this is seen to be the matrix of a reflection in a line ℓ. The slope of ℓ is given by

$$\tan\left(\frac{\varphi}{2}\right) = \pm\sqrt{\frac{1 - \cos\varphi}{1 + \cos\varphi}} = \pm 3.$$

Since φ is in the second quadrant, $\varphi/2$ is in the first quadrant and the slope of ℓ is 3. Therefore T is the reflection in the line with equation $y = 3x$.

It should be clear that any reflection in a line through the origin of $\mathscr{E}_2$

has a matrix in the form $\begin{pmatrix} 1 & 0 \\ 0 & -1 \end{pmatrix}$ with respect to some orthonormal basis, for one line (subspace) is fixed and another is reflected into itself. This result can also be obtained by considering the matrix A for T with respect to the standard basis. Since

$$A = \begin{pmatrix} \cos \varphi & \sin \varphi \\ \sin \varphi & -\cos \varphi \end{pmatrix},$$

the characteristic equation for A is $\lambda^2 = 1$. Thus ± 1 are the characteristic roots of A, and A is similar to diag $(1, -1)$. For the map in Example 3, the orthonormal basis $\{(1/\sqrt{10}, -3/\sqrt{10}), (3/\sqrt{10}, 1/\sqrt{10})\}$ yields the matrix $\begin{pmatrix} 1 & 0 \\ 0 & -1 \end{pmatrix}$ for T.

The previous argument shows that if A is a matrix for a reflection in the plane, then det $A = -1$. On the other hand, if A is the matrix of a rotation in the plane, then det $A = \cos^2\varphi + \sin^2\varphi = 1$. So the matrix of an orthogonal map in the plane has determinant ± 1. This is a general property of orthogonal matrices following from the fact that $AA^T = I_n$.

Theorem 6.14 If A is an orthogonal matrix, then det $A = \pm 1$.

Characterizing the orthogonal transformations of 3-space, $\mathscr{E}_3$, is only a little more difficult than for the plane. Suppose $T: \mathscr{E}_3 \to \mathscr{E}_3$ is orthogonal. The characteristic polynomial of T must have a nonzero real characteristic root r since it is a polynomial of odd degree. Let V_1 be a unit characteristic vector corresponding to r, and let $B = \{V_1, V_2, V_3\}$ be an orthonormal basis for $\mathscr{E}_3$. (B exists by Corollary 6.9 on page 258.) Then the matrix A of T with respect to B has the form

$$A = \begin{pmatrix} r & x & y \\ 0 & a & c \\ 0 & b & d \end{pmatrix}.$$

Since A is orthogonal, its columns are orthonormal in $\mathscr{E}_3$. Therefore $r = \pm 1$, $x = y = 0$, and $\begin{pmatrix} a & c \\ b & d \end{pmatrix}$ is orthogonal. That is, the line $\mathscr{L}\{V_1\}$ is invariant under T and its orthogonal complement, the plane $\mathscr{S} = \mathscr{L}\{V_2, V_3\}$, is sent into itself by the orthogonal, restriction map

$$T_{\mathscr{S}}(xV_2 + yV_3) = (ax + cy)V_2 + (bx + dy)V_3.$$

This already says quite a bit about an arbitrary orthogonal transformation of $\mathscr{E}_3$. But if

$$\det\begin{pmatrix} a & c \\ b & d \end{pmatrix} = 1,$$

then $T_{\mathscr{S}}$ is a rotation in the plane $\mathscr{S}$ and there is an angle φ such that

$$A = \begin{pmatrix} \pm 1 & 0 & 0 \\ 0 & \cos\varphi & -\sin\varphi \\ 0 & \sin\varphi & \cos\varphi \end{pmatrix}.$$

On the other hand, if

$$\det\begin{pmatrix} a & c \\ b & d \end{pmatrix} = -1,$$

then $T_{\mathscr{S}}$ reflects the plane $\mathscr{L}\{V_2, V_3\}$ into itself. Therefore there exist vectors V_2' and V_3' such that $B' = \{V_1, V_2', V_3'\}$ is an orthonormal basis for $\mathscr{E}_3$ and the matrix of T with respect to B' is

$$A' = \begin{pmatrix} \pm 1 & 0 & 0 \\ 0 & 1 & 0 \\ 0 & 0 & -1 \end{pmatrix}.$$

Since one of the matrices denoted by A' is similar to one denoted by A when $\varphi = \pi$, there are three different types of matrix for T. Or we have found that if $T: \mathscr{E}_3 \to \mathscr{E}_3$ is orthogonal, then T has one of the following matrices with respect to some orthonormal basis $\{U_1, U_2, U_3\}$ for $\mathscr{E}_3$:

$$\begin{pmatrix} 1 & 0 & 0 \\ 0 & \cos\varphi & -\sin\varphi \\ 0 & \sin\varphi & \cos\varphi \end{pmatrix}, \qquad \begin{pmatrix} -1 & 0 & 0 \\ 0 & 1 & 0 \\ 0 & 0 & 1 \end{pmatrix}$$

or

$$\begin{pmatrix} -1 & 0 & 0 \\ 0 & \cos\varphi & -\sin\varphi \\ 0 & \sin\varphi & \cos\varphi \end{pmatrix} = \begin{pmatrix} 1 & 0 & 0 \\ 0 & \cos\varphi & -\sin\varphi \\ 0 & \sin\varphi & \cos\varphi \end{pmatrix} \begin{pmatrix} -1 & 0 & 0 \\ 0 & 1 & 0 \\ 0 & 0 & 1 \end{pmatrix}.$$

If T has the first matrix, then T is called a *rotation* of $\mathscr{E}_3$. Thus a rotation of $\mathscr{E}_3$ is the identity on a line, $\mathscr{L}\{U_1\}$, called the *axis* of the rotation. Further the rotation T rotates every plane parallel to $\mathscr{L}\{U_2, U_3\}$ into itself, through the angle φ. This angle φ is called the *angle of rotation* of T, with the positive direction being measured from U_2 to U_3. If T has the second matrix above, then T is a *reflection* of $\mathscr{E}_3$ in the plane $\mathscr{L}\{U_2, U_3\}$. Including the third form for the matrix of T we have the following.

Theorem 6.15 An orthogonal transformation of $\mathscr{E}_3$ to itself is either a rotation, a reflection in a plane, or the composition of a rotation and a reflection in a plane.

An orthogonal transformation in $\mathscr{E}_2$ or $\mathscr{E}_3$ is a rotation if its matrix with respect to an orthonormal basis has determinant $+1$. Therefore the following definition is made.

Definition Let $T: \mathscr{E}_n \to \mathscr{E}_n$ be an orthogonal map, B an orthonormal basis for $\mathscr{E}_n$, and A the matrix of T with respect to B. Then T is a *rotation* of $\mathscr{E}_n$ if $\det A = 1$.

Can you guess how a rotation acts in 4-space, $\mathscr{E}_4$?

Example 4 Suppose $T: \mathscr{E}_3 \to \mathscr{E}_3$ is the orthogonal transformation having the matrix

$$\begin{pmatrix} 1/3 & 2/3 & -2/3 \\ -2/3 & 2/3 & 1/3 \\ 2/3 & 1/3 & 2/3 \end{pmatrix}$$

with respect to the standard basis. Determine how T acts on $\mathscr{E}_3$.

Since T is orthogonal and $\det A = 1$, T is a rotation. If U lies on the axis of T, then $T(U) = U$. This equation has the solutions $U = t(0, 1, 1)$, $t \in R$. Therefore the axis of T is $\mathscr{L}\{(0, 1, 1)\}$ and T rotates each plane with equation $y + z =$ constant into itself. To find the angle of T, let

$$B = \{(0, 1\sqrt{2}, 1\sqrt{2}), (0, -1/\sqrt{2}, 1/\sqrt{2}), (1, 0, 0)\}$$

This is an orthonormal basis for $\mathscr{E}_3$ with the first vector on the axis of rotation. Computing the matrix A of T with respect to this basis yields

$$A = \begin{pmatrix} 1 & 0 & 0 \\ 0 & 1/3 & \sqrt{8}/3 \\ 0 & -\sqrt{8}/3 & 1/3 \end{pmatrix}.$$

Therefore the angle φ of T satisfies $\cos \varphi = 1/3$ and $\sin \varphi = -\sqrt{8}/3$.

The sign of φ is meaningful only in reference to a choice of basis. If the last two vectors in B are interchanged, the sign of φ is changed, for then $\cos \varphi = 1/3$, $\sin \varphi = \sqrt{8}/3$. Thus the amount of rotation might be found without reference to the sense of the rotation. This can be done by simply finding the angle between any nonzero vector V in $\mathscr{L}\{(0, 1, 1)\}^\perp$ and its image $T(V)$. If V is taken to be $(1, 0, 0)$, then the angle φ satisfies

$$\cos \varphi = \frac{U \circ T(U)}{\|U\| \, \|T(U)\|} = (1, 0, 0) \circ (1/3, 2/3, -2/3) = 1/3.$$

One further general result should be stated to complete this section.

Theorem 6.16 The matrix of an orthogonal transformation with respect to an orthonormal basis is orthogonal.

Conversely, given an $n \times n$ orthogonal matrix A and an n-dimensional inner product space $\mathscr{E}$ with an orthonormal basis B, there exists an orthogonal map $T\colon \mathscr{E} \to \mathscr{E}$ with matrix A with respect to B.

Proof The first statement has already been derived. For the second, suppose $A = (a_{ij})$ is orthogonal and $B = \{V_1, \ldots, V_n\}$ is an orthonormal basis for $\mathscr{E}$. Define T by $T(V_j) = a_{1j}V_1 + \cdots + a_{nj}V_n$ and extend linearly to all of $\mathscr{E}$. Then T is linear and has matrix A with respect to B, so it is only necessary to show that T preserves the inner product. It is sufficient to show that $\langle T(V_i), T(V_j)\rangle = \delta_{ij}$, and

$$\langle T(V_i), T(V_j)\rangle = \langle T(V_i), T(V_j)\rangle_B = a_{1i}a_{1j} + \cdots + a_{ni}a_{nj}$$

which is the element from the ith row and jth column of the product A^TA. But $A^TA = I_n$, so $\langle T(V_i), T(V_j)\rangle = \delta_{ij}$ and T preserves the inner product of $\mathscr{E}$.

Problems

1. Determine the nature of the following orthogonal maps in $\mathscr{E}_2$.
   a. $T(x, y) = (\frac{1}{2}x + \frac{1}{2}\sqrt{3}y, \frac{1}{2}\sqrt{3}x - \frac{1}{2}y)$.
   b. $T(x, y) = (\frac{1}{2}x + \frac{1}{2}\sqrt{3}y, -\frac{1}{2}\sqrt{3}x + \frac{1}{2}y)$.
   c. $T(x, y) = (-\frac{4}{5}x + \frac{3}{5}y, -\frac{3}{5}x - \frac{4}{5}y)$.
   d. $T(x, y) = (x, y)$. e. $T(x, y) = (-y, -x)$.
2. a. Show that the composition of two orthogonal maps is orthogonal.
   b. Show that the composition of two rotations in $\mathscr{E}_n$ is a rotation.
   c. Is the composition of two reflections a reflection?
3. Suppose $T\colon \mathscr{E} \to \mathscr{E}$ is linear and preserves the norm, show that T is orthogonal.

4. Fill in the missing entries to obtain an orthogonal matrix

a. $\begin{pmatrix} 1/3 & 2/3 & a \\ b & 2/3 & c \\ 2/3 & d & 2/3 \end{pmatrix}$. b. $\begin{pmatrix} 1/\sqrt{3} & 1/\sqrt{3} & a \\ b & c & 1/\sqrt{2} \\ 1/\sqrt{6} & d & 1/\sqrt{6} \end{pmatrix}$.

5. Determine which of the following maps of $\mathscr{E}_3$ are orthogonal and give a geometric description of those that are orthogonal.

a. $T(a, b, c) = (a/\sqrt{2} + b/\sqrt{2}, b/\sqrt{2} - a/\sqrt{2}, -c)$.
b. $T(a, b, c) = (b/\sqrt{2} + c/\sqrt{2}, a, b/\sqrt{2} + c/\sqrt{2})$.
c. $T(a, b, c) = (b/\sqrt{2} - a/\sqrt{2}, a/\sqrt{2} + b/\sqrt{2}, -c)$.
d. $T(a, b, c) = (a/\sqrt{2} - c/\sqrt{2}, b, a/\sqrt{2} + c/\sqrt{2})$.
e. $T(a, b, c) = (a/\sqrt{2} + b/\sqrt{2}, b/\sqrt{2} - c/\sqrt{2}, a/\sqrt{2} + c/\sqrt{2})$.

6. Give two different proofs of the fact that the product of two $n \times n$ orthogonal matrices is orthogonal.

7. Determine how the orthogonal transformation T acts on $\mathscr{E}_3$ if the matrix of T with respect to the standard basis is

a. $\begin{pmatrix} 6/7 & 2/7 & 3/7 \\ 2/7 & 3/7 & -6/7 \\ -3/7 & 6/7 & 2/7 \end{pmatrix}$. b. $\begin{pmatrix} 6/7 & 2/7 & -3/7 \\ 2/7 & 3/7 & 6/7 \\ -3/7 & 6/7 & -2/7 \end{pmatrix}$.

c. $\begin{pmatrix} 1/3 & 2/3 & -2/3 \\ 2/3 & 1/3 & 2/3 \\ -2/3 & 2/3 & 1/3 \end{pmatrix}$. d. $\begin{pmatrix} 2/3 & 1/3 & -2/3 \\ 2/3 & -2/3 & 1/3 \\ 1/3 & 2/3 & 2/3 \end{pmatrix}$.

e. $\begin{pmatrix} -1/9 & -8/9 & 4/9 \\ -8/9 & -1/9 & -4/9 \\ 4/9 & -4/9 & -7/9 \end{pmatrix}$. f. $\begin{pmatrix} 1/9 & 8/9 & 4/9 \\ 8/9 & 1/9 & -4/9 \\ -4/9 & 4/9 & -7/9 \end{pmatrix}$.

8. Given $P \in \mathscr{E}_n$, the map T defined by $T(V) = V + P$ for all $V \in \mathscr{E}_n$ is called a *translation.*

a. Show that a translation preserves the distance between points.
b. When is a translation orthogonal?

9. Find a vector expression for a reflection of the plane $\mathscr{E}_2$ in a line that does not pass through the origin. Show that this map preserves distance between points but is not linear and does not preserve the dot product.

10. Let $T\colon \mathscr{E}_4 \to \mathscr{E}_4$ be orthogonal and suppose T has a real characteristic root.

a. Show that T has two real characteristic roots.
b. Describe the possible action of T in terms of a rotation about a plane and reflections in hyperplanes.

11. Prove Euler's theorem: If $T\colon \mathscr{E}_n \to \mathscr{E}_n$ is a rotation and n is odd, then 1 is a characteristic value for T. That is, any rotation of $\mathscr{E}_n$ fixes some line, when n is odd.

§5. Vector Spaces over Arbitrary Fields

Although the point has not been stressed, all the vector spaces considered to this point have been real vector spaces. That is, the scalars have been real numbers. But there are many situations in which it is necessary to use scalars from other "number systems." For example, we have ignored complex roots of a characteristic equation simply because of the requirement that a scalar be a real number. Because of the existence of complex roots, it will be necessary to introduce complex inner product spaces to prove an important theorem about diagonalizability.

Before considering the complex case in particular, it is worthwhile to see exactly how our definition of a vector space might be generalized. Instead of using scalars from the real number system, the scalars could be chosen from any system that shares certain properties with the real numbers. The abstract algebraic system satisfying these properties is called a field.

Definition Let F be a set on which two operations called addition and multiplication are defined such that for all $a, b \in F$, $a + b \in F$ and $a \cdot b = ab \in F$. The system $(F, +, \cdot)$ is a *field* if the following properties hold for all $a, b, c \in F$:

1. $a + (b + c) = (a + b) + c$ and $a(bc) = (ab)c$.
2. $a + b = b + a$ and $ab = ba$.
3. There exists an element $0 \in F$, such that $a + 0 = a$, and there exists and element $1 \in F$, $1 \neq 0$, such that, $a \cdot 1 = a$.
4. For each $a \in F$, there exists $-a \in F$, such that $a + (-a) = 0$, and for each $a \in F$, if $a \neq 0$, there exists $a^{-1} \in F$ such that $a \cdot a^{-1} = 1$.
5. $a(b + c) = ab + ac$.

It should be clear that the real number system is a field. In fact, the properties required of a field are the same as those listed for R on page 2 in preparation for defining the vector space $\mathscr{V}_2$. It is not difficult to show that the complex numbers $C = \{a + bi \mid a, b \in R\}$ form a field with addition and multiplication defined by

$$(a + bi) + (c + di) = a + c + (b + d)i$$

and

$$(a + bi) \cdot (c + di) = ac - bd + (ad + bc)i.$$

There are many fields besides the real and complex numbers. For example, the set of all rational numbers, denoted by Q, forms a field within R. The sum and product of rational numbers is rational, so Q is closed under addition and multiplication. The identities 0 and 1 of R are in Q, and the additive and multiplicative inverses of rational numbers are rational numbers. Thus Q is a "subfield" of R.

The fields R, C, and Q are infinite fields in that they contain an infinite number of elements. But there are also finite fields. The simplest finite fields, denoted by Z_p, are the fields of integers modulo p, where p is a prime. Z_p is the set $\{0, 1, 2, \ldots, p - 1\}$ together with addition and multiplication defined using the operations in the integers as follows:

$$a + b = c \quad \text{if} \quad a + b = kp + c \quad \text{for some} \quad k \in Z, 0 \leq c < p$$

and

$$ab = d \quad \text{if} \quad ab = hp + d \quad \text{for some} \quad h \in Z, 0 \leq d < \text{p}.$$

That is, $a + b$ and ab in Z_p are remainders obtained when $a + b$ and ab are divided by p in Z. For example, in Z_7, $2 + 3 = 5$, $6 + 5 = 4$, $5 \cdot 2 = 3$, $5 \cdot 6 = 2$, and $2 \cdot 3 = 6$. A proof that Z_p is a field is omitted since these finite fields are introduced here simply to indicate the diversity of fields. The field Z_5 is considered in problem 6 at the end of this section, and you might note that the elements of Z_5 are essentially the equivalence classes obtained in problem 3, page 221. The simplest field is Z_2, which contains only 0 and 1. In Z_2, $1 + 1 = 0$, so 1 is its own additive inverse! This fact leads to some interesting complications when statements are made for arbitrary fields.

The real number system and any field F share the properties used in defining a vector space. Therefore the concept of a vector space can be generalized by simply replacing R by F in our original definition. This yields the definition for a *vector space* $\mathscr{V}$ *over a field* F. This is an algebraic system consisting of a set $\mathscr{V}$, whose elements are called *vectors*, a field F, whose elements are called *scalars*, and operations of vector addition and scalar multiplication.

Given any field F, the set of ordered n-tuples can be made into a vector space over F for each positive integer n. Let

$$\mathscr{V}_n(F) = \{(a_1, \ldots, a_n) | a_1, \ldots, a_n \in F\}$$

and define

$$(a_1, \ldots, a_n) + (b_1, \ldots, b_n) = (a_1 + b_1, \ldots, a_n + b_n)$$

$$c(a_1, \ldots, a_n) = (ca_1, \ldots, ca_n), \qquad \text{for} \quad c \in F.$$

The proof that $\mathscr{V}_n(F)$ together with these two operations is a vector space over F is exactly like the proof that $\mathscr{V}_n$ is a real vector space. In fact, $\mathscr{V}_n(R)$ is $\mathscr{V}_n$.

Similarly, the set of all polynomials in t with coefficients from F, denoted by $F[t]$, can be turned into a vector space over the field F. The set of all $n \times m$ matrices with elements from F, denoted $\mathcal{M}_{n \times m}(F)$, forms a vector space over F, and if $\mathscr{V}$ and $\mathscr{W}$ are vector spaces over F, then so is $\text{Hom}(\mathscr{V}, \mathscr{W})$.

Example 1 Consider the vector space $\mathscr{V}_3(Z_2)$. Since the field contains only two elements, it is possible to list all the vectors in this space. They are: (0, 0, 0), (1, 0, 0), (0, 1, 0), (0, 0, 1), (1, 1, 0), (1, 0, 1), (0, 1, 1), (1, 1, 1). Therefore $\mathscr{V}_3(Z_2)$ is finite in the sense that there are only eight vectors. It must also be finite dimensional since a basis could contain at most seven vectors. However, the seven nonzero vectors are not linearly independent. For example, the set $\{(1, 1, 0), (1, 0, 1), (0, 1, 1)\}$ is linearly dependent in $\mathscr{V}_3(Z_2)$. In fact, $(1, 1, 0) + (1, 0, 1) + (0, 1, 1) = (1 + 1, 1 + 1, 1 + 1) = (0, 0, 0)$.

Example 1 implies that the basic properties of $\mathscr{V}_n(R)$ carry over to $\mathscr{V}_n(F)$. In particular, the standard basis $\{E_i\}$ for $\mathscr{V}_n(R)$ is a basis for $\mathscr{V}_n(F)$. Therefore the dimension of $\mathscr{V}_n(F)$ is n.

Example 2 Consider the vector space $\mathscr{V}_2(C)$. The vectors in this space are ordered pairs of complex numbers, such as $(3 - i, 2i - 6)$, $(1, 5)$, $(6i, i)$. Given an arbitraty vector (z, w) with $z, w \in C$, $(z, w) = z(1, 0) + w(0, 1)$, so $\{E_i\}$ spans $\mathscr{V}_2(C)$. And if

$$z(1, 0) + w(0, 1) = \mathbf{0} = (0, 0),$$

then $z = 0$ and $w = 0$ by the definition of an ordered pair, so $\{E_i\}$ is linearly independent. So $\{E_i\}$ is a basis for $\mathscr{V}_2(C)$ and the vector space has dimension 2.

Example 3 Let $T: \mathscr{V}_2(C) \to \mathscr{V}_2(C)$ be defined by

$$T(z, w) = (iz + 3w, (1 - i)w).$$

The image of any vector under T is easily computed. For example,

$$\begin{aligned} T(2 + 3i, 3 - i) &= (i(2 + 3i) + 3(3 - i), (1 - \mathrm{i})(3 - i)) \\ &= (6 - i, 4 - 4i). \end{aligned}$$

T is a linear map, for if $U = (z_1, w_1)$ and $V = (z_2, w_2)$, then

$$\begin{aligned} T(U + V) &= T(z_1 + z_2, w_1 + w_2) \\ &= (i[z_1 + z_2] + 3[w_1 + w_2], (1 - i)[w_1 + w_2]) \\ &= ([iz_1 + 3w_1] + [iz_2 + 3w_2], (1 - i)w_1 + (1 - i)w_2) \\ &= (iz_1 + 3w_1, (1 - i)w_1) + (iz_2 + 3w_2, (1 - i)w_2) \\ &= T(U) + T(V). \end{aligned}$$

Similarly, $T(zU) = zT(U)$ for any $z \in C$.

Since $T(1, 0) = (i, 0)$ and $T(0, 1) = (3, 1 - i)$, the matrix of T with respect to the standard basis $\{E_i\}$ is

$$A = \begin{pmatrix} i & 3 \\ 0 & 1 - i \end{pmatrix} \in \mathscr{M}_{2 \times 2}(C).$$

The determinant of A is $i(1 - i) = 1 + i$, so T is nonsingular and the matrix of T^{-1} with respect to $\{E_j\}$ is

$$A^{-1} = \frac{1}{1 + i}\begin{pmatrix} 1 - i & -3 \\ 0 & i \end{pmatrix} = \begin{pmatrix} -i & (3i - 3)/2 \\ 0 & (1 + i)/2 \end{pmatrix}.$$

Therefore

$$T^{-1}(a, b) = (-ia + \tfrac{1}{2}(3i - 3)b, \tfrac{1}{2}(1 + i)b).$$

You might check to see that either $T \circ T^{-1} = I$ or $T^{-1} \circ T = I$.

The characteristic polynomial of T is

$$(i - \lambda)(1 - i - \lambda) = \lambda^2 - \lambda + 1 + i.$$

Since T is defined on a complex vector space, its characteristic values may be complex numbers. In this case i and $1 - i$ are the characteristic values for T. These values are distinct, so T is diagonalizable, although its characteristic vectors will have complex components. $(1, 0)$ is a characteristic vector corresponding to i, as is $(i, 0)$ or $(z, 0)$ for any complex number $z \neq 0$, and $(3, 1 - 2i)$ is a characteristic vector corresponding to $1 - i$. Therefore the matrix of T with respect to the basis $\{(1, 0), (3, 1 - 2i)\}$ is the diagonal matrix $\begin{pmatrix} 1 & 0 \\ 0 & 1 - i \end{pmatrix}$.

Example 3 serves to show that anything that has been done in real vector spaces can be done in a complex vector space. C could have been replaced by an arbitrary field and a similar statement could be made for a vector space over any field F. Notice this means that all the results on systems of linear equations apply to systems with coefficients from an arbitrary field. However, when it comes to defining an inner product on an arbitrary vector space, the nature of the field must be taken into account. One way to see why is to consider the requirement that an inner product be positive definite. This condition requires an order relation on the elements of the field. But not all fields can be ordered, in particular the complex numbers cannot be ordered, see problem 15 below.

Problems

1. Compute the following.

a. $(2 + i, 3i - 1) - (i - 5, 2 - 4i)$. b. $[3 + i](1 + i, 2i - 4, 3 - i)$.

c. $\begin{pmatrix} 2 - i & 3i - 1 \\ 1 + i & 2i \end{pmatrix}\begin{pmatrix} 1 - i \\ 4 + i \end{pmatrix}$. d. $\begin{pmatrix} i & 3 - i \\ 2 + i & 4 + i \\ 0 & i \end{pmatrix}\begin{pmatrix} 2i & 0 & 2 + 3i \\ 1 & 3 - i & 1 + i \end{pmatrix}$.

2. a. Show that $B = \{(1 + i, 3i), (2, 3 + 2i)\}$ is a basis for $\mathscr{V}_2(C)$.
b. Find the coordinates of $(4i - 2, 6i - 9)$ and $(4i, 7i - 4)$ with respect to B.

3. a. Determine if $S = \{(i, 1 - i, 2), (2, 1, -i), (5 - 2i, 4, -1 - i)\}$ is linearly independent in $\mathscr{V}_3(C)$.
b. Determine if $(3 + i, 4, 2)$ is in the span of S.

4. Prove that $\mathscr{V}_n(F)$ is a vector space over F, noting all the points at which it is necessary to use some field property of F.

5. Define $F_n[t]$ as $R_n[t]$ was defined and show that dim $F_n[t] = n$.

6. Find the following elements in Z_5:
a. $3 + 4$, $2 + 3$, and $3 + 3$.
b. $2 \cdot 4$, $3 \cdot 4$, 4^2, and 2^6 (note $6 \notin Z_5$).
c. -1, -4, and -3.
d. 2^{-1}, 3^{-1}, and 4^{-1}.
e. Prove that Z_5 is a field.

7. Define Z_n, for any positive integer n, in the same way Z_p was defined. Are Z_4 and Z_6 fields?

8. a. Find all vectors in $\mathscr{L}\{(1, 4)\}$, if $(1, 4) \in \mathscr{V}_2(Z_5)$.
b. Is $\{(2, 3), (3, 2)\}$ a basis for $\mathscr{V}_2(Z_5)$?
c. How many vectors are there in the vector space $\mathscr{V}_2(Z_5)$?
d. Is $\{(2, 1, 3, 1), (4, 2, 1, 0), (1, 3, 4, 2)\}$ linearly independent in $\mathscr{V}_4(Z_5)$?

9. Let $T: \mathscr{V}_2(C) \rightarrow \mathscr{V}_2(C)$ be defined by $T(z, w) = (w, -z)$.
a. Show that T is linear.
b. Find the characteristic equation of T.
c. Find the characteristic values and corresponding characteristic vectors for T.
d. Find a basis for $\mathscr{V}_2(C)$ with respect to which T has a diagonal matrix.

10. Follow the directions from problem 9 for the map given by

$$T(z, w) = ([i - 1]z + [3i - 1]w, 2iz + 4w).$$

11. Let $T(a + bi, c + di) = (a + d - ci, b + (a - d)i)$. Show that T is not in $\text{Hom}(\mathscr{V}_2(C))$.

12. Let $\mathscr{V}$ be the algebraic system whose elements are ordered pairs of real numbers. Define addition as in $\mathscr{V}_2(R)$ and scalar multiplication by $z(a, b) = (za, zb)$, $z \in C$, $a, b \in R$. Determine if $\mathscr{V}$ is a vector space over C.

13. Let $\mathscr{V}$ be the algebraic system whose elements are ordered pairs of complex numbers. Define addition as in $\mathscr{V}_2(C)$ and scalar multiplication by $r(z, w) = (rz, rw)$, $r \in R$, $z, w \in C$.
a. Show that $\mathscr{V}$ is a vector space over R.
b. What is the dimension of $\mathscr{V}$?

14. Let $\mathscr{V}$ be the algebraic system whose elements are ordered pairs of real numbers. Define addition as in $\mathscr{V}_2(R)$ and scalar multiplication by $q(a, b) = (qa, qb)$, $q \in Q$, $a, b \in R$. (Q denotes the set of rationals.)
a. Show that $\mathscr{V}$ is a vector space over Q.
b. How does $\mathscr{V}$ differ from $\mathscr{V}_2(Q)$?
c. Is $\mathscr{V}$ a subspace of $\mathscr{V}_2(R)$?
d. Show that $\{(1, 0), (\sqrt{2}, 0)\}$ is linearly independent in $\mathscr{V}$.
e. Is (π, π) in the subspace of $\mathscr{V}$ given by $\mathscr{L}\{(1, 0), (0, 1)\}$?
f. What can be concluded from parts d and e about the dimension of $\mathscr{V}$?

15. A field F can be *ordered* if it contains a subset P (for positive) which is closed under addition and multiplication, and for every $x \in F$ exactly one of the following holds; $x \in P$, $x = 0$, or $-x \in P$. Show that the field of complex numbers cannot be ordered.

16. For each nonzero vector $U \in \mathscr{E}_2$, suppose θ_U is an angle satisfying

$$U = (a, b) = \|U\|(a/\|U\|, b/\|U\|) = \|U\| (\cos \theta_U, \sin \theta_U).$$

Let T_{θ_U} denote the rotation of $\mathscr{E}_2$ through the angle θ_U, and define a multiplication in $\mathscr{E}_2$ by

$$UV = \|U\| T_{\theta_U}(V) \quad \text{if} \quad U \neq \mathbf{0} \text{ and } V \in \mathscr{E}_2$$

and

$$\mathbf{0}V = \mathbf{0} \text{ when } U = \mathbf{0}.$$

That is, to multiply U times V, rotate V through the angle θ_U and then multiply by the length of U.
a. Show that the set of all ordered pairs in $\mathscr{E}_2$ together with vector addition and the above multiplication is a field.

b. Show that $\{(a, 0) | a \in R\}$ is a subfield which is isomorphic to the field of real numbers.
c. Show that $(0, 1)^2 = -1$. Therefore if we set $i = (0, 1)$, then $i^2 = -1$ using the identification given in part b.
d. Show that this field of ordered pairs is isomorphic to the field of complex numbers.

§6. Complex Inner Product Spaces and Hermitian Transformations

The dot product of $\mathscr{E}_n$ cannot be directly generalized to give an inner product on $\mathscr{V}_n(C)$ because the square of a complex number need not be real, let alone positive, e.g., $(1 - i)^2 = -2i$. However a positive definite function can be defined on $\mathscr{V}_n(C)$ using the fact that the the product of $a + bi$ and $a - bi$ is both real and nonnegative.

Definition The *conjugate* of the complex number $z = a + bi$ is $\bar{z} = a - bi$.

Thus $\overline{5 - 3i} = 5 + 3i$, $\overline{2i - 7} = -2i - 7$, and $\bar{6} = 6$.

All of the following properties of the complex conjugate are easily derived from the definition.

Theorem 6.17 Suppose z and w are complex numbers, then

1. $\overline{z + w} = \bar{z} + \bar{w}$.
2. $\overline{z \cdot w} = \bar{z} \cdot \bar{w}$.
3. $z \cdot \bar{z} = a^2 + b^2$ if $z = a + bi$, $a, b \in R$.
4. $\bar{\bar{z}} = z$.
5. $\bar{z} = z$ if and only if $z \in R$.

Definition For $Z = (z_1, \ldots, z_n)$ and $W = (w_1, \ldots, w_n)$ in $\mathscr{V}_n(C)$, let $Z \circ W = z_1 \cdot \bar{w}_1 + \cdots + z_n \cdot \bar{w}_n$.

For two ordered triples of complex numbers,

$$\begin{aligned}(3 - i, 2, 1 + i) &\circ (4, 2 + i, 3i - 1) \\ &= (3 - i)4 + 2(2 - i) + (1 + i)(-3i - 1) \\ &= 18 - 10i.\end{aligned}$$

We will call this operation the standard inner product for ordered n-tuples of complex numbers and denote the vector space $\mathscr{V}_n(C)$ together with this inner product by $\mathscr{C}_n$. But it should be quickly noted that this inner product does not satisfy all the properties of the inner product of $\mathscr{E}_n$.

Theorem 6.18 For all vectors $Z, W, U \in \mathscr{C}_n$,

1. $Z \circ W \in C$.
2. $Z \circ W = \overline{W \circ Z}$.
3. $Z \circ (W + U) = Z \circ W + Z \circ U$ and $(Z + W) \circ U = Z \circ U + W \circ U$
4. $(aZ) \circ W = a(Z \circ W)$ and $Z \circ (aW) = \bar{a}(Z \circ W)$ for all $a \in C$.
5. $Z \circ Z$ is real and $Z \circ Z > 0$ if $Z \neq \mathbf{0}$.

Proof of (2) If $Z = (z_1, \ldots, z_n)$ and $W = (w_1, \ldots, w_n)$, then

$$Z \circ W = \sum_{i=1}^{n} z_i \bar{w}_i = \sum_{i=1}^{n} \bar{w}_i z_i = \sum_{i=1}^{n} \bar{w}_i \bar{\bar{z}}_i = \sum_{i=1}^{n} \overline{w_i \bar{z}_i} = \overline{\sum_{i=1}^{n} w_i \bar{z}_i} = \overline{W \circ Z}.$$

Can you fill in a reason for each step?

Notice that the dot product in $\mathscr{E}_n$ satisfies all the preceding properties since the conjugate of a real number is the number itself. The dot product of $\mathscr{E}_n$ was said to be symmetric and bilinear. The corresponding properties for the inner product of $\mathscr{C}_n$ are numbered 2, 3, and 4 in Theorem 6.18, and the inner product of $\mathscr{C}_n$ is said to be *hermitian*. Theorem 6.18 provides the essential properties which should be required of any inner product on a complex vector space.

Definition Let $\mathscr{V}$ be a complex vector space and $\langle\ ,\ \rangle$ a function such that $\langle Z, W\rangle \in C$ for all $Z, W \in \mathscr{V}$, which satisfies

1. $\langle Z, W\rangle = \overline{\langle W, Z\rangle}$.
2. $\langle Z, aW + bU\rangle = \bar{a}\langle Z, W\rangle + \bar{b}\langle Z, U\rangle$, for all $a, b \in C$, and $Z, W, U \in \mathscr{V}$.
3. $\langle Z, Z\rangle > 0$ if $Z \neq \mathbf{0}$.

Then $\langle\ ,\ \rangle$ is a *complex inner product* on $\mathscr{V}$, and $(\mathscr{V}, \langle\ ,\ \rangle)$ is a *complex inner product space*. $\langle\ ,\ \rangle$ may also be called a *positive definite hermitian product* on $\mathscr{V}$ and $(\mathscr{V}, \langle\ ,\ \rangle)$ may be called a *unitary space*.

The second condition in this definition shows that a complex inner product is not linear in both variables. However, you might verify that the first and second conditions imply

$$\langle aZ + bW, U\rangle = a\langle Z, U\rangle + b\langle Z, U\rangle$$
$$\text{for all} \quad a, b \in C, \text{ and } Z, W, U \in \mathscr{V}.$$

Complex inner product spaces can be constructed from complex vector spaces in the same ways that real inner product spaces were obtained from real vector spaces.

Example 1 The set of all continuous complex-valued functions defined on [0, 1] might be denoted by $\mathscr{F}(C)$. This is a complex vector space with the operations defined as in $\mathscr{F}$. Define $\langle\ ,\ \rangle$ on $\mathscr{F}(C)$ by

$$\langle f, g\rangle = \int_0^1 f(x)\overline{g(x)}\,dx,$$

so that

$$\langle ix^2, 3 + ix\rangle = \int_0^1 ix^2(3 - ix)\,dx$$
$$= \int_0^1 3ix^2 + x^3\,dx = i + 1/4.$$

It can be shown that the function $\langle\ ,\ \rangle$ is a complex inner product for the space of functions $\mathscr{F}(C)$.

The definitions of terms such as norm, orthogonal, and orthonormal basis are all the same for either a real or a complex inner product space. Therefore the Gram-Schmidt process could be applied to linearly independent subsets of complex inner product spaces.

Example 2 Given

$$\mathscr{S} = \mathscr{L}\{(1 + i, 3i, 2 - i), (2 - 3i, 10 + 2i, 5 - i)\}$$

a subspace of $\mathscr{C}_3$, find an orthonormal basis for $\mathscr{S}$.

We have $\|(1 + i, 3i, 2 - i)\| = 4$, so $\frac{1}{4}(1 + i, 3i, 2 - i)$ is a unit vector. The orthogonal projection of $(2 - 3i, 10 + 2i, 5 - i)$ on this vector is

$$[(2 - 3i, 10 + 2i, 5 - i) \circ \tfrac{1}{4}(1 + i, 3i, 2 - i)]\,\tfrac{1}{4}(1 + i, 3i, 2 - i)$$
$$= [1 - 2i](1 + i, 3i, 2 - i) = (3 - i, 6 + 3i, -5i).$$

Subtracting this projection from $(2 - 3i, 10 + 2i, 5 - i)$ yields $(-1 - 2i, 4 - i, 5 + 4i)$, which has norm $\sqrt{63}$. Therefore an orthonormal basis for $\mathscr{S}$ is

$$\{\tfrac{1}{4}(1 + i, 3i, 2 - i), (1/\sqrt{63})(-1 - 2i, 4 - i, 5 + 4i)\}.$$

The primary reason for introducing complex inner product spaces here is to show that matrices of a certain type are always diagonalizable. The next step is to consider the following class of linear transformations.

Definition Let $(\mathscr{V}, \langle\ ,\ \rangle)$ be a complex inner product space and $T: \mathscr{V} \to \mathscr{V}$ a linear map. T is *hermitian* if $\langle T(U), V\rangle = \langle U, T(V)\rangle$ for all vectors $U, V \in \mathscr{V}$.

The identity map and the zero map are obviously hermitian maps, and we will soon see that there are many other examples. But first it should be shown why hermitian transformations are of interest.

Theorem 6.19 The characteristic values of a hermitian map are real.

Proof Suppose T is a hermitian transformation defined on the complex inner product space $(\mathscr{V}, \langle\ ,\ \rangle)$. Then a characteristic value z for T might well be complex. However, if W is a characteristic vector for z, then

$$\begin{aligned} z\|W\|^2 &= z\langle W, W\rangle = \langle zW, W\rangle = \langle T(W), W\rangle = \langle W, T(W)\rangle \\ &= \langle W, zW\rangle = \bar{z}\langle W, W\rangle = \bar{z}\|W\|^2. \end{aligned}$$

Since W is a nonzero vector, $z = \bar{z}$, hence the characteristic value z is real.

From this theorem we see that if $T: \mathscr{C}_n \to \mathscr{C}_n$ is hermitian, then a matrix of T with respect to any basis has only real characteristic roots. This implies that there is a collection of complex matrices that have only real characteristic roots. To determine the nature of these matrices, suppose $A = (a_{ij})$ is the matrix of T with respect to the standard basis $\{E_i\}$. Then $T(E_j) = (a_{1j}, \ldots, a_{nj})$, and since T is hermitian,

$$\begin{aligned} \overline{a_{ij}} &= E_i \circ (a_{1j}, \ldots, a_{nj}) = E_i \circ T(E_j) \\ &= T(E_i) \circ E_j = (a_{1i}, \ldots, a_{ni}) \circ E_j = a_{ji}. \end{aligned}$$

Therefore, if we set $\bar{A} = (\bar{a}_{ij})$, the matrix A satisfies $\bar{A}^T = A$.

Definition An $n \times n$ matrix A with entries from C is *hermitian* if $\bar{A}^T = A$.

Thus

$$\begin{pmatrix} 2 & 3-2i \\ 3+2i & 5 \end{pmatrix}, \quad \begin{pmatrix} 0 & 2-i & 1 \\ 2+i & 3 & 1+i \\ 1 & 1-i & 9 \end{pmatrix}, \quad \begin{pmatrix} 3 & 7 \\ 7 & 8 \end{pmatrix}$$

are examples of hermitian matrices.

Theorem 6.20 Suppose $T \in \text{Hom}(\mathscr{C}_n)$ has matrix A with respect to the standard basis. T is hermitian if and only if A is hermitian.

Proof ($\Rightarrow$) This direction has been obtained above.
($\Leftarrow$) Given that $A = (a_{ij})$ is hermitian, it is necessary to show that

$$T(U) \circ V = U \circ T(V) \qquad \text{for all} \quad U, V \in \mathscr{C}_n.$$

Let $U = (z_1, \ldots, z_n)$ and $V = (w_1, \ldots, w_n)$, then

$$\begin{aligned} T(U) \circ V &= \left(\sum_{j=1}^n a_{1j}z_j, \ldots, \sum_{j=1}^n a_{nj}z_j\right) \circ (w_1, \ldots, w_n) \\ &= \sum_{j=1}^n a_{1j}z_j\bar{w}_1 + \sum_{j=1}^n a_{2j}z_j\bar{w}_2 + \cdots + \sum_{j=1}^n a_{nj}z_j\bar{w}_n \\ &= \sum_{i=1}^n \sum_{j=1}^n a_{ij}z_j\bar{w}_i = \sum_{i=1}^n \sum_{j=1}^n \bar{a}_{ji}z_j\bar{w}_i = \sum_{j=1}^n z_j \sum_{i=1}^n \bar{a}_{ji}\bar{w}_i \\ &= \sum_{j=1}^n z_j \sum_{i=1}^n \overline{a_{ji}w_i} = \sum_{j=1}^n z_j \overline{\sum_{i=1}^n a_{ji}w_i} \\ &= (z_1, \ldots, z_n) \circ \left(\sum_{i=1}^n a_{1i}w_i, \ldots, \sum_{i=1}^n a_{ni}w_i\right) = U \circ T(V). \end{aligned}$$

Thus T is hermitian and the proof is complete.

Therefore, every hermitian matrix gives rise to a hermitian transformation on $\mathscr{C}_n$. Since a hermitian transformation has only real characteristic roots, we have:

Theorem 6.21 The characteristic values or characteristic roots of a hermitian matrix are real.

A matrix with real entries is hermitian if it equals its transpose, because conjugation does not change a real number. An $n \times n$ matrix A is said to be *symmetric* if $A^T = A$. For a real symmetric matrix, Theorem 6.21 becomes:

Theorem 6.22 The characteristic equation of a real symmetric matrix has only real roots.

This means that the characteristic roots of matrices such as

$$\begin{pmatrix} 0 & 4 \\ 4 & 3 \end{pmatrix} \quad \text{and} \quad \begin{pmatrix} 2 & 1 & 3 \\ 1 & 4 & 7 \\ 3 & 7 & 0 \end{pmatrix}$$

are all real. But even though

$$\begin{pmatrix} 1+i & 4+i \\ 4+i & 3 \end{pmatrix}$$

is symmetric, it need not have real characteristic roots.

The important fact is that not only are the characteristic roots of a real symmetric matrix real, but such a matrix is always diagonalizable. Again, the result is obtained first in the complex case.

Theorem 6.23 If $T: \mathscr{C}_n \to \mathscr{C}_n$ is hermitian, then T is diagonalizable.

Proof T is diagonalizable if it has n linearly independent characteristic vectors in $\mathscr{C}_n$. Let $\lambda_1, \ldots, \lambda_n$ be the n real characteristic values for T. Suppose V_1 is a characteristic vector corresponding to λ_1. If $n = 1$, the proof is complete, otherwise continue by induction on the dimension of $\mathscr{C}_n$.

Suppose $V_1, \ldots, V_k$, $k < n$, have been obtained, with V_i a characteristic vector corresponding to λ_i, $1 \le i \le k$, and $V_1, \ldots, V_k$ linearly independent. The proof consists in showing that there is a characteristic vector for λ_{k+1} in the orthogonal complement of $\mathscr{L}\{V_1, \ldots, V_k\}$. Therefore, suppose $\mathscr{S}_{k+1} = \mathscr{L}\{V_1, \ldots, V_k\}^\perp$ and let T_{k+1} be the restriction of T to $\mathscr{S}_{k+1}$. That is $T_{k+1}(U) = T(U)$ for each $U \in \mathscr{S}_{k+1}$. This map sends $\mathscr{S}_{k+1}$ into itself, for if $U \in \mathscr{S}_{k+1}$ and $i \le k$, then

$$T_{k+1}(U) \circ V_i = T(U) \circ V_i = U \circ T(V_i) = U \circ \lambda_i V_i = \bar{\lambda}_i(U \circ V_i) = \bar{\lambda}_i 0 = 0.$$

That is, $T_{k+1}(U)$ is in the orthogonal complement of $\mathscr{L}\{V_1, \ldots, V_k\}$, and

$T_{k+1}: \mathscr{S}_{k+1} \to \mathscr{S}_{k+1}$. As a map from $\mathscr{S}_{k+1}$ to itself, T_{k+1} has $n - k$ characteristic values, which must be $\lambda_{k+1}, \ldots, \lambda_n$. Now if V_{k+1} is a characteristic vector for T_{k+1} corresponding to λ_{k+1}, V_{k+1} is a characteristic vector for T and $V_{k+1} \in \mathscr{L}\{V_1, \ldots, V_k\}^\perp$. Therefore $V_1, \ldots, V_k, V_{k+1}$ are $k + 1$ linearly independent characteristic vectors for T, and the proof is complete by induction.

Corollary 6.24 Every hermitian matrix is diagonalizable.

This corollary refers to diagonalizability over the field of complex numbers. However, it implies that real symmetric matrices are diagonalizable over the field of real numbers.

Theorem 6.25 If A is an $n \times n$ real symmetric matrix, then there exists a real orthogonal matrix P such that $P^{-1}AP$ is a real diagonal matrix.

Proof Since A is real and symmetric, A is hermitian and hence diagonalizable over the field of complex numbers. That is, there is a matrix P with entries from C such that $P^{-1}AP$ is diag $(\lambda_1, \ldots, \lambda_n)$, where $\lambda_1, \ldots, \lambda_n$ are (real) characteristic roots of A. But the jth column of P is a solution of the system of linear equations $AX = \lambda_j X$. Since λ_j and the entries of A are real, the entries of P must also be real. Therefore, A is diagonalizable over the field of real numbers. From Theorem 6.23 we know that the columns of P are orthogonal in $\mathscr{E}_n$, therefore it is only necessary to use unit characteristic vectors in the columns of P to obtain an orthogonal matrix which diagonalizes A.

So far there has been little indication that symmetric matrices are of particular importance. Symmetric matrices are associated with inner products, as suggested in problem 9, page 247. But because of Theorem 6.25 it is often advantageous to introduce a symmetric matrix whenever possible.

Example 3 Find a rotation of the plane that transforms the polynomial $x^2 + 6xy + y^2$ into a polynomial without an xy or cross product term.

First notice that

$$x^2 + 6xy + y^2 = (x, y)\begin{pmatrix} 1 & 3 \\ 3 & 1 \end{pmatrix}\begin{pmatrix} x \\ y \end{pmatrix} = X^T AX$$

where

$$X = \begin{pmatrix} x \\ y \end{pmatrix}, \qquad A = \begin{pmatrix} 1 & 3 \\ 3 & 1 \end{pmatrix}.$$

The characteristic values for the symmetric matrix A are 4 and -2, and $(1/\sqrt{2},\ 1/\sqrt{2})$ and $(-1/\sqrt{2},\ 1/\sqrt{2})$ are corresponding unit characteristic vectors. Constructing P from these vectors yields the orthogonal matrix

$$P = \begin{pmatrix} 1/\sqrt{2} & -1/\sqrt{2} \\ 1/\sqrt{2} & 1/\sqrt{2} \end{pmatrix},$$

and

$$P^{-1}AP = \begin{pmatrix} 4 & 0 \\ 0 & -2 \end{pmatrix}.$$

Since $\det P = 1$, P is the matrix of a rotation in the plane given by

$$\begin{pmatrix} x \\ y \end{pmatrix} = P\begin{pmatrix} x' \\ y' \end{pmatrix}$$

or $X = PX'$ with

$$X' = \begin{pmatrix} x' \\ y' \end{pmatrix}.$$

Now if the given polynomial is written in terms of the new variables x' and y', we obtain

$$\begin{aligned} x^2 + 6xy + y^2 &= X^T AX = (PX')^T A(PX') = X'^T P^T APX' \\ &= X'^T P^{-1} APX' = (x', y')\begin{pmatrix} 4 & 0 \\ 0 & -2 \end{pmatrix}\begin{pmatrix} x' \\ y' \end{pmatrix} \\ &= 4x'^2 - 2y'^2. \end{aligned}$$

The polynomial in x' and y' does not have an $x'y'$ term as desired. If the graph of the equation $x^2 + 6xy + y^2 = k$ is "equivalent" to the graph of $4x'^2 - 2y'^2 = k$ for a constant k, then it would clearly be much easier to find the second graph. It may be noted that no rotation is explicitly given here; this situation is investigated in problem 10 at the end of this section.

It should be pointed out that the major theorems of this section could have been stated for an arbitrary finite-dimensional complex inner product space instead of the n-tuple space $\mathscr{C}_n$. However, since each such space is isomorphic to one of the spaces $\mathscr{C}_n$, only notational changes would be involved in the statements of such theorems.

Problems

1. Compute:
a. $(2, 1 + i, 3i)\circ(2 - i, 4i, 2 - 3i)$ in $\mathscr{C}_3$.
b. $(1 - i, 3, i + 2, 3i)\circ(5, 2i - 4, 0, i)$ in $\mathscr{C}_4$.

2. Find the orthogonal complement of $\mathscr{S}$ in $\mathscr{C}_3$ when
a. $\mathscr{S} = \mathscr{L}\{(i, 2, 1 + i)\}$.
b. $\mathscr{S} = \mathscr{L}\{(2i, i, 4), (1 + i, 0, 1 - i)\}$.

3. Show that if $Z, W, U \in \mathscr{C}_n$ and $a, b \in C$, then $Z\circ(aW + bU) = \bar{a}(Z\circ W) + \bar{b}(Z\circ U)$.

4. a. Define a complex inner product $\langle\, , \rangle$ on $\mathscr{V}_2(C)$ such that $B = \{(3 + i, 2i), (1, i - 2)\}$ is an orthonormal basis for $(\mathscr{V}_2(C), \langle\, , \rangle)$.
b. What is the norm of $(4i, 2)$ in this complex inner product space?

5. Which of the following matrices are hermitian and/or symmetric?

a. $\begin{pmatrix} 2 & 1 - i \\ 1 + i & 4i \end{pmatrix}$. b. $\begin{pmatrix} 1 & 2 \\ 2 & 3 \end{pmatrix}$. c. $\begin{pmatrix} i & 2 + i \\ 2 + i & 3 \end{pmatrix}$.

d. $\begin{pmatrix} 3 & 4 + 2i \\ 4 - 2i & 6 \end{pmatrix}$. e. $\begin{pmatrix} 7 & 4 \\ 3 & 7 \end{pmatrix}$. f. $\begin{pmatrix} 4 & 3i - 2 \\ 2 - 3i & 1 \end{pmatrix}$.

6. Suppose A is hermitian, a. Why are the diagonal entries of A real? b. Show that the determinant of A is real.

7. a. Show that the matrix $A = \begin{pmatrix} 3 & -2 \\ 9 & -3 \end{pmatrix}$ cannot be diagonalized over the field of real numbers.
b. Find a complex matrix P such that $P^{-1}AP$ is diagonal.

8. Suppose $T\colon \mathscr{C}_2 \rightarrow \mathscr{C}_2$ is given by $T(z, w) = (2z + (1 + i)w, (1 - i)z + 3w)$.
a. Find the matrix A of T with respect to the standard basis, and conclude that T is hermitian.
b. Find the characteristic values of A and a matrix P such that $P^{-1}AP$ is diagonal.

9. Do there exist matrices that cannot be diagonalized over the field of complex numbers?

10. In Example 3, page 283, it is stated that P is the matrix of a rotation, yet since no basis is mentioned, no map is defined.
a. Define a rotation $T\colon \mathscr{E}_2 \rightarrow \mathscr{E}_2$ such that P is the matrix of T with respect to the standard basis for $\mathscr{E}_2$.

b. Let $B = \{T(1, 0), T(0, 1)\}$. Show that for each $W \in \mathscr{E}_2$, if $W\colon (x, y)_{\{E_i\}}$ and $(x, y)^T = P(x', y')^T$, then $W\colon (x', y')_B$.

c. Sketch the coordinate systems associated with the bases $\{E_i\}$ and B in the plane, together with the curve that has the equation $x^2 + 6xy + y^2 = 36$ in coordinates with respect to $\{E_i\}$ and the equation $4x'^2 - 2y'^2 = 36$ in coordinates with respect to B.

11. Let G be the subset of $\mathscr{E}_2$ defined by

$$G = \{V \in \mathscr{E}_2 | V\colon (x, y)_{\{E_i\}} \text{ and } 5x^2 + 4xy + 2y^2 = 24\}.$$

a. Find an orthonormal basis B for $\mathscr{E}_2$ such that the polynomial equation defining G in coordinates with respect to B does not have a cross product term.

b. Sketch the coordinate systems representing $\{E_i\}$ and B in the plane along with the points representing the vectors in G.

12. A linear map T defined on a real inner product space $(\mathscr{V}, \langle\ ,\ \rangle)$ is *symmetric* if $\langle T(U), V\rangle = \langle U, T(V)\rangle$ for all $U, V \in \mathscr{V}$.

a. Show that the matrix of T with respect to any basis for $\mathscr{V}$ is symmetric if T is symmetric.

b. Suppose $T\colon \mathscr{V} \to \mathscr{V}$ is orthogonal and prove that T is symmetric if and only if $T^2 = T \circ T = I$.

c. What are the possible symmetric, orthogonal transformations from $\mathscr{E}_3$ into itself?

Review Problems

1. For which values of r is $\langle (a, b), (c, d)\rangle = ac + bc + ad + rbd$ an inner product on $\mathscr{V}_2$?

2. Use the Gram-Schmidt process to construct an orthonormal basis for $(\mathscr{V}_2, \langle\ ,\ \rangle)$ from the standard basis when:

a. $\langle (a, b), (c, d)\rangle = 2ac - 3ad - 3bc + 5bd$.

b. $\langle (a, b), (c, d)\rangle = (a \quad b)\begin{pmatrix} 3 & 1 \\ 1 & 2 \end{pmatrix}\begin{pmatrix} c \\ d \end{pmatrix}$.

3. Prove that $\langle U + V, U - V\rangle = 0$ if and only if $\|U\| = \|V\|$. What is the geometric interpretation of this statement?

4. Show that $\langle A, B\rangle = \operatorname{tr}(AB^T)$ defines an inner product on the space of all real $n \times n$ matrices.

5. Prove the parallelogram law:

$$\|U + V\|^2 + \|U - V\|^2 = 2\|U\|^2 + 2\|V\|^2.$$

6. Suppose $(\mathscr{V}, \langle\ ,\ \rangle)$ is a real finite-dimensional inner product space. An element of Hom $(\mathscr{V}, R)$ is called a *linear functional* on $\mathscr{V}$. Show that for every linear functional T, there exists a vector $U \in \mathscr{V}$ such that $T(V) = \langle U, V\rangle$ for all $V \in \mathscr{V}$.

7. Suppose $\mathscr{S}$ and $\mathscr{T}$ are subspaces of an inner product space $\mathscr{E}$. Prove the following statements:

a. $\mathscr{S} \subset (\mathscr{S}^\perp)^\perp$ and $\mathscr{S} = (\mathscr{S}^\perp)^\perp$ if $\mathscr{E}$ is finite dimensional.

b. $(\mathscr{S} + \mathscr{T})^\perp = \mathscr{S}^\perp \cap \mathscr{T}^\perp$.

c. $(\mathscr{S} \cap \mathscr{T})^\perp \supset \mathscr{S}^\perp + \mathscr{T}^\perp$ and $(\mathscr{S} \cap \mathscr{T})^\perp = \mathscr{S}^\perp + \mathscr{T}^\perp$ if $\mathscr{E}$ is finite dimensional.

8. If $T \in \text{Hom}(\mathscr{V})$ and there exists $T^* \in \text{Hom}(\mathscr{V})$ such that $\langle V, T^*(U)\rangle = \langle T(V), U\rangle$ for all $U, V \in \mathscr{V}$, then T^* is called the *adjoint* of T.

a. Find T^* if $T \in \text{Hom}(\mathscr{E}_2)$ is given by $T(a, b) = (2a + 3b, 5a - b)$.

b. Suppose $\mathscr{V}$ is finite dimensional with orthonormal basis B. If A is the matrix of T with respect to B, determine how the matrix of T^* with respect to B is related to A when $\mathscr{V}$ is a real inner product space; when $\mathscr{V}$ is a complex inner product space.

c. Show that $\mathscr{I}_{T^*} = \mathscr{N}_T^\perp$ if $T \in \text{Hom}(\mathscr{E}_n)$.

9. Partition $n \times n$ matrices into blocks $\left(\begin{array}{c|c} A & B \\ \hline C & D \end{array}\right)$ where $r + s = n$, A is $r \times r$, B is $r \times s$, C is $s \times r$, and D is $s \times s$.

a. Write the product $\left(\begin{array}{c|c} A_1 & B_1 \\ \hline C_1 & D_1 \end{array}\right)\left(\begin{array}{c|c} A_2 & B_2 \\ \hline C_2 & D_2 \end{array}\right)$ in block form.

b. Suppose $\left(\begin{array}{c|c} A & B \\ \hline C & D \end{array}\right)$ is symmetric and A is nonsingular. Find matrices E and F such that $\left(\begin{array}{c|c} I_r & E \\ \hline 0 & I_s \end{array}\right)^T \left(\begin{array}{c|c} A & B \\ \hline C & D \end{array}\right)\left(\begin{array}{c|c} I_r & E \\ \hline 0 & I_s \end{array}\right) = \left(\begin{array}{c|c} A & 0 \\ \hline 0 & F \end{array}\right)$.

c. Use part b to obtain a formula for the determinant of a symmetric matrix partitioned into such blocks.

10. Use the formula obtained in problem 9 to find the determinant of the following partitioned symmetric matrices.

a. $\left(\begin{array}{cc|c} 2 & 5 & 7 \\ 5 & 9 & 3 \\ \hline 7 & 3 & 4 \end{array}\right)$. b. $\left(\begin{array}{cc|cc} 2 & 3 & 1 & 4 \\ 3 & 2 & 5 & 1 \\ \hline 1 & 5 & 2 & 2 \\ 4 & 1 & 2 & 3 \end{array}\right)$. c. $\left(\begin{array}{cc|cc} 7 & 3 & 2 & 4 \\ 3 & 0 & 2 & 5 \\ \hline 2 & 2 & 3 & 5 \\ 4 & 5 & 5 & 2 \end{array}\right)$.

7

Second Degree Curves and Surfaces

Our geometric illustrations have been confined primarily to points, lines, planes, and hyperplanes, which are the graphs of linear equations or systems of linear equations. Most other types of curves and surfaces are the graphs of nonlinear equations and therefore lie outside the study of vector spaces. However, linear techniques can be used to examine second degree polynomial equations and their graphs. Such a study will not only provide a good geometric application of linear algebra, but it also leads naturally to several new concepts.

All vector spaces considered in this chapter will be over the real number field and finite dimensional.

§1. Quadratic Forms

Consider the most general second degree polynomial equation in the coordinates $x_1, \ldots, x_n$ of points in Euclidean n-space. Such an equation might be written in the form

$$\sum_{i,j=1}^{n} a_{ij}x_ix_j + \sum_{i=1}^{n} b_ix_i + c = 0, \qquad \text{where} \quad a_{ij},\ b_i,\ c \in R.$$

For example, in the equation $2x_1^2 - x_1x_2 + 9x_2x_1 - 2x_2 + 7 = 0$, $n = 2$, $a_{11} = 2$, $a_{12} = -1$, $a_{21} = 9$, $a_{22} = b_1 = 0$, $b_2 = -2$, and $c = 7$. The geometric problem is to determine the nature of the graph of the points whose coordinates satisfy such an equation. We already know that if x and y are the coordinates of points in the Cartesian plane, then the graphs of $x^2 + y^2 - 4 = 0$ and $x^2 + 4x + y^2 = 0$ are circles of radius 2. But what might the graph of $4yz + z^2 + x - 3z + 2 = 0$ look like in 3-space? Before considering such questions we will examine the second degree terms of the general equation. The first step is to write these terms, $\sum_{i,j=1}^{n} a_{ij}x_ix_j$, in a form using familiar notation. Notice that the polynomial $\sum_{i,j=1}^{n} a_{ij}x_ix_j$, can be rewritten as

$$\sum_{i=1}^{n} x_i\left(\sum_{j=1}^{n} a_{ij}x_j\right)$$

suggesting a product of matrices. In fact, if we let $X = (x_1, \ldots, x_n)^T$ and $A = (a_{ij})$, then

$$\sum_{i,j=1}^{n} a_{ij}x_ix_j = X^TAX.$$

For example, the second degree polynomial $2x_1^2 - x_1x_2 + 9x_2x_1$ can be written as

$$(x_1 \quad x_2)\begin{pmatrix} 2 & -1 \\ 9 & 0 \end{pmatrix}\begin{pmatrix} x_1 \\ x_2 \end{pmatrix} = (x_1 \quad x_2)\begin{pmatrix} 2x_1 - x_2 \\ 9x_1 \end{pmatrix} = (2x_1^2 - x_1x_2 + 9x_2x_1).$$

Although X^TAX is a 1×1 matrix, the matrix notation is dropped here for simplicity. There should be no question about the meaning of the equation $\sum_{i,j=1}^n a_{ij}x_ix_j = X^TAX$. This equation states that X^TAX is a homogeneous second degree polynomial in $x_1, \ldots, x_n$, and the coefficient of x_ix_j is the element from the ith row and the jth column of A. (In general, a polynomial is said to be *homogeneous* of degree k if every term in the polynomial is of degree k.)

Definition A *quadratic form* on an n-dimensional vector space $\mathscr{V}$, with basis B, is an expression of the form $\sum_{i,j=1}^n a_{ij}x_ix_j$ where $x_1, \ldots, x_n$ are the coordinates of an arbitrary vector with respect to B. If $X = (x_1, \ldots, x_n)^T$ and $A = (a_{ij})$, then

$$\sum_{i,j=1}^{n} a_{ij}x_ix_j = X^TAX$$

and A is the *matrix of the quadratic form with respect to B*.

Thus a quadratic form is a homogeneous second degree polynomial in coordinates, it is not simply a polynomial. In fact, any one homogeneous second degree polynomial usually yields different quadratic forms when associated with different bases.

Example 1 Consider the polynomial $2x_1^2 - x_1x_2 + 9x_2x_1$. If x_1 and x_2 are coordinates with respect to the basis $B = \{(2, 1), (-1, 0)\}$, then this polynomial determines a quadratic form on $\mathscr{V}_2$. This quadratic form assigns real numbers to each vector in $\mathscr{V}_2$. The value of the quadratic form at (2, 1) is 2 since $(2, 1)\colon (1, 0)_B$ gives $x_1 = 1$ and $x_2 = 0$. And since $(-1, 1)\colon (1, 3)_B$, the value of the quadratic form at (1, 1) is $2(1)^2 - 1(3) + 9(3)1 = 26$. For an arbitrary vector (a, b), $x_1 = b$ and $x_2 = 2b - a$ (Why?). Therefore the value of the quadratic form $2x_1^2 - x_1x_2 + 9x_2x_1$ at (a, b) is $18b^2 - 8ab$. This gives the value of the quadratic form at $(-1, 1)$ to be $18 + 8$ or 26 as before.

Since the coordinates of (a, b) with respect to the standard basis are a and b, the polynomial $18b^2 - 8ab$ is also a quadratic form on $\mathscr{V}_2$.

$2x_1^2 - x_1x_2 + 9x_2x_1$ and $18b^2 - 8ab$ are different quadratic forms in that they are different polynomials in coordinates for different bases, but they assign the same value to each vector in $\mathscr{V}_2$. That is, they both define a particular function from $\mathscr{V}_2$ to the real numbers, called a quadratic function.

Definition A *quadratic function* q on a vector space $\mathscr{V}$ with basis B is a function defined by

$$q(V) = \sum_{i,j=1}^{n} a_{ij}x_ix_j \qquad \text{for each} \quad V \in \mathscr{V}$$

where $V\colon (x_1, \ldots, x_n)_B$.

That is, a quadratic function is defined in terms of a quadratic form. The two quadratic forms above determine a quadratic function q on $\mathscr{V}_2$ for which $q(2, 1) = 2$, $q(-1, 1) = 26$, and in general $q(a, b) = 18b^2 - 8ab$, so that $q\colon \mathscr{V}_2 \to R$.

The relation between a quadratic function ($r = q(V)$) and a quadratic form ($r = X^TAX$) is much the same as the relation between a linear transformation ($W = T(V)$) and its matrix representation ($Y = AX$). But there are an infinite number of possible matrix representations for any quadratic function, in contrast to the linear case where each basis determines a unique matrix for a map. For example, the above quadratic function is given by the quadratic forms $18b^2 - 8ab$, $18b^2 - 4ab - 4ba$, and $18b^2 + 12ab - 20ba$. Therefore

$$q(a, b) = (a \quad b)\begin{pmatrix} 0 & -8 \\ 0 & 18 \end{pmatrix}\begin{pmatrix} a \\ b \end{pmatrix} = (a \quad b)\begin{pmatrix} 0 & -4 \\ -4 & 18 \end{pmatrix}\begin{pmatrix} a \\ b \end{pmatrix}$$
$$= (a \quad b)\begin{pmatrix} 0 & 12 \\ -20 & 18 \end{pmatrix}\begin{pmatrix} a \\ b \end{pmatrix}.$$

However, out of all the possible matrix representations for a quadratic function there is a best choice, for it is always possible to use a symmetric matrix. This should be clear, for if in $\sum_{i,j=1}^{n} a_{ij}x_ix_j$, $a_{hk} \neq a_{kh}$ for some h and k, then the terms $a_{hk}x_hx_k + a_{kh}x_kx_h$ can be replaced by

$$\tfrac{1}{2}(a_{hk} + a_{kh})x_hx_k + \tfrac{1}{2}(a_{hk} + a_{kh})x_kx_h$$

which have equal coefficients. For example, the polynomial $8x^2 + 7xy + 3yx + y^2$ can be written as

$$(x \quad y)\begin{pmatrix} 8 & 7 \\ 3 & 1 \end{pmatrix}\begin{pmatrix} x \\ y \end{pmatrix}$$

using a nonsymmetric matrix, but the polynomial $8x^2 + 5xy + 5yx + y^2$ yields the same values for given values of x and y and can be expressed as

$$(x \quad y)\begin{pmatrix} 8 & 5 \\ 5 & 1 \end{pmatrix}\begin{pmatrix} x \\ y \end{pmatrix}$$

using a symmetric matrix. Choosing a symmetric matrix to represent a quadratic function will make it possible to find a simple representation by a change of basis, for every symmetric matrix is diagonalizable.

Definition The *matrix of a quadratic function q with respect to a basis B* is the symmetric matrix $A = (a_{ij})$ for which

$$q(V) = \sum_{i,j=1}^{n} a_{ij}x_i x_j$$

with $V\colon (x_1, \ldots, x_n)_B$.

The quadratic form $\sum_{i,j=1}^{n} a_{ij}x_i x_j$ is *symmetric* if $a_{ij} = a_{ji}$ for all i and j.

Example 2 Recall the quadratic form considered in Example 1 given by $2x_1^2 - x_1x_2 + 9x_2x_1$ in terms of the basis B. The quadratic function q on $\mathscr{V}_2$ determined by this quadratic form is defined by

$$q(V) = (x_1 \quad x_2)\begin{pmatrix} 2 & -1 \\ 9 & 0 \end{pmatrix}\begin{pmatrix} x_1 \\ x_2 \end{pmatrix}, \qquad \text{if } V\colon (x_1, x_2)_B.$$

However, $\begin{pmatrix} 2 & -1 \\ 9 & 0 \end{pmatrix}$ is not the matrix of q with respect to B because it is not symmetric. The matrix of q with respect to B is $\begin{pmatrix} 2 & 4 \\ 4 & 0 \end{pmatrix}$, since it is symmetric, and when $V\colon (x_1, x_2)_B$,

$$q(V) = (x_1 \quad x_2)\begin{pmatrix} 2 & 4 \\ 4 & 0 \end{pmatrix}\begin{pmatrix} x_1 \\ x_2 \end{pmatrix}.$$

We have defined quadratic functions in terms of coordinates, so the definition of each quadratic function is in part a reflection of the particular basis used. But a quadratic function is a function of vectors, and as such it should be possible to obtain a description without reference to a basis. An analogous situation would be to have defined linear transformations in terms of matrix multiplication rather than requiring them to be maps that preserve the algebraic structure of a vector space. Our coordinate approach to quadratic functions followed naturally from an examination of second degree polynomial equations, but there should also be a coordinate-free characterization. We can obtain such a characterization by referring to bilinear functions. It might be noted that several inner products considered in Chapter 6 appear similar to quadratic forms. For example, the inner product in Example 6 on page 245 is expressed as a polynomial in the coordinates of vectors with respect to the standard basis.

Definition A *bilinear form* on an n-dimensional vector space $\mathscr{V}$, with basis B, is an expression of the form $\sum_{i,j=1}^{n} a_{ij}x_iy_j$ were $x_1, \ldots, x_n$ and $y_1, \ldots, y_n$ are the coordinates of two arbitrary vectors with respect to B. A bilinear form may be written as X^TAY with $X = (x_1, \ldots, x_n)^T$, $Y = (y_1, \ldots, y_n)^T$, and $A = (a_{ij})$. A is the *matrix of the bilinear form with respect to B*, and the bilinear form is *symmetric* if A is symmetric.

Thus a bilinear form is a homogeneous second degree polynomial in the coordinates of two vectors. The bilinear form suggested by Example 6, page 245, is given by $3x_1y_1 - x_1y_2 - x_2y_1 + 2x_2y_2$, where x_1, x_2 and y_1, y_2 are the coordinates of two vectors from $\mathscr{V}_2$ with respect to the standard basis. This bilinear form yields a number for each pair of vectors in $\mathscr{V}_2$ and was used to obtain an inner product on $\mathscr{V}_2$. However, a bilinear form need not determine an inner product. Consider the bilinear form $4x_1y_2 - 3x_2y_2$ defined on $\mathscr{V}_2$ in terms of coordinates with respect to $\{E_i\}$. This bilinear form defines a bilinear function from $\mathscr{V}_2$ to R, but it is neither symmetric nor positive definite.

The name "bilinear form" is justified by the fact that a bilinear form is simply a coordinate representation of a real-valued, bilinear function. That is, a function b for which:

1. $b(U, V) \in R$ for all $U, V \in \mathscr{V}$.
2. For each $W \in \mathscr{V}$, the map T given by $T(U) = b(U, W)$, for all $U \in \mathscr{V}$, is linear.
3. For each $W \in \mathscr{V}$, the map S given by $S(U) = b(W, U)$, for all $U \in \mathscr{V}$, is linear.

It is not difficult to obtain this relationship between bilinear maps and bilinear forms. It is only necessary to show that if b is a bilinear function on $\mathscr{V}$, and $B = \{U_1, \ldots, U_n\}$ is a basis for $\mathscr{V}$, then

$$b(U, V) = \sum_{i,j=1}^{n} a_{ij}x_iy_j$$

where $U\colon (x_1, \ldots, x_n)_B$, $V\colon (y_1, \ldots, y_n)_B$, and $a_{ij} = b(U_i, U_j)$.

Example 3 Construct a function b on $\mathscr{E}_2$ as follows: Suppose T is the linear map given by $T(u, v) = (2u - v, 5v - 3u)$ and set $b(U, V) = U \circ T(V)$ for all $U, V \in \mathscr{E}_2$. The linearity of T and the bilinearity of the dot product guarantees that b is bilinear. Therefore the function b gives rise to a bilinear form for each basis for $\mathscr{E}_2$. A little computation shows that

$$b((x, y), (u, v)) = 2xu - xv - 3yu + 5yv.$$

The right side of this equation is a bilinear form in coordinates with respect to $\{E_i\}$. In this case, the equations $a_{ij} = b(U_i, U_j)$ become

$$\begin{aligned} a_{11} &= b(E_1, E_1) = (1, 0) \circ T(1, 0) = (1, 0) \circ (2, 3) = 2, \\ a_{12} &= b((1, 0), (0, 1)) = -1, \\ a_{21} &= b((0, 1), (1, 0)) = -3, \\ a_{22} &= b((0, 1), (0, 1)) = 5. \end{aligned}$$

If X is set equal to Y in the bilinear form X^TAY, then the quadratic form X^TAX is obtained. This suggests that every bilinear form corresponds to a quadratic form. Since each quadratic function q is given by $q(V) = X^TAX$ with A symmetric, q should be associated with a symmetric bilinear function b, given by $b(U, V) = X^TAY$. Using the fact that a symmetric bilinear function may be defined without reference to coordinates, we have obtained a coordinate-free characterization of quadratic functions.

Theorem 7.1 If b is a symmetric, bilinear function on $\mathscr{V}$ and q is defined by $q(U) = b(U, U)$ for all $U \in \mathscr{V}$, then q is a quadratic function on $\mathscr{V}$.

Conversely, for every quadratic function q on $\mathscr{V}$, there is a unique symmetric, bilinear function b on $\mathscr{V}$ such that $q(U) = b(U, U)$ for all $U \in \mathscr{V}$.

Proof If b is a symmetric bilinear function on $\mathscr{V}$ with $b(U, V) = X^TAY$ for some basis, then X^TAY is a symmetric bilinear form on $\mathscr{V}$. To obtain the symmetry, notice that

$$X^TAY = b(U, V) = b(V, U) = Y^TAX = (Y^TAX)^T = X^TA^TY$$

implies that $A = A^T$. (Why is Y^TAX equal to its transpose?) Thus the function q defined by $q(U) = b(U, U) = X^TAX$ is a quadratic function expressed with a symmetric matrix.

For the converse, notice that if b' is a symmetric bilinear function, then

$$b'(U + V, U + V) = b'(U, U) + 2b'(U, V) + b'(V, V).$$

This suggests that given a quadratic function q, a function b might be defined by

$$b(U, V) = \tfrac{1}{2}[q(U + V) - q(U) - q(V)].$$

This function is symmetric, bilinear and $b(U, U) = q(U)$; the fact that it is symmetric is obvious and the justification of the other two properties is left as an exercise. For uniqueness, supose there are two symmetric bilinear functions b_1 and b_2 such that for all $U \in \mathscr{V}$

$$b_1(U, U) = q(U) = b_2(U, U).$$

Then for all $U, V \in \mathscr{V}$,

$$b_1(U + V, U + V) = q(U) + 2b_1(U, V) + q(V),$$
$$b_2(U + V, U + V) = q(U) + 2b_2(U, V) + q(V),$$

but both left sides equal $q(U + V)$ so $b_1(U, V) = b_2(U, V)$ and $b_1 = b_2$.

Example 4 Let b be the bilinear function defined on $\mathscr{E}_2$ by $b(U, V) = U \circ T(V)$ with $T(u, v) = (u + 4v, 4u - 3v)$. Then

$$b((x, y), (u, v)) = xu + 4xv + 4yu - 3yv = (x \quad y)\begin{pmatrix} 1 & 4 \\ 4 & -3 \end{pmatrix}\begin{pmatrix} u \\ v \end{pmatrix},$$

which shows that b is symmetric. b determines a quadratic function q on $\mathscr{E}_2$ by $q(U) = b(U, U) = U \circ T(U)$. In terms of coordinates with respect to $\{E_i\}$,

$$q(x, y) = (x \quad y)\begin{pmatrix} 1 & 4 \\ 4 & -3 \end{pmatrix}\begin{pmatrix} x \\ y \end{pmatrix} = x^2 + 8xy - 3y^2.$$

Notice that

$$\begin{pmatrix} 1 & 4 \\ 4 & -3 \end{pmatrix}$$

is the matrix of T, of b, and of q with respect to $\{E_i\}$.

Problems

1. Express the following polynomials in the form X^TAX with A symmetric.
   a. $x_1^2 + 3x_1x_2 + 3x_2x_1 - 5x_2^2$. b. $4x_1^2 + 7x_1x_2$.
   c. $x^2 + 5xy - y^2 + 6yz + 2z^2$. d. $5x^2 - 2xz + 4y^2 + yz$.

2. Find all matrices A that express the quadratic function q in the form $(a, b)\, A \begin{pmatrix} a \\ b \end{pmatrix}$ if $q(a, b) = a^2 - 5ab + 7b^2$.

3. Suppose $A = \begin{pmatrix} 3 & 1 \\ 1 & 3 \end{pmatrix}$ is the matrix of a quadratic function q with respect to a basis B. Find $q(a, b)$ if B is given by
   a. $B = \{E_i\}$. b. $B = \{(3/5, 4/5), (4/5, -3/5)\}$.
   c. $B = \{(1/\sqrt{2}, 1/\sqrt{2}), (1/\sqrt{2}, -1/\sqrt{2})\}$.
   d. $B = \{(\sqrt{2}, 1), (\sqrt{2}, -1)\}$.

4. What is the difference between a quadratic form and a quadratic function?

5. Write the double sum $\sum_{i,j=1}^n a_{ij}x_ix_j$ in four ways using two summation signs. For each of these ways, write out the sum when $n = 2$ without changing the order of the terms.

6. Show that if q is a quadratic function on $\mathscr{V}$, then $q(rV) = r^2q(V)$ for all $r \in R$ and $V \in \mathscr{V}$.

7. Let b be the bilinear function of Example 3, page 295. Find the bilinear form that arises from b when $b(U, V)$ is expressed in terms of coordinates with respect to the basis
   a. $B = \{(4, 1), (1, -2)\}$. b. $B = \{(3, 2), (1, 1)\}$.

8. Suppose $b(U, V) = \sum_{i,j=1}^n a_{ij}x_iy_j$ with $U\colon (x_1, \ldots, x_n)_B$, $V\colon (y_1, \ldots, y_n)_B$, and $B = \{U_1, \ldots, U_n\}$. Justify the equations $a_{ij} = b(U_i, U_j)$.

9. Find the quadratic function q that is obtained from the given symmetric bilinear function b.
   a. $b((x, y), (u, v)) = 2xv + 2yu - 5yv$.
   b. $b((x, y, z), (u, v, w)) = 3xu - xw + 2yv - zu + 4yw + 4zv$.

10. Find the symmetric bilinear form that is obtained from the given quadratic form.
    a. $x_1^2 - 4x_1x_2 - 4x_2x_1 + 7x_2^2$ on $\mathscr{E}_2$ with basis $\{E_i\}$.
    b. $x_1^2 + x_1x_2 + x_2x_1 - x_2^2 - 6x_2x_3 - 6x_3x_2$ on $\mathscr{E}_3$ with basis $\{E_i\}$.

11. Suppose q is a quadratic function on $\mathscr{V}$ and b is defined by
$$b(U, V) = \tfrac{1}{2}[q(U + V) - q(U) - q(V)].$$
    Prove that b is a bilinear function on $\mathscr{V}$.

12. Show that if b is a bilinear function on $\mathscr{E}_n$, then there exists a linear map $T\colon \mathscr{E}_n \to \mathscr{E}_n$ such that $b(U, V) = U \circ T(V)$.

13. a. How might the term "linear form" be defined on a vector space $\mathscr{V}$?

b. What kind of map would have a coordinate expression as a linear form?

c. Given a bilinear form $\sum_{i,j=1}^{n} a_{ij}x_i y_j$ in coordinates with respect to B, define $b(U, V) = \sum_{i,j=1}^{n} a_{ij}x_i y_j$ where $U:(x_1, \ldots, x_n)_B$ and $V:(y_1, \ldots, y_n)_B$. Use the results of parts a and b to show that b is a bilinear function.

§2. Congruence

The expression of a quadratic function in terms of a quadratic form or a bilinear function in terms of a bilinear form depends on the choice of a basis. If the basis is changed, one expects the polynomial form to change. The problem again is to determine how such a change occurs and to discover a "best" polynomial to represent a given function. That is, we have another equivalence relation to investigate, namely, the relation that states that two polynomials are equivalent if they represent the same function. Rather than work with the polynomials directly, it is simpler to study their matrices. And since a quadratic function can be defined in terms of a symmetric bilinear function, it would be simplest to consider the bilinear case first.

Suppose b is a bilinear function on $\mathscr{V}$, having matrix A with respect to the basis B. That is, $b(U, V) = X^T A Y$ for $U, V \in \mathscr{V}$ where X and Y are the coordinates of U and V with respect to B. If A' is the matrix of b with respect to another basis B', then how is A' related to A? Suppose P is the transition matrix from B to B'. Then using primes to denote coordinates with respect to B', $X = PX'$ and $Y = PY'$. Therefore,

$$b(U, V) = X^T A Y = (PX')^T A P Y' = X'^T (P^T A P) Y'.$$

But $b(U, V) = X'^T A' Y'$. Since the equality $X'^T A' Y' = X'^T (P^T A P) Y'$ holds for all X' and Y'(i.e., U and V), $A' = P^T A P$.

Definition Two $n \times n$ matrices A and B are *congruent* if there exists a nonsingular matrix P such that $B = P^T A P$.

That is, the relationship between two matrices for a bilinear function is called congruence. The converse also holds, so that two matrices are congruent if and only if they are matrices of some bilinear function with respect to two choices of basis.

Example 1 Suppose f is the bilinear function on $\mathscr{V}_2$ having

$$A = \begin{pmatrix} 1 & 2 \\ 0 & 3 \end{pmatrix}$$

as its matrix with respect to $\{E_i\}$. Does there exist a basis for which the matrix of f is diagonal?

That is, does there exist a nonisngular matrix P such that P^TAP is diagonal? If

$$P = \begin{pmatrix} a & b \\ c & d \end{pmatrix}$$

then

$$P^TAP = \begin{pmatrix} a^2 + 2ac + 3c^2 & ab + 2ad + 3cd \\ ab + 2bc + 3cd & b^2 + 2bd + 3d^2 \end{pmatrix}.$$

So P^TAP is diagonal only if $bc = ad$, since P must be nonsingular, A is not congruent to a diagonal matrix.

This example shows that congruence differs from similarity, for the matrix A is diagonalizable, that is, A is similar to the diagonal matrix diag(1, 3).

Now suppose q is a quadratic function given by $q(U) = b(U, U)$ where b is a symmetric bilinear function. If A and A' are two matrices for q with respect to two choices of basis, then A and A' are also matrices for b. Therefore, two symmetric matrices are congruent if and only if they are matrices of some quadratic function with respect to two choices of basis.

Example 2 Let $q(a, b) = 2a^2 + 8ab$. The matrix of q with respect to $\{E_i\}$ is

$$A = \begin{pmatrix} 2 & 4 \\ 4 & 0 \end{pmatrix}.$$

What is the matrix of q with respect to $B = \{(2, 1), (3, -3)\}$?

The transition matrix from $\{E_i\}$ to B is

$$P = \begin{pmatrix} 2 & 3 \\ 1 & -3 \end{pmatrix},$$

and

$$P^TAP = \begin{pmatrix} 2 & 1 \\ 3 & -3 \end{pmatrix}\begin{pmatrix} 2 & 4 \\ 4 & 0 \end{pmatrix}\begin{pmatrix} 2 & 3 \\ 1 & -3 \end{pmatrix} = \begin{pmatrix} 24 & 0 \\ 0 & -54 \end{pmatrix}.$$

Therefore

$$\begin{pmatrix} 24 & 0 \\ 0 & -54 \end{pmatrix}$$

is the matrix of q with respect to B and $q(a, b) = 24x_1^2 - 54x_2^2$ when $(a, b){:}(x_1, x_2)_B$.

Theorem 7.2 Congruence is an equivalence relation on the set of all $n \times n$ matrices.

The proof of Theorem 7.2 is like the corresponding proofs for equivalence and similarity of matrices, and as with these relations, one looks for invariants under congruence in search for basic properties of bilinear and quadratic functions. Certainly rank is invariant under congruence, but since $A' = P^T AP$ implies $\det A' = (\det P)^2 \det A$, the sign of $\det A$ is also invariant. Moreover, if the matrix P is required to satisfy $\det P = \pm 1$, as when P is orthogonal, then the value of $\det A$ is invariant. To obtain a complete set of invariants or a set of canonical forms for congruence is not an easy task, but if only real symmetric matrices are considered, then the solution is almost in hand. The restriction to real symmetric matrices is not unreasonable in that both real inner products and quadratic functions are required to have real symmetric matrices.

The first step in obtaining a canonical form for real symmetric matrices under congruence is to use the fact that every real symmetric matrix can be diagonalized with an orthogonal matrix. Given an $n \times n$ real symmetric matrix A, there exists an orthogonal matrix P such that $P^{-1}AP = \text{diag}(\lambda_1, \ldots, \lambda_n)$ where $\lambda_1, \ldots, \lambda_n$ are the n characteristic values of A. Suppose P is chosen so that the positive characteristic values, if any, are $\lambda_1, \ldots, \lambda_p$ and the negative characteristic values are $\lambda_{p+1}, \ldots, \lambda_r$, where r is the rank of A. Then if $r < n$, $\lambda_{r+1}, \ldots, \lambda_n$ are the zero characteristic values of A. Now $\text{diag}(\lambda_1, \ldots, \lambda_n)$ is congruent to the matrix

$$\text{diag}(\underbrace{1, \ldots, 1}_{p}, \underbrace{-1, \ldots, -1}_{r-p}, \underbrace{0, \ldots, 0}_{n-r}).$$

To see this, let Q be the matrix

$$\text{diag}(1/\sqrt{\lambda_1}, \ldots, 1/\sqrt{\lambda_p}, 1/\sqrt{-\lambda_{p+1}}, \ldots, 1/\sqrt{-\lambda_r}, \underbrace{1, \ldots, 1}_{n-r}).$$

Q is a nonsingular $n \times n$ matrix with $Q^T = Q$, so A is congruent to

$$\begin{aligned} Q^T &\operatorname{diag}(\lambda_1, \ldots, \lambda_n)Q \\ &= Q^T \operatorname{diag}(\lambda_1/\sqrt{\lambda_1}, \ldots, \lambda_p/\sqrt{\lambda_p}, \lambda_{p+1}/\sqrt{-\lambda_{p+1}}, \ldots, \lambda_n/\sqrt{-\lambda_n}, 0, \ldots, 0) \\ &= \operatorname{diag}(\underbrace{1, \ldots, 1}_{p}, \underbrace{-1, \ldots, -1}_{r-p}, \underbrace{0, \ldots, 0}_{n-r}). \end{aligned}$$

Thus every real symmetric matrix is congruent to a diagonal matrix with $+1$'s, -1's, and 0's on the main diagonal. However, it must be proved that no other combination is possible before this matrix can be called a canonical form. A diagonal matrix with a different number of 0's on the main diagonal could not be congruent to A because rank is not changed on multiplication by a nonsingular matrix. Therefore it is only necessary to prove the following theorem.

Theorem 7.3 If a real symmetric $n \times n$ matrix A is congruent to both

$$D_1 = \operatorname{diag}(\underbrace{1, \ldots, 1}_{p}, \underbrace{-1, \ldots, -1}_{r-p}, \underbrace{0, \ldots, 0}_{n-r})$$

and

$$D_2 = \operatorname{diag}(\underbrace{1, \ldots, 1}_{t}, \underbrace{-1, \ldots, -1}_{r-t}, \underbrace{0, \ldots, 0}_{n-r}),$$

then $p = t$.

Proof Suppose q is the quadratic function on $\mathscr{V}_n$ which has D_1 as its matrix with respect to the basis $B_1 = \{U_1, \ldots, U_n\}$, and D_2 as its matrix with respect to $B_2 = \{V_1, \ldots, V_n\}$.

If $W\colon (x_1, \ldots, x_n)_{B_1}$, then

$$q(W) = x_1^2 + \cdots + x_p^2 - x_{p+1}^2 - \cdots - x_r^2.$$

Thus q is positive on the subspace $\mathscr{S} = \mathscr{L}\{U_1, \ldots, U_p\}$. That is, if $Z \in \mathscr{S}$ and $Z \neq \mathbf{0}$, then $q(Z) > 0$. On the other hand, if $W\colon (y_1, \ldots, y_n)_{B_2}$, then

$$q(W) = y_1^2 + \cdots + y_t^2 - y_{t+1}^2 - \cdots - y_r^2.$$

Therefore q is nonpositive on the subspace

$$\mathscr{T} = \mathscr{L}\{V_{t+1}, \ldots, V_r, V_{r+1}, \ldots, V_n\}.$$

That is, if $Z \in \mathscr{T}$, then $q(Z) \leq 0$. This means that if $Z \in \mathscr{S} \cap \mathscr{T}$, then $q(Z) = 0$ and $Z = \mathbf{0}$. Thus the dimension of $\mathscr{S} + \mathscr{T}$ is the sum of the dimensions of $\mathscr{S}$ and $\mathscr{T}$, and we have

$$n \geq \dim(\mathscr{S} + \mathscr{T}) = \dim \mathscr{S} + \dim \mathscr{T} = p + (n - t).$$

Therefore $0 \geq p - t$ and $t \geq p$.

Redefining $\mathscr{S}$ and $\mathscr{T}$ yields $p \geq t$, completing the proof.

Thus there are two numerical invariants that completely determine whether two $n \times n$ real symmetric matrices are congruent. First the rank and second the number of positive or negative characteristic values. The second invariant is defined as follows.

Definition The *signature* of an $n \times n$ real symmetric matrix A is the number of positive characteristic values p minus the number of negative characteristic values $r - p$, where $r = \text{rank } A$. If the signature is denoted by s, then $s = p - (r - p) = 2p - r$.

Theorem 7.4 Two $n \times n$ real symmetric matrices are congruent if and only if they have the same rank and signature.

Proof Two real symmetric $n \times n$ matrices have the same rank r and signature s if an only if they are congruent to

$$\text{diag}(\underbrace{1, \ldots, 1}_{p}, \underbrace{-1, \ldots, -1}_{r-p}, \underbrace{0, \ldots, 0}_{n-r})$$

where $p = \frac{1}{2}(s + r)$. Since congruence is an equivalence relation, this proves the theorem.

The matrices $\text{diag}(1, \ldots, 1, -1, \ldots, -1, 0, \ldots, 0)$ are taken as the canonical forms for real symmetric matrices under congruence.

Example 3 Find the canonical form for the matrix

$$A = \begin{pmatrix} 0 & 3 & -2 \\ 3 & 0 & 5 \\ -2 & 5 & 0 \end{pmatrix}$$

under congruence.

The characteristic equation for A is $\lambda^3 - 38\lambda = 0$. This equation has roots $\pm\sqrt{38}$ and 0, so the canonical form is

$$\begin{pmatrix} 1 & 0 & 0 \\ 0 & -1 & 0 \\ 0 & 0 & 0 \end{pmatrix}$$

or diag(1, -1, 0). That is, A has rank 2 and signature 0.

Suppose A is the matrix of a quadratic function q on $\mathscr{V}_3$ with respect to the standard basis. Then q is given by

$$q(a, b, c) = 6ab - 4ac + 10bc,$$

and a basis B could be found for $\mathscr{V}_3$ for which $q(a, b, c) = x_1^2 - x_2^2$ when $(a, b, c)\colon (x_1, x_2, x_3)_B$.

Example 4 Find the canonical form for

$$A = \begin{pmatrix} 1 & -3 & 2 \\ -3 & 4 & 1 \\ 2 & 1 & 2 \end{pmatrix}$$

under congruence.

The characteristic polynomial for A is $-39 + 7\lambda^2 - \lambda^3$. Since only the signs of the characteristic roots are needed, it is not necessary to solve for them. If $\lambda_1, \lambda_2, \lambda_3$ are the three roots, then det $A = \lambda_1\lambda_2\lambda_3 = -39$, and either all three roots are negative or one is negative and two are positive. But tr $A = \lambda_1 + \lambda_2 + \lambda_3 = 7$, therefore at least one root is positive, and the canonical form is diag(1, 1, -1).

Suppose $\langle\ ,\ \rangle$ is an inner product defined on an n-dimensional, real vector space $\mathscr{V}$. As a bilinear function, $\langle\ ,\ \rangle$ may be written in the from $\langle U, V\rangle = X^TAY$ where X and Y are coordinates of U and V with respect to some basis. Since an inner product is symmetric, the matrix A must be symmetric. Now the requirement that $\langle\ ,\ \rangle$ be positive definite implies that both the rank and the signature of A are equal to n, for if not, then there exists a basis $\{U_1, \ldots, U_n\}$ in terms of which

$$\langle U, V\rangle = x_1y_1 + \cdots + x_py_p - x_{p+1}y_{p+1} - \cdots - x_ry_r.$$

If $r < n$, $\langle U_{r+1}, U_{r+1}\rangle = 0$, which contradicts positive definiteness; and if $p < n$, then $\langle U_{p+1}, U_{p+1}\rangle = -1$, which also contradicts positive definiteness. Therefore every inner product on $\mathscr{V}$ can be expressed in the form $\langle U, V\rangle = x_1y_1 + \cdots + x_ny_n$ with respect to some basis. This is not a new fact, we know that $\langle\ ,\ \rangle$ has such a coordinate expression in terms of any orthonormal basis for $(\mathscr{V}, \langle\ ,\ \rangle)$. It could however be stated in the terms of this section by saying that a matrix used in any coordinate expression for any inner product on $\mathscr{V}$ lies in the equivalence class containing the identity matrix, $[I_n]$, or that any matrix for an inner product is congruent to the identity matrix.

Problems

1. Let $q(a, b) = a^2 + 16ab - 11b^2$ for $(a, b) \in \mathscr{V}_2$.
   a. Find the matrix A of q with respect to $\{E_i\}$.
   b. Find the canonical form A' for A.
   c. Find a basis B' with respect to which A' is the matrix of q.
   d. Express q in terms of coordinates with respect to B'.
2. Follow the same directions as in problem 1 for the quadratic function defined on $\mathscr{V}_3$ by $q(a, b, c) = 9a^2 + 12ab + 12ac + 6b^2 + 12c^2$.
3. Let q be the quadratic function defined on $R_2[t]$ by $q(a + bt) = 14a^2 + 16ab + 4b^2$.
   a. Find the matrix A of q with respect to $B = \{1, t\}$.
   b. If $B' = \{2 - 5t, 3t - 1\}$, use the transition matrix from B to B' to find the matrix A' of q with respect to B'.
   c. Write q in terms of coordinates with respect to B'.
4. Prove that congruence is an equivalence relation on $\mathscr{M}_{n \times n}$.
5. Suppose A and B are $n \times n$ matrices such that $X^TAY = X^TBY$ for all X, $Y \in \mathscr{M}_{n \times 1}$. Show that $A = B$.
6. For each of the following symmetric matrices, find the rank, the signature, and the canonical form for congruence.

   a. $\begin{pmatrix} 3 & -5 \\ -5 & 2 \end{pmatrix}$. b. $\begin{pmatrix} -6 & 3 \\ 3 & -5 \end{pmatrix}$. c. $\begin{pmatrix} 3 & 1 & 4 \\ 1 & -2 & 0 \\ 4 & 0 & 5 \end{pmatrix}$. d. $\begin{pmatrix} -2 & 3 & 0 \\ 3 & -2 & 2 \\ 0 & -2 & 1 \end{pmatrix}$.

7. a. Does $b((x_1, x_2), (y_1, y_2)) = 2x_1y_1 + 6x_1y_2 + 6x_2y_1 + 18x_2y_2$ define an inner product on $\mathscr{V}_2$?

   b. Does

$$b((x_1, x_2, x_3), (y_1, y_2, y_3)) = 4x_1y_1 + 2x_1y_3 + 2x_3y_1 + 3x_2y_2 - x_2y_3 - x_3y_2 + 2x_3y_3$$

   define an inner product on $\mathscr{V}_3$?

8. Show that if A is a symmetric matrix and B is congruent to A, then B is also symmetric.

9. Suppose A and B are real symmetric $n \times n$ matrices. Show that if A is congruent to B, then there exists a quadratic function on $\mathscr{V}_n$ that has A and B as matrices with respect to two choices of basis.

§3. Rigid Motions

The coordinate transformations required to express a quadratic function in canonical form are too general to be used in studying the graph of a second degree equation in Cartesian coordinates. A transformation of Euclidean space is required to be a "rigid motion," that is, it should move figures without distortion just as rigid objects are moved in the real world. In this section we will characterize and then define the "rigid motions" of Euclidean n-space.

The intuitive idea of a "rigid motion" suggests several properties. First it should send lines into lines and second, it should be an *isometry*, that is, the distance between any two points of $\mathscr{E}_n$ should equal the distance between their images. The requirement that a transformation be an isometry, or distance preserving, implies that it is one to one. The first condition is satisfied by any linear map, and both conditions are satisfied by orthogonal transformations. However, not all orthogonal transformations would be considered "rigid motions." For example, the reflection of 3-space in a plane transforms an object such as your left hand into its mirror image, your right hand, which is not possible in the real world. Therefore a "rigid motion" should also preserve something like left or right handedness.

Definition The order of the vectors in any basis for $\mathscr{E}_n$ determines an *orientation* for $\mathscr{E}_n$. Two bases B and B' with transition matrix P determine the *same orientation* if $\det P > 0$ and *opposite orientations* if $\det P < 0$.

For example, an orientation is given for the Cartesian plane or Cartesian 3-space by the coordinate system related to the standard basis. For the plane,

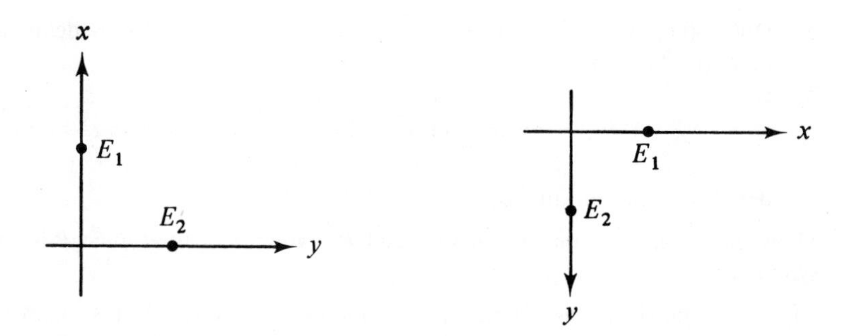

Figure 1

the orientation is usually established by choosing the second coordinate axis, or basis vector, to be 90° from the first axis, measured in the counterclockwise direction. All our diagrams in the Cartesian plane have used this orientation. Figure 1 shows two choices of coordinate systems which give opposite orientations to the plane. There are only two possible orientations for the plane; in fact, the definition of orientation implies that $\mathscr{E}_n$ can be oriented in only two ways.

The orientation for Cartesian 3-space is usually given by the "right hand rule." This rule states that if the first finger of the right hand points in the direction of the x axis, the first basis vector, and the second finger points in the direction of the y axis, second basis vector, then the thumb points in the direction of the z axis or third basis vector, see Figure 2. The other orientation for 3-space follows the "left hand rule" in which the three coordinate directions are aligned with the first two fingers and thumb of the left hand. A coordinate system which gives a left-handed orientation to 3-space is illustrated in Figure 3.

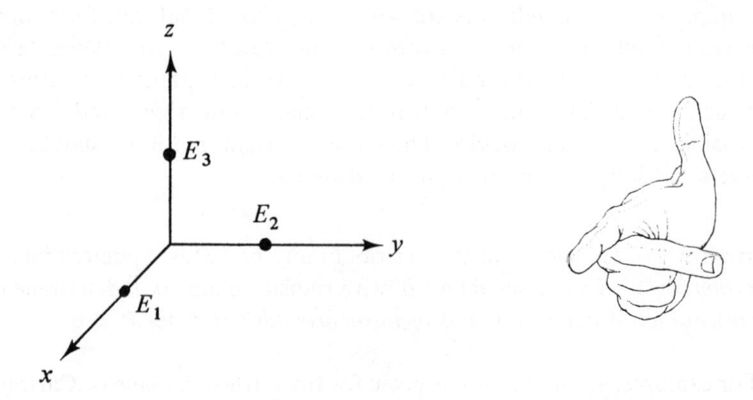

Figure 2

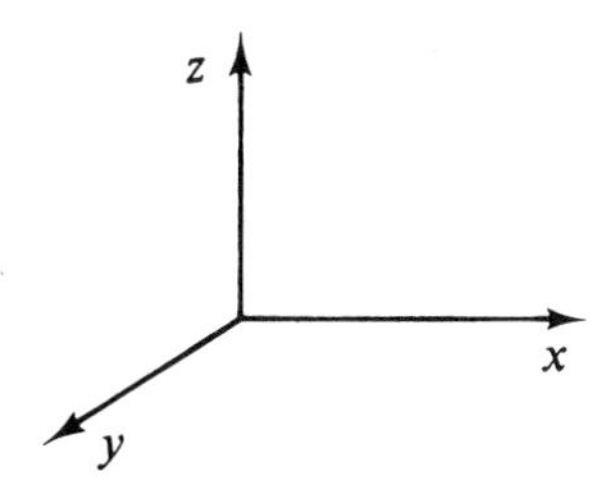

Figure 3

Definition Suppose $T: \mathscr{E}_n \to \mathscr{E}_n$ is a nonsingular linear transformation. If $B = \{U_1, \ldots, U_n\}$ is any basis for $\mathscr{E}_n$, let $B' = \{T(U_1), \ldots, T(U_n)\}$. Then T is *orientation preserving* if B and B' determine the same orientation for $\mathscr{E}_n$.

This definition could be restated as follows:

Theorem 7.5 A linear transformation $T: \mathscr{E}_n \to \mathscr{E}_n$ is orientation preserving if and only if the matrix of T with respoct to any basis for $\mathscr{E}_n$ has a positive determinant.

Proof If $B = \{U_1, \ldots, U_n\}$ is a basis for $\mathscr{E}_n$, then the matrix of T with respect to B is the transition matrix from B to $B' = \{T(U_1), \ldots, T(U_n)\}$. (Why must B' be a basis for each direction?)

Now with this definition, a reflection of 3-space in a plane is not orientation preserving, for the matrix of such a reflection with respect to an orthonormal basis has determinant -1. Therefore, the third requirement of a "rigid motion" is that it be orientation preserving. This eliminates reflections, so that the only orthogonal transformations that satisfy all three conditions are the rotations.

If the essential character of a "rigid motion" is that it move figures without distortion, then it is not necessary for the transformation to be linear. A linear map must send the origin to itself, yet from the geometric viewpoint, there is no distinguished point. It is here that Euclidean geometry differs from the study of the vector space $\mathscr{E}_n$, and nonlinear transformations must be introduced.

A translation $T: \mathscr{E}_n \to \mathscr{E}_n$ given by $T(U) = U + Z$ for $U \in \mathscr{E}_n$ and Z a fixed point in $\mathscr{E}_n$, should be admitted as a "rigid motion." It is easily seen that a translation sends lines into lines and preserves distance. However, the definition of orientation preserving must be modified to include such a nonlinear map.

Example 1 Suppose $T: \mathscr{E}_2 \to \mathscr{E}_2$ is the translation given by

$$T(a, b) = (a, b) + (2, 1) = (a + 2, b + 1).$$

How does T affect a given orientation?

Suppose $\mathscr{E}_2$ is oriented in the usual way by $\{E_i\}$. Since T moves the origin, the fact that $\{E_i\}$ is defined in relation to $\mathbf{0}$ must be taken into consideration. That is, the coordinate system is determined by the three points $\mathbf{0}$, E_1, and E_2. This coordinate system is moved by T to the coordinate system determined by $T(\mathbf{0})$, $T(E_1)$, and $T(E_2)$. These two coordinate systems are illustrated in Figure 4, with x' and y' denoting the new system. Even though the new coordinate system does not have its origin at the zero vector of $\mathscr{E}_2$, it orients the plane in the same way as $\{E_i\}$; for the second axis (y') is 90° from the first (x') measured in the counterclockwise direction. Further, the directions of the new coordinate axes are given by $T(E_1) - T(\mathbf{0})$ and $T(E_2) - T(\mathbf{0})$, that is, by E_1 and E_2. This example shows how the definition of orientation preserving might be extended to nonlinear maps.

Definition Suppose $T: \mathscr{E}_n \to \mathscr{E}_n$ is one to one and $B = \{U_1, \ldots, U_n\}$ is a basis for $\mathscr{E}_n$. Set

$$B' = \{T(U_1) - T(\mathbf{0}), \ldots, T(U_n) - T(\mathbf{0})\}.$$

T is *orientation preserving* if B and B' determine the same orientation for $\mathscr{E}_n$.

This definition essentially defines a translation to be an orientation preserving map. If T is linear, it contains nothing new, for $T(\mathbf{0}) = \mathbf{0}$. But

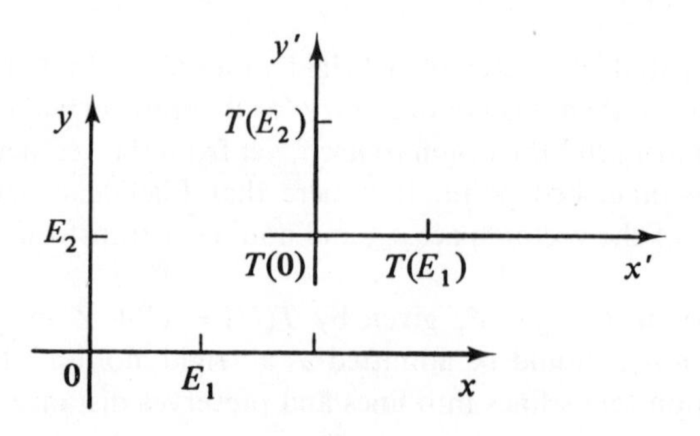

Figure 4

the definition insures that the composition of a rotation and a translation can be called orientation preserving.

Rotations and translations satisfy the three conditions deduced from the concept of a distortion-free transformation of Euclidean n-space. The characterization of a rigid motion will be complete upon showing that every such transformation is the composition of a translation and a rotation.

Theorem 7.6 Suppose $T: \mathscr{E}_n \to \mathscr{E}_n$ satisfies the following conditions:

1. T transforms lines into lines.
2. T is an isometry.
3. T is orientation preserving.

Then T is the composition of a translation and a rotation.

Proof Suppose first that $T(\mathbf{0}) \neq \mathbf{0}$. Then T may be composed with the translation which sends U to $U - T(\mathbf{0})$. This composition sends $\mathbf{0}$ to $\mathbf{0}$ and satisfies the three given conditions. Therefore it is sufficient to assume that $T(\mathbf{0}) = \mathbf{0}$ and show that T is a rotation.

By 2, T preserves distance, therefore

$$\|T(U) - T(V)\| = \|U - V\| \qquad \text{for all} \quad U, V \in \mathscr{E}_n.$$

This implies that T preserves the norm, for setting $V = \mathbf{0}$ yields $\|T(U)\| = \|U\|$.

To show that T is a linear map, first consider $T(rU)$ for some $r \in R$ and $U \in \mathscr{E}_n$. The points $\mathbf{0}$, U, and rU are collinear, so $T(\mathbf{0}) = \mathbf{0}$, $T(U)$, and $T(rU)$ are also collinear by 1. Hence $T(rU) = kT(U)$ for some scalar k. Since T preserves the norm,

$$\|T(rU)\| = \|rU\| = |r|\,\|U\| = |r|\,\|T(U)\|.$$

But $\|T(rU)\| = |k|\,\|T(U)\|$, so $k = \pm r$. If $k = -r$, then T changes the relative position of the three points, violating 2. (If $k = r = 0$, then there is nothing to prove.) Therefore $k = r$, and T preserves scalar multiplication. In a similar way, using the parallelogram rule for addition, it can be shown that T preserves addition (see problem 6 at the end of this section). Thus T is a linear map.

Since T is linear and preserves the norm, T is an orthogonal transformation (problem 3, page 269). But T is required to preserve orientation, so its matrix with respect to any basis has a positive determinant. That is, T is a rotation, completing the proof.

The preceding discussion justifies the following definition.

Definition A *rigid motion* of $\mathscr{E}_n$ is a composition of a finite number of rotations and translations.

Much of Euclidean geometry can be viewed as a search for properties of figures that remain invariant when the figure is moved by a rigid motion. Here the term "figure" simply refers to a collection of points in a Euclidean space, such as a line, a circle, or a cube.

Definition Two geometric figures in $\mathscr{E}_n$ are *congruent* if there exists a rigid motion of $\mathscr{E}_n$ that sends one onto the other.

It should be clear that congruence is an equivalence relation on any collection of figures. As a simple example, consider the set of all lines in the plane. Given any two lines, there is a rigid motion that maps one onto the other. Therefore any two lines in the plane are congruent. This means that there is no property, invariant under rigid motions, that distinguishes one line from another. Something like slope only measures relative position, it is not intrinsic to the line. Thus the set of all lines contains only one equivalence class under congruence.

As another example, consider the set of all circles in the plane. It is not difficult to show that two circles from this set are congruent if and only if they have the same radius. As with lines, circles can be in different relative positions, but two circles with the same radii are geometrically the same. The length of the radius gives a geometric invariant that distinguishes non-congruent circles. Thus the set of all circles in the plane is divided into an infinity of equivalence classes by congruence, one for each positive real number.

An invariant under congruence is an intrinsic geometric property, a property of the figure itself without reference to other objects. It is in this sense that Euclidean geometry searches for invariants under rigid motions. This is essentially what occurs in the study of congruent triangles in plane geometry. However, reflections are usually allowed in plane geometry because as 3-dimensional beings we can always pick up a triangle, turn it over, and put it back in the plane. Of course, a 4-dimensional being would view our right and left hands in just the same way.

Example 2 Suppose S is the collection of all sets $G \subset \mathscr{E}_n$ such that $G = \{C \in \mathscr{E}_n | AC^T = B\}$ for some consistent system of linear equations in n unknowns, $AX = B$. That is, S contains the solution sets for all consistent systems of linear equations in n unknowns. Examining how congruence either identifies or distinguishes between elements of S will answer three essentially identical questions: What types of geometric figures are in S? What is the nature of the equivalence classes of S under congruence? What are the geometric invariants of elements from S?

If $AX = B$ is a single linear equation, then the set of all solutions, G, is a hyperplane. As with lines (hyperplanes in $\mathscr{E}_2$), any two hyperplanes in $\mathscr{E}_n$ are congruent. Therefore one equivalence class contains all the hyperplanes in $\mathscr{E}_n$. At the other extreme, if A has rank n, then G is a single point. Since any two points are congruent, one can be sent to the other by a translation, one equivalence class contains all the points of $\mathscr{E}_n$.

In general, if the rank of A is r, then G may be expressed in terms of $n - r$ parameters, so the geometric dimension of G is $n - r$. It can be shown that two graphs from S are congruent if and only if they have the same dimension. Therefore only one invariant is needed to distinguish between elements of S, either the rank of A or the dimension of the graph G. This means there are n equivalence classes or types of solution sets. When rank $A = n$ or G has dimension 0, the equivalence class is $\mathscr{E}_n$ and the graphs are points; when rank $A = n - 1$ or G has dimension 1, the equaivalence class is the set of all lines in n-space, and so on.

Other examples can be obtained by replacing the linear system $AX = B$, defining each graph G, by some other condition on the coordinates of points. In particular, if $AX = B$ is replaced by a general second degree polynomial equation in n variables, then we are back to the problem posed at the beginning of this chapter. Now we can see that the problem is to use the geometric equivalence relation of congruence to study the possible graphs of second degree equations, and to find a complete set of geometric invariants which describe them. With the invariants will come a set of canonical forms which will amount to choosing a standard position for each type of graph.

Problems

1. Show that for any n, $\mathscr{E}_n$ can be oriented in only two ways.
2. Suppose $T: \mathscr{E}_2 \to \mathscr{E}_2$ is given by $T(a, b) = (a, b) + (-4, 3)$. Sketch the coordinate system determined by $\{E_i\}$ and the coordinate system it is sent into by T.
3. Prove that a translation of $\mathscr{E}_n$:
   a. Maps lines onto lines.
   b. Preserves the distance between points.
4. Suppose $T: \mathscr{E}_n \to \mathscr{E}_n$ sends lines onto lines and is an isometry. Prove that T preserves the order of collinear points.
5. a. Show that any two lines in $\mathscr{E}_n$ are congruent.
   b. Show that any two hyperplanes in $\mathscr{E}_n$ are congruent.

6. Prove that if $T: \mathscr{E}_n \to \mathscr{E}_n$ is an isometry, sends lines onto lines, and $T(\mathbf{0}) = \mathbf{0}$, then T preserves vector addition.

7. Determine the nature of the equivalence classes under congruence of the set of all spheres in 3-space.

8. a. Let S denote the set of all triangles in the plane. Find a complete set of geometric invariants for S under congruence. Be sure to consider orientation.

b. Do the same if S is the set of all triangles in 3-space.

§4. Conics

A general second degree polynomial in the n variables $x_1, \ldots, x_n$ can be written as the sum of a symmetric quadratic form $X^TAX = \sum_{i,j=1}^n a_{ij}x_ix_j$, a linear form $BX = \sum_{i=1}^n b_ix_i$ and a constant C. The graph G of the polynomial equation $X^TAX + BX + C = 0$ is obtained by viewing $x_1, \ldots, x_n$ as coordinates of points in Euclidean n-space with respect to some Cartesian coordinate system. This corresponds to viewing $x_1, \ldots, x_n$ as coordinates of vectors in the inner product space $\mathscr{E}_n$ with respect to some orthonormal basis. We will take $x_1, \ldots, x_n$ to be coordinates with respect to the standard basis. Coordinates with respect to any other basis or coordinate system will be primed. Thus the graph G is the subset of $\mathscr{E}_n$ given by $\{V \in \mathscr{E}_n | V{:}(x_1, \ldots, x_n)_{\{E_i\}}$ and $X^TAX + BX + C = 0$ when $X = (x_1, \ldots, x_n)^T\}$. In general, when $n = 2$, the graphs are curves in the plane called *conics* or *conic sections*. This terminology is justified in problem 5 on page 329. When $n = 3$, the graphs are surfaces in 3-space called *quadric surfaces*. It should be noted that the general second degree equation includes equations that have degenerate graphs, such as points or lines, as well as equations that have no graph at all. The latter is the case with the equation $(x_1)^2 + 1 = 0$.

Rather than consider the set of all graphs for second degree polynomial equations, and determine conditions which identify congruent graphs, it is easier to work directly with the polynomial. After determining how a rigid motion might be used to simplify an arbitrary polynomial, it is then possible to choose a set of canonical forms for the equations. Choosing such a canonical set of equations amounts to choosing "standard positions" for the various types of graph relative to a given coordinate system, and once a simple collection of equations is obtained, it is not difficult to select a set of geometric invariants for the graphs under congruence.

The first point to notice is that a translation does not affect the coefficients of the second degree terms in $X^TAX + BX + C = 0$. To see this, suppose a translation is given in coordinates by $X = X' + Z$ where Z is a fixed element of $\mathscr{M}_{n\times 1}$, that is, $x_1 = x_1' + z_1, \ldots, x_n = x_n' + z_n$. Then

$$\begin{aligned} X^TAX + BX + C &= (X' + Z)^TA(X' + Z) + B(X' + Z) + C \\ &= X'^TAX' + X'^TAZ + Z^TAX' + Z^TAZ + BX' + BZ + C \\ &= X'^TAX' + B'X' + C' \end{aligned}$$

where $B' = 2Z^TA + B$ and $C' = BZ + C$. For example, if the polynomial $x^2 + 3xy + 3yx + 7x + 2$ undergoes the translation $x = x' + 4, y = y' - 5$, then it becomes

$$\begin{aligned} (x' + 4)^2 + 3(x' + 4)(y' - 5) + 3(y' - 5)(x' + 4) + 7(x' + 4) + 2 \\ = x'^2 + 3x'y' + 3y'x' - 23x' + 24y' - 90, \end{aligned}$$

and the coefficients of the second degree terms, 1, 3, 3, 0, remain the same. You might want to work this example out in the matrix form to confirm the general expression, then $X = (x, y)^T$, $Z = (4, -5)^T$, and so on.

The other component of a rigid motion, a rotation, can always be used to obtain a polynomial without cross product terms, that is, terms of the form x_hx_k with $h \neq k$. For since A is symmetric, there exists an orthogonal matrix P with determinant $+1$, such that $P^TAP = P^{-1}AP = \text{diag}(\lambda_1, \ldots, \lambda_n)$ where $\lambda_1, \ldots, \lambda_n$ are the characteristic values of A. So the rotation of $\mathscr{E}_n$ given in coordinates by $X = PX'$ changes the polynomial $X^TAX + BX + C$ to $X'^T\text{diag}(\lambda_1, \ldots, \lambda_n)X' + BPX' + C$ in which the second degree terms are perfect squares, $\lambda_1 x'^2 + \cdots + \lambda_n x'^2$. The fact that such a polynomial can always be obtained is called the "Principal Axis Theorem." For a justification of this name, see Example 1 below.

Now consider the matrix A of coefficients for the second degree terms. Translations do not affect A and rotations transform A to a similar matrix, so any invariants of A under similarity are invariants of the polynomial under these coordinate changes. This means that similarity invariants of A are geometric invariants of the graph under rigid motions. These invariants include rank and the characteristic values, as well as the determinant, the trace, and other symmetric functions of the characteristic values. For example, the sign of the determinant of A distinguishes three broad classes of conic sections. If a second degree polynomial equation in two variables is written in the form $ax^2 + bxy + cy^2 + dx + ey + f = 0$, then

$$A = \begin{pmatrix} a & b/2 \\ b/2 & c \end{pmatrix}$$

and the determinant of A is $ac - b^2/4 = \frac{1}{4}(4ac - b^2)$. Traditionally, rather than use det A, the invariant is taken as $-4 \det A = b^2 - 4ac$ and called the *discriminant* of the equation. The graph is said to be a *hyperbola* if $b^2 - 4ac > 0$, an *ellipse* if $b^2 - 4ac < 0$, and a *parabola* if $b^2 - 4ac = 0$.

Example 1 Determine the type of curve given by

$$11x^2 + 6xy + 19y^2 - 80 = 0$$

and find a coordinate system for which the equation does not have a cross product term.

We have $b^2 - 4ac = 6^2 - 4 \cdot 11 \cdot 19 < 0$, so the curve is an ellipse. The symmetric matrix of coefficients for the second degree terms is

$$A = \begin{pmatrix} 11 & 3 \\ 3 & 19 \end{pmatrix}.$$

A has characteristic values 10 and 20, so there exist coordinates x', y' in which the equation can be written as

$$10x'^2 + 20y'^2 = 80.$$

The coordinate system and the rotation yielding these coordinates can be obtained from the characteristic vectors for A. One finds that $(3, -1)^T$ and $(1, 3)^T$ are characteristic vectors for 10 and 20, respectively, so if

$$P = \begin{pmatrix} 3/\sqrt{10} & 1/\sqrt{10} \\ -1/\sqrt{10} & 3/\sqrt{10} \end{pmatrix},$$

then P is orthogonal with det $P = +1$, and

$$P^T AP = \begin{pmatrix} 10 & 0 \\ 0 & 20 \end{pmatrix}.$$

Thus the x', y' coordinate system corresponds to the basis $B = \{(3/\sqrt{10}, -1/\sqrt{10}), (1/\sqrt{10}, 3/\sqrt{10})\}$.

The graph for the original equation is easily sketched using its equation with respect to the x', y' coordinate system. The line segments PQ and RS in Figure 5 are called the major and minor axes of the ellipse. Direction numbers for these principal axes are given by characteristic vectors for A, that is, by the columns of a matrix that diagonalized the quadratic form $X^T AX$.

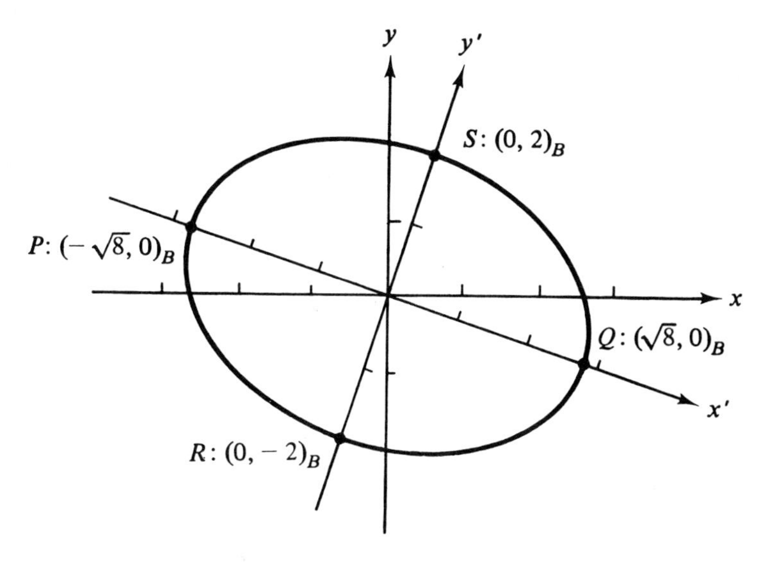

Figure 5

Now return to the general case of a second degree polynomial in n variables. Given $X^TAX + BX + C$, there is a rotation $X = PX'$ that transforms it to

$$\sum_{i=1}^{n} \lambda_i x_i'^2 + \sum_{i=1}^{n} b_i' x_i' + C,$$

where $\lambda_1, \ldots, \lambda_n$ are the characteristic values of A. It is then possible to obtain an equivalent polynomial in which no variable appears in both the quadratic and the linear parts. If for some k, λ_k and b_k' are nonzero, then completing a square gives

$$\lambda_k x_k'^2 + b_k' x_k' = \lambda_k(x_k' + \tfrac{1}{2}b_k'/\lambda_k)^2 - \tfrac{1}{4}b_k'^2/\lambda_k.$$

Therefore if in a translation $x_k'' = x_k' + \frac{1}{2}b_k'/\lambda_k$, then after the translation the terms involving $x_k'^2$ and x_k' are replaced by $\lambda_k x_k''^2$ minus a constant.

Example 2 Find a translation that transforms the polynomial $2x^2 - 5y^2 + 3x + 10y$ into a polynomial without linear terms.

Completing the squares gives

$$2(x + 3/4)^2 - 2\cdot 9/16 - 5(y - 1)^2 + 5.$$

Therefore the translation given by $(x', y') = (x + 3/4, y - 1)$ transforms the polynomial to $2x'^2 - 5y'^2 + 31/8$.

Thus, in general, given a second degree polynomial there exists a rigid motion, the composition of a rotation and a translation, that transforms it to a polynomial in a rather simple form. In this form a variable may appear either in the quadratic portion as a square or in the linear portion, but not both. From this one could choose a set of canonical forms for polynomials under coordinate changes induced by rigid motions of $\mathscr{E}_n$. However, from the geometric point of view two polynomicals should be judged equivalent if they differ by a nonzero multiple, for multiplying an equation by a nonzero constant does not affect its graph. Therefore a set of canonical forms for equations of graphs should be obtained using rigid motions and multiplication by nonzero scalars. This will be done for polynomials in two and three variables, establishing standard positions for conics and quadric surfaces.

A general second degree equation in two variables can be put into one of nine types of canonical forms by changing coordinates with a rigid motion and then multiplying the equation by a nonzero constant. For example, if the polynomial has a positive discriminant, then either $x'^2/a^2 - y'^2/b^2 = 1$ or $x'^2/a^2 - y'^2/b^2 = 0$ can be obtained where a and b represent positive constants.

Example 3 The polynomial equation

$$2x^2 - 72xy + 23y^2 + 140x - 20y + 50 = 0$$

has positive discriminant and therefore is equivalent to one of the above forms. Find the form and a coordinate transformation that produces it.

The symmetric coefficient matrix of the second degree terms $\begin{pmatrix} 2 & -36 \\ -36 & 23 \end{pmatrix}$ has 50 and -25 as characteristic values with corresponding characteristic vectors $(3, -4)^T$ and $(4, 3)^T$. The rotation given by $X = PX'$ with

$$P = \begin{pmatrix} 3/5 & 4/5 \\ -4/5 & 3/5 \end{pmatrix},$$

yields the equation

$$50x'^2 - 25y'^2 + 100x' + 100y' + 50 = 0.$$

Upon completion of the squares, the equation becomes

$$50(x' + 1)^2 - 50 - 25(y' - 2)^2 + 100 + 50 = 0.$$

Therefore under the translation $(x'', y'') = (x' + 1, y' - 2)$, it becomes

$50x''^2 - 25y''^2 = -100$. This is not quite in the first form above. But with a rotation through 90° given by $(x''', y''') = (y'', -x'')$ the equation can be written in the form

$$\frac{x'''^2}{2^2} - \frac{y'''^2}{(\sqrt{2})^2} = 1.$$

Here the a and b in the form given above are $a = 2$ and $b = \sqrt{2}$. The coordinate change required is the composition of all three transformations. Putting them together gives the relation

$$(x, y) = (4x'''/5 - 3y'''/5 + 1, 3x'''/5 + 4y'''/5 + 2).$$

Let the four coordinate systems used in obtaining the canonical form for the equation be denoted by $B = \{E_i\}$, $B' = \{(3/5, -4/5), (4/5, 3/5)\}$, B'' the translate of B' and B''' obtained from B'' by a rotation through 90°. Then Figure 6 shows the graph of each equation in relation to the appropriate coordinate system. The hyperbola is in "standard position" with respect to the x''', y''' coordinate system.

The other seven types of canonical forms for second degree equations in two variables are obtained as follows: If the discriminant is negative, the equation can be transformed to $x'^2/a^2 + y'^2/b^2 = 1$, $x'^2/a^2 + y'^2/b^2 = -1$, or $x'^2/a^2 + y'^2/b^2 = 0$, with $0 < b < a$, by using a rigid motion and multiplying by a nonzero constant. The graph in Example 1 shows an ellipse in standard position with respect to the x', y' coordinate system. For that curve, $a = \sqrt{8}$ and $b = 2$. If the discriminant is 0, the equation can be transformed to $x'^2 = 4py'$, $x'^2 = a^2$, $x'^2 = -a^2$, or $x'^2 = 0$ with p and a positive. The graph of $x'^2 = 4py'$ is the parabola in standard position with its vertex at the origin, focus at $(0, p)_{x',y'}$, and directrix the line $y' = -p$.

It is possible to determine which type of canonical form a given polynomial is equivalent to without actually finding the transformations. To see this we first express the entire polynomial in matrix form by writing the coordinates (x, y) in the form $(x, y, 1)$. Then the general second degree polynomial in two variables can be written as

$$\begin{aligned} ax^2 + bxy + cy^2 + dx + ey + f \\ = (x \quad y \quad 1)\begin{pmatrix} a & b/2 & d/2 \\ b/2 & c & e/2 \\ d/2 & e/2 & f \end{pmatrix}\begin{pmatrix} x \\ y \\ 1 \end{pmatrix}. \end{aligned}$$

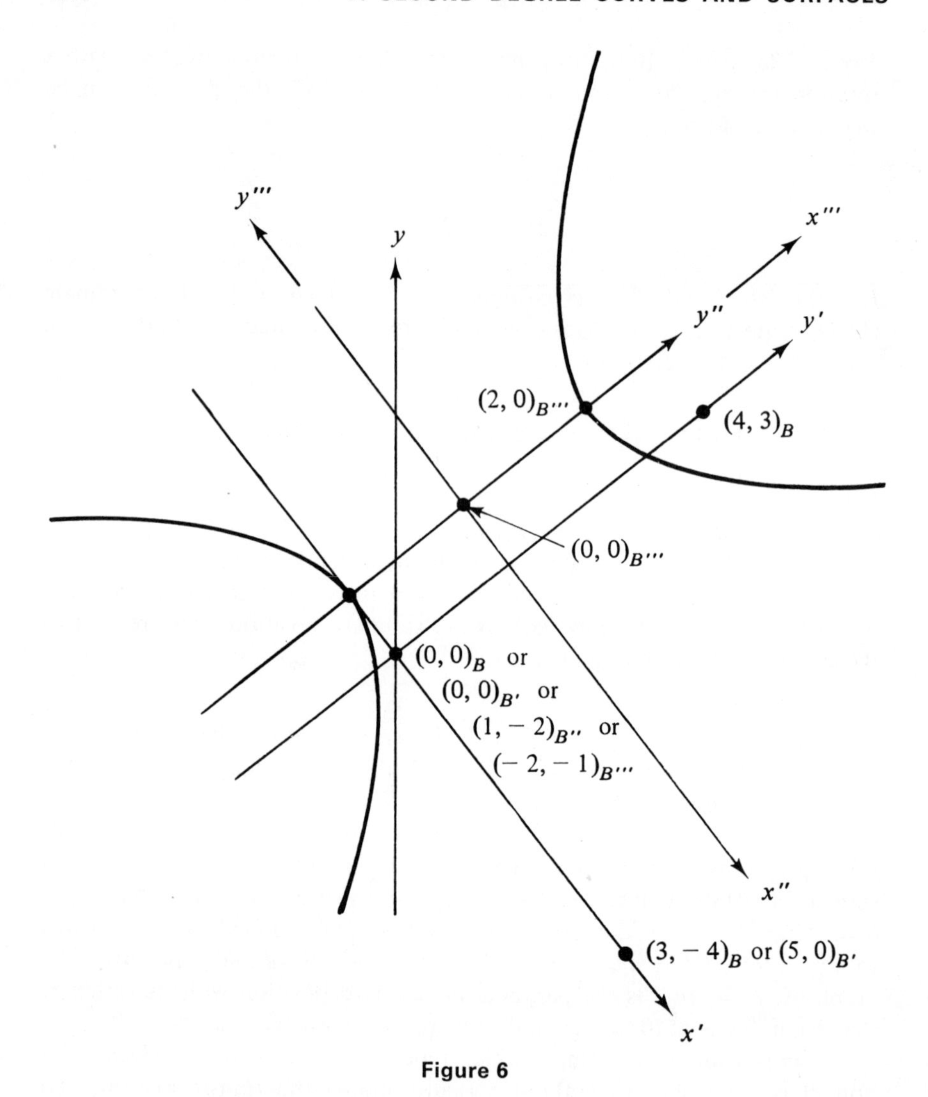

Figure 6

We will write this matrix product as $\hat{X}^T\hat{A}\hat{X}$, with $\hat{X} = (x, y, 1)^T$ and $\hat{A}$ the 3×3 matrix. Using this notation a rotation with matrix $P = (p_{ij})$ would be written as

$$\begin{pmatrix} x \\ y \\ 1 \end{pmatrix} = \begin{pmatrix} p_{11} & p_{12} & 0 \\ p_{21} & p_{22} & 0 \\ 0 & 0 & 1 \end{pmatrix} \begin{pmatrix} x' \\ y' \\ 1 \end{pmatrix}$$

or $\hat{X} = \hat{P}\hat{X}'$. Under the rotation $X = PX'$, the polynomial $X^TAX + BX + C$

$= \hat{X}^T \hat{A} \hat{X}$ becomes $(\hat{P}\hat{X}')^T \hat{A} \hat{P} \hat{X}' = \hat{X}'^T (\hat{P}^T \hat{A} \hat{P}) \hat{X}'$. This notation even makes it possible to express a translation, a nonlinear map, in matrix form. The translation $(x', y') = (x + h, y + k)$ is given by

$$\begin{pmatrix} x' \\ y' \\ 1 \end{pmatrix} = \begin{pmatrix} 1 & 0 & h \\ 0 & 1 & k \\ 0 & 0 & 1 \end{pmatrix} \begin{pmatrix} x \\ y \\ 1 \end{pmatrix}$$

and its inverse is given by

$$\begin{pmatrix} x \\ y \\ 1 \end{pmatrix} = \begin{pmatrix} 1 & 0 & -h \\ 0 & 1 & -k \\ 0 & 0 & 1 \end{pmatrix} \begin{pmatrix} x' \\ y' \\ 1 \end{pmatrix}.$$

If a translation is written as $\hat{X} = \hat{Q}\hat{X}'$, then the polynomial $\hat{X}^T \hat{A} \hat{X}$ is transformed to $(\hat{Q}\hat{X}')^T \hat{A} (\hat{Q}\hat{X}') = \hat{X}'^T (\hat{Q}^T \hat{A} \hat{Q}) \hat{X}'$.

Example 4 Perform the translation of Example 2 with matrix multiplication.

In Example 2, the translation $(x', y') = (x + 3/4, y - 1)$ was used to simplify the polynomial

$$2x^2 - 5y^2 + 3x + 10y = (x \quad y \quad 1) \begin{pmatrix} 2 & 0 & 3/2 \\ 0 & -5 & 5 \\ 3/2 & 5 & 0 \end{pmatrix} \begin{pmatrix} x \\ y \\ 1 \end{pmatrix}.$$

This translation is given by

$$\begin{pmatrix} x \\ y \\ 1 \end{pmatrix} = \begin{pmatrix} 1 & 0 & -3/4 \\ 0 & 1 & 1 \\ 0 & 0 & 1 \end{pmatrix} \begin{pmatrix} x' \\ y' \\ 1 \end{pmatrix}.$$

One can check that

$$\begin{aligned} &(x' \quad y' \quad 1) \begin{pmatrix} 1 & 0 & 0 \\ 0 & 1 & 0 \\ -3/4 & 1 & 1 \end{pmatrix} \begin{pmatrix} 2 & 0 & 3/2 \\ 0 & -5 & 5 \\ 3/2 & 5 & 0 \end{pmatrix} \begin{pmatrix} 1 & 0 & -3/4 \\ 0 & 1 & 1 \\ 0 & 0 & 1 \end{pmatrix} \begin{pmatrix} x' \\ y' \\ 1 \end{pmatrix} \\ &= (x' \quad y' \quad 1) \begin{pmatrix} 2 & 0 & 0 \\ 0 & -5 & 0 \\ 0 & 0 & 31/8 \end{pmatrix} \begin{pmatrix} x' \\ y' \\ 1 \end{pmatrix} \\ &= 2x'^2 - 5y'^2 + 31/8. \end{aligned}$$

This is the polynomial obtained in Example 2.

The composition of the rotation $X = PX'$ followed by the translation $(x'', y'') = (x' + h', y' + k')$ is given in matrix form by

$$\begin{pmatrix} x \\ y \\ 1 \end{pmatrix} = \begin{pmatrix} p_{11} & p_{12} & h \\ p_{21} & p_{22} & k \\ 0 & 0 & 1 \end{pmatrix} \begin{pmatrix} x'' \\ y'' \\ 1 \end{pmatrix}, \quad \text{with} \quad \begin{aligned} h &= -p_{11}h' - p_{12}k' \\ k &= -p_{21}h' - p_{22}k'. \end{aligned}$$

If this coordinate expression is written as $\hat{X} = \hat{P}\hat{X}''$, then the polynomial X^TAX is transformed to $\hat{X}''^T(\hat{P}^T\hat{A}\hat{P})\hat{X}''$. That is, a rigid motion transforms $\hat{A}$ to $\hat{P}^T\hat{A}\hat{P}$. Since $\hat{P}$ is a nonsingular matrix, the rank of $\hat{A}$ is an invariant of the polynomial under rigid motions of $\mathscr{E}_n$. This invariant together with the discriminant and the property of having a graph (an equation has a graph if the coordinates for some point satisfy the equation) distinguish the nine types of canonical forms for second degree polynomials in two variables. This is summarized in the Table I.

Table I

b^2–$4ac$	Rank $\hat{A}$	Graph	Canonical forms	Type of graph
>0	3	Yes	$x^2/a^2 - y^2/b^2 = 1$	Hyperbola
>0	2	Yes	$x^2/a^2 - y^2/b^2 = 0$	Intersecting lines
<0	3	Yes	$x^2/a^2 + y^2/b^2 = 1$	(Real) ellipse
<0	3	No	$x^2/a^2 + y^2/b^2 = -1$	Imaginary ellipse
<0	2	Yes	$x^2/a^2 + y^2/b^2 = 0$	Point ellipse
$=0$	3	Yes	$x^2 = 4py$	Parabola
$=0$	2	Yes	$x^2 = a^2$	Parallel lines
$=0$	2	No	$x^2 = -a^2$	Imaginary lines
$=0$	1	Yes	$x^2 = 0$	Coincident lines

Problems

1. Find the matrices A, B, and $\hat{A}$ that are used to express the following polynomials in the form $X^TAX + BX + C$ and $\hat{X}^T\hat{A}\hat{X}$.
   a. $3x^2 - 4xy + 7y + 8$. b. $8xy - 4x + 6y$.
2. Find a rotation that transforms $14x^2 + 36xy - y^2 - 26 = 0$ to $x'^2 - y'^2/2 = 1$. Sketch both coordinate systems and the graph of the two equations.

3. Find a rigid motion that transforms the equation

$$16x^2 - 24xy + 9y^2 - 200x - 100y - 500 = 0$$

to the canonical form for a parabola. Sketch the graph and the coordinate systems corresponding to the various coordinates used.

4. Write the polynomial $5x^2 - 3y^2 - 10x - 12y - 7$ in the form $\hat{X}^T\hat{A}\hat{X}$ and transform it to a polynomial without linear terms with a translation given by $\hat{X} = \hat{P}\hat{X}'$.

5. For each of the following equations, use Table I, without obtaining the canonical form, to determine the nature of its graph.

   a. $2x^2 - 6xy + 4y^2 + 2x + 4y - 24 = 0$.
   b. $4x^2 + 12xy + 9y^2 + 3x - 4 = 0$.
   c. $7x^2 - 5xy + y^2 + 8y = 0$.
   d. $x^2 + 4xy + 4y^2 + 2x + 4y + 5 = 0$.
   e. $x^2 + 3xy + y^2 + 2x = 0$.

6. The equations $X^TAX + BX + C = 0$ and $r(X^TAX + BX + C) = 0$ have the same graph for every nonzero scalar r. Should the fact that $\det A \neq \det(rA)$ have been taken into consideration when the discriminant was called a geometric invariant?

7. The line in $\mathscr{E}_2$ through (a, b) with direction numbers (α, β) can be written as $\hat{X} = t\hat{K} + \hat{U}$, $t \in R$, with

$$\hat{X} = \begin{pmatrix} x \\ y \\ 1 \end{pmatrix}, \quad \hat{U} = \begin{pmatrix} a \\ b \\ 1 \end{pmatrix}, \quad \hat{K} = \begin{pmatrix} \alpha \\ \beta \\ 0 \end{pmatrix}.$$

   a. Show that this line intersects the graph of $\hat{X}^T\hat{A}\hat{X} = 0$ at the points for which the parameter t satisfies

$$(\hat{K}^T\hat{A}\hat{K})t^2 + (2\hat{K}^T\hat{A}\hat{U})t + \hat{U}^T\hat{A}\hat{U} = 0.$$

   b. Use the equation in part a to find the points at which the line through $(-2, 3)$ with direction $(1, -2)$ intersects the graph of $x^2 + 4xy - 2y + 5 = 0$.

8. Find the canonical form for the following and sketch the graph together with any coordinate systems used.

   a. $2x^2 + 4xy + 2y^2 = 36$. b. $28x^2 + 12xy + 12y^2 = 90$.

§5. Quadric Surfaces

The general second degree equation in three variables may be written in the form $\hat{X}^T\hat{A}\hat{X} = 0$ where $\hat{A}$ is the symmetric matrix

$$\hat{A} = \begin{pmatrix} a_{11} & a_{12} & a_{13} & b_1 \\ a_{21} & a_{22} & a_{23} & b_2 \\ a_{31} & a_{32} & a_{33} & b_3 \\ b_1 & b_2 & b_3 & c \end{pmatrix}, \qquad a_{ij} = a_{ji},$$

and

$$\hat{X} = \begin{pmatrix} x_1 \\ x_1 \\ x_3 \\ 1 \end{pmatrix} \quad \text{or} \quad \begin{pmatrix} x \\ y \\ z \\ 1 \end{pmatrix}.$$

If $A = (a_{ij})$, $B = (b_1, b_2, b_3)$, and $C = (c)$, then

$$\hat{X}^T \hat{A} \hat{X} = X^T A X + 2BX + C, \qquad \text{with} \quad X = (x, y, z)^T.$$

Notice that both $\hat{A}$ and A are symmetric. If x, y, and z are viewed as Cartesian coordinates in three-space, then the graph of $\hat{X}^T \hat{A} \hat{X} = 0$ is a quadric surface. In general, this is actually a surface, but as with the conic sections some equations either have no graph or their graphs degenerate to lines or points.

We know that for any equation $\hat{X}^T \hat{A} \hat{X} = 0$ there exists a rigid motion which transforms it to an equation in which each variable that appears is either a perfect square or in the linear portion. Therefore it is not difficult to make a list of choices for canonical forms, as shown in Table II.

This list contains 17 types of equations ranked first according to the rank of $\hat{A}$ and then by the rank of A. (This relationship is shown later in Table III.) Each of the possible combinations of signatures for A, linear terms, and constant terms are included in Table II. (Recall that a constant term can be absorbed in a linear term using a translation.) Although it may appear that there is no form for an equation of the type $ax^2 + by + cz = 0$, it is possible to reduce the number of linear terms in this equation by using a rotation that fixes the x axis. To see this write $by + cz$ in the form

$$\sqrt{b^2 + c^2}\left(\frac{b}{\sqrt{b^2 + c^2}}\, y + \frac{c}{\sqrt{b^2 + c^2}}\, z\right).$$

Then the rotation given by

$$\begin{pmatrix} x' \\ y' \\ z' \end{pmatrix} = \begin{pmatrix} 1 & 0 & 0 \\ 0 & b/\sqrt{b^2 + c^2} & c/\sqrt{b^2 + c^2} \\ 0 & -c/\sqrt{b^2 + c^2} & b/\sqrt{b^2 + c^2} \end{pmatrix} \begin{pmatrix} x \\ y \\ z \end{pmatrix}$$

Table II

Type of equation	Name for graph
$x^2/a^2 + y^2/b^2 + z^2/c^2 = 1$	Ellipsoid
$x^2/a^2 + y^2/b^2 + z^2/c^2 = -1$	No graph (imaginary ellipsoid)
$x^2/a^2 + y^2/b^2 - z^2/c^2 = -1$	Hyperboloid of 2 sheets
$x^2/a^2 + y^2/b^2 - z^2/c^2 = 1$	Hyperboloid of 1 sheet
$x^2/a^2 + y^2/b^2 = 2z$	Elliptic paraboloid
$x^2/a^2 - y^2/b^2 = 2z$	Hyperbolic paraboloid
$x^2/a^2 + y^2/b^2 + z^2/c^2 = 0$	Point (imaginary elliptic cone)
$x^2/a^2 + y^2/b^2 - z^2/c^2 = 0$	Elliptic cone
$x^2/a^2 + y^2/b^2 = 1$	Elliptic cylinder
$x^2/a^2 + y^2/b^2 = -1$	No graph (imaginary elliptic cylinder)
$x^2/a^2 - y^2/b^2 = 1$	Hyperbolic cylinder
$x^2 = 4py$	Parabolic cylinder
$x^2/a^2 + y^2/b^2 = 0$	Line (imaginary intersecting planes)
$x^2/a^2 - y^2/b^2 = 0$	Intersecting planes
$x^2 = a^2$	Parallel planes
$x^2 = -a^2$	No graph (imaginary parallel planes)
$x^2 = 0$	Coincident planes

transforms $ax^2 + by + cz = 0$ to $ax'^2 + \sqrt{b^2 + c^2}y' = 0$. Thus the given equation can be put into the standard form for a parabolic cylinder.

Quadric surfaces are not as familiar as conics and at least one is rather complicated, therefore it may be useful to briefly consider the surfaces listed in Table II. Starting with the simplest, there should be no problem with those graphs composed of planes. For example, the two intersecting planes given by $x^2/a^2 - y^2/b^2 = 0$ are the planes with equations $x/a + y/b = 0$ and $x/a - y/b = 0$.

A *cylinder* is determined by a plane curve called a *directrix* and a direction. A line passing through the directrix with the given direction is called a *generator*, and the cylinder is the set of all points which lie on some generator. For the cylinders listed in Table II the generators are parallel to the z axis and the directrix may be taken as a conic section in the xy plane. These three types of cylinders are sketched in Figure 7 showing that the name of each surface is derived from the type of conic used as directrix.

The nature of the remaining quadric surfaces is most easily determined by examining how various planes interesect the surface. Given a surface and a plane, points which lie in both are subject to two constraints. Therefore a surface and a plane generally intersect in a curve. Such a curve is called a *section* of the surface, and a section made by a coordinate plane is called a *trace*. For example, the directrix is the trace in the xy plane for each of the three cylinders described above, and every section made by a plane parallel

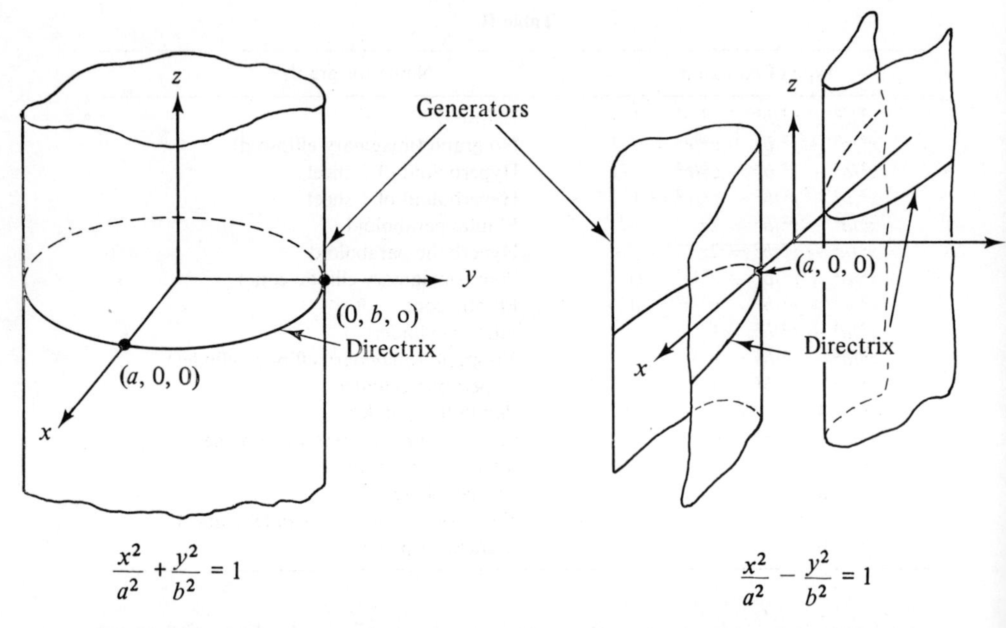

$$\frac{x^2}{a^2} + \frac{y^2}{b^2} = 1$$

Elliptic cylinder

$$\frac{x^2}{a^2} - \frac{y^2}{b^2} = 1$$

Hyperbolic cylinder

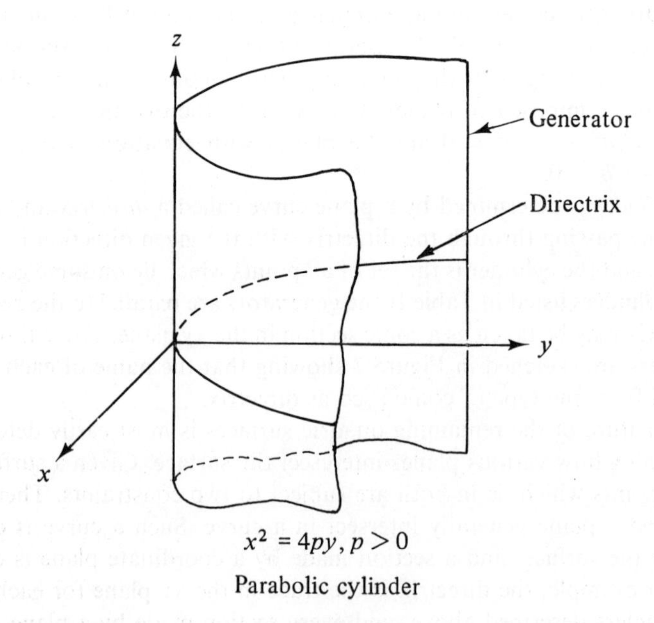

$x^2 = 4py, p > 0$

Parabolic cylinder

Figure 7

to the xy plane is congruent to the directrix. Further, if a plane parallel to the z axis intersects one of these cylinders, then the section is either one or two generators. The sections of a surface in planes parallel to the xy plane are often called *level curves*. If you are familiar with contour maps, you will recognize that the lines on a contour map are the level curves of the land surface given by parallel planes a fixed number of feet apart.

The traces of the ellipsoid $x^2/a^2 + y^2/b^2 + z^2/c^2 = 1$ in the three coordinate planes are ellipses, and a reasonably good sketch of the surface is obtained by simply sketching the three traces, Figure 8. If planes parallel to a coordinate plane intersect the ellipsoid, then the sections are ellipses, which get smaller as the distance from the origin increases. For example, the level curve in the plane $z = k$ is the ellipse $x^2/a^2 + y^2/b^2 = 1 - k^2/c^2$, $z = k$.

The elliptic cone $x^2/a^2 + y^2/b^2 - z^2/c^2 = 0$ is also easily sketched. The traces in the planes $y = 0$ and $x = 0$ are pairs of lines intersecting at the origin. And the level curve for $z = k$ is the ellipse $x^2/a^2 + y^2/b^2 = k^2/c^2$, $z = k$, as shown in Figure 9.

The hyperboloids given by $x^2/a^2 + y^2/b^2 - z^2/c^2 = \pm 1$ are so named because the sections in planes parallel to the yz plane and the xz plane are hyperbolas. (This could also be said of the elliptic cone, see problems 5 and 8 at the end of this section.) In particular, the traces in $x = 0$ and $y = 0$ are the hyperbolas given by $y^2/b^2 - z^2/c^2 = \pm 1$, $x = 0$ and $x^2/a^2 - z^2/c^2 = \pm 1$, $y = 0$, respectively. The hyperboloid of 2 sheets, $x^2/a^2 + y^2/b^2 - z^2/c^2 = -1$, has level curves in the plane $z = k$ only if $k^2 > c^2$. Such a section is an ellipse with the equation $x^2/a^2 + y^2/b^2 = -1 + k^2/c^2$, $z = k$. As k^2 increases, these level curves increase in size so that the surface is in two parts opening away from each other, Figure 10(a). For the hyperboloid of 1 sheet with the equation $x^2/a^2 + y^2/b^2 - z^2/c^2 = 1$, there is an elliptic section for every plane parallel to the xy plane. Therefore the surface is in one piece, as shown in Figure 10(b).

Sections of the paraboloids $x^2/a^2 + y^2/b^2 = 2z$ and $x^2/a^2 - y^2/b^2 = 2z$ in planes parallel to the yz and xz planes, i.e., $x = k$ and $y = k$, are parabolas. Planes parallel to the xy plane intersect the elliptic paraboloid $x^2/a^2 + y^2/b^2 = 2z$ in ellipses given by $x^2/a^2 + y^2/b^2 = 2k$, $z = k$, provided $k \geq 0$. When $k < 0$ there is no section and the surface is bowl-shaped, Figure 11(a). For the hyperbolic paraboloid with equation $x^2/a^2 - y^2/b^2 = 2z$ every plane parallel to the xy plane intersects the surface in a hyperbola. The trace in the xy plane is the degenerate hyperbola composed of two intersecting lines; while above the xy plane the hyperbolas open in the direction of the x axis and below they open in the direction of the y axis. The hyperbolic paraboloid is called a saddle surface because the region near the origin has the shape of a saddle, Figure 11(b).

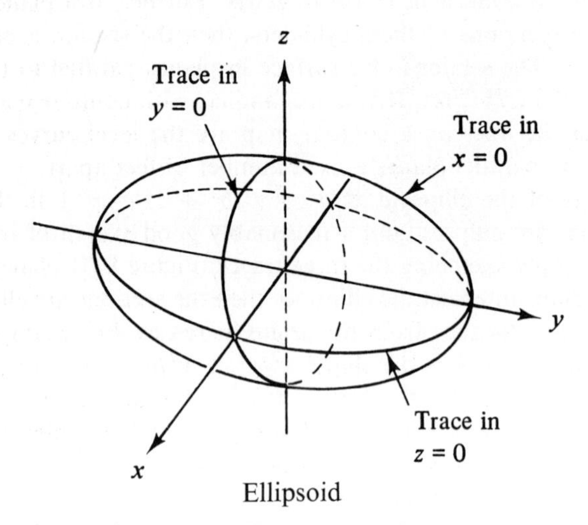

Ellipsoid

Figure 8

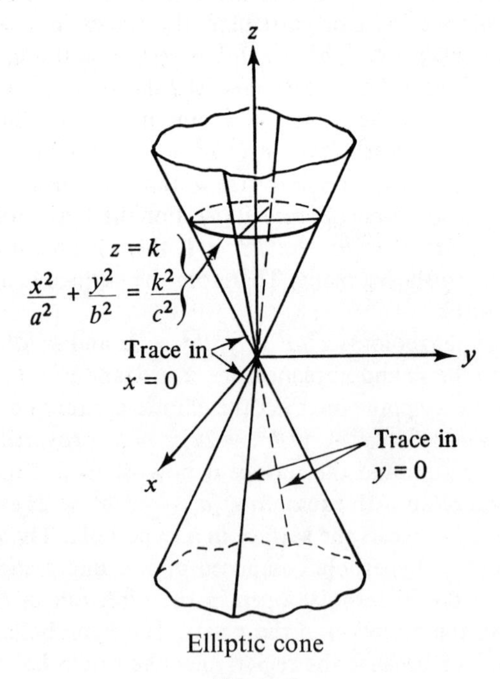

Elliptic cone

Figure 9

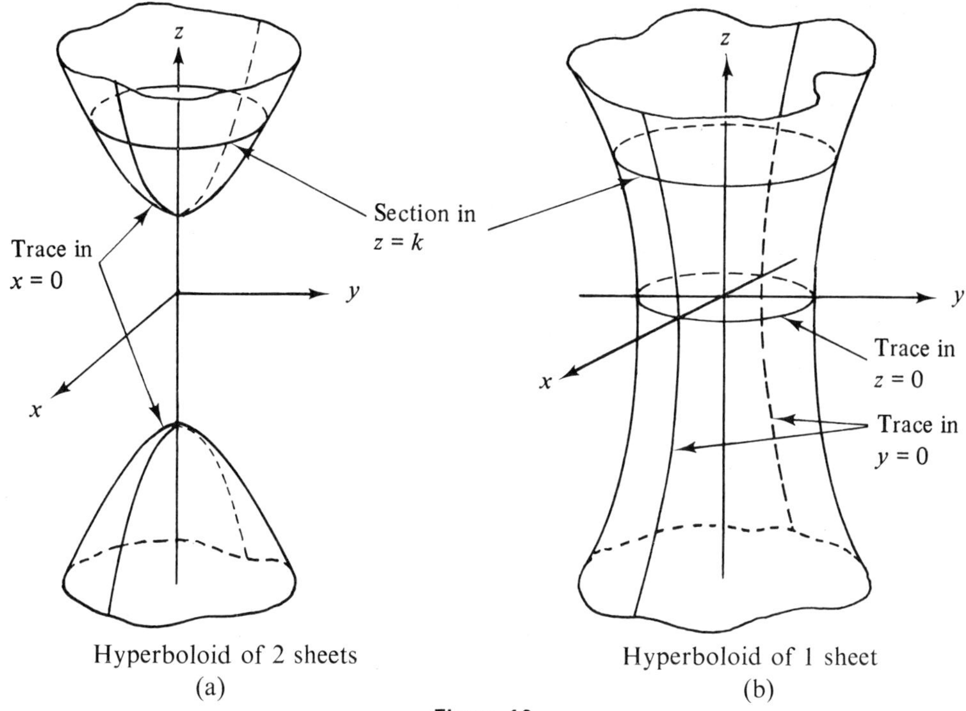

Hyperboloid of 2 sheets
(a)

Hyperboloid of 1 sheet
(b)

Figure 10

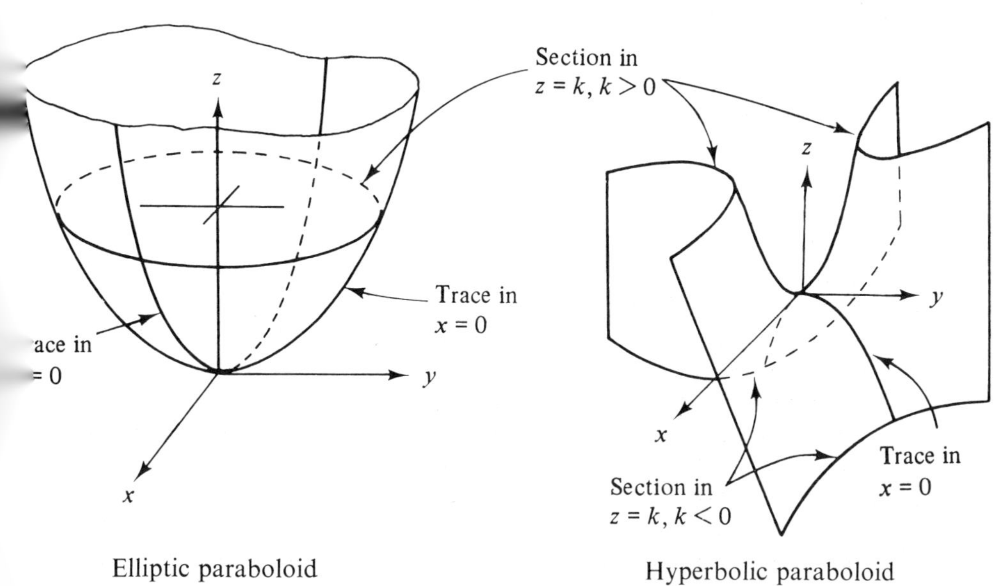

Elliptic paraboloid
(a)

Hyperbolic paraboloid
(b)

Figure 11

Given the equation of a quadric surface, $\hat{X}^T \hat{A} \hat{X} = 0$, one can determine the type of surface by obtaining values for at most five invariants. These are the rank of $\hat{A}$, the rank of A, the absolute value of the signature of A, the existence of a graph, and the sign of the determinant of $\hat{A}$. This is shown in Table III.

Table III

Rank $\hat{A}$	Rank A	Abs (sA)	Graph	Det $\hat{A}$	Type of graph
4	3	3	Yes	<0	Ellipsoid
4	3	3	No	>0	Imaginary ellipsoid
4	3	1	Yes	<0	Hyperboloid of 2 sheets
4	3	1	Yes	>0	Hyperboloid of 1 sheet
4	2	2	Yes	<0	Elliptic paraboloid
4	2	0	Yes	>0	Hyperbolic paraboloid
3	3	3	Yes		Point
3	3	1	Yes		Elliptic cone
3	2	2	Yes		Elliptic cylinder
3	2	2	No		Imaginary elliptic cylinder
3	2	0	Yes		Hyperbolic cylinder
3	1	1	Yes		Parabolic cylinder
2	2	2	Yes		Line
2	2	0	Yes		Intersecting planes
2	1	1	Yes		Parallel planes
2	1	1	No		Imaginary parallel planes
1	1	1	Yes		Coincident planes

The five invariants arise quite naturally. A rigid motion is given in coordinate form by $\hat{X}' = \hat{P}\hat{X}$ with $\hat{P}$ nonsingular. The other operation used to obtain the canonical forms is multiplication of the equation by a nonzero scalar. This has no effect on the graph or the rank of a matrix, but since A is 3×3, it can change the sign of its determinant. Therefore the ranks of $\hat{A}$ and A are invariants but the sign of det A is not. Recall that for the conics, det A gives rise to the discriminant. The signature of A is invariant under rigid motions, but it may change sign when the equation is multiplied by a negative scalar. Therefore the absolute value of the signature, abs(sA), is an invariant of the equation. These three invariants together with the existence or nonexistence of a graph are sufficient to distinguish between all types of surface except for the two hyperboloids. The hyperboloids are distinguished by the sign of det $\hat{A}$, which is an invariant because $\hat{A}$ is a 4×4 matrix.

Problems

1. For the following equations find both the canonical form and the transformation used to obtain it.
   a. $x^2 - y^2/4 - z^2/9 = 1$.
   b. $x^2/5 - z^2 = 1$.
   c. $4z^2 + 2x + 3y + 8z = 0$.
   d. $3x^2 + 6y^2 + 2z^2 + 12x + 12y + 12z + 42 = 0$.
   e. $-3x^2 + 3y^2 - 12xz + 12yz + 4x + 4y - 2z = 0$.
2. Show that every plane parallel to the z axis intersects the paraboloids $x^2/a^2 \pm y^2/b^2 = 2z$ in parabolas.
3. Show that every plane parallel to the z axis intersects the hyperboloids $x^2/a^2 + y^2/b^2 - z^2/c^2 = \pm 1$ in hyperbolas.
4. If intersecting planes, parallel planes, and coincident planes are viewed as cylinders, then what degenerate conics would be taken as the directrices?
5. Justify the name "conic section" for the graph of a second degree polynomial equation in two variables by showing that every real conic, except for parallel lines, may be obtained as the plane section of the elliptic cone $x^2/a^2 + y^2/b^2 - z^2/c^2 = 0$.
6. Find the canonical form for the following equations.
   a. $2xy + 2xz + 2yz - 8 = 0$. b. $2x^2 - 2xy + 2xz + 4yz = 0$.
7. Determine the type of quadric surface given by
   a. $x^2 + 4xy + 4y^2 + 2xz + 4yz + z^2 + 8x + 4z = 0$.
   b. $x^2 + 6xy - 4xz + yz + 4z^2 + 2x - 4z + 5 = 0$.
   c. $3x^2 + 2xy + 2y^2 + 6yz + 7z^2 + 2x + 2y + 4z + 1 = 0$.
   d. $2x^2 + 8xy - 2y^2 + 12xz + 4yz + 8z^2 + 4x + 8y + 12z + 2 = 0$.
8. Show that the elliptic cone $x^2/a^2 + y^2/b^2 - z^2/c^2 = 0$ is the "asymptotic cone" for both hyperboloids given by $x^2/a^2 + y^2/b^2 - z^2/c^2 = \pm 1$. That is, show that the distance between the cone and each hyperboloid approaches zero as the distance from the origin increases.

§6. Chords, Rulings, and Centers

There are several ways in which a line can intersect or fail to intersect a given quadric surface, $\hat{X}^T\hat{A}\hat{X} = 0$. To investigate this we need a vector equation for a line in terms of $\hat{X}$. Recall that the line through the point $U = (a, b, c)$ with direction numbers $V = (\alpha, \beta, \gamma)$ has the equation $X = tV + U$ or $(x, y, z) = t(\alpha, \beta, \gamma) + (a, b, c)$, $t \in R$. Written in terms of 4×1 matrices this equation becomes

$$\begin{pmatrix} x \\ y \\ z \\ 1 \end{pmatrix} = \hat{X} = t\hat{K} + \hat{U} = t\begin{pmatrix} \alpha \\ \beta \\ \gamma \\ 0 \end{pmatrix} + \begin{pmatrix} a \\ b \\ c \\ 1 \end{pmatrix}, \qquad t \in R.$$

The fourth entry in the direction, $\hat{K}$, must be zero to obtain an equality, therefore direction numbers and points now have different representations.

The line given by $\hat{X} = t\hat{K} + \hat{U}$ intersects the quadric surface given by $\hat{X}^T\hat{A}\hat{X} = 0$ in points for which the parameter t satisfies

$$(t\hat{K} + \hat{U})^T\hat{A}(t\hat{K} + \hat{U}) = 0.$$

Using the properties of the transpose and matrix multiplication, and the fact that for a 1×1 matrix

$$\hat{K}^T\hat{A}\hat{U} = (\hat{K}^T\hat{A}\hat{U})^T = \hat{U}^T\hat{A}^T\hat{K} = \hat{U}^T\hat{A}\hat{K},$$

this condition may be written in the form

$$(\hat{K}^T\hat{A}\hat{K})t^2 + (2\hat{K}^T\hat{A}\hat{U})t + \hat{U}^T\hat{A}\hat{U} = 0.$$

The coefficients of this equation determine exactly how the line intersects the quadric surface.

If the direction $\hat{K}$ satisfies $\hat{K}^T\hat{A}\hat{K} \neq 0$, then there are two solutions for t. These solutions may be real and distinct, real and concident, or complex conjugates. Therefore if $\hat{K}^T\hat{A}\hat{K} \neq 0$, then the line intersects the surface in two real points, either distinct or coincident, or in two imaginary points. In all three cases we will say the line *contains a chord* of the surface with the the chord being the line segment between the two points of intersection. In only one case is the chord a real line segment. However with this definition, every line with direction numbers $\hat{K}$ contains a chord of the surface $\hat{X}^T\hat{A}\hat{X} = 0$ provided $\hat{K}^T\hat{A}\hat{K} \neq 0$.

Example 1 Show that every line contains a chord of an arbitrary ellipsoid.

It is sufficient to consider an ellipsoid given by $x^2/a^2 + y^2/b^2 + z^2/c^2 = 1$ (Why?). For this equation $\hat{A} = \text{diag}(1/a^2, 1/b^2, 1/c^2, -1)$, so if $\hat{K} = (\alpha, \beta, \gamma, 0)^T$

is an arbitrary direction, then $\hat{K}^T\hat{A}\hat{K} = \alpha^2/a^2 + \beta^2/a^2 + \gamma^2/c^2$. This is zero only if $\alpha = \beta = \gamma = 0$, but as direction numbers, $\hat{K} \neq \mathbf{0}$. Therefore, $\hat{K}^T\hat{A}\hat{K} \neq 0$ for all directions $\hat{K}$.

A line near the origin intersects this ellipsoid in two real points, far from the origin it intersects the surface in two imaginary points, and somewhere inbetween for each direction there are (tangent) lines which intersect the ellipsoid in two coincident points.

If the line ℓ given by $\hat{X} = t\hat{K} + \hat{U}$ intersects $\hat{X}^T\hat{A}\hat{X} = 0$ when $t = t_1$ and $t = t_2$, then the point for which $t = \frac{1}{2}(t_1 + t_2)$ is the *midpoint* of the chord. This midpoint is always real, for even if t_1 and t_2 are not real, they are complex conjugates. Thus if $\hat{K}^T\hat{A}\hat{K} \neq 0$, then every line with direction $\hat{K}$ contains the real midpoint of a chord. It is not difficult to show that the locus of all these midpoints is a plane. For suppose $\hat{M}$ is the midpoint of the chord on the line ℓ. If ℓ is parameterized so that $\hat{X} = \hat{M}$ when $t = 0$, then the equation for ℓ is $\hat{X} = t\hat{K} + \hat{M}$. ℓ intersects the surface $\hat{X}^T\hat{A}\hat{X} = 0$ at the points for which t satisfies $\hat{K}^T\hat{A}\hat{K}t^2 + 2\hat{K}^T\hat{A}\hat{M}t + \hat{M}^T\hat{A}\hat{M} = 0$. Since these points are the endpoints of the chord, the coefficient of t in this equation must vanish. That is, if ℓ intersects the surface when $t = t_1$ and $t = t_2$, then $\frac{1}{2}(t_1 + t_2) = 0$ implies that $t_1 = -t_2$. Therefore the midpoint $\hat{M}$ satisfies $\hat{K}^T\hat{A}\hat{M} = 0$. Conversely, if $\hat{X} = s\hat{K} + \hat{U}$ is another parameterization with $\hat{K}^T\hat{A}\hat{U} = 0$, then ℓ intersects the surface when $s = \pm s_1$ for some s_1 and $\hat{U}$ is the midpoint of the chord. Thus we have proved the following.

Theorem 7.7 If $\hat{K}^T\hat{A}\hat{K} \neq 0$ for the quadric surface given by $\hat{X}^T\hat{A}\hat{X} = 0$, then the locus of midpoints of all the chords with direction $\hat{K}$ is the plane with equation $\hat{K}^T\hat{A}\hat{X} = 0$.

Definition The plane that bisects a set of parallel chords of a quadric surface is a *diametral plane* of the surface.

Example 2 Find the equation of the diametral plane for the ellipsoid $x^2/9 + y^2 + z^2/4 = 1$ that bisects the chords with direction numbers $(3, 5, 4)$.

$$\begin{aligned}
\hat{K}^T\hat{A}\hat{X} &= (3, 5, 4, 0)\ \mathrm{diag}(1/9, 1, 1/4, -1)(x, y, z, 1)^T \\
&= \mathrm{diag}(1/3, 5, 1, 0)(x, y, z, 1)^T \\
&= x/3 + 5y + z.
\end{aligned}$$

Therefore, the diametral plane is given by $x/3 + 5y + z = 0$. Notice that the direction $\hat{K}$ is not normal to the diametrial plane in this case.

Suppose the line $\hat{X} = t\hat{K} + \hat{U}$ does not contain a chord of $\hat{X}^T\hat{A}\hat{X} = 0$, then $\hat{K}^T\hat{A}\hat{K} = 0$ and the nature of the intersection depends on the values of $\hat{K}^T\hat{A}\hat{U}$ and $\hat{U}^T\hat{A}\hat{U}$. If $\hat{K}^T\hat{A}\hat{U} \neq 0$, then the line intersects the surface in only one point. (Why isn't this a tangent line?)

Example 3 Determine how lines with direction numbers (0, 2, 3) intersect the hyperboloid of two sheets given by $x^2 + y^2/4 - z^2/9 = -1$.

For this surface $\hat{A} = \text{diag}(1, 1/4, -1/9, 1)$. Therefore, with $\hat{K} = (0, 2, 3, 0)^T$, $\hat{K}^T\hat{A}\hat{K} = 4/4 - 9/9 = 0$, and lines with direction $\hat{K}$ do not contain chords. For an arbitrary point $\hat{U} = (a, b, c, 1)^T$, $\hat{K}^T\hat{A}\hat{U} = 2b/4 - 3c/9 = b/2 - c/3$. This is nonzero if (a, b, c) is not in the plane with equation $3y - 2z = 0$. Therefore, every line with direction numbers (0, 2, 3) that is not in the plane $3y - 2z = 0$ intersects the hyperboloid in exactly one point. Each line with direction (0, 2, 3) that is in this plane fails to intersect the surface. For such a line has the equation $\hat{X} = t\hat{K} + \hat{U}$ with $\hat{U} = (a, 0, 0, 1)^T$. In this case, the equation

$$\hat{K}^T\hat{A}\hat{K}t^2 + 2\hat{K}^T\hat{A}\hat{U}t + \hat{U}^T\hat{A}\hat{U} = 0$$

becomes $0t^2 + 0t + a^2 + 1 = 0$, which has no solution. In particular when $a = 0$ the line is in the asymptotic cone, $x^2 + y^2/4 - z^2/9 = 0$, for the surface (see problem 8, page 329).

This example gives a third relationship between a line and a quadric surface. If $\hat{K}^T\hat{A}\hat{K} = \hat{K}^T\hat{A}\hat{U} = 0$ but $\hat{U}^T\hat{A}\hat{U} \neq 0$, then there are no values for t, real or complex, that satisfy the equation $\hat{K}^T\hat{A}\hat{K}t^2 + 2\hat{K}^T\hat{A}\hat{U}t + \hat{U}^T\hat{A}\hat{U} = 0$. So the line $\hat{X} = t\hat{K} + \hat{U}$ fails to intersect the surface $\hat{X}^T\hat{A}\hat{X} = 0$. The last possibility is that $\hat{K}^T\hat{A}\hat{K} = \hat{K}^T\hat{A}\hat{U} = \hat{U}^T\hat{A}\hat{U} = 0$ for which all $t \in R$ satisfy the equation and the line lies in the surface.

Definition A line that lies entirely in a surface is a *ruling* of the surface.

Example 4 Show that a line with direction numbers (0, 0, 1) is either a ruling of the cylinder $x^2/4 - y^2/9 = 1$ or it does not intersect the surface.

For this surface $\hat{A} = \text{diag}(1/4, -1/9, 0, -1)$. For the direction $\hat{K} = (0, 0, 1, 0)^T$ and any point $\hat{U}$, $\hat{K}^T\hat{A}\hat{K} = \hat{U}^T\hat{A}\hat{U} = 0$ and $\hat{U}^T\hat{A}\hat{U} = 0$ if and only if $\hat{U}$ is on the surface. That is, $\hat{X} = t\hat{K} + \hat{U}$ is either a ruling of the cylinder or it does not intersect the surface. Since $\hat{U}$ could be taken as an arbitrary point on the cylinder, this shows that there is a ruling through every point of the hyperbolic cylinder.

Definition A quadric surface is said to be *ruled* if every point on the surface lies on a ruling.

The definition of a cylinder implies that all cylinders are ruled surfaces; are any other quadric surfaces ruled? Certainly an ellipsoid is not since every line contains a chord, but a cone appears to be ruled. For the cone given by $x^2/a^2 + y^2/b^2 - z^2/c^2 = 0$, $\hat{A} = \text{diag}(1/a^2, 1/b^2, -1/c^2, 0)$. So if $\hat{K}$ is the direction given by $\hat{K} = (\alpha, \beta, \gamma, 0)^T$, then $\hat{K}^T\hat{A}\hat{K} = \alpha^2/a^2 + \beta^2/b^2 - \gamma^2/c^2$. This vanishes if an only if the point (α, β, γ) is on the cone. Therefore, if $\hat{U} = (\alpha, \beta, \gamma, 1)^T$ is a point on the cone, not the origin, then $\hat{K}^T\hat{A}\hat{K} = \hat{K}^T\hat{A}\hat{U} = \hat{U}^T\hat{A}\hat{U} = 0$ and the line $\hat{X} = t\hat{K} + \hat{U}$ is a ruling through $\hat{U}$. Since all these rulings pass through the origin, every point lies on a ruling and the elliptic cone is a ruled surface.

Although the figures do not suggest it, there are two other ruled quadric surfaces. Consider first the two hyperboloids with equations of the from $x^2/a^2 + y^2/b^2 - z^2/c^2 = \mp 1$. For these surfaces $\hat{A} = \text{diag}(1/a^2, 1/b^2, -1/c^2, \pm 1)$. Suppose $\hat{U} = (x, y, z, 1)^T$ is an arbitrary point on either of these surfaces, i.e., $\hat{U}^T\hat{A}\hat{U} = 0$, and consider an arbitrary line throuth $\hat{U}$, $\hat{X} = t\hat{K} + \hat{U}$ with $\hat{K} = (\alpha, \beta, \gamma, 0)^T$. These surfaces are ruled if for each point $\hat{U}$ there exists a direction $\hat{K}$ such that $\hat{K}^T\hat{A}\hat{K} = \hat{K}^T\hat{A}\hat{U} = \hat{U}^T\hat{A}\hat{U} = 0$. Suppose these conditions are satisfied, then

$$\hat{K}^T\hat{A}\hat{K} = \alpha^2/a^2 + \beta^2/b^2 - \gamma^2/c^2 = 0.$$

$$\hat{K}^T\hat{A}\hat{U} = \alpha x/a^2 + \beta y/b^2 - \gamma z/c^2 = 0,$$

$$\hat{U}^T\hat{A}\hat{U} = x^2/a^2 + y^2/b^2 - z^2/c^2 \pm 1 = 0.$$

So

$$\begin{aligned}\alpha x/a^2 + \beta y/b^2 &= \gamma z/c^2 = (\gamma/c)(z/c)\\ &= \pm\sqrt{\alpha^2/a^2 + \beta^2/b^2}\,\sqrt{x^2/a^2 + y^2/b^2 \pm 1}.\end{aligned}$$

Squaring both sides and combining terms yields

$$(\alpha y - \beta x)^2 \pm (\alpha^2 b^2 + \beta^2 a^2) = 0.$$

Since a and b are nonzero and $\alpha = \beta = 0$ implies $\gamma = 0$, there is no solution

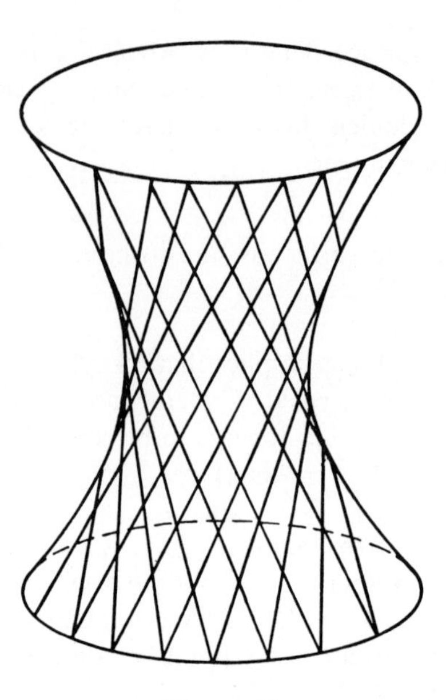

Figure 12

for $\hat{K}$ using the plus sign, which means that the hyperboloid of 2 sheets is not a ruled surface. Thus the question becomes, can α and β be found so that $(\alpha y - \beta x)^2 - (\alpha^2 b^2 + \beta^2 a^2) = 0$, if so, then the hyperboloid of 1 sheet is a ruled surface. If $|y| \neq b$, view this condition as a quadratic equation for α is terms of β,

$$(y^2 - b^2)\alpha^2 - (2xy\beta)\alpha + \beta^2(x^2 - a^2) = 0.$$

Then the discriminant in the quadratic formula is $4\beta^2(a^2y^2 + x^2b^2 - a^2b^2)$. This is always nonnegative for $x^2/a^2 + y^2/b^2 = 1$ is the level curve of the surface in $z = 0$ and all other level curves are larger. That is, if (x, y, z) is on the surface, then $x^2/a^2 + y^2/b^2 \geq 1$. Therefore if $\hat{U}$ is not on the trace in $z = 0$ and $|y| \neq b$, then there are two different solutions for α in terms of β. Further, for each solution there is only one value for γ satisfying $\hat{K}^T\hat{A}\hat{K} = \hat{K}^T\hat{A}\hat{U} = 0$, so in this case there are two rulings through $\hat{U}$. If you check the remaining cases, you will find that there are two distinct rulings through every point on the surface. Thus not only is the hyperboloid of 1 sheet a ruled surface, but it is "doubly ruled." This is illustrated in Figure 12, but it is much better shown in a three-dimensional model.

Now turn to the two types of paraboloids given by $x^2/a^2 \pm y^2/b^2 = 2z$.

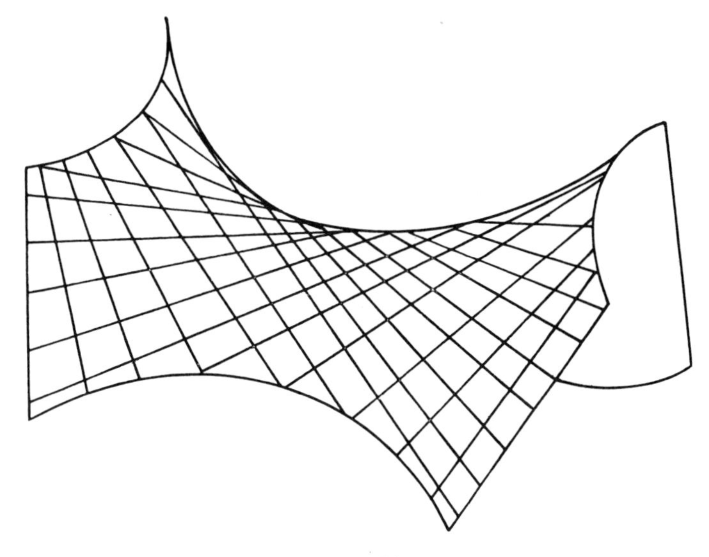

Figure 13

For these surfaces

$$\hat{A} = \begin{pmatrix} 1/a^2 & 0 & 0 & 0 \\ 0 & \pm 1/b^2 & 0 & 0 \\ 0 & 0 & 0 & -1 \\ 0 & 0 & -1 & 0 \end{pmatrix}.$$

So if $\hat{K} = (\alpha, \beta, \gamma, 0)^T$, then $\hat{K}^T\hat{A}\hat{K} = \alpha^2/a^2 \pm \beta^2/b^2$. For the plus sign this vanishes only if $\alpha = \beta = 0$. But $\hat{K}^T\hat{A}\hat{U} = \alpha x/a^2 \pm \beta y/b^2 - \gamma$. Therefore there cannot be a ruling through $\hat{U}$ if $\hat{U}$ lies on the elliptic paraboloid. For the hyperbolic paraboloid, α, β, γ need only satisfy $\alpha^2/a^2 - \beta^2/b^2 = 0$ and $\alpha x/a^2 - \beta y/b^2 - \gamma = 0$. Again there are two rulings through $\hat{U}$ with the directions $\hat{K} = (a, b, x/a - y/b, 0)^T$ and $\hat{K} = (a, -b, x/a + y/b, 0)^T$. Thus the hyperbolic paraboloid is also a doubly ruled surface, as shown in Figure 13. Since a portion of a hyperbolic paraboloid can be made of straight lines, the surface can be manufactured without curves. This fact has been used in the design of roofs for some modern buildings, Figure 14.

We will conclude this section with a determination of when a quadric surface has one or more centers.

Definition A point U is a *center* of a quadric surface S if every line through U either

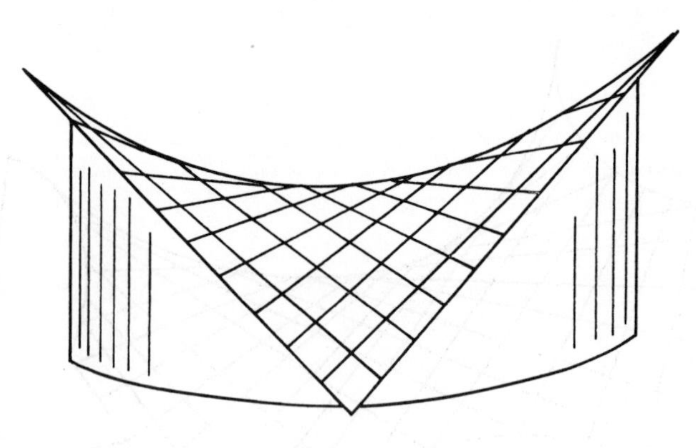

Figure 14

1. contains a chord of S with midpoint U, or
2. fails to intersect S, or
3. is a ruling of S.

That is, a center is a point about which the quadric surface is symmetric.

If $\hat{U}$ is a center for the surface given by $\hat{X}^T\hat{A}\hat{X} = 0$ and $\hat{K}$ is some direction, then each of the above three conditions implies that $\hat{K}^T\hat{A}\hat{U} = 0$. For $\hat{K}^T\hat{A}\hat{U}$ to vanish for all directions $\hat{K}$, implies that the first three entries of $\hat{A}\hat{U}$ must be zero, or $\hat{A}\hat{U} = (0, 0, 0, *)^T$. Conversely, suppose $\hat{P}$ is a point for which $\hat{A}\hat{P} = (0, 0, 0, *)^T$. Then for any direction $\hat{K} = (\alpha, \beta, \gamma, 0)^T$, $\hat{K}^T\hat{A}\hat{P} = 0$. Therefore, the line $\hat{X} = t\hat{K} + \hat{P}$ (1) contains a chord with midpoint $\hat{P}$ if $\hat{K}^T\hat{A}\hat{K} \neq 0$, or (2) does not intersect the surface if $\hat{K}^T\hat{A}\hat{K} = 0$ and $\hat{U}^T\hat{A}\hat{U} \neq 0$, or (3) is a ruling of the surface if $\hat{K}^T\hat{A}\hat{K} = \hat{U}^T\hat{A}\hat{U} = 0$. This proves:

Theorem 7.8 The point $\hat{U}$ is a center of the quadric surface given by $\hat{X}^T\hat{A}\hat{X} = 0$ if and only if $\hat{A}\hat{U} = (0, 0, 0, *)^T$.

Example 5 Determine if the quadric surface given by

$$2x^2 + 6xy + 6y^2 - 2xz + 2z^2 - 2x + 4z + 5 = 0$$

has any centers.

For this polynomial,

$$\hat{A} = \begin{pmatrix} 2 & 3 & -1 & -1 \\ 3 & 6 & 0 & 0 \\ -1 & 0 & 2 & 2 \\ -1 & 0 & 2 & -5 \end{pmatrix}.$$

Therefore x, y, z are coordinates of a center provided

$$\hat{A}\hat{X} = \begin{pmatrix} 2 & 3 & -1 & -1 \\ 3 & 6 & 0 & 0 \\ -1 & 0 & 2 & 2 \\ -1 & 0 & 2 & -5 \end{pmatrix}\begin{pmatrix} x \\ y \\ z \\ 1 \end{pmatrix} = \begin{pmatrix} 0 \\ 0 \\ 0 \\ * \end{pmatrix}$$

or

$$\begin{pmatrix} 2 & 3 & -1 \\ 3 & 6 & 0 \\ -1 & 0 & 2 \end{pmatrix}\begin{pmatrix} x \\ y \\ z \end{pmatrix} = \begin{pmatrix} 1 \\ 0 \\ -2 \end{pmatrix}.$$

One finds that this system of equations is consistent, so the surface has at least one center. In fact, there is a line of centers, for the solution is given by $x = 2t + 2$, $y = -t - 1$, and $z = t$ for any $t \in R$. This quadric surface is an elliptic cylinder.

Example 6 Determine if the hyperboloid of 2 sheets given by $x^2 - 3xz + 4y^2 + 2yz + 10x - 12y + 4z = 0$ has any centers.

For this equation

$$\hat{A} = \begin{pmatrix} 1 & 0 & -3 & 5 \\ 0 & 4 & 1 & -6 \\ -3 & 1 & 0 & 2 \\ 5 & -6 & 2 & 0 \end{pmatrix} \quad \text{and} \quad \begin{pmatrix} 1 & 0 & -3 \\ 0 & 4 & 1 \\ -3 & 1 & 0 \end{pmatrix}\begin{pmatrix} x \\ y \\ z \end{pmatrix} = \begin{pmatrix} -5 \\ 6 \\ -2 \end{pmatrix}$$

is the system of linear equations given by $\hat{A}\hat{X} = (0, 0, 0, *)^T$. This system has the unique solution $x = y = 1$, $z = 2$. Therefore this hyperboloid has exactly one center.

Definition A quadric surface with a single center is a *central quadric*.

The system of equations $\hat{A}\hat{X} = (0, 0, 0, *)^T$ has a unique solution if and only if the rank of A is 3. Therefore, ellipsoids, hyperboloids, and cones are the only central quadric surfaces. A quadric surface has at least one center if the rank of A equals the rank of the first three rows of $\hat{A}$. That is, when the system $\hat{A}\hat{X} = (0, 0, 0, *)^T$ is consistent. Further, a surface with at least one center has a line of centers if rank $A = 2$ and a plane of centers if rank $A = 1$.

Problems

1. Determine how the line intersects the quadric surface. If the line contains a chord, find the corresponding diametral plane.
   a. $(x, y, z) = t(1, -1, 2) + (-2, -2, 2);$
   $$2x^2 + 2xy + 6yz + z^2 + 2x + 8 = 0.$$
   b. $(x, y, z) = t(1, -1, 1) + (0, 1, -2);$
   $$2xy + 2xz - 2yz - 2z^2 + 8x + 6z + 2 = 0.$$
   c. $(x, y, z) = t(2, \sqrt{2}, 0);$
   $$x^2 - 2xz + 2y^2 + 8yz + 8z^2 + 4y + 1 = 0.$$
2. Determine how the family of lines with direction numbers $(1, -2, 1)$ intersects the quadric surface given by
   $$4xy + y^2 - 2yz + 6x + 4z + 5 = 0.$$
3. Show that every diametral plane for a sphere is perpendicular to the chords it bisects.
4. For each surface find the direction numbers of the rulings through the given point.
   a. $x^2/5 + y^2/3 - z^2/2 = 0;\ \ (5, 3, 4).$
   b. $x^2 - y^2/8 = 2z;\ \ (4, 4, 7).$
   c. $4xy + 2xz - 6yz + 2y - 4 = 0;\ \ (1, 0, 2).$
   d. $x^2 - 8xy + 2yz - z^2 - 6x - 6y - 6 = 0;\ \ (-1, 0, 1).$
5. Find all the centers of the following quadric surfaces.
   a. $3x^2 + 4xy + y^2 + 10xz + 6yz + 8z^2 + 4y - 6z + 2 = 0.$
   b. $x^2 + 2xy + y^2 + 2xz + 2yz + z^2 + 2x + 2y + 2z = 0.$
   c. $x^2 + 6xy - 4yz + 3z^2 - 8x + 6y - 14z + 13 = 0.$
   d. $4xy + 6y^2 - 8xz - 12yz + 8x + 12y = 0.$
6. Suppose $\hat{U}$ is a point on the surface given by $\hat{X}^T\hat{A}\hat{X} = 0.$
   a. What conditions should a direction $\hat{K}$ satisfy for the line $\hat{X} = t\hat{K} + \hat{U}$ to be tangent to the surface?
   b. Why should $\hat{U}$ not be a center if there is to be a tangent line through $\hat{U}$? A point of a quadric surface which is not a center is called a *regular point* of the surface.
   c. What is the equation of the tangent plane to $\hat{X}^T\hat{A}\hat{X} = 0$ at a regular point $\hat{U}$?
7. If the given point is a regular point for the surface, find the equation of the tangent plane at that point, see problem 6.
   a. $x^2 + y^2 + z^2 = 1;\ \ (0, 0, 1).$
   b. $x^2 + 6xy + 2y^2 + 4yz + z^2 - 8x - 6y - 2z + 6 = 0;\ \ (1, 1, -1).$
   c. $4xy - 7y^2 + 3z^2 - 6x + 8y - 2 = 0;\ \ (1, 2, -2).$
   d. $3x^2 - 5xz + y^2 + 3yz - 8x + 2y - 6z = 0;\ \ (2, 4, 5).$
8. Find the two lines in the plane $x - 4y + z = 0$, passing through the point $(2, 1, 2)$, that are tangent to the ellipsoid given by $x^2/4 + y^2 + z^2/4 = 1.$

Review Problems

1. Find a symmetric matrix A and a matrix B that express the given polynomial as the sum of a symmetric bilinear form X^TAX and a linear form BX.

a. $3x^2 - 5xy + 2y^2 + 5x - 7y$.
b. $x^2 + 4xy + 6y$.
c. $4x^2 + 4xy - y^2 + 10yz + 4x - 9z$.
d. $3xy + 3y^2 - 4xz + 2yz - 5x + y - 2z$.

2. Prove that if $S, T \in \text{Hom}(\mathscr{V}, R)$, then the function b defined by $b(U, V) = S(U)T(V)$ is bilinear. Is such a function ever symmetric and positive definite?

3. Determine which of the following functions define inner products.

a. $b((x_1, x_2), (y_1, y_2)) = 5x_1y_1 + 6x_1y_2 + 6x_2y_1 + 9x_2y_2$.
b. $b((x_1, x_2), (y_1, y_2)) = 6x_1y_1 + 7x_1y_2 + 3x_2y_1 + 5x_2y_2$.
c. $b((x_1, x_2), (y_1, y_2)) = 4x_1y_1 + 3x_1y_2 + 3x_2y_1 + 2x_2y_2$.

4. Find the signature and the canonical form under congruence for

a. $\begin{pmatrix} -4 & 1 \\ 1 & -4 \end{pmatrix}$. b. $\begin{pmatrix} 3 & 9 \\ 9 & 4 \end{pmatrix}$. c. $\begin{pmatrix} 5 & 2 & 3 \\ 2 & 1 & 1 \\ 3 & 1 & 2 \end{pmatrix}$.

5. Transform the following equations for conics into canonical form. Sketch the curve and the coordinate systems used.

a. $x^2 - 4y^2 + 4x + 24y - 48 = 0$.
b. $x^2 - y^2 - 10x - 4y + 21 = 0$.
c. $9x^2 - 24xy + 16y^2 - 4x - 3y = 0$.
d. $x^2 + 2xy + y^2 - 8 = 0$.
e. $9x^2 - 4xy + 6y^2 + 12\sqrt{5}x + 4\sqrt{5}y - 20 = 0$.

6. a. Define center for a conic.
b. Find a system of equations that gives the centers of conics.
c. Which types of conics have centers?

7. Find the centers of the following conics, see problem 6.

a. $2x^2 + 2xy + 5y^2 - 12x + 12y + 4 = 0$.
b. $3x^2 + 12xy + 12y^2 + 2x + 4y = 0$.
c. $4x^2 - 10xy + 7y^2 + 2x - 4y + 9 = 0$.

8. Identify the quadric surface given by

a. $2x^2 + 2xy + 3y^2 - 4xz + 2yz + 2x - 2y + 4z = 0$.
b. $x^2 + 2xy + y^2 + 2xz + 2yz + z^2 + 2x + 2y + 2z - 3 = 0$.
c. $x^2 + 4xy + 3y^2 - 2xz - 3z^2 + 2y - 4z + 2 = 0$.

9. Given the equation for an arbitrary quadric surface, which of the 17 types would it most likely be?

10. Find all diametral planes for the following quadric surfaces, if their equation is in standard form.

a. Real elliptic cylinder.

b. Elliptic paraboloid.
c. Hyperbolic paraboloid.

11. How do lines with direction numbers (3, 6, 0) intersect the quadric surface with equation $4xz - 2yz + 6x = 0$?

12. Find a rigid motion of the plane, in the form $\hat{X} = \hat{P}\hat{X}'$, which transforms $52x^2 - 72xy + 73y^2 - 200x + 100y = 0$ into a polynomial equation without cross product or linear terms. What is the canonical form for this equation?

Canonical Forms Under Similarity

In Chapter 5 we found several invariants for similarity, but they were insufficient to distinguish between similar and nonsimilar matrices. Therefore it was not possible to define a set of canonical forms for matrices under similarity. However, there are good reasons for taking up the problem again. From a mathematical point of view our examination of linear transformations is left open by not having a complete set of invariants under similarity. On the other hand, the existence of a canonical form that is nearly diagonal permits the application of matrix algebra to several problems. Some of these applications will be considered in the final section of this chapter.

Throughout this chapter $\mathscr{V}$ will be a finite-dimensional vector space over an arbitrary field F.

§1. Invariant Subspaces

A set of canonical forms was easily found for matrices under equivalence. The set consisted of all diagonal matrices with 1's and then 0's on the main diagonal. It soon became clear that diagonal matrices could not play the same role for similarity. However, we can use the diagonalizable case to show how we should proceed in general.

Since two matrices are similar if and only if they are matrices for some map $T \in \operatorname{Hom}(\mathscr{V})$, it will be sufficient to look for properties that characterize maps in $\operatorname{Hom}(\mathscr{V})$, rather than $n \times n$ matrices. Therefore suppose $T \in \operatorname{Hom}(\mathscr{V})$ is a diagonalizable map. Let $\lambda_1, \ldots, \lambda_k$, $k \leq \dim \mathscr{V}$, be the distinct characteristic values of T. For each characteristic value λ_i, the subspace $\mathscr{S}(\lambda_i) = \{V \in \mathscr{V} | T(V) = \lambda_i V\}$ is invariant under T. To say that $\mathscr{S}(\lambda_i)$ is a T-invariant subspace means that if $U \in \mathscr{S}(\lambda_i)$, then $T(U) \in \mathscr{S}(\lambda_i)$. Call $\mathscr{S}(\lambda_i)$ the *characteristic space* of λ_i (it is also called the eigenspace of λ_i.) Then the characteristic spaces of the diagonalizable map T yield a direct sum decomposition of $\mathscr{V}$ in the following sense.

Definition Let $\mathscr{S}_1, \ldots, \mathscr{S}_h$ be subspaces of $\mathscr{V}$. The *sum* of these spaces is

$$\mathscr{S}_1 + \cdots + \mathscr{S}_h = \{U_1 + \cdots + U_h | U_i \in \mathscr{S}_i, 1 \leq i \leq h\}.$$

The sum is *direct*, written $\mathscr{S}_1 \oplus \cdots \oplus \mathscr{S}_h$, if $U_1 + \cdots + U_h = U_1' + \cdots + U_h'$ implies $U_i = U_i'$ when $U_i, U_i' \in \mathscr{S}_i$, $1 \leq i \leq h$. $\mathscr{S}_1 \oplus \cdots \oplus \mathscr{S}_h$ is a *direct sum decomposition of* $\mathscr{V}$ if $\mathscr{V} = \mathscr{S}_1 \oplus \cdots \oplus \mathscr{S}_h$.

The following theorem gives another characterization of when a sum is direct. The proof is not difficult and is left to the reader.

Theorem 8.1 The sum $\mathscr{S}_1 + \cdots + \mathscr{S}_h$ is direct if and only if for $U_i \in \mathscr{S}_i$, $1 \leq i \leq h$, $U_1 + \cdots + U_h = \mathbf{0}$ implies $U_1 = \cdots = U_h = \mathbf{0}$.

Now the above statement regarding characteristic spaces becomes:

Theorem 8.2 If $T \in \operatorname{Hom}(\mathscr{V})$ is diagonalizable and $\lambda_1, \ldots, \lambda_k$ are the distinct characteristic values of T, then $\mathscr{V} = \mathscr{S}(\lambda_1) \oplus \cdots \oplus \mathscr{S}(\lambda_k)$.

Proof Since $\mathscr{V}$ has a basis B of characteristic vectors, $\mathscr{V} \subset \mathscr{S}(\lambda_1) + \cdots + \mathscr{S}(\lambda_k)$, therefore $\mathscr{V} = \mathscr{S}(\lambda_1) + \cdots + \mathscr{S}(\lambda_k)$.

To show the sum is direct suppose $U_1 + \cdots + U_k = \mathbf{0}$ with $U_i \in \mathscr{S}(\lambda_i)$. Each U_i is a linear combination of the characteristic vectors in B corresponding to λ_i. Characteristic vectors corresponding to different characteristic values are linearly independent, therefore replacing each U_i in $U_1 + \cdots + U_k$ with this linear combination of characteristic vectors yields a linear combination of the vectors in B. Since B is a basis, all the coefficients vanish and $U_i = \mathbf{0}$ for each i. So by Theorem 8.1 the sum is direct.

Example 1 Suppose $T \in \operatorname{Hom}(\mathscr{V}_3)$ is given by

$$T(a, b, c) = (3a - 2b - c, -2a + 6b + 2c, 4a - 10b - 3c).$$

T has three distinct characteristic values, 1, 2, and 3, and so is diagonalizable. Therefore $\mathscr{V}_3 = \mathscr{S}(1) \oplus \mathscr{S}(2) \oplus \mathscr{S}(3)$. In fact one can show that

$$\mathscr{S}(1) = \mathscr{L}\{(1, -2, 6)\}, \quad \mathscr{S}(2) = \mathscr{L}\{(0, 1, -2)\}, \quad \mathscr{S}(3) = \mathscr{L}\{(1, -2, 4)\}$$

and $\{(1, -2, 6), (0, 1, -2), (1, -2, 4)\}$ is a basis for $\mathscr{V}_3$.

Return to a diagonalizable map T with distinct characteristic values $\lambda_1, \ldots, \lambda_k$. Then T determines the direct sum decomposition $\mathscr{V} = \mathscr{S}(\lambda_1) \oplus \cdots \oplus \mathscr{S}(\lambda_k)$ and this decomposition yields a diagonal matrix for T as follows: Choose bases arbitrarily for $\mathscr{S}(\lambda_1), \ldots, \mathscr{S}(\lambda_k)$ and let B denote the basis for $\mathscr{V}$ composed of these k bases. Then the matrix of T with respect to B is the diagonal matrix with the characteristic values $\lambda_1, \ldots, \lambda_k$ on the main diagonal. In Example 1 such a basis is given by

$B = \{(1, -2, 6), (0, 1, -2), (1, -2, 4)\}$ and the matrix of T with respect to B is diag(1, 2, 3).

How might this be generalized if T is not diagonalizable? Suppose enough T-invariant subspaces could be found to express $\mathscr{V}$ as a direct sum of T-invariant subspaces. Then a basis B could be constructed for $\mathscr{V}$ from bases for the T-invariant subspaces in the decomposition. Since T sends each of these subspaces into itself, the matrix for T with respect to B will have its nonzero entries confined to blocks along the main diagonal. To see this suppose $\mathscr{S}_1, \ldots, \mathscr{S}_h$ are T-invariant subspaces, although not necessarily characteristic spaces, and $\mathscr{V} = \mathscr{S}_1 \oplus \cdots \oplus \mathscr{S}_h$. If dim $\mathscr{S}_i = a_i$, choose a basis $B_i = \{U_{i1}, \ldots, U_{ia_i}\}$ for each $\mathscr{S}_i$, and let

$$B = \{U_{11}, \ldots, U_{1a_1}, U_{21}, \ldots, U_{2a_2}, \ldots, U_{h1}, \ldots, U_{ha_h}\}.$$

Then the matrix A of T with respect to B has its nonzero entries confined to blocks, $A_1, \ldots, A_h$, which contain the main diagonal of A. The block A_i being the matrix of the restriction of T to $\mathscr{S}_i$ with respect to the basis B_i. (Since $\mathscr{S}_i$ is T-invariant, $T\colon \mathscr{S}_i \to \mathscr{S}_i$.)

Thus

$$A = \begin{pmatrix} A_1 & & & 0 \\ & A_2 & & \\ & & \ddots & \\ 0 & & & A_h \end{pmatrix}.$$

We will call A a *diagonal block matrix* and denote it by diag$(A_1, \ldots, A_h)$. A diagonal block matrix is also called a "direct sum of matrices" and denoted by $A_1 \oplus \cdots \oplus A_h$, but this notation will not be used here.

Example 2 Suppose $T\colon \mathscr{E}_3 \to \mathscr{E}_3$ is rotation with axis $\mathscr{L}\{U\} = \mathscr{S}$ and angle $\theta \neq 0$. Let $\mathscr{T} = \mathscr{S}^\perp$, then $\mathscr{T}$ is not a characteristic space for T, but $\mathscr{S}$ and $\mathscr{T}$ are T-invariant subspaces and $\mathscr{E}_3 = \mathscr{S} \oplus \mathscr{T}$. Now suppose $\{U, V, W\}$ is an orthonormal basis for $\mathscr{E}_3$ with $\mathscr{S} = \mathscr{L}\{V, W\}$. Then the matrix of T with respect to $\{U, V, W\}$ is

$$\begin{pmatrix} A_1 & 0 \\ 0 & A_2 \end{pmatrix} = \left(\begin{array}{c|cc} 1 & 0 & 0 \\ \hline 0 & \cos\theta & -\sin\theta \\ 0 & \sin\theta & \cos\theta \end{array}\right).$$

Thus a decomposition of $\mathscr{V}$ into a direct sum of T-invariant subspaces gives rise to a diagonal block matrix. The procedure for obtaining canonical forms under similarity now becomes one of selecting T-invariant subspaces in some well-defined way and then choosing bases which make the blocks relatively simple. The problem is in the selection of the subspaces, for it is easy to generate T-invariant subspaces and obtain bases in the process.

For any nonzero $U \in \mathscr{V}$, there is an integer $k \leq \dim \mathscr{V}$ such that the set $\{U, T(U), \ldots, T^{k-1}(U)\}$ is linearly independent and the set $\{U, T(U), \ldots, T^{k-1}(U), T^k(U)\}$ is linearly dependent. Set $\mathscr{S}(U, T) = \mathscr{L}\{U, T(U), \ldots, T^{k-1}(U)\}$. Then it is a simple matter to show that $\mathscr{S}(U, T)$ is a T-invariant subspace called the *T-cyclic subspace generated by U*. Moreover the set $\{U, T(U), \ldots, T^{k-1}(U)\}$ is a good choice as a basis for $\mathscr{S}(U, T)$. Problem 4 at the end of this section provides an example of how a direct sum of T-cyclic subspaces leads to a simple matrix for T.

There is a polynomial implicit in the definition of the T-cyclic subspace $\mathscr{S}(U, T)$ which arises when a matrix is found for T using the vectors in $\{U, T(U), \ldots, T^{k-1}(U)\}$. This polynomial is of fundamental importance in determining how we should select the T-invariant subspaces for a direct sum decomposition of $\mathscr{V}$. Since $T^k(U) \in \mathscr{S}(U, T)$, there exist scalars $a_{k-1}, \ldots, a_1, a_0$ such that

$$T^k(U) + a_{k-1}T^{k-1}(U) + \cdots + a_1 T(U) + a_0 U = \mathbf{0}$$

or

$$(T^k + a_{k-1}T^{k-1} + \cdots + a_1 T + a_0 I)U = \mathbf{0}.$$

Therefore U determines a polynomial

$$a(x) = x^k + a_{k-1}x^{k-1} + \cdots + a_1 x + a_0$$

such that when the symbol x is replaced by the map T, a linear map $a(T)$ is obtained which sends U to $\mathbf{0}$. This implies that $a(T)$ sends all of $\mathscr{S}(U, T)$ to $\mathbf{0}$ or that $\mathscr{S}(U, T)$ is contained in $\mathscr{N}_{a(T)}$, the null space of $a(T)$.

Example 3 Consider the map $T \in \mathrm{Hom}(\mathscr{V}_3)$ given by $T = 2I$. For any nonzero vector $U \in \mathscr{V}_3$, $T(U) = 2U$. Therefore the polynomial associated with U is $a(x) = x - 2$. Then $a(T) = T - 2I$ and

$$a(T)U = (T - 2I)U = T(U) - 2U = 2U - 2U = \mathbf{0}.$$

Since $T = 2I$, $\mathscr{S}(U, T) = \mathscr{L}\{U\}$ and $\mathscr{N}_{a(T)} = \mathscr{V}_3$. Thus the null space contains the T-cyclic space, but they need not be equal.

The null space of $a(T)$ in Example 3 is all of $\mathscr{V}_3$ and, therefore, a T-

invariant subspace. But the fact is that the null space of such a polynomial in T is always T-invariant.

Theorem 8.3 Suppose $T \in \text{Hom}(\mathscr{V})$ and $b_m, \ldots, b_1, b_0 \in F$. Let $b(T) = b_m T^m + \cdots + b_1 T + b_0 I$. Then $b(T) \in \text{Hom}(\mathscr{V})$ and $\mathscr{N}_{b(T)}$ is a T-invariant subspace of $\mathscr{V}$.

Proof Since $\text{Hom}(\mathscr{V})$ is a vector space and closed under composition, a simple induction argument shows that $b(T) \in \text{Hom}(\mathscr{V})$.

To show that $\mathscr{N}_{b(T)}$ is T-invariant, let $W \in \mathscr{N}_{b(T)}$, then

$$\begin{aligned} b(T)T(W) &= (b_m T^m + \cdots + b_1 T + b_0 I)T(W) \\ &= b_m T^m(T(W)) + \cdots + b_1 T(T(W)) + b_0 I(T(W)) \\ &= b_m T(T^m(W)) + \cdots + b_1 T(T(W)) + b_0 T(I(W)) \\ &= T(b_m T^m(W) + \cdots + b_1 T(W) + b_0 I(W)) \\ &= T(b(T)W) = T(\mathbf{0}) = \mathbf{0}. \end{aligned}$$

Therefore $T(W) \in \mathscr{N}_{b(T)}$ and $\mathscr{N}_{b(T)}$ is T-invariant.

Example 4 Let T be the map in Example 1 and $b(x) = x^2 - 3x$. Find $\mathscr{N}_{b(T)}$.

$b(T) = T^2 - 3T$. A little computation shows that

$$b(T)(a, b, c) = (-2b - c, -4a + 2b + 2c, 8a - 8b - 6c),$$

and that $\mathscr{N}_{b(T)} = \mathscr{L}\{(1, -2, 4)\}$. In Example 1 we found that $(1, -2, 4)$ is a characteristic vector for T, which confirms that $\mathscr{N}_{b(T)}$ is T-invariant.

Our procedure for obtaining canonical forms under similarity is to obtain a direct sum decomposition of $\mathscr{V}$ into T-invariant subspaces and choose bases for these subspaces in a well-defined way. The preceding discussion shows that every nonzero vector U generates a T-invariant subspace $\mathscr{S}(U, T)$ and each polynomial $a(x)$ yields a T-invariant subspace $\mathscr{N}_{a(T)}$. Although these two methods of producing T-invariant subspaces are related, Example 3 shows that they are not identical. Selecting a sequence of T-invariant subspaces will use both methods; it will be based on a polynomial associated with T and basic properties of polynomials. These properties closely parallel properties of the integers and would be obtained in a general study of rings.

However, we cannot proceed without them. Therefore the needed properties are derived in the next section culminating in the unique factorization theorem for polynomials. It would be possible to go quickly through this discussion and then refer back to it as questions arise.

Problems

1. Suppose

$$\begin{pmatrix} 3 & 1 & 0 \\ 1 & 3 & 2 \\ 1 & 1 & 2 \end{pmatrix}$$

is the matrix of $T \in \text{Hom}(\mathscr{V}_3)$ with respect to $\{E_i\}$, for the given vector U, find $\mathscr{S}(U, T)$ and the polynomial $a(x)$ associated with U.

a. $U = (1, 0, 0)$. b. $(1, -1, 0)$. c. $(-2, 0, 1)$. d. $(0, 1, 0)$.

2. Let $T \in \text{Hom}(\mathscr{V}_3)$ have matrix

$$\begin{pmatrix} 2 & -1 & 2 \\ 1 & 3 & 2 \\ 1 & 2 & 3 \end{pmatrix}$$

with respect to $\{E_i\}$, find $\mathscr{S}(U, T)$ and the polynomial $a(x)$ associated with U when:

a. $U = (-1, 0, 1)$. b. $U = (2, 0, -1)$. c. $U = (1, 1, 1)$.
d. What is the characteristic polynomial of T?

3. Show that $\mathscr{S}(U, T)$ is the smallest T-invariant subspace of $\mathscr{V}$ that contains U. That is, if $\mathscr{T}$ is a T-invariant subspace and $U \in \mathscr{T}$, then $\mathscr{S}(U, T) \subset \mathscr{T}$.

4. Suppose

$$\begin{pmatrix} -2 & 1 & -1 & -1 & 0 \\ 1 & 1 & 1 & 1 & 0 \\ 2 & -3 & 1 & 1 & 3 \\ 0 & 1 & 0 & 0 & -3 \\ 1 & 0 & 1 & 2 & 2 \end{pmatrix}$$

is the matrix of $T \in \text{Hom}(\mathscr{V}_5)$ with respect to $\{E_i\}$.

a. Show that $\mathscr{V}_5 = \mathscr{S}(U_1, T) \oplus \mathscr{S}(U_2, T) \oplus \mathscr{S}(U_3, T)$ when

$$U_1 = (0, 0, -1, 1, 0),\ U_2 = (1, 0, -1, 0, 0),\ U_3 = (-1, 2, 3, -1, 0).$$

b. Find the diagonal block matrix of T with respect to the basis

$$B = \{U_1, T(U_1), U_2, U_3, T(U_3)\}.$$

5. Prove that for a diagonal block matrix,

$$\det \operatorname{diag}(A_1, A_2, \ldots, A_k) = (\det A_1)(\det A_2) \cdots (\det A_k).$$

6. Suppose $T \in \text{Hom}(\mathscr{V}_3)$ has

$$\begin{pmatrix} 2 & 1 & -2 \\ 2 & -2 & 1 \\ 1 & 2 & 2 \end{pmatrix}$$

as its matrix with respect to $\{E_i\}$. Find the null space $\mathscr{N}_{p(T)}$ when

a. $p(x) = x + 3$. b. $p(x) = x^2 - 5x + 9$.
c. $p(x) = x^2 - 4$. d. $p(x) = (x + 3)(x^2 - 5x + 9)$.

7. Suppose $T \in \text{Hom}(\mathscr{V}_3)$ has

$$\begin{pmatrix} 1 & 0 & 1 \\ 3 & 2 & 4 \\ 1 & 0 & 1 \end{pmatrix}$$

as its matrix with respect to $\{E_i\}$. Find $\mathscr{N}_{p(T)}$ when
a. $p(x) = x - 2$. b. $p(x) = (x - 2)^2$. c. $p(x) = x$.
d. $p(x) = x(x - 2)$. e. $p(x) = x(x - 2)^2$.

8. Suppose $T \in \text{Hom}(\mathscr{V})$ and $p(x)$ and $q(x)$ are polynomials such that $\mathscr{V} = \mathscr{N}_{p(T)} + \mathscr{N}_{q(T)}$. What can be concluded about the map $p(T) \circ q(T)$?

9. Suppose $T \in \text{Hom}(\mathscr{V}_3)$ has

$$\begin{pmatrix} 2 & 1 & 0 \\ -2 & 1 & 1 \\ 0 & 1 & 2 \end{pmatrix}$$

as its matrix with respect to $\{E_i\}$. Find $\mathscr{N}_{p(T)}$ when
a. $p(x) = x - 2$. b. $p(x) = (x - 2)^2$. c. $p(x) = x^2 - 3x + 3$.
d. $p(x) = x - 1$. e. Find $p(x)$ such that $p(T) = \mathbf{0}$.

§2. Polynomials

We have encountered polynomials in two situations, first as vectors in the vector space $R[t]$, for which t is an indeterminant symbol never allowed to be a member of the field R, and then in polynomial equations used to find characteristic values, for which the symbol λ was regarded an unknown variable in the field. A general study of polynomials must combine both roles for the symbol used. Therefore suppose that the symbol x is an indeterminant, which may be assigned values from the field F, and denote by $F[x]$ the set of all polynomials in x with coefficients from F. That is, $p(x) \in F[x]$ if $p(x) = a_n x^n + \cdots + a_1 x + a_0$ with $a_n, \ldots, a_1, a_0 \in F$.

Definition Two polynomials from $F[x]$, $p(x) = a_n x^n + \cdots + a_1 x + a_0$ and $q(x) = b_m x^m + \cdots + b_1 x + b_0$ are *equal* if $n = m$ and $a_i = b_i$ for all i, $1 \leq i \leq n$. The *zero polynomial* $0(x)$ of $F[x]$ satisfies the condition that if $0(x) = a_n x^n + \cdots + a_1 x + a_0$, then $a_n = \cdots = a_1 = a_0 = 0$.

The two different roles for x are nicely illustrated by the two statements $p(x) = 0(x)$ and $p(x) = 0$. The first means that $p(x)$ is the zero polynomial, while the second will be regarded as an equation for x as an element in F, that is, $p(x)$ is the zero element of F. The distinciton is very clear over a finite field where a polynomial may vanish for every element and yet not be the zero polynomial, see problem 1 at the end of this section.

Two operations are defined on the polynomials of $F[x]$. Addition in $F[x]$ is defined as in the vector space $F[t]$, and multiplication is defined as follows: If $p(x) = a_n x^n + \cdots + a_1 x + a_0$ and $q(x) = b_m x^m + \cdots b_1 x + b_0$, then

$$p(x)q(x) = c_{n+m}x^{n+m} + \cdots + c_1 x + c_0$$

with

$$c_k = \sum_{i+j=k} a_i b_j, \quad 1 \le k \le n + m.$$

For example, if $p(x) = x^3 - x^2 + 3x - 2$ and $q(x) = 2x^2 + 4x + 3$, then

$$\begin{aligned} p(x)q(x) &= 1\cdot 2x^5 + (1\cdot 4 - 1\cdot 2)x^4 + (1\cdot 3 - 1\cdot 4 + 3\cdot 2)x^3 \\ &\quad + (-1\cdot 3 + 3\cdot 4 - 2\cdot 3)x^2 + (3\cdot 3 - 2\cdot 4)x - 2\cdot 3 \\ &= 2x^5 + 2x^4 + 5x^3 + 3x^2 + x - 6. \end{aligned}$$

The algebraic system composed of $F[x]$ together with the operations of addition and multiplication of polynomials will also be denoted by $F[x]$. Thus $F[x]$ is not a vector space but is an algebraic system with the same type of operations as the field F. In fact, $F[x]$ satisfies all but one of the defining properties of a field—in general polynomials do not have multiplicative inverses in $F[x]$.

Theorem 8.4 $F[x]$ is a commutative ring with identity in which the cancellation law for multiplication holds.

Proof The additive properties for a ring are the same as the additive properties for the vector space $F[t]$ and therefore carry over to $F[x]$. This leaves six properties to be established. Let $p(x)$, $q(x)$, and $r(x)$ be arbitrary elements from $F[x]$, then the remaining properties for a ring are:

1. $p(x)q(x) \in F[x]$
2. $p(x)(q(x)r(x)) = (p(x)q(x))r(x)$
3. $p(x)(q(x) + r(x)) = p(x)q(x) + p(x)r(x)$ and
$(p(x) + q(x))r(x) = p(x)r(x) + q(x)r(x)$.

$F[x]$ is commutative if

4. $p(x)q(x) = q(x)p(x)$.

$F[x]$ has an identity if

5. there exists $1(x) \in F[x]$ such that $p(x)1(x) = 1(x)p(x) = p(x)$. The cancellation law for multiplication holds in $F[x]$ if

6. $p(x)q(x) = p(x)r(x)$ and $p(x) \neq 0(x)$, then $q(x) = r(x)$ and if $q(x)p(x) = r(x)p(x)$ and $p(x) \neq 0(x)$, then $q(x) = r(x)$.

Parts 1 and 4 are obvious from the definition of multiplication and the corresponding properties of F. Thus only half of parts 3 and 6 need be established, and this is left to the reader. For 5, take $1(x)$ to be the multiplicative identity of F. Parts 2 and 3 follow with a little work from the corresponding properties in F. This leaves 6, which may be proved as follows:

Suppose $p(x)q(x) = p(x)r(x)$, then $p(x)(q(x) - r(x)) = 0(x)$ and it is sufficient to show that if $p(x)s(x) = 0(x)$ and $p(x) \neq 0(x)$, then $s(x) = 0(x)$. Therefore, suppose $p(x)s(x) = 0(x)$, $p(x) = a_n x^n + \cdots + a_1 x + a_0$ with $a_n \neq 0$, so $p(x) \neq 0(x)$, and $s(x) = b_m x^m + \cdots + b_1 x + b_0$. Assume that $s(x) \neq 0(x)$ so that $b_m \neq 0$. Then $p(x)s(x) = a_n b_m x^{n+m}$ + terms in lower powers of x. $p(x)s(x) = 0(x)$ implies that $a_n b_m = 0$, but $a_n \neq 0$ and F is a field, so $b_m = 0$, contradicting the assumption that $s(x) \neq 0(x)$. Therefore $s(x) = 0(x)$, and the cancellation law holds in $F[x]$.

Definition An *integral domain* is a commutative ring with identity in which the cancellation law for multiplication holds.

Hence a polynomial ring $F[x]$ is an integral domain. The integers together with the arithmetic operations of addition and multiplication also comprise an integral domain. Actually the term integral domain is derived from the fact that such an algebraic system is an abstraction of the integers. Our main objective in this section is to prove that there is a unique way to factor any polynomial just as there is a unique factorization of any integer into primes. Since every element of F is a polynomial in $F[x]$, and F is a field, there are many factorizations that are not of interest. For example, $2x + 3 = 4(x/2 + 3/4) = \frac{1}{5}(10x + 15)$.

Definition Let $p(x) = a_n x^n + \cdots + a_1 x + a_0 \in F[x]$. If $p(x) \neq 0(x)$ and $a_n \neq 0$, then a_n is the *leading coefficient* of $p(x)$ and $p(x)$ has *degree* n denoted $\deg p = n$. If the leading coefficient of $p(x)$ is 1, then $p(x)$ is *monic*. For $a(x), b(x) \in F[x]$, $b(x)$ is a *factor* of $a(x)$ or $b(x)$ *divides* $a(x)$, denoted $b(x)|a(x)$, if there exists $q(x) \in F[x]$ such that $a(x) = b(x)q(x)$.

There is no number that can reasonably be taken for the degree of the zero polynomial. However it is useful if the degree of $0(x)$ is taken to be minus infinity.

The nonzero elements of F are polynomials of degree 0 in $F[x]$. F is a field so that any nonzero element of F is a factor of any polynomial in $F[x]$. However, a polynomial such as $2x + 3$ above can be factored into the product of an element from F and a monic polynomial in only one way aside from order; $2x + 3 = 2(x + 3/2)$. Thus, given a monic polynomial, we will only look for monic factors.

In deriving our main result on factorization we will need two properties of the integers which are equivalent to induction, see problem 13 at the end of this section.

Theorem 8.5 The set of positive integers is *well ordered*. That is, every nonempty subset of the positive integers contains a smallest element.

Proof The proof is by induction. Suppose A is a nonempty subset of positive integers. Assume A has no smallest element and let $S = \{n|n$ is a positive integer and $n < a$ for all $a \in A\}$. Then $1 \in S$ for if not then $1 \in A$ and A has a smallest element. If $k \in S$ and $k + 1 \notin S$, then $k + 1 \in A$ and again A has a smallest element. Therefore $k \in S$ implies $k + 1 \in S$ and S is the set of all positive integers by induction. This contradicts the fact that A is nonempty, therefore A has a smallest element as claimed.

Many who find induction difficult to believe, regard this principle of well ordering as "obvious." For them it is a paradox that the two principles are equivalent.

There are two principles of induction for the positive integers. The induction used in the above proof may be called the "first principle of induction."

Theorem 8.6 (Second Principle of Induction) Suppose B is a set of positive integers such that

1. $1 \in B$ and
2. If m is a positive integer such that $k \in B$ for all $k < m$, then $m \in B$.

Then B is the set of all positive integers.

Proof Let $A = \{n|n$ is a positive integer and $n \notin B\}$. If A is nonempty, then by well ordering A has a smallest element, call it m. $m \notin B$ so $m \neq 1$ and if k is a positive integer less than m, then $k \in B$. Therefore, by 2, $m \in B$ and the assumption that A is nonempty leads to a contradiction. Thus B is the set of all positive integers.

As with the first principle of induction, both the well ordering property and the second principle of induction can be applied to any set of integers obtained by taking all integers greater than some number.

Theorem 8.7 (Division Algorithm for Polynomials) Let $a(x)$, $b(x) \in F[x]$ with $b(x) \neq 0(x)$. Then there exist polynomials $q(x)$, called the *quotient*, and $r(x)$, called the *remainder*, such that

$$a(x) = b(x)q(x) + r(x) \qquad \text{with} \quad \deg r < \deg b.$$

Moreover the polynomials $q(x)$ and $r(x)$ are uniquely determined by $a(x)$ and $b(x)$.

Proof If $\deg a < \deg b$, then $q(x) = 0(x)$ and $r(x) = a(x)$ satisfy the existence part of the theorem. Therefore, suppose $\deg a \geq \deg b$ and apply the second principle of induction to the set of non-negative integers $B = \{n | \text{if } \deg a = n, \text{ then } q(x) \text{ and } r(x) \text{ exist}\}$.

If $\deg a = 0$, there is nothing to prove for $a(x)$ and $b(x)$ are in F and we can take $q(x) = a(x)/b(x)$, $r(x) = 0(x)$. So $0 \in B$.

Now assume $k \in B$ if $k < m$, that is, that $q(x)$ and $r(x)$ exist for $c(x)$ if $\deg c = k$. Let $a(x) = a_m x^m + \cdots + a_1 x + a_0$ and $b(x) = b_n x^n + \cdots + b_1 x + b_0$ with $m > n$. Consider the polynomial $a'(x) = a(x) - (a_m/b_n)x^{m-n}b(x)$. Deg $a' < m$, so by the induction assumption there exist polynomials $q'(x)$ and $r(x)$ such that $a'(x) = b(x)q'(x) + r(x)$ with $\deg r < \deg b$. Therefore, if $q(x) = q'(x) + (a_m/b_n)x^{m-n}$, then $a(x) = b(x)q(x) + r(x)$ with $\deg r < \deg b$. Therefore $k < m$ and $k \in B$ implies $m \in B$, and by the second principle of induction, B is the set of all non-negative integers. That is, $q(x)$ and $r(x)$ exist for all polynomials $a(x)$ such that $\deg a \geq \deg b$.

The proof of uniqueness is left to problem 5 at the end of this section. In practice the quotient and remainder are easily found by long division.

Example 1 Find $q(x)$ and $r(x)$ if $a(x) = 4x^3 + x^2 - x + 2$ and $b(x) = x^2 + x - 2$.

$$\begin{array}{r|l}
 & \phantom{4x^3 + x^2 - {}} 4x - 3 \\
\cline{2-2}
x^2 + x - 2 & 4x^3 + x^2 - x + 2 \\
 & 4x^3 + 4x^2 - 8x \\
\cline{2-2}
 & \quad -3x^2 + 7x + 2 \\
 & \quad -3x^2 - 3x + 6 \\
\cline{2-2}
 & \qquad\qquad 10x - 4
\end{array}$$

Therefore $q(x) = 4x - 3$ and $r(x) = 10x - 4$.

Corollary 8.8 (Remainder Theorem) Let $c \in F$ and $p(x) \in F[x]$. Then the remainder of the divison $p(x)/(x - c)$ is $p(c)$.

Proof Since $x - c \in F[x]$, there exist polynomials $q(x)$ and $r(x)$ such that $p(x) = (x - c)q(x) + r(x)$ with $\deg r(x) < \deg (x - c)$, but $\deg (x - c) = 1$ so $r(x)$ is an element of F. Since $r(x)$ is constant, its value can be obtained by setting x equal to c; $p(c) = (c - c)q(c) + r(c)$. Thus the remainder $r(x) = p(c)$.

Corollary 8.9 (Factor Theorem) Let $p(x) \in F[x]$ and $c \in F$, $x - c$ is a factor of $p(x)$ if and only if $p(c) = 0$.

Definition Given $p(x) \in F[x]$ and $c \in F$. c is a *root* of $p(x)$ if $x - c$ is a factor of $p(x)$.

A third corollary is easily proved using the first principle of induction.

Corollary 8.10 If $p(x) \in F[x]$ and $\deg p(x) = n > 0$, then $p(x)$ has at most n roots in F.

Of course a polynomial need have no roots in a particular field. For example, $x^2 + 1$ has no roots in R. However, every polynomial in $R[x]$ has at least one root in C, the complex numbers. This follows from the Fundamental Theorem of Algebra, since $F[x] \subset C[x]$. This theorem is quite difficult to prove, and will only be stated here.

Theorem 8.11 (Fundamental Theorem of Algebra) If $p(x) \in C[x]$, then $p(x)$ has at least one root in C.

Definition A field F is *algebraically closed* if every polynomial of positive degree in $F[x]$ has a root in F.

The essential part of this definition is the requirement that the root be in F. Thus the complex numbers are algebraically closed while the real numbers are not. We will find that the simplest canonical form under similarity is obtained when the matrix is considered to be over an algebraically closed field.

If the polynomial $p(x)$ has c as a root, then $x - c$ is a factor and there exists a polynomial $q(x)$ such that $p(x) = (x - c)q(x)$. Since $\deg q =$

$\deg p - 1$, an induction argument can be used to extend Theorem 8.11 as follows.

Corollary 8.12 If $p(x) \in C[x]$ and $\deg p = n > 0$, then $p(x)$ has exactly n roots. (The n roots of $p(x)$ need not all be distinct.)

Thus an nth degree polynomial $p(x) \in C[x]$ can be factored into the product of n linear factors of the form $x - c_i$, $1 \leq i \leq n$, with $p(c_i) = 0$. A polynomial $p(x)$ in $R[x]$ is also in $C[x]$ and it can be shown that any complex roots of $p(x)$ occur in conjugate pairs. Therefore $p(x)$ can be expressed as a product of linear and quadratic factors in $R[x]$.

Before obtaining a factorization for every polynomial in $F[x]$, we must investigate greatest common divisors of polynomials. This is done with the aid of subsets called ideals.

Definition An *ideal* in $F[x]$ is a nonempty subset S of $F[x]$ such that

1. if $p(x), q(x) \in S$, then $p(x) - q(x) \in S$, and
2. if $r(x) \in F[x]$ and $p(x) \in S$, then $r(x)p(x) \in S$.

Example 2 Show that $S = \{p(x) \in R[x] | 4 \text{ is a root of } p(x)\}$ is an ideal in $R[x]$.

If $p(x)$ and $q(x) \in S$, then $(x - 4)|p(x)$ and $(x - 4)|q(x)$ implies that $(x - 4)|(p(x) - q(x))$, so $p(x) - q(x) \in S$, and $(x - 4)|r(x)p(x)$ for any $r(x) \in R[x]$, so $r(x)p(x) \in S$. Therefore both conditions are satisfied and S is an ideal.

Any set of polynomials can be used to generate an ideal in $F[x]$. For example, given $a(x)$ and $b(x)$ in $F[x]$, define S by

$$S = \{u(x)a(x) + v(x)b(x) | u(x), v(x) \in F[x]\}.$$

It is not difficult to check that S is an ideal in $F[x]$ containing both $a(x)$ and $b(x)$. Notice that if c is a root of both $a(x)$ and $b(x)$, then c is a root of every polynomial in S. In fact this ideal is determined by the roots or factors common to $a(x)$ and $b(x)$. This is shown in the next two theorems.

Theorem 8.13 $F[x]$ is a *principal ideal domain*. That is, if S is an ideal in $F[x]$, then there exists a polynomial $d(x)$ such that $S = \{q(x)d(x) | q(x) \in F[x]\}$.

Proof If $S = \{0(x)\}$, there is nothing to prove, therefore suppose $S \neq \{0(x)\}$. Then the well-ordering principle implies that S contains a polynomial of least finite degree. If $d(x)$ denotes such a polynomial, then it is only necessary to show that $d(x)$ divides every element of S.

If $p(x) \in S$, then there exist polynomials $q(x)$ and $r(x)$ such that $p(x) = d(x)q(x) + r(x)$ with $\deg r < \deg d$. But $r(x)$ is in the ideal since $r(x) = p(x) - q(x)d(x)$ and $p(x), d(x) \in S$. Now since $d(x)$ has the minimal finite degree of any polynomial in S and $\deg r < \deg d$, $r(x) = 0(x)$ and $d(x)|p(x)$ as desired.

The ideal in Example 2 is principal with $d(x) = x - 4$ as a single generator. For an ideal generated by two polynomials, a single generator guaranteed by Theorem 8.13 is also a greatest common divisor.

Theorem 8.14 Any two nonzero polynomials $a(x), b(x) \in F[x]$ have a *greatest common divisor* (g.c.d.) $d(x)$ satisfying

1. $d(x)|a(x)$ and $d(x)|b(x)$
2. if $p(x)|a(x)$ and $p(x)|b(x)$, then $p(x)|d(x)$.

Moreover there exist polynomials $s(x), t(x) \in F[x]$ such that

$$d(x) = s(x)a(x) + t(x)b(x).$$

Proof The set

$$S = \{u(x)a(x) + v(x)b(x)|u(x), v(x) \in F[x]\}$$

is an ideal in $F[x]$ containing $a(x)$ and $b(x)$. Therefore, by Theorem 8.13, there exists a nonzero polynomial $d(x) \in S$ such that $d(x)$ divides every element of S. In particular, $d(x)|a(x)$ and $d(x)|b(x)$.

Since $d(x) \in S$, there exist polynomials $s(x), t(x) \in F[x]$ such that $d(x) = s(x)a(x) + t(x)b(x)$, and it only remains to show that 2 is satisfied. But if $p(x)|a(x)$ and $p(x)|b(x)$, then $p(x)$ divides $s(x)a(x) + t(x)b(x) = d(x)$.

Two polynomials will have many g.c.d.'s but they have a unique monic g.c.d. In practice a g.c.d. is most easily found if the polynomials are given in a factored form. Thus the monic g.c.d. of $a(x) = 4(x - 2)^3(x^2 + 1)$ and $b(x) = 2(x - 2)^2(x^2 + 1)^5$ is $(x - 2)^2(x^2 + 1)$.

Definition Two nonzero polynomials in $F[x]$ are *relatively prime* if their only common factors in $F[x]$ are elements of F.

Thus $x^2 - 4$ and $x^2 - 9$ are relatively prime even though each has polynomial factors of positive degree.

Theorem 8.15 If $a(x)$ and $b(x)$ are relatively prime in $F[x]$, then there exist $s(x), t(x) \in F[x]$ such that $1 = s(x)a(x) + t(x)b(x)$.

Proof Since $a(x)$ and $b(x)$ are relatively prime, 1 is a g.c.d.

We are finally ready to define the analogue in $F[x]$ to a prime number and to prove the analogue to the Fundamental Theorem of Arithmatic. Recall that this theorem states that every integer greater than 1 may be expressed as a product of primes in a unique way, up to order.

Definition A polynomial $p(x) \in F[x]$ of positive degree is *irreducible over* F if $p(x)$ is relatively prime to all nonzero polynomials in $F[x]$ of degree less than the degree of $p(x)$.

That is, if $\deg p > 0$, then $p(x)$ is irreducible over F if its only factors from $F[x]$ are constant multiples of $p(x)$ and constants from F.

Examples

1. All polynomials of degree 1 in $F[x]$ are irreducible over F. (Why?)
2. The only irreducible polynomials in $C[x]$ are polynomials of degree 1. (Why?)
3. $ax^2 + bx + c \in R[x]$ is irreducible over R if the discriminant $b^2 - 4ac < 0$.
4. $x^2 + 1$ is irreducible over R but not over C.
5. $x^2 - 2$ is irreducible over the field of rationals but not over R or C.

Theorem 8.16 Suppose $a(x), b(x), p(x) \in F[x]$ with $p(x)$ irreducible over F. If $p(x) | a(x)b(x)$, then either $p(x) | a(x)$ or $p(x) | b(x)$.

Proof If $p(x) | a(x)$ we are done. Therefore, suppose $p(x) \nmid a(x)$. Then $p(x)$ and $a(x)$ are relatively prime so there exist $s(x), t(x) \in F[x]$ such that $1 = s(x)p(x) + t(x)a(x)$. Multiply this equation by $b(x)$ to obtain $b(x) = (b(x)s(x))p(x) + t(x)(a(x)b(x))$. Now $p(x)$ divides the right side of this equation so $p(x) | b(x)$, completing the proof.

Theorem 8.17 (Unique Factorization Theorem) Every polynomial of positive degree in $F[x]$ can be expressed as an element of F times a product of irreducible monic polynomials. This expression is unique except for the order of the factors.

Proof Every polynomial in $F[x]$ can be expressed in a unique way as $ca(x)$ where $c \in F$ and $a(x)$ is monic. Therefore it is sufficient to consider $a(x)$ a monic polynomial of positive degree. The existence proof uses the second principle of induction on the degree of $a(x)$. If $\deg a = 1$, then $a(x)$ is irreducible, therefore an expression exists when $a(x)$ has degree 1. So assume every monic polynomial of degree less than m has an expression in terms of irreducible monic polynomials and suppose $\deg a = m$, $m > 1$. If $a(x)$ is irreducible over F, then it is expressed in the proper form. If $a(x)$ is not irreducible over F, then $a(x) = p(x)q(x)$ where $p(x)$ and $q(x)$ are monic polynomials of positive degree. Since $\deg a = \deg p + \deg q$ (see problem 2 at the end of this section), $p(x)$ and $q(x)$ have degrees less than m. So by the induction assumption $p(x)$ and $q(x)$ may be expressed as products of irreducible monic polynomials. This gives such an expression for $a(x)$, which has degree m. Therefore, by the second principle of induction, every monic polynomial of positive degree may be expressed as a product of irreducible monic polynomials.

For uniqueness, suppose $a(x) = p_1(x)p_2(x)\cdots p_h(x)$ and $a(x) = q_1(x)q_2(x)\cdots q_k(x)$ with $p_1(x), \ldots, p_h(x)$, $q_1(x), \ldots, q_k(x)$ irreducible monic polynomials. $p_1(x)|a(x)$ so $p_1(x)|q_1(x)\cdots q_k(x)$. Since $p_1(x)$ is irreducible, an extension of Theorem 8.16 implies that $p_1(x)$ divides one of the q's. Say $p_1(x)|q_1(x)$, then since both are irreducible and monic, $p_1(x) = q_1(x)$, and by the cancellation law, $p_2(x)\cdots p_h(x) = q_2(x)\cdots q_k(x)$. Continuing this process with an induction argument yields $h = k$ and $p_1(x), \ldots, p_k(x)$ equal to $q_1(x), \ldots, q_k(x)$ in some order.

Corollary 8.18 If $a(x) \in F[x]$ has positive degree, then there exist distinct irreducible (over F) monic polynomials $p_1(x), \ldots, p_k(x)$, positive integers $e_1, \ldots, e_k$ and an element $c \in F$ such that $a(x) = cp_1^{e_1}(x)\,p_2^{e_2}(x)\cdots p_k^{e_k}(x)$. This expression is unique up to the order of the polynomials $p_1, \ldots, p_k$.

Definition If $a(x) = cp_1^{e_1}(x)\cdots p_k^{e_k}(x)$ with $c \in F$, $p_1(x), \ldots, p_k(x)$ distinct irreducible monic polynomials, and $e_1, \ldots, e_k$ positive integers, then e_i is the *multiplicity* of $p_i(x)$, $1 \le i \le k$.

Finding roots for an arbitrary polynomial from $R[x]$ or $C[x]$ is a difficult task. The quadratic fomula provides complex roots for any second degree polynomial, and there are rather complicated formulas for third and fourth degree polynomials. But there are no formulas for general polynomials of degree larger than four. Therefore, unless roots can be found by good guesses, approximation techniques are necessary. One should not expect an arbitrary polynomial to have rational roots, let alone integral roots. In fact, for a polynomial with rational coefficients there are few rational numbers which could be roots, see problem 10.

Problems

1. Suppose $p(x) \in F[x]$ satisfies $p(c) = 0$ for all $c \in F$. Show that if F is a finite field, then $p(x)$ need not be $0(x)$. What if F were infinite?

2. Show that if $a(x), b(x) \in F[x]$, then $\deg ab = \deg a + \deg b$. (Take $r - \infty = -\infty$ for any $r \in R$.)

3. Let F be a field. a. Show that the cancellation law for multiplication holds in F. b. Suppose $a, b \in F$ and $ab = 0$, show that either $a = 0$ or $b = 0$.

4. Find the quotient and remainder when $a(x)$ is divided by $b(x)$.
a. $a(x) = 3x^4 + 5x^3 - 3x^2 + 6x + 1$, $b(x) = x^2 + 2x - 1$.
b. $a(x) = x^2 + 2x + 1$, $b(x) = x^3 - 3x$.
c. $a(x) = x^5 + x^3 - 2x^2 + x - 2$, $b(x) = x^2 + 1$.
d. $a(x) = x^4 - 2x^3 - 3x^2 + 10x - 8$, $b(x) = x - 2$.

5. Show that the quotient and remainder given by the division algorithm for polynomials are unique.

6. Prove that any two nonzero polynomials $a(x), b(x) \in F[x]$ have a *least commom multiple* (l.c.m.) $m(x)$ satisfying
1. $a(x)|m(x)$ and $b(x)|m(x)$
2. if $a(x)|p(x)$ and $b(x)|p(x)$, then $m(x)|p(x)$.

7. Find the monic g.c.d. and the monic l.c.m. of $a(x)$ and $b(x)$.
a. $a(x) = x(x-1)^4(x+2)^2$, $b(x) = (x-1)^3(x+2)^4(x+5)$.
b. $a(x) = (x^2 - 16)(5x^3 - 5)$, $b(x) = 5(x-4)^3(x-1)$.

8. Prove Euclid's algorithm: If $a(x), b(x) \in F[x]$ with $\deg a \geq \deg b > 0$, and the polynomials $q_i(x), r_i(x) \in F[x]$ satisfy

$$\begin{aligned}
a(x) &= b(x)q_1(x) + r_1(x) && 0 < \deg r_1 < \deg b \\
b(x) &= r_1(x)q_2(x) + r_2(x) && 0 < \deg r_2 < \deg r_1 \\
r_1(x) &= r_2(x)q_3(x) + r_3(x) && 0 < \deg r_3 < \deg r_2 \\
&\vdots \\
r_{k-2}(x) &= r_{k-1}(x)q_k(x) + r_k(x) && 0 \leq \deg r_k < \deg r_{k-1} \\
r_{k-1}(x) &= r_k(x)q_{k+1}(x) + 0(x),
\end{aligned}$$

then $r_k(x)$ is a g.c.d. of $a(x)$ and $b(x)$.

9. Use Euclid's algorithm to find a g.c.d. of

a. $a(x) = 2x^5 - 2x^4 + x^3 - 4x^2 - x - 2$, $b(x) = x^4 - x^3 - x - 1$.

b. $a(x) = 2x^4 + 5x^3 - 6x^2 - 6x + 22$, $b(x) = x^4 + 2x^3 - 3x^2 - x + 8$.

c. $a(x) = 6x^4 + 7x^3 + 2x^2 + 8x + 4$, $b(x) = 6x^3 - x^2 + 4x + 3$.

10. Suppose $p(x) = a_n x^n + \cdots + a_1 x + a_0 \in R[x]$ has integral coefficients. Show that if the rational number c/d is a root of $p(x)$, then $c|a_0$ and $d|a_n$. How might this be extended to polynomials with rational coefficients?

11. Using problem 10, list all possible rational roots of

a. $2x^5 + 3x^4 - 6x + 6$.

b. $x^n + 1$.

c. $x^4 - 3x^3 + x - 8$.

d. $9x^3 + x - 1$.

12. Is the field of integers modulo 2, that is the field containing only 0 and 1, algebraically closed?

13. Prove that the first principle of induction, the second principle of induction, and the well-ordering principle are equivalent properties of the positive integers.

14. Let $\mathscr{V}$ be a vector space over F, $U \in \mathscr{V}$, $T \in \text{Hom}(\mathscr{V})$ and show that $\{p(T)U | p(x) \in F[x]\}$ is the T-cyclic subspace $\mathscr{S}(U, T)$.

15. Let $T \in \text{Hom}(\mathscr{V})$ and $U \in \mathscr{V}$.

a. Show that $S = \{q(x) \in F[x] | q(T)U = \mathbf{0}\}$ is an ideal in $F[x]$.

b. Find the polynomial that generates S as a principal ideal if $S \neq \{0(x)\}$.

16. You should be able to prove any of the following for Z, the integral domain of integers.

a. The division algorithm for integers.

b. Z is a principal ideal domain.

c. If $a, b \in Z$ are relatively prime, then there exist $s, t \in Z$ such that $1 = sa + tb$.

d. The Fundamental Theorem of Arithmetic as stated on page 356.

17. Extend the definition of $F[x]$ to $F[x, y]$ and $F[x, y, z]$, the rings of polynomials in two and three variables, respectively.

a. Show that $F[x, y]$ is not a principal ideal domain. For example, consider the ideal generated by xy and $x + y$.

b. Suppose S is the ideal in $R[x, y, z]$ generated by $a(x, y, z) = x^2 + y^2 - 4$ and $b(x, y, z) = x^2 + y^2 - z$, and let G be the graph of the intersection of $a(x, y, z) = 0$ and $b(x, y, z) = 0$. Show that S can be taken as an algebraic description of G in that every element of S is zero on G.

c. Show that the polynomials $z - 4$, $x^2 + y^2 + (z - 4)^2 - 4$, and $4x^2 + 4y^2 - z^2$ are in the ideal of part b. Note that the zeros of these polynomials give the plane, a sphere, and a cone through G.

d. Give a geometric argument to show that the ideal S in part b is not principal.

§3. Minimum Polynomials

Now return to the polynomial associated with the T-cyclic subspace $\mathscr{S}(U, T)$. Recall that if $\mathscr{S}(U, T) = \mathscr{L}\{U, T(U), \ldots, T^{k-1}(U)\}$ and $T^k(U) + a_{k-1}T^{k-1}(U) + \cdots + a_1T(U) + a_0U = \mathbf{0}$, then this equation may be written as $a(T)U = \mathbf{0}$ where $a(x) = x^k + a_{k-1}x^{k-1} + \cdots + a_1x + a_0$. Notice that $a(x)$ is monic and since $\mathscr{S}(T, U)$ has dimension k, there does not exist a nonzero polynomial $b(x)$ of lower degree for which $b(T)U = \mathbf{0}$. Therefore we will call $a(x)$ the *T-minimum polynomial of U.*

Theorem 8.19 If $a(x)$ is the T-minimum polynomial of U and $b(T)U = \mathbf{0}$, then $a(x)|b(x)$. The monic polynomial $a(x)$ is uniquely determined by U.

Proof By the division algorithm there exist polynomials $q(x)$ and $r(x)$ such that $b(x) = a(x)q(x) + r(x)$ with deg $r <$ deg a. But $r(T)U = \mathbf{0}$ for

$$r(T)U = (b(T) - q(T) \circ a(T))U = b(T)U - q(T)(a(T)U).$$

If $r(x) \neq 0(x)$, then $r(x)$ is a nonzero polynomial of degree less than deg a and $r(T)U = \mathbf{0}$ which contradicts the remarks preceding this theorem. Therefore $a(x)$ divides $b(x)$. This also gives the uniqueness of $a(x)$, for if deg b = deg a, then they differ by a constant multiple. Since $a(x)$ is monic, it is uniquely determined by U.

The T-minimum polynomial of U determines a map that sends the entire subspace $\mathscr{S}(U, T)$ to $\mathbf{0}$. There is a corresponding polynomial for the vector space that plays the central role in obtaining a canonical form under similarity. We see that there is a polynomial in T that maps all of $\mathscr{V}$ to $\mathbf{0}$ as follows: If dim $\mathscr{V} = n$, then dim Hom$(\mathscr{V}) = n^2$ and the set $\{I, T, T^2, \ldots, T^{n^2}\}$ is linearly dependent. That is, there is a linear combination of these maps (a polynomial in T) that is the zero map on $\mathscr{V}$.

Theorem 8.20 Given a nonzero map $T \in$ Hom$(\mathscr{V})$, there exists a unique monic polynomial $m(x)$ such that $m(T) = \mathbf{0}_{\text{Hom}(\mathscr{V})}$ and if $p(T) = \mathbf{0}_{\text{Hom}(\mathscr{V})}$ for some polynomial $p(x)$ then $m(x)|p(x)$.

Proof By the above remarks there exists a polynomial $p(x)$ such that $\deg p \leq n^2$ and $p(T) = \mathbf{0}$. Therefore the set

$$B = \{k | \text{there exists } p(x) \in F[x], \deg p = k > 0 \text{ and } p(T) = \mathbf{0}\}$$

is nonempty, and since the positive integers are well ordered, B has a least element. Let $m(x)$ be a monic polynomial of least positive degree such that $m(T) = \mathbf{0}$. This proof is completed in the same manner as the corresponding properties were obtained for the T-minimum polynomial of U.

Is it clear that the symbol $\mathbf{0}$ in this proof denotes quite a different vector from $\mathbf{0}$ in the preceding proof?

Definition The *minimum polynomial* $m(x)$ of a nonzero map $T \in \text{Hom}(\mathscr{V})$ is the unique monic polynomial of lowest positive degree such that $m(T) = \mathbf{0}_{\text{Hom}(\mathscr{V})}$.

If the minimum polynomial of T is the key to obtaining a canonical from under similarity, then a method of computation is required. Fortunately it is not necessary to search through all $n^2 + 1$ maps $I, T, \ldots, T^{n^2}$ for a dependency. We will find that the degree of the minimum polynomial of T cannot exceed the dimension of $\mathscr{V}$, Corollary 8.28, page 378. However, this does leave $n + 1$ maps or matrices to consider.

Example 1 Find the minimum polynomial of $T \in \text{Hom}(\mathscr{V}_3)$ if the matrix of T with respect to $\{E_i\}$ is

$$A = \begin{pmatrix} 1 & 3 & 2 \\ -1 & 1 & -2 \\ 0 & -1 & 1 \end{pmatrix}.$$

Multiplication gives

$$A^2 = \begin{pmatrix} -2 & 4 & -2 \\ -2 & 0 & -6 \\ 1 & -2 & 3 \end{pmatrix} \quad \text{and} \quad A^3 = \begin{pmatrix} -6 & 0 & -14 \\ -2 & 0 & -10 \\ 3 & -2 & 9 \end{pmatrix}.$$

It is clear that the sets $\{I_3, A\}$ and $\{I_3, A, A^2\}$ are linearly independent, and a little computation shows that $\{I_3, A, A^2, A^3\}$ is linearly dependent with $A^3 = 3A^2 - 4A + 4I_3$. Therefore $\{I, T, T^2\}$ is linearly independent and $T^3 - 3T^2 + 4T - 4I = \mathbf{0}$, so the minimum polynomial of T is $x^3 - 3x^2 + 4x - 4$.

Another approach to finding the minimum polynomial of a map is given in the next two theorems.

Theorem 8.21 Suppose $T \in \text{Hom}(\mathscr{V})$ and $U \in \mathscr{V}$. If $m(x)$ is the minimum polynomial of T and $a(x)$ is the T-minimum polynomial of U, then $a(x)|m(x)$.

Proof By the division algorithm there exist polynomials $q(x)$ and $r(x)$ such that $m(x) = a(x)q(x) + r(x)$ with $\deg r < \deg a$. Since $m(T)U = a(T)U = \mathbf{0}$ and $r(T) = m(T) - q(T) \circ a(T)$, we have $r(T)U = \mathbf{0}$. But $a(x)$ is the T-minimum polynomial of U and $\deg r < \deg a$. Therefore $r(x) = 0(x)$ and $a(x)|m(x)$.

Theorem 8.22 Let $\{U_1, \ldots, U_n\}$ be a basis for $\mathscr{V}$ and $T \in \text{Hom}(\mathscr{V})$. If $a_1(x), \ldots, a_n(x)$ are the T-minimum polynomials of $U_1, \ldots, U_n$, respectively, then the minimum polynomial of T is the monic least common multiple of $a_1(x), \ldots, a_n(x)$.

Proof Suppose $m(x)$ is the minimum polynomial of T and $s(x)$ is the monic l.c.m. of $a_1(x), \ldots, a_n(x)$. Each polynomial $a_i(x)$ divides $s(x)$ so $s(T)U_i = \mathbf{0}$ for $1 \le i \le n$. Therefore $s(T) = \mathbf{0}$ and $m(x)|s(x)$.

To show that $m(x) = s(x)$, suppose $p_1(x), \ldots, p_k(x)$ are the distinct irreducible monic factors of the polynomials $a_1(x), \ldots, a_n(x)$. Then

$$a_j(x) = p_1^{e_{1j}}(x)p_2^{e_{2j}}(x)\cdots p_k^{e_{kj}}(x) \qquad \text{for} \quad 1 \le j \le n$$

with $e_{ij} = 0$ allowed. Since $s(x)$ is the l.c.m. of $a_1(x), \ldots, a_n(x)$,

$$s(x) = p_1^{e_1}(x)p_2^{e_2}(x)\cdots p_k^{e_k}(x)$$

where $e_i = \max\{e_{i1}, e_{i2}, \ldots, e_{ik}\}$. Now $m(x)|s(x)$ implies that there exist integers $f_1, \ldots, f_k$ such that

$$m(x) = p_1^{f_1}(x)p_2^{f_2}(x)\cdots p_k^{f_k}(x).$$

$m(x)$ divides $s(x)$ and both are monic, so if $m(x) \ne s(x)$, then $f_h < e_h$ for some index h and $f_h < e_{hj}$ for some index j. But then $a_j(x) \nmid m(x)$, contradicting Theorem 8.21. Therefore $m(x) = s(x)$ as claimed.

Example 2 Use these theorems to find the minimum polynomial of the map in Example 1.

Consider the basis vector E_1; $E_1 = (1, 0, 0)$, $T(E_1) = (1, -1, 0)$, and $T^2(E_1) = (-2, -2, 1)$ are linearly independent. $T^3(E_1) = (-6, -2, 3)$ and $T^3(E_1) = 3T^2(E_1) - 4T(E_1) + 4E_1$. Therefore the T-minimum polynomial of E_1 is $x^3 - 3x^2 - 4x + 4$. This polynomial divides the minimum polynomial of T, $m(x)$ and $\deg m \le 3$, therefore $m(x) = x^3 - 3x^2 + 4x - 4$.

Example 3 Find the minimum polynomial of $T \in \text{Hom}(\mathscr{V}_3)$ if the matrix of T with respect to $\{E_i\}$ is

$$\begin{pmatrix} 3 & 2 & -1 \\ -2 & -2 & 2 \\ 3 & 6 & -1 \end{pmatrix}.$$

$E_1 = (1, 0, 0)$, $T(E_1) = (3, -2, 3)$, $T^2(E_1) = (2, 4, -6)$, and $T^2(E_1) = -2T(E_1) + 8E_1$. Since E_1 and $T(E_1)$ are linearly independent, the T-minimum polynomial of E_1 is $x^2 + 2x - 8$. Further computation shows that E_2 and E_3 also have $x^2 + 2x - 8$ as their T-minimum polynomial, so this is the minimum polynomial of T.

The map T in Example 3 was considered in Example 4, page 229. In that example we found that the characteristic polynomial of T is $-(x - 2)^2(x + 4)$. Since the minimum polynomial of T is $(x - 2)(x + 4)$, the minimum polynomial of a map need not equal its characteristic polynomial. However, the minimum polynomial of a map always divides its characteristic polynomial. This is a consequence of the Cayley-Hamilton theorem, Theorem 8.31, page 380.

Recall that our approach to obtaining canonical forms under similarity relies on finding a direct sum decomposition of $\mathscr{V}$ into T-invariant subspaces. This decomposition is accomplished in two steps with the first given by a factorization of the minimum polynomial of T.

Theorem 8.23 (Primary Decomposition Theorem) Suppose $m(x)$ is the minimum polynomial of $T \in \text{Hom}(\mathscr{V})$ and $m(x) = p_1^{e_1}(x) \cdots p_k^{e_k}(x)$ where $p_1(x), \ldots, p_k(x)$ are the distinct irreducible monic factors of $m(x)$ with multiplicities $e_1, \ldots, e_k$, respectively. Then

$$\mathscr{V} = \mathscr{N}_{p_1^{e_1}(T)} \oplus \mathscr{N}_{p_2^{e_2}(T)} \oplus \cdots \oplus \mathscr{N}_{p_k^{e_k}(T)}.$$

Proof There are two parts to the proof: (1) Show that the sum of the null spaces equals $\mathscr{V}$ and (2) show that the sum is direct.

Define $q_i(x)$ for $1 \le i \le k$ by $m(x) = p_i^{e_i}(x)q_i(x)$. Then for any $U \in \mathscr{V}$, $\mathbf{0} = m(T)U = p_i^{e_i}(T)(q_i(T)U)$, therefore $q_i(T)U \in \mathscr{N}_{p_i^{e_i}(T)}$. This fact is used to obtain (1) as follows: The monic g.c.d. of $q_1(x), \ldots, q_k(x)$ is 1, so by an extension of Theorem 8.15 there exist polynomials $s_1(x), \ldots, s_k(x)$ such that

$$1 = s_1(x)q_1(x) + \cdots + s_k(x)q_k(x).$$

Therefore

$$I = s_1(T) \circ q_1(T) + \cdots + s_k(T) \circ q_k(T)$$

and applying this map to the vector U yields

$$U = I(U) = s_1(T)(q_1(T)U) + \cdots + s_k(T)(q_k(T)U).$$

Now $q_i(T)U \in \mathscr{N}_{p_i^{e_i}(T)}$ implies $s_i(T)(q_i(T)U) \in \mathscr{N}_{p_i^{e_i}(T)}$, therefore

$$U \in \mathscr{N}_{p_1^{e_1}(T)} + \mathscr{N}_{p_2^{e_2}(T)} + \cdots + \mathscr{N}_{p_k^{e_k}(T)},$$

which establishes (1).

For (2), suppose $W_1 + \cdots + W_k = \mathbf{0}$ with $W_i \in \mathscr{N}_{p_i^{e_i}(T)}$ for $1 \le i \le k$. $q_i(x)$ and $p_i^{e_i}(x)$ are relatively prime, so there exist $s(x)$ and $t(x)$ such that $1 = s(x)q_i(x) + t(x)p_i^{e_i}(x)$. Applying the map

$$I = s(T) \circ q_i(T) + t(T) \circ p_i^{e_i}(T)$$

to $W_i \in \mathscr{N}_{p_i^{e_i}(T)}$ yields

$$W_i = s(T) \circ q_i(T)W_i + t(T) \circ p_i^{e_i}(T)W_i = s(T) \circ q_i(T)W_i + \mathbf{0}.$$

But $p_j(x) | q_i(x)$ if $j \ne i$, so $q_i(T)W_j = \mathbf{0}$ if $i \ne j$, and we have

$$\mathbf{0} = q_i(T)\mathbf{0} = q_i(T)(W_1 + \cdots + W_k) = q_i(T)W_i.$$

Therefore, $W_i = s(T)(q_i(T)W_i) = s(T)\mathbf{0} = \mathbf{0}$, proving that the sum is direct and completing the proof of the theorem.

It should be clear that none of the direct summands $\mathscr{N}_{p_i^{e_i}(T)}$ in this decomposition can be the zero subspace.

This decomposition theorem provides a diagonal block matrix for T

as follows. Suppose a basis B_i is chosen for each null space $\mathscr{N}_{p_i^{e_i}(T)}$ and these bases are combined in order to give a basis B for $\mathscr{V}$. Then the matrix of T with respect to B is diag$(A_1, \ldots, A_k)$, where A_i is the matrix of the restriction

$$T: \mathscr{N}_{p_i^{e_i}(T)} \to \mathscr{N}_{p_i^{e_i}(T)}$$

with respect to the basis B_i. In general this cannot be regarded as a canonical form because no invariant procedure is given for choosing the bases B_i. The second decomposition theorem will provide such a choice for each B_i. However there is one case for which choosing the bases presents no problems, namely, when the irreducible factors are all linear and of multiplicity 1.

Suppose the minimum polynomial of T factors to $p_1(x)\cdots p_k(x)$ with $p_i(x) = x - \lambda_i$, $1 \leq i \leq k$, and $\lambda_1, \ldots, \lambda_k$ distinct elements from the field F. Then the null space $\mathscr{N}_{p_i(T)}$ is the characteristic space $\mathscr{S}(\lambda_i)$. Therefore if B_i is any basis for $\mathscr{N}_{p_i(T)}$, then the block A_i is diag$(\lambda_i, \ldots, \lambda_i)$ and the diagonal block matrix diag$(A_1, \ldots, A_k)$ is a diagonal matrix. This establishes half of the following theorem.

Theorem 8.24 $T \in \text{Hom}(\mathscr{V})$ is diagonalizable if and only if the minimum polynomial of T factors into linear factors, each of multiplicity 1.

Proof ($\Rightarrow$) If T is diagonalizable, then $\mathscr{V}$ has a basis $B = \{U_1, \ldots, U_n\}$ of characteristic vectors for T. Let $\lambda_1, \ldots, \lambda_k$ be the distinct characteristic values for T and consider the polynomial

$$q(x) = (x - \lambda_1)(x - \lambda_2)\cdots(x - \lambda_k).$$

Since each vector $U_j \in B$ is a characteristic vector, $q(T)U_j = \mathbf{0}$. (Multiplication of polynomials is commutative, thus the factor involving the characteristic value for U_j can be written last in $q(T)$.) Therefore $q(T) = \mathbf{0}$ and the minimum polynomial of T divides $q(x)$. But this implies the minimum polynomial has only linear factors of multiplicity 1.

($\Leftarrow$) This part of the proof was obtained above.

Example 4 Suppose the matrix of $T \in \text{Hom}(\mathscr{V}_3)$ with respect to $\{E_i\}$ is

$$\begin{pmatrix} 3 & -2 & 0 \\ -1 & 3 & -1 \\ -5 & 7 & -1 \end{pmatrix}.$$

Determine if T is diagonalizable by examining the minimum polynomial of T.

Choosing E_1 arbitrarily, we find that the T-minimum polynomial of E_1 is $x^3 - 5x^2 + 8x - 4 = (x - 2)^2(x - 1)$. Since this polynomial divides the minimum polynomial of T, the minimum polynomial of T contains a linear factor with multiplicity greater than 1. Therefore T is not diagonalizable. The same conclusion was obtained in Example 5, page 230.

Problems

1. Suppose the matrix of $T \in \operatorname{Hom}(\mathscr{V}_3)$ with respect to $\{E_i\}$ is

$$\begin{pmatrix} -3 & -1 & -3 \\ 2 & 2 & 1 \\ 6 & 1 & 6 \end{pmatrix}.$$

Find the T-minimum polynomial of the following vectors.
a. $(1, 0, 0)$. b. $(1, 0, -2)$. c. $(-3, 1, 3)$. d. $(1, -1, -1)$.

2. Suppose $T \in \operatorname{Hom}(\mathscr{V}_2)$ is given by $T(a, b) = (3a - 2b, 7a - 4b)$.
a. Find the T-minimum polynomial of $(1, 0)$ and $(0, 1)$.
b. Show that every nonzero vector in $\mathscr{V}_2$ has the same T-minimum polynomial.
c. If T is viewed as a map in $\operatorname{Hom}(\mathscr{V}_2(C))$, does every nonzero vector have the same T-minimum polynomial?

3. Find the minimum polynomial for $T \in \operatorname{Hom}(\mathscr{V}_n)$ given the matrix for T with respect to the standard basis for $\mathscr{V}_n$.

a. $\begin{pmatrix} -3 & 1 \\ 6 & 2 \end{pmatrix}$. b. $\begin{pmatrix} -6 & -4 \\ 4 & 2 \end{pmatrix}$. c. $\begin{pmatrix} 0 & 0 & 1 \\ 1 & 0 & -3 \\ 0 & 1 & 3 \end{pmatrix}$. d. $\begin{pmatrix} 0 & -16 & 0 \\ 1 & -8 & 0 \\ 0 & 0 & -5 \end{pmatrix}$.

e. $\begin{pmatrix} 2 & -1 & 2 \\ 2 & -1 & 1 \\ -4 & 1 & -4 \end{pmatrix}$. f. $\begin{pmatrix} 0 & 0 & 0 & -9 \\ 1 & 0 & 0 & 0 \\ 0 & 1 & 0 & -6 \\ 0 & 0 & 1 & 0 \end{pmatrix}$. g. $\begin{pmatrix} 7 & 10 & -2 \\ -5 & -8 & 2 \\ -5 & -10 & 4 \end{pmatrix}$.

4. Use the minimum polynomial to determine which of the maps in problem 3 are diagonalizable.

5. For each map T in problem 3, find the direct sum decomposition of $\mathscr{V}_n$ given by the primary decomposition theorem.

6. Let T be the map in problem 3e, then the decomposition of $\mathscr{V}_3$ obtained in problem 5 is of the form $\mathscr{S}(U_1, T) \oplus \mathscr{S}(U_2, T)$ for some $U_1, U_2 \in \mathscr{V}_3$. Find the matrix of T with respect to the basis $\{U_1, T(U_1), U_2\}$.

7. Suppose $\dim \mathscr{V} = n$, $T \in \operatorname{Hom}(\mathscr{V})$, and the minimum polynomial of T is irreducible over F and of degree n.
a. Show that if $U \in \mathscr{V}$, $U \neq \mathbf{0}$, then $\mathscr{V} = \mathscr{S}(U, T)$.
b. Find the matrix of T with respect to the basis $\{U, T(U), \ldots, T^{n-1}(U)\}$.
c. What is this matrix if $\mathscr{V} = \mathscr{V}_2$ and the minimum polynomial of T is $x^2 - 5x + 7$?

8. For the map T of Example 4, page 365, the T-minimum polynomial of E_1 is $(x - 2)^2(x - 1)$. Use this fact to obtain a vector from E_1 which has as its T-minimum polynomial
a. $(x - 2)^2$. b. $x - 1$. c. $x - 2$.

9. Suppose $T \in \text{Hom}(\mathscr{V}_n(C))$ and every nonzero vector in $\mathscr{V}_n(C)$ has the same T-minimum polynomial. Show that this implies that $T = zI$ for some $z \in C$.

10. Suppose $T \in \text{Hom}(\mathscr{V})$ and the T-minimum polynomial of U is irreducible over F. Prove that $\mathscr{S}(U, T)$ cannot be decomposed into a direct sum of nonzero T-invariant subspaces.

11. Let $T \in \text{Hom}(\mathscr{V})$ and suppose $p_i(x)$ is the T-minimum polynomial of U_i, $U_i \in \mathscr{V}$, for $i = 1, 2$. Prove that $p_1(x)$ and $p_2(x)$ are relatively prime if and only if $p_1(x)p_2(x)$ is the T-minimum polynomial of $U_1 + U_2$.

12. A map $T \in \text{Hom}(\mathscr{V})$ is a *projection* (*idempotent*) if $T^2 = T$.
a. Find the three possible minimum polynomials for a projection.
b. Show that every projection is diagonalizable.

13. Show that $T \in \text{Hom}(\mathscr{V})$ is a projection if and only if there exist subspaces $\mathscr{S}$ and $\mathscr{T}$ such that $\mathscr{V} = \mathscr{S} \oplus \mathscr{T}$ and $T(V) = U$ when $V = U + W$ with $U \in \mathscr{S}$, $W \in \mathscr{T}$.

14. Show that $\mathscr{V} = \mathscr{S}_1 \oplus \ldots \oplus \mathscr{S}_k$ if and only if there exist k projection maps $T_1, \ldots, T_k$ such that
i. $T_i(\mathscr{V}) = \mathscr{S}_i$.
ii. $T_i \circ T_j = \mathbf{0}$ when $i \neq j$.
iii. $I = T_1 + \cdots + T_k$.

15. Prove the primary decomposition theorem using the result of problem 14.

§4. Elementary Divisors

Given $T \in \text{Hom}(\mathscr{V})$, the primary decomposition theorem expresses $\mathscr{V}$ as a direct sum of the null spaces $\mathscr{N}_{p_i^{e_i}(T)}$, where $p_i(x)$ is an irreducible factor of the minimum polynomial of T with multiplicity e_i. The next problem is to decompose each of these null spaces into a direct sum of T-invariant subspaces in some well-defined way. There are several approaches to this problem, but none of them can be regarded as simple. The problem is somewhat simplified by assuming the field F is algebraically closed. In that case each $p_i(x)$ is linear, and the most useful canonical form is obtained at once. However this approach does not determine canonical forms for matrices over an arbitrary field, such as R, and so will not be followed here. Instead we will choose a sequence of T-cyclic subspaces for each null space in such way that each subspace is of maximal dimension.

A decomposition must be found for each of the null spaces in the primary decomposition theorem. Therefore it will be sufficient to begin by assuming that the minimum polynomial of $T \in \text{Hom}(\mathscr{V})$ is $p^e(x)$ with $p(x)$ irreducible over F. Then $\mathscr{N}_{p^e(T)} = \mathscr{V}$. It is easy to show that $\mathscr{V}$ contains a T-cyclic subspace of maximal dimension. Since $p^e(x)$ is the minimum polynomial of T, there exists a vector $U_1 \in \mathscr{V}$ such that $p^{e-1}(T)U_1 \neq \mathbf{0}$. Therefore, the minimum polynomial of U_1 is $p^e(x)$. (Why?) Suppose the degree of $p(x)$ is b, then $\deg p^e = eb$ and $\dim \mathscr{S}(U_1, T) = eb$. Since no element of $\mathscr{V}$ can have a T-minimum polynomial of degree greater than eb, $\mathscr{S}(U_1, T)$ is a T-cyclic subspace of maximal dimension in $\mathscr{V}$. If $\mathscr{V} = \mathscr{S}(U_1, T)$, then we have obtained the desired decomposition.

Example 1 Suppose

$$\begin{pmatrix} 1 & -1 & 4 \\ 1 & -1 & -4 \\ -1 & 2 & 6 \end{pmatrix}$$

is the matrix of $T \in \text{Hom}(\mathscr{V}_3)$ with respect to the standard basis. The minimum polynomial of T is found to be $x^3 - 6x^2 + 12x - 8 = (x - 2)^3$ which is of the form $p^e(x)$ with $p(x) = x - 2$ and $e = 3$. The T-minimum polynomial of E_1 is also $(x - 2)^3$, so we may take $U_1 = (1, 0, 0)$. Then $\dim \mathscr{S}(U_1, T) = 3 \cdot 1 = 3$ and $\mathscr{S}(U_1, T) = \mathscr{V}_3$. Although T is not diagonalizable, the matrix of T with respect to the basis $\{U_1, T(U_1), T^2(U_1)\}$ has the simple form

$$\begin{pmatrix} 0 & 0 & 8 \\ 1 & 0 & -12 \\ 0 & 1 & 6 \end{pmatrix}.$$

Notice that the elements in the last column come from the minimum polynomial of T, for

$$T(T^2(U_1)) = T^3(U_1) = 6T^2(U_1) - 12T(U_1) + 6U_1.$$

Therefore this matrix for T does not depend on the choice of U_1, provided the T-minimum polynomial of U_1 is $(x - 2)^3$.

Definition The *companion matrix* of the monic polynomial $q(x)$

$= x^h + a_{h-1}x^{h-1} + \cdots + a_1x + a_0$ is the $h \times h$ matrix

$$C(q) = \begin{pmatrix} 0 & 0 & & 0 & 0 & -a_0 \\ 1 & 0 & & 0 & 0 & -a_1 \\ 0 & 1 & & 0 & 0 & -a_2 \\ & & \ddots & & & \\ 0 & 0 & & 1 & 0 & -a_{h-2} \\ 0 & 0 & & 0 & 1 & -a_{h-1} \end{pmatrix}.$$

That is, the companion matrix of $q(x)$ has 1's on the "subdiagonal" below the main diagonal and negatives of coefficients from $q(x)$ in the last column.

If the minimum polynomial of $T \in \text{Hom}(\mathscr{V})$ is $p^e(x)$ and $\deg p^e = \dim \mathscr{V}$, then the companion matrix of $p^e(x)$ is the matrix of T with respect to a basis for $\mathscr{V}$. In this case $C(p^e)$ can be taken as the canonical form for the matrices of T. However, it is not always so simple. For example, if $T \in \text{Hom}(\mathscr{V}_3)$ has

$$\begin{pmatrix} -2 & -4 & 2 \\ 2 & 4 & -1 \\ -4 & -4 & 4 \end{pmatrix}$$

as its matrix with respect to $\{E_i\}$, then the minimum polynomial of T is $(x - 2)^2$. Therefore T is not diagonalizable, and if U_1 is chosen as above, the dimension of $\mathscr{S}(U_1, T)$ is only 2. So a T-cyclic subspace of maximal dimension in $\mathscr{V}_3$ is a proper subspace of $\mathscr{V}_3$.

Returning to the general case, suppose $\mathscr{S}(U_1, T) \neq \mathscr{V}$. Then we must find a vector U_2 that generates a T-cyclic subspace of maximal dimension satisfying $\mathscr{S}(U_1, T) \cap \mathscr{S}(U_2, T) = \{\mathbf{0}\}$. The first step in searching for such a vector is to observe that if $V \notin \mathscr{S}(U_1, T)$, then when k is sufficiently large, $p^k(T)V \in \mathscr{S}(U_1, T)$. In particular, $p^e(T)V \in \mathscr{S}(U_1, T)$. Thus for each $V \notin \mathscr{S}(U_1, T)$, there exists an integer k, $1 \leq k \leq e$, such that $p^k(T)V \in \mathscr{S}(U_1, T)$ and $p^{k-1}(T)V \notin \mathscr{S}(U_1, T)$. Let r_2 be the largest of all such integers k. Then $1 \leq r_2 \leq e$ and there exists a vector $W \in \mathscr{V}$ such that $p^{r_2}(T)W \in \mathscr{S}(U_1, T)$ and $p^{r_2-1}(T)W \notin \mathscr{S}(U_1, T)$. If $p^{r_2}(T)W = \mathbf{0}$, then we may set $U_2 = W$ and we will be able to show that the sum $\mathscr{S}(U_1, T) + \mathscr{S}(U_2, T)$ is direct. However, if $p^{r_2}(T)W \neq \mathbf{0}$, then we must subtract a vector from W to obtain a choice for U_2. To find this vector, notice that $p^{r_2}(T)V \in \mathscr{S}(U_1, T)$ means that there exists a polynomial $f(x) \in F[x]$ such that

$$p^{r_2}(T)W = f(T)U_1. \tag{1}$$

(See problem 14, page 359.) Applying the map $p^{e-r_2}(T)$ to Eq. (1) gives

$$p^{e-r_2}(T) \circ f(T)U_1 = p^{e-r_2}(T) \circ p^{r_2}(T)W = p^e(T)W = \mathbf{0}.$$

Since $p^e(x)$ is the T-minimum polynomial of U_1, $p^e(x)|p^{e-r_2}(x)f(x)$ and therefore $p^{r_2}(x)|f(x)$. Suppose $f(x) = p^{r_2}(x)q(x)$, then Eq. (1) becomes

$$p^{r_2}(T)W = f(T)U_1 = p^{r_2}(T) \circ q(T)U_1.$$

Therefore if we set $U_2 = W - q(T)U_1$, then

$$p^{r_2}(T)U_2 = p^{r_2}(T)W - p^{r_2}(T) \circ q(T)U_1 = \mathbf{0}$$

as desired, and it is only necessary to check that $p^{r_2-1}(T)U_2 \notin \mathscr{S}(U_1, T)$. But

$$\begin{aligned} p^{r_2-1}(T)U_2 &= p^{r_2-1}(T)(W - q(T)U_1) \\ &= p^{r_2-1}(T)W - p^{r_2-1}(T) \circ q(T)U_1. \end{aligned}$$

So if $p^{r_2-1}(T)U_2 \in \mathscr{S}(U_1, T)$, then so is $p^{r_2-1}(T)W$, contradicting the choice of W. Therefore there exists a vector U_2 with minimum polynomial $p^{r_2}(x)$, $1 \le r^2 \le e$, such that $p^{r_2-1}(T)U_2 \notin \mathscr{S}(U_1, T)$ and r_2 is the largest integer such that $p^{r_2-1}(T)V \notin \mathscr{S}(U_1, T)$ for some $V \in \mathscr{V}$.

It remains to be shown that $\mathscr{S}(U_1, T) \cap \mathscr{S}(U_2, T) = \{\mathbf{0}\}$. Therefore suppose $V \neq \mathbf{0}$ is in the intersection. Then $V = g(T)U_2$ for some $g(x) \in F[x]$. Suppose the g.c.d. of $g(x)$ and $p^{r_2}(x)$ is $p^a(x)$. Then there exist polynomials $s(x)$ and $t(x)$ such that

$$p^a(x) = s(x)g(x) + t(x)p^{r_2}(x).$$

Replacing x by T and applying the map $p^a(T)$ to U_2 yields

$$p^a(T)U_2 = s(T) \circ g(T)U_2 + t(T) \circ p^{r_2}(T)U_2 = s(T)V + \mathbf{0}.$$

Since V is in the intersection, $p^a(T)U_2 \in \mathscr{S}(U_1, T)$ and $a \ge r_2$. But $a < r_2$ for $g(T)U_2 = V \neq \mathbf{0}$ implies $p^{r_2}(x)$, the T-minimum polynomial of U_2, does not divide $g(x)$. This contradiction shows that the intersection only contains $\mathbf{0}$, and so the sum $\mathscr{S}(U_1, T) + \mathscr{S}(U_2, T)$ is direct. If $\mathscr{V} = \mathscr{S}(U_1, T) \oplus \mathscr{S}(U_2, T)$, we have obtained the desired decomposition; otherwise this procedure may be continued to prove the following.

Theorem 8.25 (Secondary Decomposition Theorem) Suppose the minimum polynomial of $T \in \mathrm{Hom}(\mathscr{V})$ is $p^e(x)$ where $p(x)$ is irreducible over

F. Then there exist h vectors $U_1, \ldots, U_h$ such that

$$\mathscr{V} = \mathscr{S}(U_1, T) \oplus \mathscr{S}(U_2, T) \oplus \cdots \oplus \mathscr{S}(U_h, T)$$

and $\mathscr{S}(U_i, T)$ is a T-cyclic subspace of maximal dimension contained in

$$\mathscr{S}(U_i, T) \oplus \mathscr{S}(U_{i+1}, T) \oplus \cdots \oplus \mathscr{S}(U_h, T), \qquad 1 \leq i < h.$$

Further if $p^{r_1}(x), \ldots, p^{r_h}(x)$ are the T-minimum polynomials of $U_1, \ldots, U_h$, respectively, then $e = r_1 \geq \cdots \geq r_h \geq 1$.

It is too soon to claim that we have a method for choosing a canonical form under similarity since the decomposition given by Theorem 8.25 is not unique. But we will show that the T-minimum polynomials $p^{r_1}(x), \ldots, p^{r_h}(x)$ are uniquely determined by T and thereby obtain a canonical form. Both these points are illustrated in Example 2 below. The candidate for a canonical form is the diagonal block matrix given by this decomposition. Call $\{U_i, T(U_i), \ldots, T^{r_ib-1}(U_i)\}$ a *cyclic basis* for the T-cyclic space $\mathscr{S}(U_i, T)$, $1 \leq i \leq h$, where $\deg p = b$. Then the restriction of T to $\mathscr{S}(U_i, T)$ has the companion matrix $C(p^{r_i})$ as its matrix with respect to this cyclic basis. Therefore, if we build a *cyclic basis* B for $\mathscr{V}$ by combining these h cyclic bases in order, then the matrix of T with respect to B is $\operatorname{diag}(C(p^{r_1}), C(p^{r_2}), \ldots, C(p^{r_h}))$.

Example 2 If

$$\begin{pmatrix} 6 & -9 & 6 \\ -1 & 6 & -2 \\ -3 & 9 & -3 \end{pmatrix}$$

is the matrix of $T \in \operatorname{Hom}(\mathscr{V}_3)$ with respect to $\{E_i\}$, then $(x - 3)^2$ is the minimum polynomial of T. Find the matrix of T with respect to a cyclic basis for $\mathscr{V}_3$ given by the secondary decomposition theorem.

The degree of $(x - 3)^2$ is 2, so U_1 must be a vector that generates a T-cyclic subspace of dimension 2. Clearly any of the standard basis vectors could be used. (Why?) If we take $U_1 = (1, 0, 0)$, then $\mathscr{S}(U_1, T) = \mathscr{L}\{(1, 0, 0), (6, -1, 3)\}$. Since the dimension of $\mathscr{V}_3$ is 3, U_2 must generate a 1-dimensional T-cyclic subspace. Therefore U_2 can be any characteristic vector for the characteristic value 3 which is not in $\mathscr{S}(U_1, T)$. The characteristic space is

$$\mathscr{S}(3) = \{(a, b, c) | a - 3b + 2c = 0\}$$

and $\mathscr{S}(3) \cap \mathscr{S}(U_1, T) = \mathscr{L}\{(3, -1, -3)\}$. Therefore we may set $U_2 = (3, 1, 0)$

and

$$\mathscr{V}_3 = \mathscr{S}((1, 0, 0), T) \oplus \mathscr{S}((3, 1, 0), T)$$

is a decomposition of $\mathscr{V}_3$ given by the secondary decomposition theorem. This decomposition determines the cyclic basis $B = \{U_1, T(U_1), U_2\}$.

Now the T-minimum polynomial of U_1 is $(x - 3)^2$, so the restriction of T ot $\mathscr{S}(U_1, T)$ has $C((x - 3)^2)$ as its matrix with respect to $\{U_1, T(U_1)\}$. And the T-minimum polynomial of U_2 is $x - 3$, so $C(x - 3)$ is the matrix of T on $\mathscr{S}(U_2, T)$. Therefore the matrix of T on $\mathscr{V}_3$ with respect to the cyclic basis B is

$$\operatorname{diag}(C((x - 3)^2), C(x - 3)) = \begin{pmatrix} 0 & -9 & 0 \\ 1 & 6 & 0 \\ 0 & 0 & 3 \end{pmatrix}.$$

Note that almost any vector in $\mathscr{V}_3$ could have been chosen for U_1 in this example. That is, U_1 could have been any vector not in the 2-dimensional characteristic space $\mathscr{S}(3)$. And then U_2 could have been any vector in $\mathscr{S}(3)$ not on a particular line. So the vectors U_1 and U_2 are far from uniquely determined. However, no matter how they are chosen, their T-minimum polynomials will always be $(x - 3)^2$ and $x - 3$.

The secondary decomposition theorem provides the basis for the claim made in the last section that the degree of the minimum polynomial for $T \in \operatorname{Hom}(\mathscr{V})$ cannot exceed the dimension of $\mathscr{V}$. For if $p^e(x)$ is the minimum polynomial of T, then

$$\dim \mathscr{V} \geq \dim \mathscr{S}(U_1, T) = \deg p^{r_1} = \deg p^e.$$

In Example 2, $p^e(x) = (x - 3)^2$ has degree 2 and $\mathscr{V} = \mathscr{V}_3$ has dimension 3.

Example 3 If the matrix of $T \in \operatorname{Hom}(\mathscr{V}_5)$ is

$$\begin{pmatrix} 14 & 8 & -1 & -6 & 2 \\ -12 & -4 & 2 & 8 & -1 \\ 8 & -2 & 0 & -9 & 0 \\ 8 & 8 & 0 & 0 & 2 \\ -8 & -4 & 0 & 4 & 0 \end{pmatrix}$$

with respect to $\{E_i\}$, then the minimum polynomial of T is $p^3(x) = (x - 2)^3$. Find a direct sum decomposition for $\mathscr{V}_5$ following the procedure given by the secondary decomposition theorem, and find the corresponding diagonal block matrix.

U_1 must be a vector such that $p^2(T)U_1 \neq \mathbf{0}$. Choosing E_1 arbitrarily (the definition of T does not suggest a choice for U_1) we find that

$$p^2(T)E_1 = (T^2 - 4T + 4I)E_1 = (-24, 16, 32, -32, 0) \neq \mathbf{0}.$$

Therefore we set $U_1 = E_1$ and obtain

$$\begin{aligned}\mathscr{S}(U_1, T) &= \mathscr{L}\{(1, 0, 0, 0, 0), (14, -12, 8, 8, -8), (28, -32, 64, 0, -32)\}.\\ &= \mathscr{L}\{(1, 0, 0, 0, 0), (0, 1, 2, -2, 0), (0, 0, -4, 2, 1)\}.\end{aligned}$$

In looking for a vector W, dimensional restrictions leave only two possibilities; either $p(T)V \in \mathscr{S}(U_1, T)$ for all $V \notin \mathscr{S}(U_1, T)$ or there exists a vector V such that $p(T)V \notin \mathscr{S}(U_1, T)$. Again we arbitrarily choose a vector not in $\mathscr{S}(U_1, T)$, say E_2. Then

$$p(T)E_2 = (T - 2I)E_2 = (8, -6, -2, 8, -4),$$

and it is easily shown that $p(T)E_2 \notin \mathscr{S}(U_1, T)$. Therefore set $W = E_2$ and $r_2 = 2$. Since $p^2(T)W = (-6, 4, 8, -8, 0)$ is not $\mathbf{0}$, we cannot let U_2 equal W. [Note that $p^2(T)W = \frac{1}{4}T^2(U_1) - T(U_1) + U_1$, showing that $p^2(T) \in \mathscr{S}(U_1, T)$ as expected. Is it clear that $r_2 = 2$ is maximal? What would $p^2(T)W \notin \mathscr{S}(U_1, T)$ and $r_2 \geq 3$ imply about the dimension of $\mathscr{V}_5$?] Still following the proof for the secondary decomposition theorem, it is now necessary to subtract an element of $\mathscr{S}(U_1, T)$ from W. Equation (1) in this case is $p^2(T)W = f(T)U_1$ with $f(x) = \frac{1}{4}x^2 - x + 1$. Therefore the polynomial $q(x)$ satisfying $f(x) = p^{r_2}(x)q(x)$ is the constant polynomial $q(x) = 4$ and we should take

$$U_2 = W - q(T)U_1 = E_2 - 4E_1 = (-4, 1, 0, 0, 0).$$

Then since $\dim \mathscr{S}(U_1, T) = 3$, $\dim \mathscr{S}(U_2, T) = 2$, and $\dim \mathscr{V}_5 = 5$, we have

$$\mathscr{V}_5 = \mathscr{S}((1, 0, 0, 0, 0), T) \oplus \mathscr{S}((-4, 1, 0, 0, 0), T).$$

The cyclic basis given by this decomposition is $B = \{U_1, T(U_1), T^2(U_1), U_2, T(U_2)\}$ and the matrix of T with respect to B is

$$\operatorname{diag}(C(p^3), C(p^2)) = \left(\begin{array}{ccc|cc} 0 & 0 & 8 & 0 & 0\\ 1 & 0 & -12 & 0 & 0\\ 0 & 1 & 6 & 0 & 0\\ \hline 0 & 0 & 0 & 0 & -4\\ 0 & 0 & 0 & 1 & 4 \end{array}\right).$$

The two previous examples suggest quite strongly that the T-minimum polynomials associated with $\mathscr{S}(U_1, T), \ldots, \mathscr{S}(U_h, T)$ are independent of $U_1, \ldots, U_h$ and that they are determined entirely by the map T. If so, then the matrix of T with respect to a cyclic basis for $\mathscr{V}$ can be taken as the canonical form for the equivalence class under similarity containing all possible matrices for T.

Theorem 8.26 The numbers $h, r_1, \cdots r_h$ in the secondary deposition theorem depend only on T.

That is, if $\mathscr{V} = \mathscr{S}(U_1', T) \oplus \cdots \oplus \mathscr{S}(U_{h'}', T)$ is another decomposition for $\mathscr{V}$ satisfying the secondary decomposition theorem and the T-minimum polynomial of U_i' is $p^{r'_i}(x)$, then $h' = h$ and $r_1' = r_1, \ldots, r_h' = r_h$.

Proof The proof uses the second principle of induction on the dimension of $\mathscr{V}$.

If $\dim \mathscr{V} = 1$, the theorem is clearly true.

Therefore suppose the theorem holds for all T-invariant spaces of dimension less than n and let $\dim \mathscr{V} = n$, $n > 1$. Consider the null space of $p(T)$, $\mathscr{N}_{p(T)}$. If $\mathscr{N}_{p(T)} = \mathscr{V}$, then $e = 1$ and the dimension of each T-cyclic space is deg p. In this case $hp = \dim \mathscr{V} = h'p$ and the theorem is true. If $\mathscr{N}_{p(T)} \neq \mathscr{V}$, let $V \in \mathscr{N}_{p(T)}$. Then $V \in \mathscr{S}(U_1, T) \oplus \cdots \oplus \mathscr{S}(U_h, T)$ implies there exist polynomials $f_1(x), \ldots, f_h(x)$ such that $V = f_1(T)U_1 + \cdots + f_h(T)U_h$ and

$$\mathbf{0} = p(T)V = p(T) \circ f_1(T)U_1 + \cdots + p(T) \circ f_h(T)U_h.$$

Since there is only one expression for $\mathbf{0}$ in a direct sum, $p(T) \circ f_i(T)U_i = \mathbf{0}$ for $1 \leq i \leq h$. Therefore the minimum polynomial of U_i, $p^{r_i}(x)$, divides $p(x)f_i(x)$, implying that $p^{r_i-1}(x)$ divides $f_i(x)$. Thus there exist polynomials $q_1(x), \ldots, q_h(x)$ such that

$$f_i(x) = q_i(x)p^{r_i-1}(x).$$

Therefore

$$V = q_1(T) \circ p^{r_1-1}(T)U_1 + \cdots + q_h(T) \circ p^{r_h-1}(T)U_h,$$

which means that

$$\mathscr{N}_{p(T)} \subset \mathscr{S}(p^{r_1-1}(T)U_1, T) \oplus \cdots \oplus \mathscr{S}(p^{r_h-1}(T)U_h, T).$$

The reverse inclusion is immediate so that

$$\mathcal{N}_{p(T)} = \mathcal{S}(p^{r_1-1}(T)U_1, T) \oplus \cdots \oplus \mathcal{S}(p^{r_h-1}(T)U_h, T).$$

Similarly

$$\mathcal{N}_{p(T)} = \mathcal{S}(p^{r'_1-1}(T)U_1, T) \oplus \cdots \oplus \mathcal{S}(p^{r'_{h'}-1}(T)U_{h'}, T).$$

These direct sum decompositions satisfy the secondary decomposition theorem and $\mathcal{N}_{p(T)}$ is a T-invariant space of dimension less than n, so the induction assumption applies and $h' = h$.

Now consider the image space $\mathcal{I}_{p(T)} = p(T)[\mathcal{V}]$. By a similar agrument it can be shown that

$$\mathcal{I}_{p(T)} = \mathcal{S}(p(T)U_1, T) \oplus \cdots \oplus \mathcal{S}(p(T)U_k, T)$$

where k is the index such that $r_k > 1$ and $r_{k+1} = 1$ or $k = h$ with $r_h > 1$. Then the T-minimum polynomial of $p(T)U_i$ is $p^{r_i-1}(x)$, $1 \leq i \leq k$. Also,

$$\mathcal{I}_{p(T)} = \mathcal{S}(p(T)U'_1, T) \oplus \cdots \oplus \mathcal{S}(p(T)U'_{k'}, T)$$

with k' defined similarly. Now $\mathcal{I}_{p(T)}$ is also a T-invariant space, and since $\mathcal{N}_{p(T)} \neq \{\mathbf{0}\}$, $\mathcal{I}_{p(T)}$ is a proper subspace of $\mathcal{V}$. Therefore the induction assumption yields $k' = k$ and $r'_1 = r_1, \ldots, r'_k = r_k$ with $r'_{k+1} = \cdots = r'_h = r_{k+1} = \cdots = r_h = 1$, which completes the proof of the theorem.

Definition If the minimum polynomial of $T \in \text{Hom}(\mathcal{V})$ is $p^e(x)$ with $p(x)$ irreducible over F, then the polynomials $p^e(x) = p^{r_1}(x), p^{r_2}(x), \ldots, p^{r_h}(x)$ from the secondary decomposition theorem are the *$p(x)$-elementary divisors of T*.

From the comments that have been made above, it should be clear that if $p^e(x)$ is the minimum polynomial of T, then a diagonal block matrix can be given for T once the $p(x)$-elementary divisors are known.

Example 4 If $T \in \text{Hom}(\mathcal{V}_8)$ has $p(x)$-elementary divisors $(x^2 - x + 3)^2$, $x^2 - x + 3$, $x^2 - x + 3$, then the diagonal block matrix of T with respect to a cyclic basis for $\mathscr{V}_8$ is

$$\operatorname{diag}(C(p^2), C(p), C(p)) = \begin{pmatrix} 0 & 0 & 0 & -9 & & & & 0 \\ 1 & 0 & 0 & 6 & & & & \\ 0 & 1 & 0 & -7 & & & & \\ 0 & 0 & 1 & 2 & & & & \\ & & & & 0 & -3 & & \\ & & & & 1 & 1 & & \\ & & & & & & 0 & -3 \\ 0 & & & & & & 1 & 1 \end{pmatrix}.$$

Here $p(x) = x^2 - x + 3$ and $p^2(x) = x^4 - 2x^3 + 7x^2 - 6x + 9$.

Problems

1. Find the companion matrices of the following polynomials.
   a. $x - 5$. b. $(x - 5)^2$. c. $(x - 5)^3$. d. $x^2 - 3x + 4$.
   e. $(x^2 - 3x + 4)^2$.

2. Suppose the companion matrix $C(q) \in \mathscr{M}_{h\times h}$ is the matrix of $T \in \operatorname{Hom}(\mathscr{V})$ with respect to some basis B.
   a. Show that $q(x)$ is the minimum polynomial of T.
   b. Show that $(-1)^h q(x)$ is the characteristic polynomial of T.

3. Find a cyclic basis B for $\mathscr{V}_3$ and the matrix of T with respect to B if the matrix of T with respect to $\{E_i\}$ is
   a. $\begin{pmatrix} 0 & -4 & -1 \\ 1 & 4 & 1 \\ 0 & 0 & 2 \end{pmatrix}$. b. $\begin{pmatrix} -2 & -4 & 2 \\ 2 & 4 & -1 \\ -4 & -4 & 4 \end{pmatrix}$.

4. Suppose $T \in \operatorname{Hom}(\mathscr{V})$ and B is a basis for $\mathscr{V}$. Let $U_1, \ldots, U_h$ be vectors satisfying the secondary decomposition theorem with $U_2, \ldots, U_h$ obtained from $W_2, \ldots, W_h$ as in the proof.
   a. Why is there always a vector in B that can be used as U_1?
   b. Show that $W_2, \ldots, W_h$ can always be chosen from B.

5. Suppose $T \in \operatorname{Hom}(\mathscr{V}_6)$ and $p^e(x)$ is the minimum polynomial of T. List all possible diagonal block matrices for T (up to order) composed of companion matrices if
   a. $p^e(x) = (x + 3)^2$. b. $p^e(x) = x - 4$. c. $p^e(x) = (x^2 + x + 1)^2$.
   d. $p^e(x) = (x^2 + 4)^3$. e. $p^e(x) = (x - 1)^3$.

6. Suppose $(x - c)^r$ is the T-minimum polynomial of U, $c \in F$.
   a. Show that $\mathscr{S}(U, T)$ contains a characteristic vector for c.
   b. Show that $\mathscr{S}(U, T)$ cannot contain two linearly independent characteristic vectors for c.

7. Find the $(x + 1)$-elementary divisors of $T \in \operatorname{Hom}(\mathscr{V}_4)$ given the matrix of T with respect to $\{E_i\}$ and the fact that its minimum polynomial is $(x + 1)^2$.

a. $\begin{pmatrix} 1 & 2 & 2 & 4 \\ -1 & -2 & -1 & -2 \\ -1 & -1 & -2 & -2 \\ 0 & 0 & 0 & -1 \end{pmatrix}$. b. $\begin{pmatrix} 5 & -6 & 4 & 2 \\ 0 & 0 & 0 & -1 \\ -9 & 10 & -7 & -4 \\ 0 & 1 & 0 & -2 \end{pmatrix}$.

8. Find a diagonal block matrix for a map with the given elementary divisors.

a. $x^2 + x + 1$, $x^2 + x + 1$, $x^2 + x + 1$.

b. $(x + 3)^3$, $(x + 3)^2$, $(x + 3)^2$.

c. $(x^2 - 4x + 7)^2$, $x^2 - 4x + 7$.

9. Combine the primary and secondary decomposition theorems to find a diagonal block matrix composed of companion matrices similar to

a. $\begin{pmatrix} -6 & 1 & -6 & 11 \\ 1 & -2 & 2 & -1 \\ 1 & -1 & 1 & -2 \\ -4 & 0 & -4 & 7 \end{pmatrix}$. b. $\begin{pmatrix} 10 & 4 & 10 & -14 \\ 1 & 4 & 2 & -1 \\ -2 & -4 & -2 & 2 \\ 6 & 0 & 6 & -10 \end{pmatrix}$.

§5. Canonical Forms

The primary and secondary decomposition theorems determine a direct sum decomposition of $\mathscr{V}$ for any map $T \in \operatorname{Hom}(\mathscr{V})$. This decomposition yields a diagonal block matrix for T, which is unique up to the order of the blocks, and thus may be used as a canonical representation for the equivalence class of matrices for T.

Theorem 8.27 Suppose $m(x)$ is the minimum polynomial of $T \in \operatorname{Hom}(\mathscr{V})$ and $m(x) = p_1^{e_1}(x) \cdots p_k^{e_k}(x)$ is a factorization for which $p_1(x), \ldots, p_k(x)$ are distinct monic polynomials, irreducible over F. Then there exist unique positive integers h_i, r_{ij} with $e_i = r_{i1} \geq \cdots \geq r_{ih_i} \geq 1$ for $1 \leq i \leq k$ and $1 \leq j \leq h_i$ and there exist (not uniquely) vectors $U_{ij} \in \mathscr{V}$ such that $\mathscr{V}$ is the direct sum of the T-cyclic subspaces $\mathscr{S}(U_{ij}, T)$ and the T-minimum polynomial of U_{ij} is $p_i^{r_{ij}}(x)$ for $1 \leq i \leq k$, $1 \leq j \leq h_i$.

Proof The factorization $m(x) = p_1^{e_1}(x) \cdots p_k^{e_k}(x)$ is unique up to the order of the irreducible factors, and the primary decomposition theorem gives

$$\mathscr{V} = \mathscr{N}_{p_1^{e_1}(T)} \oplus \cdots \oplus \mathscr{N}_{p_k^{e_k}(T)}. \tag{1}$$

Now consider the restriction of T to each T-invariant space $\mathscr{N}_{p_i^{e_i}(T)}$. The minimum polynomial of this restriction map is $p_i^{e_i}(x)$. So from the previous section we know that there exist unique positive integers h_i, $e_i = r_{i1} \geq \cdots \geq r_{ih_i} \geq 1$ and there exist vectors $U_{i1}, \ldots, U_{ih_i} \in \mathscr{N}_{p_i^{e_i}(T)} \subset \mathscr{V}$ with T-minimum polynomials $p_i^{r_{i1}}(x), \ldots, p_i^{r_{ih_i}}(x)$, respectively, such that

$$\mathscr{N}_{p_i^{e_i}(T)} = \mathscr{S}(U_{i1}, T) \oplus \cdots \oplus \mathscr{S}(U_{ih_i}, T).$$

Replacing each null space in the sum (1) by such direct sums of T-cyclic subspaces completes the proof of the theorem.

Corollary 8.28 The degree of the minimum polynomial of $T \in \text{Hom}(\mathscr{V})$ cannot exceed the dimension of $\mathscr{V}$.

Proof

$$\dim \mathscr{V} = \sum_{i=1}^{k} \sum_{j=1}^{h_i} \dim \mathscr{S}(U_{ij}, T) \geq \sum_{i=1}^{k} \dim \mathscr{S}(U_{i1}, T) = \sum_{i=1}^{k} \deg p_i^{e_i}$$
$$= \deg m.$$

The collection of all $p_i(x)$-elementary divisors of T, $1 \leq i \leq k$, will be referred to as the *elementary divisors* of T. These polynomials are determined by the map T and therefore must be reflected in any matrix for T. Since two matrices are similar if and only if they are matrices for some map, we have obtained a set of invariants for matrices under similarity. It is a simple matter to extend our terminology to $n \times n$ matrices with entries from the field F. Every matrix $A \in \mathscr{M}_{n\times n}(F)$, determines a map from the vector space $\mathscr{M}_{n\times 1}(F)$ to itself with $A(V) = AV$ for all $V \in \mathscr{M}_{n\times 1}(F)$. Therefore we can speak of the minimum polynomial of A and of the elementary divisors of A. In fact, the examples and problems of the two previous sections were stated in terms of matrices for simplicity and they show that the results for a map were first obtained for the matrix.

Theorem 8.29 Two matrices from $\mathscr{M}_{n\times n}(F)$ are similar if and only if they have the same elementary divisors.

Proof ($\Rightarrow$) If A is the matrix of T with respect to the basis B, then $Y = AX$ is the matrix equation for $W = T(V)$. Therefore, the elementary divisors of A equal the elementary divisors of T. If A and A' are similar, then they are matrices for some map T and hence they have the same elementary divisors.

($\Leftarrow$) Let A be an $n \times n$ matrix with elementary divisors $p_i^{r_{ij}}(x)$ for $1 \leq j \leq h_i$, $1 \leq i \leq k$. Suppose A is the matrix of $T \in \operatorname{Hom}(\mathscr{V})$ with respect to some basis. Then there exists a basis for $\mathscr{V}$ with respect to which T has the matrix

$$D = \operatorname{diag}(C(p_1^{r_{11}}), \ldots, C(p_k^{r_{kh_k}})).$$

Thus A is similar to D. If $A' \in \mathscr{M}_{n \times n}(F)$ has the same elementary divisors as A, then by the same argument A' is similar to D and hence similar to A.

This theorem shows that the elementary divisors of a matrix form a complete set of invariants under similarity, and it provides a canonical form for each matrix.

Definition If $A \in \mathscr{M}_{n \times n}(F)$ has elementary divisors $p_i^{r_{ij}}(x)$, $1 \leq j \leq h_i$ and $1 \leq i \leq k$, then the diagonal block matrix

$$\operatorname{diag}(C(p_1^{r_{11}}), \ldots, C(p_1^{r_{1h_1}}), \ldots, C(p_k^{r_{k1}}), \ldots, C(p_k^{r_{kh_k}}))$$

is a *rational canonical form* for A.

Two rational canonical forms for a matrix A differ only in the order of the blocks. It is possible to specify an order, but there is little need to do so here. It should be noticed that the rational canonical form for a matrix A depends on the nature of the field F. For example, if $A \in \mathscr{M}_{n \times n}(R)$, then A is also in $\mathscr{M}_{n \times n}(C)$, and two quite different rational canonical forms can be obtained for A by changing the base field. This is the case in Example 1 below.

The following result is included in the notion of a canonical form but might be stated for completeness.

Theorem 8.30 Two matrices from $\mathscr{M}_{n \times n}(F)$ are similar if and only if they have the same rational canonical form.

Example 1 Find a rational canonical form for

$$A = \begin{pmatrix} 0 & -1 & 0 & -1 & 0 \\ 0 & 0 & 4 & 0 & -4 \\ -2 & 0 & 3 & -2 & -2 \\ 1 & 0 & -1 & 2 & 1 \\ -2 & 1 & 3 & -2 & -2 \end{pmatrix} \in \mathscr{M}_{5 \times 5}(R).$$

Suppose $\{E_i\}$ denotes the standard basis for $\mathscr{M}_{n\times 1}(R)$. Then a little computation shows that the A-minimum polynomial of E_1 and E_4 is $(x-1)^2$, the A-minimum polynomial of E_2 is x^2+4 and the A-minimum polynomial of E_3 and E_5 is $(x-1)^2(x^2+4)$. Therefore the minimum polynomial of A is $m(x) = (x-1)^2(x^2+4)$. The polynomials $p_1(x) = x-1$ and $p_2(x) = x^2+4$ are irreducible over R and $m(x) = p_1^2(x)p_2(x)$. Thus $p_1^2(x)$ and $p_2(x)$ are two elementary divisors of A. The degrees of the elementary divisors of A must add up to 5, so A has a third elementary divisor of degree 1. This must be $x-1$. Since the elementary divisors of A are $p_1^2(x)$, $p_1(x)$, and $p_2(x)$, a rational canonical form for A is

$$D = \operatorname{diag}(C(p_1^2), C(p_1), C(p_2)) = \begin{pmatrix} 0 & -1 & & & 0 \\ 1 & 2 & & & \\ & & 1 & & \\ & & & 0 & -4 \\ 0 & & & 1 & 0 \end{pmatrix}.$$

If the matrix of this example were given as a member of $\mathscr{M}_{5\times 5}(C)$, then the minimum polynomial would remain the same but its irreducible factors would change. Over the complex numbers,

$$m(x) = (x-1)^2(x+2i)(x-2i).$$

Then $(x-1)^2$, $x+2i$, and $x-2i$ are elementary divisors of A, and again there must be another elementary divisor of degree 1. Why is it impossible for the fourth elementary divisor to be either $x+2i$ or $x-2i$? With elementary divisors $(x-1)^2$, $x-1$, $x+2i$, and $x-2i$, a rational canonical form for $A \in \mathscr{M}_{5\times 5}(C)$ is obtained from D by replacing the block $\begin{pmatrix} 0 & -4 \\ 1 & 0 \end{pmatrix}$ with $\begin{pmatrix} 2i & 0 \\ 0 & -2i \end{pmatrix}$.

Theorem 8.31 (Cayley-Hamilton Theorem) An $n \times n$ matrix [or a map in Hom($\mathscr{V}$)] satisfies its characteristic equation.

That is, if $q(\lambda) = 0$ is the characteristic equation for the matrix A, then $q(A) = \mathbf{0}$.

Proof Suppose the rational canonical form for A is $D = \operatorname{diag}(C(p_1^{r_{11}}), \ldots, C(p_k^{r_{kh_k}}))$. Then since A is similar to D, $q(\lambda) = \det(D - \lambda I_n)$.

The determinant of a diagonal block matrix is the product of the determinants of the blocks (problem 5, page 347), so if we set $b_{ij} = \deg p_i^{r_{ij}}$, then

$$q(\lambda) = \det(C(p_1^{r_{11}}) - \lambda I_{b_{11}}) \cdots \det(C(p_k^{r_{kh_k}}) - \lambda I_{b_{kh_k}}).$$

But $\det(C(p_i^{r_{ij}}) - \lambda I_{b_{ij}})$ is the characteristic polynomial of $C(p_i^{r_{ij}})$, and by problem 2, page 376 this polynomial is $(-1)^{b_{ij}} p_i^{r_{ij}}(\lambda)$. So $q(\lambda) = (-1)^n p_1^{r_{11}}(\lambda) \cdots p_k^{r_{kh_k}}(\lambda)$. But if $m(x)$ is the minimum polynomial of A, then $m(x) = p_1^{r_{11}}(x) p_2^{r_{21}}(x) \cdots p_k^{r_{k1}}(x)$. Therefore $q(\lambda) = m(\lambda)s(\lambda)$ for some polynomial $s(\lambda)$, and $m(A) = \mathbf{0}$ implies $q(A) = \mathbf{0}$ as desired.

Corollary 8.32 If $p_1^{r_{11}}(x), \ldots, p_k^{r_{kh_k}}(x)$ are the elementary divisors for an $n \times n$ matrix A, then the characteristic polynomial of A is $(-1)^n p_1^{r_{11}}(\lambda) \cdots p_k^{r_{kh_k}}(\lambda)$.

Although we have used the rational canonical form to prove the Caley-Hamilton theorem, the form itself leaves something to be desired. As a diagonal block matrix most of its entries are 0, but a general companion matrix is quite different from a diagonal matrix. This is true even in an algebraically closed field, for example, the companion matrix of $(x - a)^4$ is

$$\begin{pmatrix} 0 & 0 & 0 & -a^4 \\ 1 & 0 & 0 & 4a^3 \\ 0 & 1 & 0 & -6a^2 \\ 0 & 0 & 1 & 4a \end{pmatrix}.$$

For practical purposes it would be better to have a matrix that is more nearly diagonal. One might hope to obtain a diagonal block matrix similar to the companion matrix $C(p^e)$ which has e blocks of $C(p)$ on the diagonal. Although this is not possible, there is a similar matrix, which comes close to this goal.

Consider a map $T \in \mathrm{Hom}(\mathscr{V})$ for which $\mathscr{V} = \mathscr{S}(U, T)$ and $p^e(x)$ is the T-minimum polynomial of U. Then $C(p^e)$ is the matrix of T with respect to the cyclic basis $\{U, T(U), \ldots, T^{eb-1}(U)\}$, where b is the degree of $p(x)$. We will obtain a simpler matrix for T by choosing a more complicated basis. Suppose $p(x) = x^b + a_{b-1}x^{b-1} + \cdots + a_1 x + a_0$. Replace x by T and apply the resulting map to U and obtain

$$p(T)U = T^b(U) + a_{b-1}T^{b-1}(U) + \cdots + a_1 T(U) + a_0 U.$$

Now repeatedly apply the map $p(T)$ to obtain the following relations:

$$p^2(T)U = p(T)T^b(U) + a_{b-1}p(T)T^{b-1}(U) + \cdots + a_1p(T)T(U) + a_0p(T)U$$

$$p^3(T)U = p^2(T)T^b(U) + a_{b-1}p^2(T)T^{b-1}(U) + \cdots + a_1p^2(T)T(U) + a_0p^2(T)U$$

$$\vdots$$

$$p^{e-1}(T)U = p^{e-2}(T)T^b(U) + a_{b-1}p^{e-2}(T)T^{b-1}(U) + \cdots + a_1p^{e-2}(T)T(U) + a_0p^{e-2}(T)U.$$

Notice that the entries from the last column of $C(p)$ appear in each equation, and the $p + 1$ vectors on the right side of each equation are like the vectors of a cyclic basis. Also the vector on the left side of each equation appears in the next equation. Therefore, consider the following set of vectors:

$$\begin{array}{ccccc} U & T(U) & T^2(U) & \cdots & T^{b-1}(U) \\ p(T)U & p(T)T(U) & p(T)T^2(U) & \cdots & p(T)T^{b-1}(U) \\ p^2(T)U & p^2(T)T(U) & p^2(T)T^2(U) & \cdots & p^2(T)T^{b-1}(U) \\ & & & \vdots & \\ p^{e-1}(T)U & p^{e-1}(T)T(U) & p^{e-1}(T)T^2(U) & \cdots & p^{e-1}(T)T^{b-1}(U). \end{array}$$

There are eb vectors in this set, so if they are linearly independent, they form a basis for $\mathscr{V}$. A linear combination of these vectors may be written as $q(T)U$ for some polynomial $q(x) \in F[x]$. Since the highest power of T occurs in $p^{e-1}(T)T^{b-1}(U)$, the degree of $q(x)$ cannot exceed $eb - 1$. Therefore, if the set is linearly dependent, there exists a nonzero polynomial $q(x)$ of degree less than $p^e(x)$ for which $q(T)U = \mathbf{0}$. This contradicts the fact that $p^e(x)$ is the T-minimum polynomial of U, so the set is a basis for $\mathscr{V}$. Order these vectors from left to right, row by row, starting at the top, and denote the basis by B_c. We will call B_c a *canonical basis* for $\mathscr{V}$. Let A_c be the matrix of T with respect to this canonical basis B_c. The first $b - 1$ vectors of B_c are the same as the first $b - 1$ vectors of the cyclic basis, but

$$T(T^{b-1}(U)) = T^b(U) = p(T)U - a_{b-1}T^{b-1}(U) - \cdots - a_1T(U) - a_0U.$$

Therefore the $b \times b$ submatrix in the upper left corner of A_c is $C(p)$, and the entry in the $p + 1$st row and pth column is 1. This pattern is repeated; each equation above gives rise to a block $C(p)$ and each equation except for the last adds a 1 on the subdiagonal between two blocks. For example, the second row of elements in B_c begins like a cyclic basis with $T(p(T)U) = p(T)T(U)$, $T(p(T)T(U)) = p(T)T^2(U)$, $T(p(T)T^2(U)) = p(T)T^3(U)$, and so on. Then using the second equation above, we have

$$T(p(T)T^{b-1}(U)) = p(T)T^b(U)$$
$$= p^2(T)U - a_{b-1}p(T)T^{b-1}(U) - \cdots - a_0p(T)U.$$

Therefore

$$A_c = \begin{pmatrix} C(p) & & & & 0 \\ 1 & C(p) & & & \\ & 1 & C(p) & & \\ & & 1 & \ddots & \\ 0 & & & 1 & C(p) \end{pmatrix}.$$

Call this matrix the *hypercompanion* matrix of $p^e(x)$ and denote it by $C_h(p^e)$.

Example 2 If $p(x) = x^2 - 2x + 3$, then $p(x)$ is irreducible over R. Find the companion matrix of $p^3(x)$, $C(p^3)$, and the hypercompanion matrix of $p^3(x)$, $C_h(p^3)$.

$$C(p^3) = \begin{pmatrix} 0 & 0 & 0 & 0 & 0 & -27 \\ 1 & 0 & 0 & 0 & 0 & 54 \\ 0 & 1 & 0 & 0 & 0 & -63 \\ 0 & 0 & 1 & 0 & 0 & 44 \\ 0 & 0 & 0 & 1 & 0 & -21 \\ 0 & 0 & 0 & 0 & 1 & 6 \end{pmatrix}$$

and

$$C_h(p^3) = \begin{pmatrix} 0 & -3 & 0 & 0 & 0 & 0 \\ 1 & 2 & 0 & 0 & 0 & 0 \\ 0 & 1 & 0 & -3 & 0 & 0 \\ 0 & 0 & 1 & 2 & 0 & 0 \\ 0 & 0 & 0 & 1 & 0 & -3 \\ 0 & 0 & 0 & 0 & 1 & 2 \end{pmatrix}$$

If the companion matrices of a rational canonical form are replaced by hypercompanion matrices for the same polynomials, then a new canonical form is obtained, which has its nonzero entries clustered more closely about the main diagonal.

Definition If $A \in \mathscr{M}_{n \times n}(F)$ has elementary divisors $p_i^{r_{ij}}(x)$, $1 \leq j \leq h_i$

and $1 \le i \le k$, then the diagonal block matrix

$$\operatorname{diag}(C_h(p_1^{r_{11}}), \ldots, C_h(p_1^{r_{1h_1}}), \ldots, C_h(p_k^{r_{k1}}), \ldots, C_h(p_k^{r_{kh_k}}))$$

is a *classical canonical form* for A.

Example 3 Suppose $T \in \operatorname{Hom}(\mathscr{V}_6)$ has elementary divisors $(x^2 + x + 2)^2$ and $(x - 5)^2$. Find a matrix D_1 for T in rational canonical form and a matrix D_2 for T in classical canonical form.

$$D_1 = \begin{pmatrix} 0 & 0 & 0 & -4 & & 0 \\ 1 & 0 & 0 & -4 & & \\ 0 & 1 & 0 & -5 & & \\ 0 & 0 & 1 & -2 & & \\ & & & & 0 & -25 \\ 0 & & & & 1 & 10 \end{pmatrix}, \quad D_2 = \begin{pmatrix} 0 & -2 & 0 & 0 & & 0 \\ 1 & -1 & 0 & 0 & & \\ 0 & 1 & 0 & -2 & & \\ 0 & 0 & 1 & -1 & & \\ & & & & 5 & 0 \\ 0 & & & & 1 & 5 \end{pmatrix}.$$

If $p(x)$ is a first degree polynomial, say $p(x) = x - c$, then the hypercompanion matrix $C_h(p^e)$ is particularly simple with the characteristic value c on the main diagonal and 1's on the subdiagonal. Thus if all irreducible factors of the minimum polynomial for A are linear, then the nonzero entries in a classical canonical form for A are confined to the main and subdiagonals. If A is not diagonalizable, then such a form is the best that can be hoped for in computational and theoretical applications. In particular, this always occurs over an algebraically closed field such as the complex numbers.

Definition If the minimum polynomial (or characteristic polynomial) of $A \in \mathscr{M}_{n \times n}(F)$ factors into powers of first degree polynomials over F, then a classical canonical form for A is called a *Jordan canonical form* or simply a Jordan form.

Every matrix with complex entries has a Jordan form. So every matrix with real entries has a Jordan form if complex entries are permitted. However, not every matrix in $\mathscr{M}_{n \times n}(R)$ has a Jordan form in $\mathscr{M}_{n \times n}(R)$. For example, the equivalence class of 6×6 real matrices containing D_1 and D_2 of Example 3 does not contain a matrix in Jordan form.

Example 4 Suppose $T \in \operatorname{Hom}(\mathscr{V}_5)$ has elementary divisors $(x - 2)^3$ and $(x - 2)^2$, as in Example 3, page 372. Then a matrix for T in classical canonical form is a Jordan form, and there is a canonical basis for $\mathscr{V}_5$ with respect to which the matrix of T is

$$\left(\begin{array}{ccc:cc} 2 & 0 & 0 & & 0 \\ 1 & 2 & 0 & & \\ 0 & 1 & 2 & & \\ \hdashline & & & 2 & 0 \\ 0 & & & 1 & 2 \end{array}\right) = \operatorname{diag}(C_h(p^3), C_h(p^2)), \qquad p(x) = x - 2.$$

Example 5 Let

$$A = \begin{pmatrix} -3 & -3 & 8 & -8 \\ 0 & 0 & 1 & -24 \\ -1 & -1 & 3 & -16 \\ 0 & 1 & -1 & 0 \end{pmatrix}.$$

Find a classical canonical form for A as a member of $\mathcal{M}_{4\times 4}(R)$ and a Jordan form for A as a member of $\mathcal{M}_{4\times 4}(C)$.

The A-minimum polynomial of $(1/8, -1/8, 0, 0)^T$ is easily found to be $x^4 + 8x^2 + 16$. Since the degree of the minimum polynomial for A cannot exceed 4, $x^4 + 8x^2 + 16$ is the minimum polynomial for A. Over the real numbers this polynomial factors into $(x^2 + 4)^2$, whereas over the complex numbers it factors into $(x - 2i)^2(x + 2i)^2$. Therefore the classical canonical form for A is

$$\left(\begin{array}{cc:cc} 0 & -4 & 0 & 0 \\ 1 & 0 & 0 & 0 \\ \hdashline 0 & 1 & 0 & -4 \\ 0 & 0 & 1 & 0 \end{array}\right) \quad \text{in} \quad \mathcal{M}_{4\times 4}(R)$$

and

$$\left(\begin{array}{cc:cc} 2i & 0 & 0 & 0 \\ 1 & 2i & 0 & 0 \\ \hdashline 0 & 0 & -2i & 0 \\ 0 & 0 & 1 & -2i \end{array}\right) \quad \text{in} \quad \mathcal{M}_{4\times 4}(C).$$

The second matrix is a Jordan form as required.

If a map T has a matrix in Jordan form, then this fact can be used to express T as the sum of a diagonalizable map and a map which becomes zero when raised to some power.

Definition A map $T \in \mathrm{Hom}(\mathscr{V})$ is *nilpotent* if $T^k = \mathbf{0}$ for some positive integer k. The *index* of a nilpotent map is 1 if $T = \mathbf{0}$ and the integer k if $T^k = \mathbf{0}$ but $T^{k-1} \neq \mathbf{0}$, $k > 1$.

Nilpotent matrices are defined similary.

Theorem 8.33 If the minimum polynomial of $T \in \mathrm{Hom}(\mathscr{V})$ has only linear irreducible factors, then there exists a nilpotent map T_n and a diagonalizable map T_d such that $T = T_n + T_d$.

Proof Since the minimum polynomial has only linear factors, there exists a basis B_c for $\mathscr{V}$ such that if A is the matrix for T with respect to B_c, then A is in Jordan canonical form. Write $A = N + D$ where D is the diagonal matrix containing the diagonal entries of A and N is the "subdiagonal matrix" containing the subdiagonal entries of A. It is easily shown that N is a nilpotent matrix, problem 14 below. So if $T_n \in \mathrm{Hom}(\mathscr{V})$ has N as its matrix with respect to B_c and $T_d \in \mathrm{Hom}(\mathscr{V})$ has D as its matrix with respect to B_c, then $T = T_n + T_d$.

Theorem 8.33 may be proved without reference to the existence of a Jordan form. This fact can then be used to derive the existence of a Jordan form over an algebraically closed field. See problem 16 below.

Problems

1. Given the elementary divisors of $A \in \mathscr{M}_{n \times n}(R)$, find a rational canonical form D_1 for A and a classical canonical form D_2 for A.
   a. $(x - 3)^3$, $x - 3$, $x - 3$.
   b. $(x^2 + 4)^2$, $(x + 2)^2$.
   c. $(x^2 + 16)^2$, $x^2 + 16$.
   d. $(x^2 + 2x + 2)^3$.
2. Suppose the matrices of problem 1 are in $\mathscr{M}_{n \times n}(C)$. Determine the elementary divisors of A and find a rational canonical form D_3 for A and a Jordan canonical form D_4 for A.
3. Suppose the matrix of $T \in \mathrm{Hom}(\mathscr{V}_4)$ with respect to $\{E_i\}$ is
$$\begin{pmatrix} 2 & -1 & 1 & -2 \\ 1 & 0 & 0 & -2 \\ 0 & 0 & 3 & -1 \\ 0 & 1 & 0 & 3 \end{pmatrix}.$$
   a. Show that the minimum polynomial of T is $(x - 2)^4$.
   b. Find the canonical basis B_c for $\mathscr{V}_4$ which begins with the standard basis vector E_1.
   c. Show that the matrix of T with respect to B_c is a Jordan canonical form.

4. Find a classical canonical form for A as a member of $\mathscr{M}_{4\times 4}(R)$ and a Jordan form for A as a member of $\mathscr{M}_{4\times 4}(C)$ when

$$A = \begin{pmatrix} -1 & 0 & 1 & -9 \\ -1 & 0 & 1 & -18 \\ -19 & 9 & 1 & -9 \\ -1 & 1 & 0 & 0 \end{pmatrix}.$$

5. Determine conditions which insure that the minimum polynomial of an $n \times n$ matrix A equals $(-1)^n$ times the characteristic polynomial of A.

6. Use the characteristic polynomial to find the minimum polynomial and the Jordan form of the following matrices from $\mathscr{M}_{3\times 3}(C)$.

a. $\begin{pmatrix} 4 & -2 & 3 \\ 0 & 2 & -1 \\ -1 & 1 & 0 \end{pmatrix}$. b. $\begin{pmatrix} 3 & -1 & 2 \\ -1 & 3 & -2 \\ -1 & 1 & 0 \end{pmatrix}$. c. $\begin{pmatrix} -1 & 2 & 5 \\ -4 & 5 & 10 \\ 2 & -2 & -3 \end{pmatrix}$.

7. Suppose the characteristic polynomial of A is $(x - 2)^6$.
   a. Find all possible sets of elementary divisors for A.
   b. Which sets of elementary divisors are determined entirely by the number of linearly independent characteristic vectors for 2?

8. Compute $\begin{pmatrix} a & 0 \\ 1 & a \end{pmatrix}^n$ for any positive integer n.

9. For the following matrices in Jordan form, determine by inspection their elementary divisors and the number of linearly independent characteristic vectors for each characteristic value.

a. $\begin{pmatrix} 2 & 0 & 0 & 0 \\ 1 & 2 & 0 & 0 \\ 0 & 0 & 2 & 0 \\ 0 & 0 & 1 & 2 \end{pmatrix}$. b. $\begin{pmatrix} 2 & 0 & 0 & 0 \\ 1 & 2 & 0 & 0 \\ 0 & 1 & 2 & 0 \\ 0 & 0 & 0 & 2 \end{pmatrix}$. c. $\begin{pmatrix} 2 & 0 & 0 & 0 \\ 1 & 2 & 0 & 0 \\ 0 & 0 & 2 & 0 \\ 0 & 0 & 0 & 2 \end{pmatrix}$.

10. Suppose the matrix of T with respect to $\{E_i\}$ is

$$\begin{pmatrix} -1 & 0 & 1 & 1 \\ -1 & 0 & 2 & 1 \\ 4 & -3 & 0 & 0 \\ -5 & 2 & 1 & 1 \end{pmatrix}.$$

   a. Show that the minimum polynomial of T is $(x^2 + 2)^2$.
   b. With $T \in \text{Hom}(\mathscr{V}_4(R))$, find a canonical basis B_c for $\mathscr{V}_4$ begining with $(0, 0, 1, -1)$ and find the matrix of T with respect to B_c.
   c. With $T \in \text{Hom}(\mathscr{V}_4(C))$, find a canonical basis B_c for $\mathscr{V}_4(C)$ and the matrix of T with respect to B_c.

11. a. Show that every characteristic value of a nilpotent map is zero.
   b. Suppose $T \in \text{Hom}(\mathscr{V})$ is nilpotent, and $\dim \mathscr{V} = 4$. List all posible sets of elementary divisors for T and find a matrix in Jordan form for each.
   c. Prove that if T is nilpotent, then T has a matrix with 0's off the subdiagonal and 1's as the only nonzero entries on the subdiagonal, for some choice of basis.

12. Show that the following matrices are nilpotent and find a Jordan form for each.

a. $\begin{pmatrix} 6 & -4 & 3 \\ 6 & -4 & 1 \\ -3 & 2 & -2 \end{pmatrix}$. b. $\begin{pmatrix} 1 & -1 & 1 \\ -1 & 1 & -1 \\ -1 & 1 & -1 \end{pmatrix}$. c. $\begin{pmatrix} 1 & 1 & -1 & 2 \\ -2 & -2 & 2 & -4 \\ 5 & 3 & -1 & 2 \\ 3 & 2 & -1 & 2 \end{pmatrix}$.

13. Find à nilpotent map T_n and a diagonalizable map T_d such that $T = T_n + T_d$ when

a. $T(a, b) = (5a - b, a + 3b)$.

b. $T(a, b, c) = (3a - b - c, 4b + c, a - 3b)$.

14. Suppose A is an $n \times n$ matrix with all entries on and above the main diagonal equal to 0. Show that A is nilpotent.

15. Suppose the mimimum polynomial of $T \in \text{Hom}(\mathscr{V})$ has only linear irreducible factors. Let $T_1, \ldots, T_k$ be the projection maps that determine a primary decomposition of $\mathscr{V}$, as obtained in problem 15 on page 367. Thus if $\lambda_1, \ldots, \lambda_k$ are the k distinct characteristic values of T, then $T_i \colon \mathscr{V} \to \mathscr{N}_{(T - \lambda_i I)^{e_i}}$.

a. Prove that $T_d = \lambda_1 T_1 + \cdots + \lambda_k T_k$ is diagonalizable.

b. Prove that $T_n = T - T_d$ is nilpotent. Suggestion: Write $T = T \circ I = T \circ T_1 + \cdots + T \circ T_k$ and show that for any positive integer r,

$$T_n^r = (T - \lambda_1 I)^r \circ T_1 + \cdots + (T - \lambda_k I)^r \circ T_k.$$

16. Combine problems 11c and 15 to show how the decomposition $T = T_d + T_n$ yields a matrix in Jordan form. (The statement of 11c can be strengthened to specify how the 1's appear, and then proved without reference to a Jordan form. See problem 17.)

17. Suppose $T \in \text{Hom}(\mathscr{V})$ is nilpotent of index r and nullity s.

a. Show that $\{0\} = \mathscr{N}_{T^0} \subset \mathscr{N}_T \subset \mathscr{N}_{T^2} \subset \cdots \subset \mathscr{N}_{T^r} = \mathscr{V}$.

b. Show that $\mathscr{V}$ has a basis

$$B_1 = \{U_1, \ldots, U_{a_1}, U_{a_1+1}, \ldots, U_{a_2}, \ldots, U_{a_{r-1}+1}, \ldots, U_{a_r}\},$$

with $\{U_1, \ldots, U_{a_i}\}$ a basis for $\mathscr{N}_{T^i}$ for $i = 1, \ldots, r$.

c. Prove that the set $\{U_1, \ldots, U_{a_{i-1}}, T(U_{a_i+1}), \ldots, T(U_{a_{i+1}})\}$ may be extended to a basis for $\mathscr{N}_{T^i}$ for $i = 1, \ldots, r - 1$.

d. Use part c to obtain a basis $B_2 = \{W_1, \ldots, W_{a_1}, \ldots, W_{a_{r-1}+1}, \ldots, W_{a_r}\}$ such that $\{W_1, \ldots, W_{a_i}\}$ is a basis for $\mathscr{N}_{T^i}$ for each i, and if $W \in B_2$ is in the basis for $\mathscr{N}_{T^i}$, but not the basis for $\mathscr{N}_{T^{i-1}}$, then

$$T(W) \in \{W_{a_{i-2}+1}, \ldots, W_{a_{i-1}}\}, \ T^2(W) \in \{W_{a_{i-3}+1}, \ldots, W_{a_{i-2}}\},$$

and so on.

e. Find a basis B_3 for $\mathscr{V}$ such that the matrix of T with respect to B_3 is $\text{diag}(N_1, \ldots, N_s)$, where N_j has 1's on the subdiagonal and 0's elsewhere, and if N_j is $b_j \times b_j$, then $r = b_1 \geq b_2 \geq \cdots \geq b_s \geq 1$.

§6. Applications of the Jordan Form

Our examples have been chosen primarily from Euclidean geometry to provide clarification of abstract concepts and a context in which to test new ideas. However, there are no simple geometric applications of the Jordan form for a nondiagonalizable matrix. On the other hand, there are many places where the Jordan form is used, and a brief discussion of them will show how linear techniques are applied in other fields of study. Such an application properly belongs in an orderly presentation of the field in question, so no attempt will be made here to give a rigorous presentation. We will briefly consider applications to differential equations, probability, and numerical analysis.

Differential equations: Suppose we are given the problem of finding $x_1, \ldots, x_n$ as functions of t when

$$D_t x_i = dx_i/dt = a_{i1}x_1 + \cdots + a_{in}x_n, \qquad x_i(t_0) = c_i$$

for each i, $1 \leq i \leq n$. If the a_{ij}'s are all scalars, then this system of equations is called a *homogeneous system of linear first-order differential equations with constant coefficients*. The conditions $x_i(t_0) = c_i$ are the *initial conditions* at $t = t_0$. Such a system may be written in matrix form by setting $A = (a_{ij})$, $X = (x_1, \ldots, x_n)^T$, and $C = (c_1, \ldots, c_n)^T$. Then the system becomes $D_t X = AX$ or $D_t X(t) = AX(t)$ with $X(t_0) = C$.

Example 1 Such a system is given by

$$D_t x = 4x - 3y - z, \qquad x(0) = 2$$

$$D_t y = x - z, \qquad y(0) = 1$$

$$D_t z = -x + 2y + 3z, \qquad z(0) = 4.$$

This system is written $D_t X = AX$, with initial conditions at $t = 0$ given by $X(0) = C$, if

$$A = \begin{pmatrix} 4 & -3 & -1 \\ 1 & 0 & -1 \\ -1 & 2 & 3 \end{pmatrix}, \qquad X = \begin{pmatrix} x \\ y \\ z \end{pmatrix}, \qquad C = \begin{pmatrix} 2 \\ 1 \\ 4 \end{pmatrix}.$$

A *solution* for the system $D_t X = AX$ with initial conditions $X(t_0) = C$ is a set on n functions $f_1, \ldots, f_n$ such that $x_1 = f_1(t), \ldots, x_n = f_n(t)$ satisfies the equations and $f_1(t_0) = c_1, \ldots, f_n(t_0) = c_n$. Usually $D_t X = AX$ cannot be solved by inspection or by simple computation of antiderivatives. However, the system would be considerably simplified if the coefficient matrix A could be replaced by its Jordan form. Such a change in the coefficient matrix amounts to a change in variables.

Theorem 8.34 Given the system of differential equations $D_tX = AX$ with initial conditions $X(t_0) = C$, let $X = PX'$ define a new set of variables, P nonsingular. Then X is a solution of $D_tX = AX$, $X(t_0) = C$ if and only if X' is a solution of $D_tX' = P^{-1}APX'$, $X'(t_0) = P^{-1}C$.

Proof Since differentiation is a linear operation and P is a matrix of constants, $P(D_tX') = D_t(PX')$. Therefore $D_tX' = P^{-1}APX'$ if and only if $D_t(PX') = A(PX')$, that is, if and only if $X = PX'$ satisfies the equation $D_tX = AX$.

Example 2 Find x, y, z if

$$D_tX = D_t\begin{pmatrix} x \\ y \\ z \end{pmatrix} = \begin{pmatrix} 8 & 2 & -8 \\ 13 & 3 & -14 \\ 10 & 2 & -10 \end{pmatrix}\begin{pmatrix} x \\ y \\ z \end{pmatrix} = AX \quad \text{and} \quad X(0) = \begin{pmatrix} 4 \\ 1 \\ 3 \end{pmatrix} = C.$$

For this problem $t_0 = 0$.

The characteristic values of A are 1 and ± 2 so A is diagonalizable. In fact, if

$$P = \begin{pmatrix} 2 & 1 & 1 \\ 1 & 1 & 3 \\ 2 & 1 & 2 \end{pmatrix},$$

then

$$P^{-1}AP = \begin{pmatrix} 1 & 0 & 0 \\ 0 & 2 & 0 \\ 0 & 0 & -2 \end{pmatrix}.$$

Therefore if new variables x', y', z' are defined by $X = PX'$, then it is necessary to solve $D_tX' = \operatorname{diag}(1, 2, -2)X'$ with initial conditions

$$X'(0) = P^{-1}C = \begin{pmatrix} -1 & -1 & 2 \\ 4 & 2 & -5 \\ -1 & 0 & 1 \end{pmatrix}\begin{pmatrix} 4 \\ 1 \\ 3 \end{pmatrix} = \begin{pmatrix} 1 \\ 3 \\ -1 \end{pmatrix}.$$

The equations in the new coordinates are simply $D_tx' = x'$, $D_ty' = 2y'$, and $D_tz' = -2z'$, which have general solutions $x' = ae^t$, $y' = be^{2t}$, $z' = ce^{-2t}$ where a, b, c are arbitrary constants. But the initial conditions yield $x' = e^t$,

$y' = 3e^{2t}$, and $z' = -e^{-2t}$, therefore the solution of the original system is

$$X = PX' = \begin{pmatrix} 2 & 1 & 1 \\ 1 & 1 & 3 \\ 2 & 1 & 2 \end{pmatrix} \begin{pmatrix} e^t \\ 3e^{2t} \\ -e^{-2t} \end{pmatrix}$$

or

$$x = 2e^t + 3e^{2t} - e^{-2t}$$
$$y = e^t + 3e^{2t} - 3e^{-2t}$$
$$z = 2e^t + 3e^{2t} - 2e^{-2t}.$$

Example 3 Consider the system given in Example 1. A Jordan form for A is given by $P^{-1}AP$ with

$$P = \begin{pmatrix} 1 & -1 & 2 \\ 1 & -1 & 1 \\ 0 & 1 & -1 \end{pmatrix} \quad \text{and} \quad P^{-1}AP = \begin{pmatrix} 2 & 0 & 0 \\ 1 & 2 & 0 \\ 0 & 0 & 3 \end{pmatrix}.$$

Thus A is not diagonalizable. If new variables are given by $X = PX'$, then it is necessary to solve the system

$$D_t \begin{pmatrix} x' \\ y' \\ z' \end{pmatrix} = \begin{pmatrix} 2 & 0 & 0 \\ 1 & 2 & 0 \\ 0 & 0 & 3 \end{pmatrix} \begin{pmatrix} x' \\ y' \\ z' \end{pmatrix}$$

with

$$\begin{pmatrix} x'(0) \\ y'(0) \\ z'(0) \end{pmatrix} = P^{-1}C = \begin{pmatrix} 0 & 1 & 1 \\ 1 & -1 & 1 \\ 1 & -1 & 0 \end{pmatrix} \begin{pmatrix} 2 \\ 1 \\ 4 \end{pmatrix} = \begin{pmatrix} 5 \\ 5 \\ 1 \end{pmatrix}.$$

The first and last equations are easily solved: $x' = 5e^{2t}$ and $z' = e^{3t}$. The second equation is $D_t y' = x' + 2y'$ or $D_t y' - 2y' = 5e^{2t}$. Using e^{-2t} as an integrating factor,

$$(D_t y' - 2y')e^{-2t} = 5e^{2t}e^{-2t} \quad \text{or} \quad D_t(y'e^{-2t}) = D_t(5t).$$

Therefore $y'e^{-2t} = 5t + b$, but $y' = 5$ when $t = 0$ so $y' = 5(t + 1)e^{2t}$, and the solution for the original system is

$$\begin{pmatrix} x \\ y \\ z \end{pmatrix} = X = PX' = \begin{pmatrix} 1 & -1 & 2 \\ 1 & -1 & 1 \\ 0 & 1 & -1 \end{pmatrix} \begin{pmatrix} 5e^{2t} \\ 5(t+1)e^{2t} \\ e^{3t} \end{pmatrix}.$$

The general solution of $D_t x = ax$ is $x = ce^{at}$ where c is an arbitrary constant. The surprising fact is that if e^{tA} is properly defined, then the general solution of $D_t X = AX$ is given by $e^{tA}C$ where C is a column matrix of constants. There is no point in proving this here, however computation of e^{tA} is most easily accomplished using a Jordan form for A.

Computation of e^A *and* e^{tA}: The matrix e^A is defined by following the pattern set for e^x in the Maclaurin series expansion:

$$e^x = 1 + x + \frac{x^2}{2!} + \frac{x^3}{3!} + \cdots = \sum_{k=0}^{\infty} \frac{x^k}{k!}.$$

Given any $n \times n$ matrix A, we would set

$$e^A = I_n + A + \frac{1}{2!}A^2 + \frac{1}{3!}A^3 + \cdots = \sum_{k=0}^{\infty} \frac{1}{k!}A^k, \qquad A^0 = I_n,$$

if this infinite sum of matrices can be made meaningful.

Definition Suppose $A_k \in \mathcal{M}_{n \times n}$ for $k = 0, 1, 2, \ldots$, and $A_k = (a_{ij,k})$. If $\lim_{k \to \infty} a_{ij,k}$ exists for all i and j, then we write

$$\lim_{k \to \infty} A_k = B \quad \text{where} \quad B = (b_{ij}) \quad \text{and} \quad b_{ij} = \lim_{k \to \infty} a_{ij,k}.$$

$S_k = \sum_{i=0}^{k} A_i$ is a *partial sum* of the *infinite series* of matrices $\sum_{k=0}^{\infty} A_i$. This series *converges* if $\lim_{k \to \infty} S_k$ exists. If $\lim_{k \to \infty} S_k = S$, then we write $\sum_{k=0}^{\infty} A_k = S$.

Theorem 8.35 The series

$$\sum_{k=0}^{\infty} \frac{1}{k!} A^k$$

converges for every $A \in \mathcal{M}_{n \times n}$.

Proof Let $A = (a_{ij})$ and $A^k = (a_{ij,k})$. The entries in A must be bounded, say by $M = \max\{|a_{ij}| | 1 \leq i, j \leq n\}$. Then the entries in A^2 are bounded by nM^2 for

$$|a_{ij,2}| = \left|\sum_{h=1}^{n} a_{ih}a_{hj}\right| \leq \sum_{h=1}^{n} |a_{ih}a_{hj}| \leq \sum_{h=1}^{n} M^2 = nM^2.$$

An induction argument gives $|a_{ij,k}| \leq n^{k-1}M^k$. Therefore,

$$\left|\frac{1}{k!} a_{ij,k}\right| \leq \frac{1}{k!} n^{k-1} M^k < \frac{(nM)^k}{k!}.$$

But $(nM)^k/k!$ is the $k + 1$st term in the series expansion for e^{nM}, which always converges. Therefore, by the comparison test the seires $\sum_{k=0}^{\infty} (1/k!)a_{ij,k}$ converges absolutely, hence it converges for all i and j. That is, the series $\sum_{k=0}^{\infty} (1/k!)A^k$ converges.

Thus we may define

$$e^A = \sum_{k=0}^{\infty} \frac{1}{k!} A^k \qquad \text{and} \qquad e^{tA} = \sum_{k=0}^{\infty} \frac{t^k}{k!} A^k.$$

If A is diagonal, then e^A is easily computed for

$$e^{\operatorname{diag}(d_1, \ldots, d_n)} = \operatorname{diag}(e^{d_1}, \ldots, e^{d_n}).$$

But if A is not diagonal, then e^A can be computed using a Jordan form for A.

Theorem 8.36 If $B = P^{-1}AP$, then $e^B = P^{-1}e^AP$.

Proof

$$e^B = \sum_{k=0}^{\infty} \frac{1}{k!} B^k = \sum_{k=0}^{\infty} \frac{1}{k!} (P^{-1}AP)^k = \sum_{k=0}^{\infty} \frac{1}{k!} P^{-1}A^kP$$
$$= P^{-1}\left(\sum_{k=0}^{\infty} \frac{1}{k!} A^k\right)P = P^{-1}e^AP.$$

Example 4 Let A and P be as in Example 2. Then $P^{-1}AP =$

diag(1, 2, −2) and

$$\begin{aligned} e^{tA} &= P(e^{t\,\mathrm{diag}(1,2,-2)})P^{-1} \\ &= P\,\mathrm{diag}(e^{t}, e^{2t}, e^{-2t})P^{-1} \end{aligned}$$

Now with C as in Example 2,

$$e^{tA}C = P\,\mathrm{diag}(e^{t}, e^{2t}, e^{-2t})P^{-1}C$$

is the solution obtained for the system of differential equations given in Example 2.

The calculation of e^{A} is more difficult if A is not diagonalizable. Suppose $J = \mathrm{diag}(J_1, \ldots, J_h)$ is a Jordan form for A where each block J_i contains 1's on the subdiagonal and a characteristic value of A on the diagonal. Then $e^{J} = \mathrm{diag}(e^{J_1}, \ldots, e^{J_h})$. Thus it is sufficient to consider J an $n \times n$ matrix with λ on the diagonal and 1's on the subdiagonal. Write $J = \lambda I_n + N$, then N is nilpotent of index n. It can be shown that $e^{A+B} = e^{A}e^{B}$ provided that $AB = BA$, therefore

$$e^{J} = e^{\lambda I_n}e^{N} = e^{\lambda}\begin{pmatrix} 1 & 0 & \cdots & 0 \\ 1 & 1 & \cdots & 0 \\ 1/2! & 1 & \cdots & 0 \\ & & \vdots & \\ 1/(n-1)! & 1/(n-2)! & \cdots & 1 \end{pmatrix}.$$

and

$$e^{tJ} = e^{\lambda t}\begin{pmatrix} 1 & 0 & \cdots & 0 \\ t & 1 & \cdots & 0 \\ t^2/2! & t & \cdots & 0 \\ & & \vdots & \\ t^{n-1}/(n-1)! & t^{n-2}/(n-2)! & \cdots & 1 \end{pmatrix}.$$

Example 5 Let A and P be the matrices of Example 3. Then

$$P^{-1}AP = \begin{pmatrix} 2 & 0 & 0 \\ 1 & 2 & 0 \\ 0 & 0 & 3 \end{pmatrix}.$$

Consider

$$J = \begin{pmatrix} 2 & 0 \\ 1 & 2 \end{pmatrix} = 2I_2 + N \qquad \text{with} \quad N = \begin{pmatrix} 0 & 0 \\ 1 & 0 \end{pmatrix}.$$

Then

$$e^{tJ} = e^{2tI_2}\, e^{tN} = e^{2t}\begin{pmatrix} 1 & 0 \\ t & 1 \end{pmatrix},$$

and

$$e^{tA} = P\begin{pmatrix} e^{2t} & 0 & 0 \\ te^{2t} & e^{2t} & 0 \\ 0 & 0 & e^{3t} \end{pmatrix}P^{-1}.$$

With

$$C = \begin{pmatrix} 2 \\ 1 \\ 4 \end{pmatrix} \qquad \text{and} \qquad P^{-1}C = \begin{pmatrix} 5 \\ 5 \\ 1 \end{pmatrix},$$

$X = e^{tA}C$ gives the solution obtained for the system of differential equations in Example 3.

Markov chains: In probability theory, a stochastic process concerns events that change with time. If the probability that an object will be in a certain state at time t depends only on the state it occupied at time $t - 1$, then the process is called a *Markov process*. If, in addition, there are only a finite number of states, then the process is called a *Markov chain*. Possible states for a Markow chain might be positions an object can occupy, or qualities that change with time, as in Example 6.

If a Markov chain involves n states, let a_{ij} denote the probability that an object in the ith state at time t was in the jth state at time $t - 1$. Then a_{ij} is called a *transition probability*. A Markov chain with n states is then characterized by the set of all n^2 transition probabilities and the initial distribution in the various states. Suppose $x_1, \ldots, x_n$ give the distribution at $t = 0$, and set $X = (x_1, \ldots, x_n)^T$, then the distribution at time $t = 1$ is given by the product AX, where $A = (a_{ij})$ is the $n \times n$ matrix of transition probabilities. When $t = 2$, the distribution is given by A^2X, and in general, the distribution when $t = k$ is given by the product A^kX.

The matrix A of transition probabilities is called a *stochastic matrix*. The sum of the elements in any column of a stochastic matrix must equal 1. This follows from the fact that the sum $a_{1j} + a_{2j} + \cdots + a_{nj}$ is the probability that an element in the jth state at time t was in some state at time $t - 1$.

If A is the stochastic matrix for a Markov process, then the ij entry in A^k is the probability of moving from the jth state to the ith state in k steps. If $\lim_{k\to\infty} A^k$ exists, then the process approaches a "steady-state" distribution. That is, if $B = \lim_{k\to\infty} A^k$, then the distribution $Y = BX$ is unchanged by the process, and $AY = Y$. This is the situation in Example 6.

If A is a diagonal matrix, then it is a simple matter to determine if a steady state is approached. For if $A = \text{diag}(d_1, \ldots, d_n)$, then $A^k = \text{diag}(d_1^k, \ldots, d_n^k)$. So for a diagonal matrix, $\lim_{k\to\infty} A$ exists provided $|d_i| < 1$ or $d_i = 1$ for all the diagonal entries d_i. This condition can easily be extended to diagonalizable matrices.

Theorem 8.37 If $B = P^{-1}AP$, and $\lim_{k\to\infty} A^k$ exists, then

$$\lim_{k\to\infty} B^k = P^{-1}\left(\lim_{k\to\infty} A^k\right)P.$$

Proof

$$\lim_{k\to\infty} B^k = \lim_{k\to\infty} (P^{-1}AP)^k = \lim_{k\to\infty} P^{-1}A^kP = P^{-1}\left(\lim_{k\to\infty} A^k\right)P.$$

Using Theorem 8.37, we have the following condition.

Theorem 8.38 If A is diagonalizable and $|\lambda| < 1$ or $\lambda = 1$ for each characteristic value λ of A, then $\lim_{k\to\infty} A^k$ exists.

Example 6 Construct a Markov chain as follows: Call being brown-eyed, state 1, and blue-eyed, state 2. Then a transition occurs from one generation to the next. Suppose the probability of a brown-eyed parent having a brown-eyed child is 5/6, so $a_{11} = 5/6$; and the probability of having a blue-eyed child is 1/6, so $a_{21} = 1/6$. Suppose further that the probability of a blue-eyed parent having a brown-eyed child is $1/3 = a_{12}$; and the probability of having a blue-eyed child is $2/3 = a_{22}$.

If the initial population is half brown-eyed and half blue-eyed, then

$X = (1/2, 1/2)^T$ gives the initial distribution. Therefore, the product

$$AX = \begin{pmatrix} 5/6 & 1/3 \\ 1/6 & 2/3 \end{pmatrix}\begin{pmatrix} 1/2 \\ 1/2 \end{pmatrix} = \begin{pmatrix} 7/12 \\ 5/12 \end{pmatrix}$$

indicates that after one generation, 7/12th of the population should be brown-eyed and 5/12th blue-eyed. After k generations, the proportions would be given by $A^k(1/2, 1/2)^T$.

The stochastic matrix for this Markov chain is diagonalizable. If

$$P = \begin{pmatrix} 2 & -1 \\ 1 & 1 \end{pmatrix},$$

then

$$P^{-1}AP = \begin{pmatrix} 1 & 0 \\ 0 & 1/2 \end{pmatrix}.$$

Therefore the hypotheses of Theorem 8.38 are satisfied, and

$$\lim_{k\to\infty} A^k = P\left[\lim_{k\to\infty}\begin{pmatrix} 1 & 0 \\ 0 & 1/2 \end{pmatrix}^k\right]P^{-1} = P\begin{pmatrix} 1 & 0 \\ 0 & 0 \end{pmatrix}P^{-1} = \begin{pmatrix} 2/3 & 2/3 \\ 1/3 & 1/3 \end{pmatrix}.$$

Hence this chain approaches a steady-state distribution. In fact,

$$\begin{pmatrix} 2/3 & 2/3 \\ 1/3 & 1/3 \end{pmatrix}\begin{pmatrix} 1/2 \\ 1/2 \end{pmatrix} = \begin{pmatrix} 2/3 \\ 1/3 \end{pmatrix},$$

so that, in time, the distribution should approach two brown-eyed individuals for each blue-eyed individual. Notice that, in this case, the steady-state distribution is independent of the initial distribution. That is, for any a with $0 \le a \le 1$,

$$\begin{pmatrix} 2/3 & 2/3 \\ 1/3 & 1/3 \end{pmatrix}\begin{pmatrix} 1 - \mathrm{a} \\ a \end{pmatrix} = \begin{pmatrix} 2/3 \\ 1/3 \end{pmatrix}.$$

Numerical analysis: A stochastic matrix A must have 1 as a characteristic value (see problem 8 at the end of this section), therefore $\lim_{k\to\infty} A^k$ cannot be $\mathbf{0}$. In contrast, our application of the Jordan form to numerical analysis requires that A^k approach the zero matrix.

Theorem 8.39 Suppose A is a matrix with complex entries. $\text{Lim}_{k \to \infty} A^k = \mathbf{0}$ if and only if $|\lambda| < 1$ for every characteristic value λ of A.

Proof If $J = \text{diag}(J_1, \ldots, J_h)$ is a Jordan form for A, then $\lim_{k \to \infty} A^k = \mathbf{0}$ if and only if $\lim_{k \to \infty} J^k = \mathbf{0}$. But $J^k = \text{diag}(J_1^k, \ldots, J_h^k)$, so it is sufficient to consider a single Jordan block $\lambda I_n + N$ where λ is a characteristic value of A and N contains 1's on the subdiagonal and 0's elsewhere. That is, it must be shown that

$$\lim_{k \to \infty} (\lambda I_n + N)^k = \mathbf{0} \qquad \text{if and only if} \quad |\lambda| < 1.$$

If $k \geq n$, then since the index of N is n,

$$(\lambda I_n + N)^k = \lambda^k I_n + \binom{k}{1} \lambda^{k-1} N + \binom{k}{2} \lambda^{k-2} N^2 + \cdots + \binom{k}{n-1} \lambda^{k-n+1} N^{n-1}$$

where $\binom{k}{h}$ is the binomial coefficient $\dfrac{k!}{h!(k-h)!}$. Therefore

$$(\lambda I_n + N)^k = \begin{bmatrix} \lambda^k & 0 & \cdots & 0 \\ \binom{k}{1}\lambda^{k-1} & \lambda^k & \cdots & 0 \\ \binom{k}{2}\lambda^{k-2} & \binom{k}{1}\lambda^{k-1} & \cdots & 0 \\ & & \vdots & \\ \binom{k}{n-1}\lambda^{k-n+1} & \binom{k}{n-2}\lambda^{k-n+2} & \cdots & \lambda^k \end{bmatrix}.$$

If $\lim_{k \to \infty} (\lambda I_n + N)^k = \mathbf{0}$, then $|\lambda| < 1$ since λ^k is on the diagonal. Conversely, if $|\lambda| < 1$, then

$$\lim_{k \to \infty} \lambda^k = 0 \qquad \text{and} \qquad \lim_{k \to \infty} \binom{k}{i} \lambda^{k-i} = 0$$

follows from repeated applicalions of L'Hopital's rule for $1 \leq i \leq n$, so that $\lim_{k \to \infty} (\lambda I_n + N)^k = \mathbf{0}$.

Using this result we can show that the series $\sum_{k=0}^{\infty} A^k$ behaves very much like a geometric series of the form $\sum_{n=0}^{\infty} ar^n$, a and r scalars.

Theorem 8.40 The series $\sum_{k=0}^{\infty} A^k$ converges if and only if $\lim_{k\to\infty} A^k = \mathbf{0}$. Further when the series converges,

$$\sum_{k=0}^{\infty} A^k = (I_n - A)^{-1}.$$

Proof Let S_k be the partial sum $\sum_{h=0}^{k} A^h$.

($\Rightarrow$) If $\sum_{k=0}^{\infty} A^k$ converges, say $\sum_{k=0}^{\infty} A^k = S$, then for $k > 1$, $A^k = S_k - S_{k-1}$ and

$$\lim_{k\to\infty} A^k = \lim_{k\to\infty} (S_k - S_{k-1}) = S - S = \mathbf{0}.$$

($\Leftarrow$) Suppose $\lim_{k\to\infty} A^k = \mathbf{0}$ and consider

$$S_k = I_n + A + A^2 + \cdots + A^k.$$

Then

$$AS_k = A + A^2 + A^3 + \cdots + A^k + A^{k+1} = S_k - I_n + A^{k+1},$$

or

$$I_n - A^{k+1} = (I_n - A)S_k.$$

Taking the limit gives

$$I_n = \lim_{k\to\infty}(I_n - A^{k+1}) = \lim_{k\to\infty} (I_n - A)S_k = (I_n - A) \lim_{k\to\infty} S_k.$$

Therefore the limit of the partial sums exists and is equal to $(I_n - A)^{-1}$, that is,

$$\sum_{k=0}^{\infty} A^k = (I_n - A)^{-1}.$$

Corollary 8.41 $\sum_{k=0}^{\infty} A^k = (I_n - A)^{-1}$ if and only if $|\lambda| < 1$ for every characteristic value λ of A.

The above theorems are employed in iterative techniques. For example, suppose $EX = B$ is a system of n linear equations in n unknowns. If $A = I_n - E$, then $X = AX + B$ is equivalent to the original system. Given a nonzero vector X_0, define X_k by $X_k = AX_{k-1} + B$ for $k = 1, 2, \ldots$. If the sequence of vectors $X_0, X_1, X_2, \ldots$ converges, then it converges to a solution of the given system. For if $\lim_{h\to\infty} X_h = C$, then

$$C = AC + B = (I_n - E)C + B,$$

or $EC = B$. This is a typical iterative technique of numerical analysis.

Theorem 8.42 The sequence $X_0, X_1, X_2, \ldots$ converges to a solution for $EC = B$ if $|\lambda| < 1$ for every characteristic value λ of $A = I_n - E$.

Proof

$$X_1 = AX_0 + B,$$

$$X_2 = AX_1 + B = A^2X_0 + AB + B,$$

$$X_3 = AX_2 + B = A^3X_0 + (A^2 + A + I_n)B,$$

and by induction

$$X_h = A^hX_0 + \left(\sum_{k=0}^{h-1} A^k\right)B.$$

With the given assumption,

$$\lim_{h\to\infty} X_h = \left(\lim_{h\to\infty} A^h\right)X_0 + \left(\lim_{h\to\infty} \sum_{k=0}^{h-1} A^k\right)B = (I_n - A)^{-1}B.$$

$(I_n - A)^{-1}B$ is a solution for $EX = B$ for

$$E((I_n - A)^{-1}B) = (I_n - A)(I_n - A)^{-1}B = B.$$

Notice that the solution obtained above agrees with Cramer's rule because

$$(I_n - A)^{-1} = (I_n - (I_n - E))^{-1} = E^{-1}.$$

What if E were singular?

Because of our choice of problems and examples, it may not be clear why one would elect to use such an iterative technique to solve a system of linear equations. But suppose the coefficients of a system were given only approximately, say to four or five decimal places. Then this technique might provide a method for approximating a solution involving only matrix multiplication and addition. This is obviously not something to be done by hand, but a computer can be programmed to perform such a sequence of operations.

Problems

1. Solve the following systems of differential equations by making a change of variables.

a. $D_t x = x - 6y, \quad x(0) = 1$
$D_t y = 2x - 6y, \quad y(0) = 5.$

b. $D_t x = 7x - 4y, \quad x(0) = 2$
$D_t y = 9x - 5y, \quad y(0) = 5.$

c. $D_t x = x - 2y + 2z, \quad x(0) = 3$
$D_t y = 5y - 4z, \quad y(0) = 1$
$D_t z = 6y - 5z, \quad z(0) = 2.$

d. $D_t x = 5y - z, \quad x(0) = 1$
$D_t y = -x + 5y - z, \quad y(0) = 2$
$D_t z = -x + 3y + z, \quad z(0) = 1.$

2. Compute e^A and e^{tA} if A is

a. $\begin{pmatrix} -1 & 4 \\ -2 & 5 \end{pmatrix}$. b. $\begin{pmatrix} 3 & 0 \\ 1 & 3 \end{pmatrix}$. c. $\begin{pmatrix} 4 & 0 & 0 \\ 0 & 5 & 0 \\ 0 & 1 & 5 \end{pmatrix}$. d. $\begin{pmatrix} 3 & 0 & 0 & 0 \\ 1 & 3 & 0 & 0 \\ 0 & 1 & 3 & 0 \\ 0 & 0 & 1 & 3 \end{pmatrix}$.

3. For the systems $D_t X = AX$, $X(0) = \dot{C}$, of problem 1, show that $e^{tA}C$ is the solution obtained by making a change of variables.

4. An nth order linear homogeneous differential equation with constant coefficients has the form

$$D_t^n x + a_{n-1} D_t^{n-1} x + \cdots + a_1 D_t x + a_0 x = 0 \tag{1}$$

Define $x_1 = x$, $x_2 = D_t x$, $x_3 = D_t^2 x, \ldots, x_n = D_t^{n-1} x$, and consider the homogeneous system of linear first-order differential equations

$$\begin{aligned} D_t x_1 &= x_2 \\ D_t x_2 &= x_3 \\ &\vdots \\ D_t x_{n-1} &= x_n \\ D_t x_n &= -a_{n-1} x_n - a_{n-1} x_{n-2} - \cdots - a_1 x_2 - a_0 x_1. \end{aligned} \tag{2}$$

a. What is the form of the matrix A if the system (2) is written as $D_t X = AX$?

b. Show that $x = f(t)$ is a solution of (1) if and only if $x_1 = f(t)$ is included in a solution of (2).

5. Find x as a function of t if

a. $D_t^2x - 5D_tx + 6x = 0$, $x(0) = 2$, $D_tx(0) = -1$.
b. $D_t^2x - 8D_tx + 16x = 0$, $x(0) = 7$, $D_tx(0) = 34$.
c. $D_t^3x + 2D_t^2x - D_tx - 2x = 0$, $x(0) = 2$, $D_tx(0) = 1$, $D_t^2x(0) = -1$.

6. Find the $\lim_{k\to\infty} A^k$ if it exists when A is

a. $\begin{pmatrix} 1 & 0 \\ 0 & 3/5 \end{pmatrix}$. b. $\begin{pmatrix} 1 & 0 \\ 0 & 4/3 \end{pmatrix}$. c. $\begin{pmatrix} -1 & 1 \\ -2 & 11/6 \end{pmatrix}$. d. $\begin{pmatrix} 7 & -4 \\ 9 & -5 \end{pmatrix}$.

e. $\begin{pmatrix} 1/2 & 0 & 0 \\ 3 & 1 & 0 \\ -1 & 0 & 1 \end{pmatrix}$. f. $\begin{pmatrix} 1/5 & 0 & 0 & 0 \\ 3/5 & 1 & 0 & 0 \\ 0 & 0 & 2/3 & 0 \\ 2 & 0 & -4 & 1 \end{pmatrix}$. g. $\begin{pmatrix} 1 & 4 & -3 \\ -1/2 & 3 & -3/2 \\ -1/2 & 2 & -1/2 \end{pmatrix}$.

7. a. Show that if $(x - 1)^2$ is an elementary divisor of A, then $\lim_{k\to\infty} A^k$ does not exist.
b. Show that if $\lim_{k\to\infty} A^k = \mathbf{0}$, then $\det A < 1$. Is the converse true?

8. Let A be a stochastic matrix and $B = (1, 1, \ldots, 1)$. Show that 1 is a characteristic value of A by considering the product BA.

9. Does $\lim_{k\to\infty} A^k$ exist for the following stochastic matrices?

a. $\begin{pmatrix} 3/4 & 1/3 \\ 1/4 & 2/3 \end{pmatrix}$. b. $\begin{pmatrix} 1/4 & 1/3 & 2/3 \\ 2/4 & 1/3 & 0 \\ 1/4 & 1/3 & 1/3 \end{pmatrix}$.

10. Show that if E is singular, then $A = I_n - E$ has a characteristic value λ such that $|\lambda| \geq 1$.

Review Problems

1. Suppose $\begin{pmatrix} 4 & 9 & 6 \\ -1 & 1 & 0 \\ -2 & -9 & -5 \end{pmatrix}$ is the matrix of $T \in \operatorname{Hom}(\mathscr{V}_3)$ with respect to $\{E_i\}$.

a. Show that $x^3 - 1$ is the T-minimum polynomial of E_3.
b. Find the vectors $U_1 = (T - I)(E_3)$ and $U_2 = (T^2 + T + I)(E_3)$.
c. Find the T-minimum polynomials of U_1 and U_2.
d. Find the matrix of T with respect to the basis $\{U_1, T(U_1), U_2\}$.
e. Is there a matrix for T in Jordan form?

2. Prove Theorem 8.1: $\mathscr{S}_1 + \cdots + \mathscr{S}_h$ is a direct sum if and only if $U_1 + \cdots + U_h = \mathbf{0}$, with $U_i \in \mathscr{S}_i$ for $1 \leq i \leq h$, implies $U_i = \mathbf{0}$ for $1 \leq i \leq h$.

3. Suppose $\begin{pmatrix} -3 & 0 & 10 & -2 \\ 2 & -1 & 1 & 1 \\ 0 & 0 & 2 & 0 \\ 5 & 0 & 3 & 4 \end{pmatrix}$ is the matrix of $T \in \operatorname{Hom}(\mathscr{V}_4)$ with respect to the standard basis.

a. Find the T-minimum polynomial of E_3.
b. What is the minimum polynomial of T?

c. Find a matrix for T in rational canonical form.
d. Find a matrix for T in classical canonical form.

4. Given $T \in \text{Hom}(\mathscr{V})$, suppose there exist n T-invariant subspaces $\mathscr{S}_1 \subset \mathscr{S}_2 \subset \cdots \subset \mathscr{S}_{n-1} \subset \mathscr{S}_n = \mathscr{V}$, with $\dim \mathscr{S}_i = i$, $1 \leq i \leq n$. (Such a sequence of subspaces exists if $\mathscr{V}$ is over an algebraically closed field F.)

a. Show that there exists a basis $B = \{U_1, \ldots, U_n\}$ for $\mathscr{V}$ such that for each i, $1 \leq i \leq n$, $\{U_1, \ldots, U_i\}$ is a basis for $\mathscr{S}_i$.
b. Determine the form of the matrix A of T with respect to B.
c. Show that the entries on the main diagonal of A are the characteristic values of T.
d. Suppose $p(x) \in F[x]$. Prove that μ is a characteristic value for $p(T)$ if and only if $\mu = p(\lambda)$ for some characteristic value λ for T.

5. Find the companion matrix and the hypercompanion matrix of

a. $(x + 2)^3$. b. $(x^2 + x + 1)^2$.

6. Find rational and classical canonical forms for

a. $\begin{pmatrix} 2 & 0 & 0 & 0 \\ 1 & 1 & 0 & 0 \\ 1 & 0 & 2 & 0 \\ -2 & 1 & 1 & 1 \end{pmatrix}$. b. $\begin{pmatrix} 2 & 0 & 0 & 0 \\ 2 & 1 & 0 & 0 \\ -1 & 1 & 1 & 0 \\ 7 & -3 & -1 & 2 \end{pmatrix}$.

7. Suppose $(x + 4)^3(x + 1)^2$ is the minimum polynomial of a matrix A. Determine the possible elementary divisors of A and the corresponding number of independent characteristic vectors if

a. A is 5×5. b. A is 7×7.

8. Suppose the minimum polynomial of A is $(x^2 + 4)^3$. Find a classical canonical form for A if

a. $A \in \mathscr{M}_{6\times 6}(R)$. b. $A \in \mathscr{M}_{6\times 6}(C)$.

9. Suppose the minimum polynomial of A is $(x - 3)^3$. Find the possible Jordan forms for A if A is a. 3×3. b. 5×5.

10. Suppose T is nilpotent and A is a matrix in Jordan form for T. Show that the number of 0's on the subdiagonal of A is one less than the nullity of T.

11. Find the Jordan form for the nilpotent map $T \in \text{Hom}(\mathscr{V}_n)$ if

a. $n = 6$, nullity of $T = 4$, and the index is 3.
b. $n = 7$, nullity of $T = 4$, and the index is 2.
c. Show there are two possible forms if $n = 7$, nullity $=$ index $= 3$.

12. a. Solve $D_t\begin{pmatrix} x \\ y \\ z \end{pmatrix} = \begin{pmatrix} 1 & 1 & -5 \\ 2 & 0 & -5 \\ 2 & -2 & -3 \end{pmatrix}\begin{pmatrix} x \\ y \\ z \end{pmatrix} = AX$ if $X(0) = \begin{pmatrix} 1 \\ -2 \\ 1 \end{pmatrix} = C$.

b. Show that the solution is given by $X = e^{tA}C$.

13. How might Theorem 8.38 be extended to include nondiagonalizable matrices?

APPENDIX

Determinants

A computational definition for the determinant was given in Chapter 3. This approach was sufficient for the purposes of computation, but it did not provide a proof of the important fact that a determinant may be expanded along any row or any column. Now that we are familiar with the concepts of linearity and bilinearity, it is reasonable to take an abstract approach and view the determinant as a multilinear function. We have also introduced arbitrary fields since Chapter 3. So the determinant should be defined for matrices in $\mathscr{M}_{n \times n}(F)$, that is, for matrices with entries from an arbitrary field F.

§1. A Determinant Function

We begin by redefining determinants using as a guide the properties of the determinant obtained in Chapter 3. By definition, det is a function that assigns a real number to each real $n \times n$ matrix. However, because of the simple effect of elementary row and column operations on a determinant, it may well be viewed as a function of the n rows or the n columns. For example, suppose $A_1, \ldots, A_n$ are the n columns of the $n \times n$ matrix A, and write $A = (A_1, \ldots, A_n)$. Then $\det A = \det(A_1, \ldots, A_n)$, and det takes on the appearance of a function of n (vector) variables. Now the property relating to an elementary column operation of type II takes the form

$$\det(A_1, \ldots, rA_k, \ldots, A_n) = r \det(A_1, \ldots, A_k, \ldots, A_n).$$

That is, as a function of n columns, the function det preserves scalar multiplication in the kth variable. Or the function given by

$$V \to \det(A_1, \ldots, A_{k-1}, V, A_{k+1}, \ldots, A_n)$$

from $\mathscr{M}_{n \times 1}(R)$ to R preserves scalar multiplication. We could go on to show that this function is linear or that, as a function of the n columns, det is linear in each variable.

Example 1 Suppose

$$A = \begin{pmatrix} 3 & -2 & 2 \\ 2 & 1 & 1 \\ 5 & 3 & 0 \end{pmatrix},$$

then

$$A_1 = \begin{pmatrix} 3 \\ 2 \\ 5 \end{pmatrix}, \quad A_2 = \begin{pmatrix} -2 \\ 1 \\ 3 \end{pmatrix}, \quad A_3 = \begin{pmatrix} 2 \\ 1 \\ 0 \end{pmatrix}.$$

If the second column of A is multiplied by 4, then the determinant of the new matrix is

$$\det(A_1, 4A_2, A_3) = \det\begin{pmatrix} 3 & -8 & 2 \\ 2 & 4 & 1 \\ 5 & 12 & 0 \end{pmatrix} = 4 \det\begin{pmatrix} 3 & -2 & 2 \\ 2 & 1 & 1 \\ 5 & 3 & 0 \end{pmatrix}$$
$$= 4 \det(A_1, A_2, A_3).$$

And the function $V \to \det(A_1, V, A_3)$ is given by

$$\begin{pmatrix} x \\ y \\ z \end{pmatrix} \to \det\begin{pmatrix} 3 & x & 2 \\ 2 & y & 1 \\ 5 & z & 0 \end{pmatrix} = 5x - 10y + z.$$

This function is clearly a linear map from $\mathscr{M}_{3 \times 1}(R)$ to R.

Definition Let $\mathscr{V}$ be a vector space and f a function of n variables $U_1, \ldots, U_n \in \mathscr{V}$. f is *multilinear* if the function

$$V \to f(U_1, \ldots, U_{k-1}, V, U_{k+1}, \ldots, U_n)$$

is linear for each k, $1 \le k \le n$, and for each choice of vectors

$$U_1, \ldots, U_{k-1}, U_{k+1}, \ldots, U_n \in \mathscr{V}.$$

Therefore the claim is that det is a multilinear function of the n columns in an arbitrary matrix. We will not prove this, but simply note that this result was obtained for 2×2 matrices in problem 6, page 246.

We found that an elementary column operation of type I, interchanging columns, changes the sign of det A. This property will be introduced with the following concept.

Definition A multilinear function f is *alternating* if $f(U_1, \ldots, U_n) = 0$ whenever at least two of the variables $U_1, \ldots, U_n$ are equal.

If two columns of a real matrix are equal, then the determinant is zero so if the function det is multilinear, then it is alternating.

We are now prepared to redefine determinants.

Definition A *determinant function* D from $\mathcal{M}_{n\times n}(F)$ to F is an alternating multilinear function of the n columns for which $D(I_n) = 1$.

Therefore a determinant function D must satisfy the following conditions for all $A_1, \ldots, A_n \in \mathcal{M}_{n\times 1}(F)$ and all k, $1 \le k \le n$:

1. $D(A_1, \ldots, A_k + A_k', \ldots, A_n)$
$= D(A_1, \ldots, A_k, \ldots, A_n) + D(A_1, \ldots, A_k', \ldots, A_n)$.
2. $D(A_1, \ldots, cA_k, \ldots, A_n) = cD(A_1, \ldots, A_k, \ldots, A_n)$, for all $c \in F$.
3. $D(A_1, \ldots, A_n) = 0$ if at least two columns of $A = (A_1, \ldots, A_n)$ are equal.
4. $D(I_n) = D(E_1, \ldots, E_n) = 1$ where $E_1, \ldots, E_n$ is the standard basis for $\mathcal{M}_{n\times 1}(F)$. These vectors are actually $E_1^T, \ldots, E_n^T$, but the transpose notation will be dropped for simplicity.

So little is required of a determinant function that one might easily doubt that D could equal det. In fact there is no assurance that a determinant function D exists. One cannot say, "of course one exists, det is obviously a determinant function," for the proof of this statement relies on the unproved result about expansions along arbitrary columns.

Assuming that a determinant function exists, is it clear why the condition $D(I_n) = 1$ is assumed? Suppose there are many alternating multilinear functions of the columns of a matrix. This is actually the case, and such a condition is required to obtain uniqueness and equality with det.

We will proceed by assuming that a determinant function exists and search for a characterization in terms of the entries from the matrix A. The first step is to note that D must be *skew-symmetric*. That is, interchanging two variables changes the sign of the functional value.

Theorem A1 If D exists, then it is skew-symmetric. That is, interchanging two columns of A changes $D(A)$ to $-D(A)$.

Proof It must be shown that if $1 \le h < k \le n$, then

$$D(A_1, \ldots, A_h, \ldots, A_k, \ldots, A_n) = -D(A_1, \ldots, A_k, \ldots, A_h, \ldots, A_n).$$

For the matrix with $A_h + A_k$ in both its hth and kth columns,

$$\begin{aligned}0 &= D(A_1, \ldots, A_h + A_k, \ldots, A_h + A_k, \ldots, A_n)\\ &= D(A_1, \ldots, A_h, \ldots, A_h, \ldots, A_n) + D(A_1, \ldots, A_h, \ldots, A_k, \ldots, A_n)\\ &\quad + D(A_1, \ldots, A_k, \ldots, A_h, \ldots, A_n) + D(A_1, \ldots, A_k, \ldots, A_k, \ldots, A_n)\\ &= D(A_1, \ldots, A_h, \ldots, A_k, \ldots, A_n) + D(A_1, \ldots, A_k, \ldots, A_h, \ldots, A_n).\end{aligned}$$

This shows that the two functional values are additive inverses in F, establishing the theorem.

Skew-symmetry is not equivalent to alternating. However, in a field satisfying $1 + 1 \neq 0$, a skew-symmetric multilinear function is alternating, see problem 6 at the end of this section.

To find an expression for $D(A)$, write each column A_k of A in the form

$$A_k = \sum_{i=1}^{n} a_{ik} E_i.$$

For example, if

$$A = \begin{pmatrix} 4 & 9 & 2 \\ 7 & 2 & 8 \\ 1 & 0 & 5 \end{pmatrix},$$

then

$$A_3 = \begin{pmatrix} 2 \\ 8 \\ 5 \end{pmatrix} = 2\begin{pmatrix} 1 \\ 0 \\ 0 \end{pmatrix} + 8\begin{pmatrix} 0 \\ 1 \\ 0 \end{pmatrix} + 5\begin{pmatrix} 0 \\ 0 \\ 1 \end{pmatrix} = 2E_1 + 8E_2 + 5E_3.$$

When every column is written in this way it will be necessary to keep track of which column contributes which vector E_i. Therefore we introduce a second index and write

$$A_k = \sum_{i_k=1}^{n} a_{i_k k} E_{i_k}.$$

So in the above example, $A_3 = 2E_{1_3} + 8E_{2_3} + 5E_{3_3}$. With this notation, A

takes the form

$$
\begin{aligned}
A &= (A_1, \ldots, A_n) \\
&= \left(\sum_{i_1=1}^{n} a_{i_1 1} E_{i_1}, \sum_{i_2=1}^{n} a_{i_2 2} E_{i_2}, \ldots, \sum_{i_n=1}^{n} a_{i_n n} E_{i_n} \right).
\end{aligned}
$$

Now if a determinant function D exists, it is linear in each variable. Therefore, writing A in this way shows that

$$
D(A) = \sum_{i_1=1}^{n} a_{i_1 1} \sum_{i_2=1}^{n} a_{i_2 2} \cdots \sum_{i_n=1}^{n} a_{i_n n} D(E_{i_1}, E_{i_2}, \ldots, E_{i_n}).
$$

Example 2 Use the above formula to compute $D(A)$ when A is the 2×2 matrix

$$
\begin{pmatrix} a_{11} & a_{12} \\ a_{21} & a_{22} \end{pmatrix}.
$$

In this case the expression

$$
A = \left(\sum_{i_1=1}^{n} a_{i_1 1} E_{i_1}, \sum_{i_2=1}^{n} a_{i_2 2} E_{i_2} \right)
$$

becomes

$$
A = \left(a_{11}\begin{pmatrix} 1 \\ 0 \end{pmatrix} + a_{21}\begin{pmatrix} 0 \\ 1 \end{pmatrix}, a_{12}\begin{pmatrix} 1 \\ 0 \end{pmatrix} + a_{22}\begin{pmatrix} 0 \\ 1 \end{pmatrix} \right).
$$

Therefore

$$
E_{1_1} = E_{2_1} = \begin{pmatrix} 1 \\ 0 \end{pmatrix} = E_1 \text{ and } E_{1_2} = E_{2_2} = \begin{pmatrix} 0 \\ 1 \end{pmatrix} = E_2,
$$

and

$$
\begin{aligned}
D(A) &= \sum_{i_1=1}^{2} a_{i_1 1} \sum_{i_2=1}^{2} a_{i_2 2} D(E_{i_1}, E_{i_2}) \\
&= a_{11}a_{12}D(E_1, E_1) + a_{11}a_{22}D(E_1, E_2) \\
&\quad + a_{21}a_{12}D(E_2, E_1) + a_{21}a_{22}D(E_2, E_2).
\end{aligned}
$$

But D must be skew-symmetric and $D(E_1, E_2) = 1$. Therefore,

$$D(A) = a_{11}a_{22} - a_{21}a_{12}.$$

That is, if D exists, then it equals det on 2×2 matrices.

This example suggests the general situation, but it may be useful to compute $D(A)$ for an arbitrary 3×3 matrix using the above formula. You should find that $D(A)$ is again equal to det A when A is 3×3.

There are now two ways to proceed in showing that a determinant function exists. The classical approach is to show that such a function can only assign one value to each matrix of the form $(E_{i_1}, \ldots, E_{i_n})$. Of course, if two columns are equal, then the value must be 0. But if no two columns are equal, then all the vectors $E_1, \ldots, E_n$ are in the matrix. So using enough interchanges, the sequence $E_{i_1}, \ldots, E_{i_n}$ can be rearranged to $E_1, \ldots, E_n$. That is

$$D(E_{i_1}, \ldots, E_{i_n}) = \pm D(E_1, \ldots, E_n) = \pm 1.$$

In the next section we will show that this sign is determined only by the order of the indices $i_1, \ldots, i_n$. A more abstract approach to the problem of existence will be considered in Section 3. Using this approach we will ignore this sign problem and obtain an expression for $D(A)$ in terms of cofactors.

Before proceeding to show that a determinant function exists, it should be pointed out that if such a function exists, then it must be unique, for a function is single valued. So if D exists, then $D(E_{i_1}, \ldots, E_{i_n})$ has a unique value for each choice of $i_1, \ldots, i_n$. This value depends only on the indices $i_1, \ldots, i_n$ and the value of $D(E_1, \ldots, E_n)$. Since $D(E_1, \ldots, E_n)$ is required to be 1, we have:

Theorem A2 If a determinant function exists, then it is unique.

Problems

1. Assuming a determinant function exists, let $\{E_i\}$ be the standard basis in $\mathscr{M}_{3 \times 1}(F)$ and find
   a. $D(E_1, E_3, E_2)$.
   b. $D(E_2, E_1, E_2)$.
   c. $D(E_3, E_2, E_1)$.
   d. $D(E_3, E_1, E_2)$.
2. Assuming a determinant function exists, let $\{E_i\}$ be the standard basis in $\mathscr{M}_{4 \times 1}(F)$ and find
   a. $D(E_1, E_3, E_4, E_2)$.
   b. $D(E_3, E_2, E_1, E_3)$.
   c. $D(E_2, E_4, E_1, E_3)$.
   d. $D(E_1, E_1, E_1, E_1)$.

3. Show that the function $W \to \det(U_1, \ldots, U_{k-1}, W, U_{k+1}, \ldots, U_n)$ is a linear function from $\mathscr{M}_{n \times 1}(R)$ to R for any $n - 1$ vectors $U_1, \ldots, U_{k-1}, U_{k+1}, \ldots, U_n$ from $\mathscr{M}_{n \times 1}(R)$.
4. Show that conditions 1 and 2 following the definition of a determinant function are equivalent to conditions 1* and 2 if 1* is given by:
   1.* $D(A_1, \ldots, A_h + cA_k, \ldots, A_n) = D(A_1, \ldots, A_h, \ldots, A_n)$
   for any two columns A_h and A_k, $h \neq k$, and any $c \in F$.
5. Prove that if $U_1, \ldots, U_n$ are linearly dependent in $\mathscr{M}_{n \times 1}(F)$ and a determinant function exists, then $D(U_1, \ldots, U_n) = 0$.
6. Suppose f is a multilinear function of n variables, and F is a field in which $1 + 1 \neq 0$. Prove that if f is skew-symmetric, then f is alternating.
7. Could the definition of a determinant function have been made in terms of rows instead of columns?
8. Suppose f is a multilinear function such that $f(U_1, \ldots, U_n) = 0$ whenever two adjacent columns are equal. Prove that f is alternating.
9. Suppose f is a skew-symmetric function of n variables. Prove that if f is linear in the first variable, then f is multilinear.

§2. Permutations and Existence

Definition A *permutation* is a one to one transformation of a finite set onto itself.

We are intersted in permutations of the first n positive integers and will let $\mathscr{P}_n$ denote the set of all permutations of $\{1, \ldots, n\}$. For example if $\sigma(1) = 3$, $\sigma(2) = 4$, $\sigma(3) = 1$, $\sigma(4) = 5$, and $\sigma(5) = 2$, then σ is a member of $\mathscr{P}_5$. Another element of $\mathscr{P}_5$ is given by $\tau(1) = 2$, $\tau(2) = 1$, $\tau(3) = 3$, $\tau(4) = 5$, $\tau(5) = 4$. Since permutations are $1 - 1$ maps, their compositions are $1 - 1$, so two permutations may be composed to obtain a permutation. For σ and τ above, $\sigma \circ \tau$ is the permutation $\mu \in \mathscr{P}_5$ given by $\mu(1) = 4$, $\mu(2) = 3$, $\mu(3) = 1$, $\mu(4) = 2$, and $\mu(5) = 5$. Find $\tau \circ \sigma$, and note that it is not μ.

Theorem A3 The set of permutations $\mathscr{P}_n$ together with the operation of composition is a group. If $n > 2$, then the group is not commutative.

Proof Closure of $\mathscr{P}_n$ under composition was obtained above. Associativity is a general property of composition. The identity map I is the identity in $\mathscr{P}_n$. And finally since permutations are $1 - 1$, they have inverses, so $\mathscr{P}_n$ is a group. The proof of noncommutativity is left as a problem.

In the previous section we found that $D(A)$ is expressed in terms of $D(E_{i_1}, \ldots, E_{i_n})$ and the value of $D(E_{i_1}, \ldots, E_{i_n})$ depends on the arrangement of the integers $i_1, \ldots, i_n$. If this arrangement contains all n integers, then it is a permutation in $\mathscr{P}_n$ given by $\sigma(1) = i_1, \ldots, \sigma(n) = i_n$. Therefore the expression obtained for $D(A)$ might be written as

$$D(A) = \sum_{\sigma \in \mathscr{P}_n} a_{\sigma(1)1}a_{\sigma(2)2} \cdots a_{\sigma(n)n}D(E_{\sigma(1)}, E_{\sigma(2)}, \ldots, E_{\sigma(n)}).$$

Still assuming such a function D exists, the problem now is to determine how the sign of $D(E_{\sigma(1)}, \ldots, E_{\sigma(n)})$ depends on σ.

Definition A *transposition* is a permutation that interchanges two elements and sends every other element to itself.

For a given permutation $\sigma \in \mathscr{P}_n$, or $i_1, \ldots, i_n$, there is a sequence of transpositions that rearranges $i_1, \ldots, i_n$ into the normal order $1, \ldots, n$. That is, there is a sequence of transpositions which when composed with the permutation σ yields the identity permutation I.

Example 1 Use a sequence of transpositions to rearrange the permutation 4, 2, 1, 5, 3 into 1, 2, 3, 4, 5.

This may be done with three transpositions.

$$\begin{aligned} 4, 2, 1, 5, 3 &\to 1, 2, 4, 5, 3 && \text{interchange 4 and 1} \\ &\to 1, 2, 4, 3, 5 && \text{interchange 5 and 3} \\ &\to 1, 2, 3, 4, 5 && \text{interchange 4 and 3.} \end{aligned}$$

But this sequence of transpositions is not unique, for example:

$$4, 2, 1, 5, 3 \xrightarrow[2,1]{} 4, 1, 2, 5, 3 \xrightarrow[4,1]{} 1, 4, 2, 5, 3 \xrightarrow[4,2]{} 1, 2, 4, 5, 3$$
$$\xrightarrow[5,3]{} 1, 2, 4, 3, 5 \xrightarrow[4,3]{} 1, 2, 3, 4, 5$$

where the numbers under each arrow indicate the pair interchanged by a transposition.

Since both sequences in Example 3 contain an odd number of transpositions, either sequence would give

$$D(E_4, E_2, E_1, E_5, E_3) = -D(E_1, E_2, E_3, E_4, E_5) = -1.$$

But if 4, 2, 1, 5, 3 can be transformed to 1, 2, 3, 4, 5 with an even number of transpositions, then the determinant function D does not exist. Therefore it is necessary to prove that only an odd number of transpositions will do the job. The general proof of such an "invariance of parity" relies on a special polynomial.

The product of all factors $x_i - x_j$ for $1 \leq i < j \leq n$ is written as

$$\prod_{i<j}^{n} (x_i - x_j).$$

Denote this polynomial by $P(x_1, \ldots, x_n)$. Then

$$P(x_1, x_2) = x_1 - x_2, \qquad P(x_1, x_2, x_3) = (x_1 - x_2)(x_1 - x_3)(x_2 - x_3),$$

and so on.

For $\sigma \in \mathscr{P}_n$, apply σ to the indices in the polynomial $P(x_1, \ldots, x_n)$. Then for the indices in $x_{\sigma(i)} - x_{\sigma(j)}$, either $\sigma(i) < \sigma(j)$ or $\sigma(j) < \sigma(i)$ and $x_{\sigma(i)} - x_{\sigma(j)} = -(x_{\sigma(j)} - x_{\sigma(i)})$. Therefore the factors in $\prod_{i<j}^{n} (x_{\sigma(i)} - x_{\sigma(j)})$ differ from the factors in $\prod_{i<j}^{n} (x_i - x_j)$ by at most multiples of -1 and

$$P(x_{\sigma(1)}, \ldots, x_{\sigma(n)}) = \pm P(x_1, \ldots, x_n).$$

Definition A permutation $\sigma \in \mathscr{P}_n$ is *even* if

$$P(x_{\sigma(1)}, \ldots, x_{\sigma(n)}) = P(x_1, \ldots, x_n)$$

and *odd* if

$$P(x_{\sigma(1)}, \ldots, x_{\sigma(n)}) = -P(x_1, \ldots, x_n).$$

Example 2 Let $\sigma \in \mathscr{P}_3$ be given by $\sigma(1) = 3, \sigma(2) = 2$, and $\sigma(3) = 1$, then σ is odd for

$$\begin{aligned} P(x_{\sigma(1)}, x_{\sigma(2)}, x_{\sigma(3)}) &= (x_{\sigma(1)} - x_{\sigma(2)})(x_{\sigma(1)} - x_{\sigma(3)})(x_{\sigma(2)} - x_{\sigma(3)}) \\ &= (x_3 - x_2)(x_3 - x_1)(x_2 - x_1) \\ &= (-1)^3(x_2 - x_3)(x_1 - x_3)(x_1 - x_2) \\ &= -P(x_1, x_2, x_3). \end{aligned}$$

Theorem A4 Every transposition is an odd permutation.

Proof Suppose $\sigma \in \mathscr{P}_n$ is a transposition interchanging h and k with $h < k$. Then there are $2(k - h - 1) + 1$ (an odd number) factors in $\prod_{i<j}^n (x_{\sigma(i)} - x_{\sigma(j)})$ that must be multiplied by -1 to obtain $\prod_{i<j}^n (x_i - x_j)$. If $h < i < k$, then $k - h - 1$ of these factors are

$$x_{\sigma(h)} - x_{\sigma(i)} = x_{\mathrm{k}} - x_i = -(x_i - x_k).$$

Another $k - h - 1$ factors in the wrong order are

$$x_{\sigma(i)} - x_{\sigma(\mathrm{k})} = x_i - x_h = -(x_h - x_i),$$

and finally

$$x_{\sigma(h)} - x_{\sigma(k)} = x_k - x_h = -(x_h - x_k).$$

These are all the factors that must be multiplied by -1 since all other choices for $i < j$ yield $\sigma(i) < \sigma(j)$. Thus σ is an odd permutation.

Theorem A5 If $i_1, \ldots, i_n$ gives an even (odd) permutation in $\mathscr{P}_n$ and there exist m transpositions that transform $i_1, \ldots, i_n$ into $1, \ldots, n$, then m is even (odd).

Proof Suppose σ is an even permutation given by $\sigma(1) = i_1, \ldots, \sigma(n) = i_n$ and $\tau_1, \ldots, \tau_m$ are transpositions such that $\tau_m \circ \cdots \circ \tau_1 \circ \sigma = I$. Since I is an even permutation, applying the permutation $\tau_m \circ \cdots \circ \tau_1 \circ \sigma$ to the indicies in $P(x_1, \ldots, x_n)$ leaves the polynomial unchanged. But σ leaves $P(x_1, \ldots, x_n)$ unchanged, and each transposition changes the sign. Therefore the composition changes the polynomial to $(-1)^m P(x_1, \ldots, x_n)$ and m must be even. Similarly m is odd if σ is an odd permutation.

Definition The *sign* or *signum* of a permutation σ, denoted sgn σ, is 1 if σ is even and -1 if σ is odd.

If there exist m transpositions that transform a permutation into the normal order, then sgn $\sigma = (-1)^m$, and we can write

$$P(x_{\sigma(1)}, \ldots, x_{\sigma(n)}) = (\operatorname{sgn} \sigma) P(x_1, \ldots, x_n).$$

A value for m can be obtained by counting the number of inversions in the permutation σ. An *inversion* in a permutation occurs whenever an integer precedes a smaller integer. Thus the permutation 2, 1, 5, 4, 3 contains four inversions. Each inversion in a permutation can be turned around with a single transposition. Therefore if σ has m inversions, then $\operatorname{sgn} \sigma = (-1)^m$.

Example 3 The permutation given in Example 1 contains five inversions, therefore $\operatorname{sgn}(4, 2, 1, 5, 3) = (-1)^5 = -1$. The second sequence of transpositions used to transform 4, 2, 1, 5, 3 to 1, 2, 3, 4, 5 is a sequence that inverts each of the five inversions.

Now if $E_{i_1}, \ldots, E_{i_n}$ is a rearrangement of $E_1, \ldots, E_n$, then the number of interchanges required to obtain the standard order depends on whether $i_1, \ldots, i_n$ is an even or odd permutation. Therefore

$$\begin{aligned} D(E_{i_1}, \ldots, E_{i_n}) &= \operatorname{sgn}(i_1, \ldots, i_n) D(E_1, \ldots, E_n) \\ &= \operatorname{sgn}(i_1, \ldots, i_n), \end{aligned}$$

and the formula for $D(A)$ obtained on page 410 becomes

$$D(A) = \sum_{\sigma \in \mathscr{P}_n} (\operatorname{sgn} \sigma) a_{\sigma(1)1} a_{\sigma(2)2} \cdots a_{\sigma(n)n}.$$

So to prove the existence of a determinant function it is only necessary to show that this formula does in fact define such a function.

Theorem A6 The function $D\colon \mathscr{M}_{n \times n}(F) \to F$ defined by

$$D(A) = \sum_{\sigma \in \mathscr{P}_n} (\operatorname{sgn} \sigma) a_{\sigma(1)1} \cdots a_{\sigma(n)n} \quad \text{for} \quad A = (a_{ij})$$

is a determinant function.

Proof Showing that D is multilinear and that $D(I_n) = 1$ is not difficult and is left to the reader.

In proving that D is alternating we will use the *alternating group* of even permutations, $\mathscr{A}_n = \{\mu \in \mathscr{P}_n | \operatorname{sgn} \mu = 1\}$. If τ is a transposition in $\mathscr{P}_n$, then

$$\mathscr{P}_n = \mathscr{A}_n \cup \{\tau \circ \mu | \mu \in \mathscr{A}_n\},$$

see problem 8 below.

To show that D is alternating, suppose the hth and kth columns of A are equal. Then if τ is the transposition interchanging h and k,

$$D(A) = \sum_{\mu \in \mathscr{A}_n} (\operatorname{sgn} \mu) a_{\mu(1)1} \cdots a_{\mu(n)n} + \sum_{\mu \in \mathscr{A}_n} (\operatorname{sgn} \tau \circ \mu) a_{\tau\circ\mu(1)1} \cdots a_{\tau\circ\mu(n)n}.$$

For each $\mu \in \mathscr{A}_n$, $\operatorname{sgn} \mu = 1$ and $\operatorname{sgn} \tau \circ \mu = -1$. Further since the hth and kth columns of A are equal,

$$a_{\tau\circ\mu(1)1} \cdots a_{\tau\circ\mu(n)n} = a_{\mu(1)} \cdots a_{\mu(n)n}.$$

Therefore $D(A) = 0$, and D is alternating.

Problems

1. Write out all the permutations in $\mathscr{P}_3$ and $\mathscr{P}_4$ and determine which are even and which are odd.

2. Let $\sigma, \tau \in \mathscr{P}_5$ be given by $\sigma(1) = 4$, $\sigma(2) = 2$, $\sigma(3) = 5$, $\sigma(4) = 1$, $\sigma(5) = 3$ and $\tau(1) = 5$, $\tau(2) = 1$, $\tau(3) = 2$, $\tau(4) = 4$, $\tau(5) = 3$.
Find a. $\sigma \circ \tau$. b. $\tau \circ \sigma$. c. σ^{-1}. d. τ^{-1}.

3. Suppose $\sigma \in \mathscr{P}_3$ satisfies $\sigma(1) = 2$, $\sigma(2) = 3$, and $\sigma(3) = 1$. Find all permutations in $\mathscr{P}_3$ that commute with σ, that is, all $\tau \in \mathscr{P}_3$ for which $\sigma \circ \tau = \tau \circ \sigma$.

4. Prove that if $n > 2$, then $\mathscr{P}_n$ is not a commutative group.

5. Show that $\operatorname{sgn} \sigma \circ \tau = (\operatorname{sgn} \sigma)(\operatorname{sgn} \tau)$ for all $\sigma, \tau \in \mathscr{P}_n$.

6. a. Write out $P(x_1, x_2, x_3, x_4)$.
b. Use this polynomial to find the sign of σ and τ if σ is given by $\sigma(1) = 4$, $\sigma(2) = 2$, $\sigma(3) = 1$, $\sigma(4) = 3$ and τ is given by 1, 4, 3, 2.

7. How many elements are there in $\mathscr{P}_n$?

8. a. Show that $\mathscr{A}_n = \{\mu \in \mathscr{P}_n | \mu \text{ is even}\}$ is a subgroup of $\mathscr{P}_n$.
b. Suppose $\tau \in \mathscr{P}_n$ is a transposition. Show that each odd permutation $\sigma \in \mathscr{P}_n$ can be expressed as $\tau \circ \mu$ for some $\mu \in \mathscr{A}_n$.

9. Find the number of inversions in the following permutations. Use the number of inversions to determine the sign of each.
a. 5, 4, 3, 2, 1. c. 5, 1, 4, 2, 3.
b. 6, 5, 4, 3, 2, 1. d. 4, 3, 6, 2, 1, 5.

10. Suppose $A = (A_1, \ldots, A_n)$ and $B = (b_{ij})$ are $n \times n$ matrices.
a. Show that the product may be written as

$$AB = \left(\sum_{i=1}^{n} b_{i1} A_i, \ldots, \sum_{i=1}^{n} b_{in} A_i\right).$$

b. Use the definition given for D in Thoerem A6 and part a to prove that $D(AB) = D(A)D(B)$.

11. Suppose f is an alternating multilinear function of k variables and $\sigma \in \mathscr{P}_k$. Prove that $f(V_{\sigma(1)}, \ldots, V_{\sigma(k)}) = (\text{sgn } \sigma) f(V_1, \ldots, V_k)$.

§3. Cofactors, Existence, and Applications

We will obtain another definition for $D(A)$ by again assuming that a determinant function exists. First, since D is to be linear in the first variable,

$$D(A) = D(A_1, A_2, \ldots, A_n) = D\left(\sum_{k=1}^{n} a_{k1} E_k, A_2, \ldots, A_n\right)$$
$$= \sum_{k=1}^{n} a_{k1} D(E_k, A_2, \ldots, A_n\biggr).$$

This expresses $D(A)$ in terms of $D(E_k, A_2, \ldots, A_n)$. But since D is alternating, $D(E_k, A_2, \ldots, A_n)$ does not involve the kth entries of $A_2, \ldots, A_n$. Therefore, if $A_h^{(k)}$ denotes the hth column of A with the kth entry deleted, $A_h^{(k)} = \sum_{i \neq k} a_{ih} E_i$, and $D(E_k, A_2, \ldots, A_n)$ differs from $D(A_2^{(k)}, \ldots, A_n^{(k)})$ by at most a sign. Write

$$D(E_k, A_2, \ldots, A_n) = cD(A_2^{(k)}, \ldots, A_n^{(k)}).$$

The value of c can be found by setting the columns of A equal to basis vectors. For

$$\begin{aligned} &D(E_k, E_1, \ldots, E_{k-1}, E_{k+1}, \ldots, E_n) \\ &\qquad = cD(E_1^{(k)}, \ldots, E_{k-1}^{(k)}, E_{k+1}^{(k)}, \ldots, E_n^{(k)}) \\ &\qquad = cD(I_{(n-1)\times(n-1)}) = c. \end{aligned}$$

On the other hand, D must be skew-symmetric, so

$$\begin{aligned} &D(E_k, E_1, \ldots, E_{k-1}, E_{k+1}, \ldots, E_n) \\ &\qquad = (-1)^{k-1} D(E_1, \ldots, E_{k-1}, E_k, E_{k+1}, \ldots, E_n) \\ &\qquad = (-1)^{k-1} D(I_n) = (-1)^{k-1}. \end{aligned}$$

This gives $c = (-1)^{k-1}$, but it will be better to write c in the form $(-1)^{k+1}$.

Therefore

$$D(E_k, A_2, \ldots, A_n) = (-1)^{k+1}D(A_1^{(k)}, \ldots, A_n^{(k)}).$$

Now since $(A_2^{(k)}, \ldots, A_n^{(k)})$ is the $(n-1) \times (n-1)$ matrix obtained from A by deleting the kth row and the 1st column, we can give an inductive definition of D just as was done in Chapter 3.

Theorem A7 Define $D: \mathscr{M}_{n\times n}(F) \to F$ inductively by $D(A) = a_{11}$ when $n = 1$ and $A = (a_{11})$; and

$$\begin{aligned} D(A) &= D(A_1, A_2, \ldots, A_n) \\ &= \sum_{k=1}^{n} a_{k1}(-1)^{k+1}D(A_2^{(k)}, \ldots, A_n^{(k)}) \qquad \text{when} \quad n > 1. \end{aligned}$$

Then D is a determinant function for each n.

Proof It is clear from the definition that D is linear in the first variable, and a straightfoward induction argument shows that D is linear in every other variable.

To prove that D is alternating, first note that the definition gives

$$D\begin{pmatrix} a & b \\ c & d \end{pmatrix} = a(-1)^{1+1}D(d) + c(-1)^{1+2}D(b) = ad - cb.$$

So

$$D\begin{pmatrix} a & a \\ c & c \end{pmatrix} = ac - ca$$

and D is alternating for 2×2 matrices. Therefore assume D is alternating for $(n-1) \times (n-1)$ matrices, $n > 2$. If A is an $n \times n$ matrix with two columns equal, then there are two cases.

Case I: Neither of the equal columns is the first column of A In this case two columns of $(A_2^{(k)}, \ldots, A_n^{(k)})$ are equal, so by the induction assumption $D(A_2^{(k)}, \ldots, A_n^{(k)}) = 0$ for each k and $D(A) = 0$.

Case II: The first column of A equals some other column. Since case I and the multilinearity of D imply that D is skew-symmetric in all variables but the first, it is sufficient to assume the first two columns of A are equal.

Thus

$$D(A) = D(A_1, A_1, A_3, \ldots, A_n)$$
$$= \sum_{k=1}^{n} a_{k1}(-1)^{k+1} D(A_1^{(k)}, A_3^{(k)}, \ldots, A_n^{(k)}).$$

To evaluate the terms in this expression, let $A_j^{(k,h)}$ denote the jth column of A with the hth and kth entries deleted, so

$$A_j^{(k,h)} = \sum_{i \neq k,h} a_{ij} E_i.$$

Then

$$D(A_1^{(k)}, A_3^{(k)}, \ldots, A_n^{(k)})$$
$$= \sum_{h=1}^{k-1} a_{h1}(-1)^{h+1} D(A_3^{(k,h)}, \ldots, A_n^{(k,h)})$$
$$+ \sum_{h=k+1}^{n} a_{h1}(-1)^{h} D(A_3^{(k,h)}, \ldots, A_n^{(k,h)}),$$

and

$$D(A) = D(A_1, A_1, A_3, \ldots, A_n)$$
$$= \sum_{k=1}^{n} \sum_{h=1}^{k-1} a_{k1} a_{h1}(-1)^{k+h+1} D(A_3^{(k,h)}, \ldots, A_n^{(k,h)})$$
$$+ \sum_{k=1}^{n} \sum_{h=k+1}^{n} a_{k1} a_{h1}(-1)^{k+h} D(A_3^{(k,h)}, \ldots, A_n^{(k,h)}).$$

The product $a_{i1}a_{j1}$ appears twice in this expression for each $i \neq j$, and since the coefficients are $(-1)^{i+j+1} D(A_3^{(i,j)}, \ldots, A_n^{(i,j)})$ and $(-1)^{j+i} D(A_3^{(j,i)}, \ldots, A_n^{(j,i)})$, $D(A) = 0$. Therefore D is alternating.

To complete the proof it is necessary to show that $D(I_n) = 1$. This is proved by induction, noting that if $k \neq 1$, then the multilinearity of D implies that $D(E_2^{(k)}, \ldots, E_n^{(k)}) = 0$.

Since there is only one determinant function, the functions defined in Theorems A6 and A7 must be equal. The first step in showing that this function equals det is to generalize the idea used above. If we write the hth column of A as $\sum_{k=1}^{n} a_{kh}E_k$, then

$$D(A) = D\left(A_1, \ldots, A_{h-1}, \sum_{k=1}^{n} a_{kh}E_k, A_{h+1}, \ldots, A_n\right)$$

$$= \sum_{k=1}^{n} a_{kh}D(A_1, \ldots, A_{h-1}, E_k, A_{h+1}, \ldots, A_n)$$

$$= \sum_{k=1}^{n} a_{kh}(-1)^{k+h}D(A_1^{(k)}, \ldots, A_{h-1}^{(k)}, A_{h+1}^{(k)}, \ldots, A_n^{(k)}).$$

For the last equality, note that if $k < h$, then

$$D(E_1, \ldots, E_{k-1}, E_{k+1}, \ldots, E_{h-1}, E_h, E_k, E_{h+1}, \ldots, E_n)$$
$$= (-1)^{h-(k+1)+1}D(E_1, \ldots, E_{k-1}, E_k, E_{k+1}, \ldots, E_{h-1}, E_{h+1}, \ldots, E_n)$$
$$= (-1)^{h-k}D(I_n) = (-1)^{h+k}.$$

The case $k > h$ is handled similarly. Now the matrix

$$(A_1^{(k)}, \ldots, A_{h-1}^{(k)}, A_{h+1}^{(k)}, \ldots, A_n^{(k)})$$

is obtained from A by deleting the kth row and the hth column. Therefore we will denote $(-1)^{k+h}D(A_1^{(k)}, \ldots, A_{h-1}^{(k)}, A_{h+1}^{(k)}, \ldots, A_n^{(k)})$ by A_{kh} and call it the *cofactor* of a_{kh}. This terminology will agree with that of Chapter 3 when we show that the functions D and det are equal. We will say that $D(A) = \sum_{k=1}^{n} a_{kh}A_{kh}$ is the *expansion by cofactors of* $D(A)$ *along the* hth *column of* A.

Writing $D(A) = \sum_{k=1}^{n} a_{kh}A_{kh}$ shows that all the terms containing a_{ij} for any i and j are in the product $a_{ij}A_{ij}$. But each term in $D(A)$ contains exactly one element from each row. Therefore if the terms in $D(A)$ are grouped according to which element they contain from the ith row, then $D(A)$ must have the form

$$a_{i1}A_{i1} + a_{i2}A_{i2} + \cdots + a_{in}A_{in}.$$

The equality $D(A) = \sum_{j=1}^{n} a_{ij}A_{ij}$ gives the *expansion by cofactors of* $D(A)$ *along the* ith *row of* A.

Theorem A8 The functions D and det are equal on $\mathcal{M}_{n\times n}(R)$.

Proof By induction on n. D and det are equal for 1×1 real matrices. Therefore assume they are equal for $(n-1) \times (n-1)$ real matrices, $n > 1$, and let A be a real $n \times n$ matrix. Then expanding $D(A)$ along the first row gives

$$\begin{aligned} D(A) &= \sum_{j=1}^{n} a_{1j}A_{1j} \\ &= \sum_{j=1}^{n} a_{1j}(-1)^{1+j}D(A_1^{(1)}, \ldots, A_{j-1}^{(1)}, A_{j+1}^{(1)}, \ldots, A_n^{(1)}) \\ &= \sum_{j=1}^{n} a_{1j}(-1)^{1+j}\det(\text{minor of } a_{1j}) \qquad \text{by induction assumption} \\ &= \det A. \end{aligned}$$

We will now drop the notation $D(A)$ and write det A for the determinant of $A \in \mathscr{M}_{n\times n}(F)$. Theorem A8 shows that our two definitions of cofactor are equivalent, and it supplies a proof for Theorem 3.13 which stated that det A could be expanded by cofactors along any row or any column. We might now prove that the basic properties of determinants hold for matrices with enties from an arbitrary field. The fact that $\det A^T = \det A$ is still most easily proved by induction, but the other properties can be obtained quite easily using the fact that det is an alternating multilinear function. For example, the fact that the determinant of a product is the product of the determinants (Theorem 4.26) might be proved over an arbitrary field as follows:

Let $A = (a_{ij}) = (A_1, \ldots, A_n)$ and $B = (b_{ij})$ be $n \times n$ matrices. Then the jth column of AB has the form

$$\begin{pmatrix} \sum_{k=1}^{n} a_{1k}b_{kj} \\ \vdots \\ \sum_{k=1}^{n} a_{nk}b_{kj} \end{pmatrix} = \sum_{i=1}^{n}\left(\sum_{k=1}^{n} a_{ik}b_{kj}\right)E_i = \sum_{k=1}^{n} b_{kj}\left(\sum_{i=1}^{n} a_{ik}E_i\right) = \sum_{k=1}^{n} b_{kj}A_k.$$

Therefore

$$\det AB = \det\left(\sum_{k=1}^{n} b_{k1}A_k, \ldots, \sum_{k=1}^{n} b_{kn}A_k\right).$$

Since the function det is multilinear and skew symmetric,

$$\det B = \det\left(\sum_{k=1}^{n} b_{k1}E_k, \ldots, \sum_{k=1}^{n} b_{kn}E_k\right) = \det B \det(E_1, \ldots, E_n).$$

Then replacing the columns of E's with A's gives

$$\det\left(\sum_{k=1}^{n} b_{k1}A_k, \ldots, \sum_{k=1}^{n} b_{kn}A_k\right) = \det B \det(A_1, \ldots, A_n).$$

So $\det AB = \det B \det A = \det A \det B$.

We may also prove Cramer's rule (Theorem 4.22) directly from the properties of the determinant function. Suppose $AX = B$ is a system of n linear equations in n unknowns. If $A = (A_1, \ldots, A_n)$, then the system can be written in the form

$$x_1A_1 + \cdots + x_nA_n = B \qquad \text{or} \qquad \sum_{i=1}^{n} x_iA_i = B.$$

Now since det is an alternating multilinear function

$$\det\left(A_1, \ldots, A_{j-1}, \sum_{i=1}^{n} x_iA_i, A_{j+1}, \ldots, A_n\right)$$
$$= x_j \det(A_1, \ldots, A_{j-1}, A_j, A_{j+1}, \ldots, A_n)$$
$$= x_j \det A.$$

But $\sum_{i=1}^{n} x_iA_i = \text{B}$, so

$$\det(A_1, \ldots, A_{j-1}, B, A_{j+1}, \ldots, A_n) = x_j \det A.$$

Therefore if $\det A \neq 0$,

$$x_j = \frac{\det(A_1, \ldots, A_{j-1}, B, A_{j+1}, \ldots, A_n)}{\det A}.$$

It is now possible to derive the adjoint formula for the inverse of a matrix using Cramer's rule, see problem 4 at the end of this section.

As a final application of the determinant function we will redefine the cross product in $\mathscr{E}_3$ and prove that

$$\|U \times V\| = \|U\| \, \|V\| \sin \theta.$$

Definition The *cross product* of $U, V \in \mathscr{E}_3$, denoted by $U \times V$, is the vector in $\mathscr{E}_3$ that satisfies

$$(U \times V) \circ W = \det(U^T, V^T, W^T) \qquad \text{for all} \quad W \in \mathscr{E}_3.$$

The existence of a vector $U \times V$ for each pair of vectors from $\mathscr{E}_3$ is a consequence of problem 6, page 286. Since this is a new definition for the cross product, it is necessary to show that it yields the computational formula given in Chapter 3.

Theorem A9 If $U = (x_1, x_2, x_3)$ and $V = (y_1, y_2, y_3)$, then

$$U \times V = \left(\det\begin{pmatrix} x_2 & x_3 \\ y_2 & y_3 \end{pmatrix}, -\det\begin{pmatrix} x_1 & x_3 \\ y_1 & y_3 \end{pmatrix}, \det\begin{pmatrix} x_1 & x_2 \\ y_1 & y_2 \end{pmatrix}\right).$$

Proof This follows at once from the fact that the components of $U \times V$ are given by $U \times V \circ E_i$ when $i = 1, 2, 3$. For example, when $i = 1$,

$$U \times V \circ E_1 = \det\begin{pmatrix} x_1 & y_1 & 1 \\ x_2 & y_2 & 0 \\ x_3 & y_3 & 0 \end{pmatrix} = \det\begin{pmatrix} x_2 & y_2 \\ x_3 & y_3 \end{pmatrix} = \det\begin{pmatrix} x_2 & x_3 \\ y_2 & y_3 \end{pmatrix}.$$

This theorem shows that if the standard basis vectors are treated as numbers, then

$$U \times V = \det\begin{pmatrix} x_1 & y_1 & E_1 \\ x_2 & y_2 & E_2 \\ x_3 & y_3 & E_3 \end{pmatrix} = \det\begin{pmatrix} E_1 & E_2 & E_3 \\ x_1 & x_2 & x_3 \\ y_1 & y_2 & y_3 \end{pmatrix}.$$

The determinants of these two matrices are equal because the second matrix is obtained from the first by interchanging two pairs of columns and then taking the transpose.

The basic properties of the cross product are now immediate consequences of the fact that det is an alternating multilinear function.

Theeorm A10 If $U, V, W \in \mathscr{E}_3$ and $a, b \in R$, then

1. $U \times V$ is orthogonal to U and V.
2. The cross product is skew-symmetric, that is, $U \times V = -V \times U$.
3. The cross product is bilinear, that is,

$$U \times (aV + bW) = aU \times V + bU \times W$$

and

$$(aU + bV) \times W = aU \times W + bV \times W.$$

We are also in a position to show that the cross product is invariant under rotations, which is equivalent to showing that the cross product of two vectors can be computed using their coordinates with respect to any orthonormal basis having the same orientation as the standard basis.

Theorem A11 Suppose $T: \mathscr{E}_3 \to \mathscr{E}_3$ is a rotation. Then

$$T(U) \times T(V) = T(U \times V) \qquad \text{for all} \quad U, V \in \mathscr{E}_3.$$

Proof Let P be the matrix of T with respect to the standard basis. If the transpose notation is omitted for simplicity, then we can write $T(W) = PW$ for any $W \in \mathscr{E}_3$ and

$$\begin{aligned} T(U) \times T(V) \circ W &= \det(T(U), T(V), T(T^{-1}(W))) \\ &= \det(PU, PV, PT^{-1}(W)) \\ &= \det P \det(U, V, T^{-1}(W)) \\ &= U \times V \circ T^{-1}(W) && \text{Since } \det P = 1 \\ &= T(U \times V) \circ T(T^{-1}(W)) && \text{Since } T \text{ is orthogonal} \\ &= T(U \times V) \circ W. \end{aligned}$$

Since this holds for all $W \in \mathscr{E}_3$, it implies that $T(U) \times T(V) = T(U \times V)$.

Theorem A12 If U and V are orthogonal unit vectors, then $U \times V$ is a unit vector.

Proof There exists a rotation T that sends U to E_1 and V to E_2. therefore, using the fact that T is orthogonal we have

$$\|U \times V\| = \|T(U \times V)\| = \|T(U) \times T(V)\| = \|E_1 \times E_2\| = \|E_3\| = 1.$$

Now the fact that $\|U \times V\| = \|U\| \, \|V\| \sin \theta$ follows from the next theorem.

Theorem A13 Suppose θ is the angle between U and V. Then there exists a unit vector N, orthogonal to U and V, such that

$$U \times V = \|U\| \, \|V\| \sin \theta \, N.$$

Proof If U and V are linearly dependent, say $V = rU$, then $U \times V \circ W = \det(U, rU, W) = 0$ for all $W \in \mathscr{E}_3$, and $U \times V = \mathbf{0}$. In this case, $\sin \theta = 0$ and the equation holds for any vector N.

Therefore suppose U and V are linearly independent. Use the Gram-Schmidt process to build an orthonormal basis $\{X, Y\}$ for $\mathscr{L}\{U, V\}$ from the basis $\{U, V\}$. Then $X = U/\|U\|$ and the angle between V and Y is either $90° - \theta$ or $\theta - 90°$. Since $\{X, Y\}$ is an orthonormal basis,

$$\begin{aligned} V &= (V \circ X)X + (V \circ Y)Y \\ &= (V \circ X)X + \|V\| \, \|Y\| \cos \pm(\theta - 90°)Y \\ &= (V \circ X)X + \|V\| \sin \theta \, Y. \end{aligned}$$

Therefore

$$\begin{aligned} U \times V &= (\|U\|X) \times ((V \circ X)X + \|V\| \sin \theta \, Y) \\ &= \|U\| \, \|V\| \sin \theta \, X \times Y. \end{aligned}$$

$X \times Y$ is a unit vector by Theorem A12. It is orthogonal to X and Y by Theorem A10, and so is orthogonal to U and V. Therefore, setting $N = X \times Y$ gives the desired equation. Notice that when U and V are linearly independent, $N = U \times V/\|U \times V\|$, as expected.

Corollary A14 The set $\{U, V\}$ is linearly dependent in $\mathscr{E}_3$ if and only if $U \times V = \mathbf{0}$.

Problems

1. Show that the formulas in Theorems A6 and A7 give the same expression for $D(A)$ when A is an arbitrary 3×3 matrix.
2. Using the definition for D given in Theorem A7,
   a. Prove that D is multilinear.
   b. Prove that $D(I_n) = 1$.
3. Show that when $k > h$,
$$\begin{aligned} &D(A_1, \ldots, A_{h-1}, E_k, A_{h+1}, \ldots, A_n) \\ &\quad = (-1)^{k+h} D(A_1^{(k)}, \ldots, A_{h-1}^{(k)}, A_{h+1}^{(k)}, \ldots, A_n^{(k)}). \end{aligned}$$
4. Use Cramer's rule to derive the adjoint formula for the inverse of a nonsingular matrix.
5. Show that $U \times V$ is orthogonal to U and V.

6. Prove that the cross product is a. skew-symmetric. b. bilinear.

7. a. Show that the parallelogram with vertices **0**, U, V, $U + V$ in $\mathscr{E}_3$ has area $\|U \times V\|$.

b. Find the area of the parallelogram with vertices **0**, (2, 3, 1), (4, −5, −2), and (6, −2, −1).

c. Use part a to find a geneneral formula for the area of the triangle with vertices $A, B, C \in \mathscr{E}_3$.

d. Find the area of the triangle with vertices (3, −2, 6), (1, 5, 7), and (2, −4, 1).

8. Consider the set $\mathscr{E}_3$ together with the operation given by the cross product. How close does $(\mathscr{E}_3, \times)$ come to satisfying the definition of a group?

9. a. Extend the definition of cross product to $\mathscr{E}_4$.

b. Suppose U, V, and W are linearly independent in $\mathscr{E}_4$. Show that $U \times V \times W$ is normal to the hyperplane spanned by U, V, and W.

c. Find (1, 0, 0, 0) × (0, 1, 0, 0) × (0, 0, 1, 0).

d. Find (2, 0, 1, 0) × (0, 3, 0, 2) × (2, 0, 0, 1).

10. Suppose $\mathscr{V}$ is an n-dimensional vector space over F. A *p-vector* on $\mathscr{V}$ is an expression of the form $rU_1 \wedge U_2 \wedge \cdots \wedge U_p$ where $r \in F$ and $U_1, \ldots, U_p \in \mathscr{V}$, for $2 \leq p \leq n$. p-vectors are required to satisfy

i. $U_1 \wedge \cdots \wedge (aU_i + bU_i') \wedge \cdots \wedge U_p = aU_1 \wedge \cdots \wedge U_i \wedge \cdots \wedge U_p + bU_1 \wedge \cdots \wedge U_i' \wedge \cdots \wedge U_p$ for $i = 1, \ldots, p$ and $a, b \in F$.

ii. $U_1 \wedge \cdots \wedge U_p = \mathbf{0}$ if any two vectors U_i and U_j are equal, $i \neq j$.

$U \wedge W$, read "U wedge W," is called the *exterior product* of U and W. The vector space of all p-vectors on $\mathscr{V}$, denoted by $\Lambda^p\mathscr{V}$, is the set of all finite sums of p-vectors on $\mathscr{V}$.

a. Show that in $\Lambda^2\mathscr{V}_2$, $(a, b) \wedge (c, d) = [ad - bc](1, 0) \wedge (0, 1)$.

b. Show that $\dim \Lambda^2\mathscr{V}_3 = 3$ and that $\dim \Lambda^2\mathscr{V}_4 = 6$.

c. Prove that $\dim \Lambda^n\mathscr{V} = 1$ by showing that if $\{V_1, \ldots, V_n\}$ is a basis for $\mathscr{V}$, then $\{V_1 \wedge \cdots \wedge V_n\}$ is a basis for $\Lambda^n\mathscr{V}$.

d. For $T \in \text{Hom}(\mathscr{V})$, define $\Lambda T: \Lambda^n\mathscr{V} \rightarrow \Lambda^n\mathscr{V}$ by

$$\Lambda T(\Sigma\, rU_1 \wedge \cdots \wedge U_n) = \Sigma\, rT(U_1) \wedge \cdots \wedge T(U_n).$$

Show that there exists a constant $|T|$ such that

$$T(W_1) \wedge \cdots \wedge T(W_n) = |T| W_1 \wedge \cdots \wedge W_n \quad \text{for all} \quad W_1, \ldots, W_n \in \mathscr{V}.$$

Call $|T|$ the *determinant* of T. Notice that $|T|$ is defined without reference to a matrix.

e. Prove that if $S, T \in \text{Hom}(\mathscr{V})$, then $|S \circ T| = |S|\,|T|$.

f. Prove that if A is the matrix of T with respect to some basis, then $|T| = \det A$.

Answers and Suggestions

Chapter 1

§1 Page 4

1. a. Seven is a real number. b. $\sqrt{-6}$ is not a real number.
c. The set containing only 0 and 5.
d. The set of all x such that x is a negative real number.
e. The set of all real numbers x such that $x^2 = -1$. Notice that this set is empty.

2. a. $1 + 1 = 2$. b. Consider the product $(2n_1 + 1)(2n_2 + 1)$.

3. a. Closed under multiplication but not addition. b. Neither.
c. Closed under addition but not multiplication. d. Both. e. Both.

4. Addition in R is *commutative* if $r + s = s + r$ for all $r, s \in R$.

5. $(r + s)t = rt + st$ can proved using the given distributive law and the fact that multiplication is commutative. Can you prove this?

§2 Page 6

1. a. $(5, -3)$. b. $(-1, 0)$. c. $\mathbf{0}$. d. $(1/4, 7/3)$.

2. a. $(2, -5/2)$. b. $\mathbf{0}$. c. $(6, -1)$. d. $(-3, 6)$.

3. a. $(2, 10)$. b. $(-1/3, -1/3)$. c. $(5, -1)$. d. $(13/2, 1)$.

4. Suppose U is the ordered pair (a, b), then

$$\begin{aligned} U + \mathbf{0} &= (a, b) + (0, 0) && \text{Definition of } \mathbf{0} \\ &= (a + 0, b + 0) && \text{Definition of addition in } \mathscr{V}_2 \\ &= (a, b) && \text{Zero is the additive identity in } R \\ &= U. \end{aligned}$$

5. Set $U = (a, b)$, $V = (c, d)$, and $W = (e, f)$ and perform the indicated operations, thus

$$\begin{aligned} U + (V + W) &= (a, b) + ((c, d) + (e, f)) \\ &= (a, b) + (c + e, d + f) \qquad \text{Reason?} \end{aligned}$$

Continue until you obtain $(U + V) + W$, it should take four more steps, with reasons, and the fact that addition is associative in R must be used.

6. No. The operation yields scalars (numbers) and vectors are ordered pairs of numbers.

7. a. Vectors in $\mathscr{V}_2$ are ordered pairs.
b. Definition of vector addition in $\mathscr{V}_2$.
c. By assumption $U + W = U$.
d. Definition of equality of ordered pairs.

e. Property of $0 \in R$ and the fact that equations have unique solutions in R. (The latter can be derived from the properties of addition.)

f. Definition of the zero vector.

8. For $U = (a, b) \in \mathscr{V}_2$, assume there is a vector $T \in \mathscr{V}_2$ such that $U + T = \mathbf{0}$ and show that $T = (-a, -b)$.

9. All these properties are proved using properties of real numbers. To show that $(r + s)U = rU + sU$, let $U = (a, b)$, then

$$\begin{aligned}
[r + s]U &= [r + s](a, b) \\
&= ([r + s]a, [r + s]b) && \text{Definition of scalar multiplication in } \mathscr{V}_2 \\
&= (ra + sa, rb + sb) && \text{Distributive law in } R \\
&= (ra, rb) + (sa, sb) && \text{Definition of addition in } \mathscr{V}_2 \\
&= r(a, b) + s(a, b) && \text{Definition of scalar multiplication in } \mathscr{V}_2 \\
&= rU + sU.
\end{aligned}$$

10. b. If $U = (a, b)$, then from problem 8, $-U = (-a, -b)$. But $(-1)U = (-1)(a, b) = ((-1)a, (-1)b) = (-a, -b)$.

11. Subtraction would be commutative if $x - y = y - x$ for all $x, y \in R$. This does not hold for $2 - 5 = -3$ while $5 - 2 = 3$. This shows why we use signed numbers.

§3 Page 12

1. b. If (x, y) is a point on the line representing all scalar multiples of $(3, -4)$, then $(x, y) = r(3, -4)$ for some $r \in R$. Therefore $x = 3r$ and $y = -4r$, so $x/3 = r = y/-4$ and $y = (-4/3)x$.

3. The three points are collinear with the origin.

4. Remember that there are many equations for each line.

a. $P = (t, t), t \in R$. b. $P = t(1, 5), t \in R$. c. $P = t(3, 2), t \in R$.
d. $P = (t, 0), t \in R$. e. $P = t(1, 5) + (1, -1), t \in R$.
f. $P = (4, t), t \in R$. g. $P = t(1, 1/2) + (1/2, 1), t \in R$.

5. a. $(1, 1)$. b. $(1, 5)$. c. $(3, 2)$. d. $(1, 0)$. e. $(1, 5)$.
f. $(0, 1)$. g. $(2, 1)$.

6. a. $P = t(2, 3) + (5, -2), t \in R$. b. $P = t(1, 1) + (4, 0), t \in R$.
c. $P = t(0, 1) + (1, 2), t \in R$.

7. a. $x - 3y + 2 = 0$. b. $y = x$. c. $4x - y - 2 = 0$.

8. The equations have the same graph, ℓ, and the vectors V and B are on ℓ. Therefore there exist scalars s_0 and t_0 such that $V = s_0A + B$ and $B = t_0U + V$. This leads to $U = (-s_0/t_0)A$. What if $t_0 = 0$?

§4 Page 18

1. a. 2. b. 0. c. 0. d. -23.

2. a. 1. b. 1. c. 5. d. $\sqrt{10}$.

3. Let $U = (a, b)$ and compute the length of rU, remember that $\sqrt{r^2} = |r|$.

4. a. It has none. b. Use the same procedure as when proving the properties of addition in $\mathscr{V}_2$.

6. a. $P = t(1, -3) + (2, -5)$, $t \in R$. b. $P = t(1, 0) + (4, 7)$, $t \in R$.
c. $P = t(3, -7) + (6, 0)$, $t \in R$.

7. a. $(1/2, -1/2)$. b. $(7, 7)$. c. $(9/2, 7/2)$.

8. $Q = \frac{1}{5}A + \frac{4}{5}B$.

11. Notice that if A, B, C, and D are the vertices of a parallelogram, with B opposite D, then $A - B = D - C$. Write out what must be proved in vector form and use this relation.

§5 Page 21

1. a. $90°$. b. $\cos^{-1} -2/\sqrt{5}$. c. $30°$.

2. a. $P = t(3, 1)$, $t \in R$. b. $P = t(1, -1) + (5, 4)$, $t \in R$.
c. $P = t(1, 0) + (2, 1)$, $t \in R$.

7. If A, B, C are the vertices of the triangle with A and B at the ends of the diameter and $\mathbf{0}$ is the center, then $A = -B$. Consider $(A - C) \circ (B - C)$.

§6 Page 28

1. a. $(8, 1, 6)$, $\mathscr{V}_3$ or $\mathscr{E}_3$.
b. $(5, 0, 10, 18)$, $\mathscr{V}_4$ or $\mathscr{E}_4$.
c. 0, $\mathscr{E}_3$.
d. $(0, -1, -1, -6, -4)$, $\mathscr{V}_5$ or $\mathscr{E}_5$
e. 1, $\mathscr{E}_3$.
f. -6, $\mathscr{E}_4$.

2. Since $U, V \in \mathscr{V}_3$, let $U = (a, b, c)$, $V = (d, e, f)$ and proceed as in the corresponding proof for $\mathscr{V}_2$.

4. a. $\sqrt{3}/2$. b. $\sqrt{14}$. c. 1.

5. Choose a coordinate system with a vertex of the cube at the origin and the coordinate axes along edges of the cube. $\cos^{-1} 1/\sqrt{3}$.

6. a. $P = t(5, 3, -8) + (2, -1, 4)$, $t \in R$.
d. $P = t(2, 8, -5, -7, -5) + (1, -1, 4, 2, 5)$, $t \in R$.

8. a. The intersection of the two walls. b. The floor. c. The plane parallel to and 4 feet above the floor. d. The plane parallel to and 3 feet in front of the wall representing the yz coordinate plane. e. The line on the wall representing the yz coordinate plane which intersects the floor and the other wall in points 3 feet from the corner. f. The plane parallel to and 1 foot behind the wall representing the xz coordinate plane. g. The plane which bisects the angle between the two walls. h. The intersection of the plane in part g with the floor. i. The plane which intersects the floor and two walls in an equilateral triangle with side $3\sqrt{2}$.

9. a. $[P - (1, 2, 0)] \circ (0, 0, 1) = 0$, $z = 0$.
b. $[P - (4, 0, 8)] \circ (-3, 2, 1) = 0$, $3x - 2y - z = 4$.
c. $[P - (1, 1)] \circ (2, -3) = 0$, $2x - 3y = -1$.
d. $[P - (-7, 0)] \circ (1, 0) = 0$, $x = -7$.
e. $[P - (2, 1, 1, 0)] \circ (3, 1, 0, 2) = 0$, $3x + y + 2w = 7$, $P = (x, y, z, w)$.

11. a. Yes at (0, 0, 0). b. Yes at (2, −2, 4) since the three equations $t = s + 4$, $-t = 3s + 4$, and $2t = -2s$ are all satisfied by $t = -s = 2$. c. No for a solution for any two of the equations does not satisfy the third. d. No. e. Yes at (1, 1, 1).

12. a. $[P - (0, -3)] \circ (4, -1) = 0$. d. $[P - (1, 1)] \circ (-3, 1) = 0$.
b. $[P - (0, 7)] \circ (0, 1) = 0$. e. $P \circ (1, -1) = 0$.
c. $[P - (6, 1)] \circ (-2, 3) = 0$.

13. b. $\pm(2/3, 2/3, 1/3)$, $\pm(3/5, 4/5)$, $\pm(1/\sqrt{3}, 1/\sqrt{3}, 1/\sqrt{3})$.

14. The angle between U and the positive x axis is the angle between U and (1, 0, 0).

15. a. $(-1/3, 2/3, 2/3)$ or $(1/3, -2/3, -2/3)$.
b. $\pm(1/\sqrt{2}, 0, 1/\sqrt{2})$. c. $\pm(0, 1, 0)$.

16. $(a, b, 0)$, $(a, 0, c)$, $(0, b, c)$.

17. As points on a directed line.

Chapter 2

§1 Page 36

1. a. $(7, -2, -4, 4, -8)$. b. $2 + 2t + 6t^2$.
c. $3i - 5$. d. $\cos x + 4 - \sqrt{x}$.

2. a. $(-2, 4, 0, -2, -1)$, $(-5, -2, 4, -2, 9)$. b. $3t^4 - 6t^2 - 2$, $-2t - 3t^4$. c. $i - 2$, $7 - 4i$. d. $-\cos x$, $\sqrt{x} - 4$.

5. a. Definition of addition in $\mathscr{F}$; Definition of **0**; Property of zero in R.
b. $f + (-f) = \mathbf{0}$ if $(f + (-f))(x) = \mathbf{0}(x)$ for all $x \in [0, 1]$.

$$(f + (-f))(x) = f(x) + (-f)(x) = f(x) + (-f(x)) = 0 = \mathbf{0}(x).$$

What is the reason for each step?

c. The fact that $\mathscr{F}$ is closed under scalar multiplication is proven in any introductory calculus course. To show that $r(f + g) = rf + rg$, for all $r \in R$ and $f, g \in \mathscr{F}$, consider the value of $[r(f + g)](x)$.

$[r(f + g)](x) = r[(f + g)(x)]$	Definition of scalar multiplication in $\mathscr{F}$
$= r[f(x) + g(x)]$	Definition of addition in $\mathscr{F}$
$= r[f(x)] + r[g(x)]$	Distributive law in R
$= [rf](x) + [rg](x)$	Definition of scalar multiplication in $\mathscr{F}$
$= [rf + rg](x)$	Definition of addition in $\mathscr{F}$.

Therefore the functions $r(f + g)$ and $rf + rg$ are equal.

6. a. (0, 0, 0). b. $\mathbf{0}(x) = 0$, for all $x \in [0, 1]$. c. $0 = 0 + 0i$.
d. $0 \in R$. e. (0, 0, 0, 0, 0). f. $\{a_n\}$ where $a_n = 0$ for all n.
g. The origin.

7. a. The distributive law $(r + s)U = rU + sU$ fails to hold.
b. If $b \neq 0$, then $1(a, b) \neq (a, b)$.
c. Addition is neither commutative nor associative.
d. Not closed under addition or scalar multiplication. There is no zero element and there are no additive inverses.

8. Define $\begin{pmatrix} a & b \\ c & d \end{pmatrix} + \begin{pmatrix} e & f \\ g & h \end{pmatrix} = \begin{pmatrix} a+e & b+f \\ c+g & d+h \end{pmatrix}$ and $r\begin{pmatrix} a & b \\ c & d \end{pmatrix} = \begin{pmatrix} ra & rb \\ rc & rd \end{pmatrix}$.

§2 Page 42

1. a. Call the set $\mathscr{S}$. $(4, 3, 0) \in \mathscr{S}$, so $\mathscr{S}$ is nonempty. If $U, V \in \mathscr{S}$, then $U = (a, b, 0)$ and $U = (c, d, 0)$ for some scalars $a, b, c, d \in R$. Since the third component of $U + V = (a + c, b + d, 0)$ is zero, $U + V$ is in $\mathscr{S}$, and $\mathscr{S}$ is closed under addition. Similarly, $\mathscr{S}$ is closed under scalar multiplication. (How does $\mathscr{S}$ compare with $\mathscr{V}_2$?)
b. Call the set $\mathscr{T}$. If $U \in \mathscr{T}$, then $U = (a, b, c)$ with $a = 3b$ and $a + b = c$. Suppose $rU = (x, y, z)$, then $(x, y, z) = (ra, rb, rc)$. Now $x = ra = r(3b) = 3(rb) = 3y$ and $x + y = ra + rb = r(a + b) = rc = z$. Thus $rU \in \mathscr{T}$ and $\mathscr{T}$ is closed under scalar multiplication.

2. If N is normal to the plane, then the problem is to show that $\{U \in \mathscr{E}_3 | U \circ N = 0\}$ is a subspace of $\mathscr{E}_3$.

3. The question is, does the vector equation

$$x(1, 1, 1) + y(0, 1, 1) + z(0, 0, 1) = (a, b, c)$$

have a solution for x, y, z in terms of a, b, c, for all possible scalars a, b, c? That is, is $(a, b, c) \in \mathscr{S}$ for all $(a, b, c) \in \mathscr{V}_3$? The answer is yes, for $x = a$, $y = b - a$ and $z = c - b$ is a solution.

4. a. Improper: $(a, b, c) = a(1, 0, 0) + b(0, 1, 0) + c(0, 0, 1)$.
b. Proper. For example, it need not contain (7, 0, 0).
c. Proper. If $b \neq 0$, then (a, b, c) need not be a member.
d. Improper: $2(a, b, c) = [a + b - c](1, 1, 0) + [a - b + c](1, 0, 1) + [-a + b + c](0, 1, 1)$.

5. a. $(a, b, a + 4b)$ for some $a, b \in R$.
b. $(x, 5y, \frac{3}{2}x, y)$ for some $x, y \in R$.
c. $(-t, 2t, t)$ for some $t \in R$. d. $(-4t, 2t, t)$ for some $t \in R$.

6. Only one set forms a subspace.

8. Show that if $\mathscr{S}$ is a subspace of $\mathscr{V}_1$, and $\mathscr{S}$ contains a nonzero vector, then $\mathscr{V}_1 \subset \mathscr{S}$.

9. a. $\mathscr{S}$ is the subspace of all constant functions, it is essentially the vector space R.

b. The function $f(x) = -5/3$ is in $\mathscr{S}$, so $\mathscr{S}$ is nonempty. But $\mathscr{S}$ is neither closed under addition nor scalar multiplication.

c. $\mathscr{S}$ is subspace. d. $\mathscr{S}$ is not a subspace.

10. a. Since $\mathscr{S}$ and $\mathscr{T}$ are nonempty so is the sum $\mathscr{S} + \mathscr{T}$. If $A, B \in (\mathscr{S} + \mathscr{T})$, then $A = U_1 + V_1$ and $B = U_2 + V_2$ for some $U_1, U_2 \in \mathscr{S}$ and some $V_1, V_2 \in \mathscr{T}$. (Why is this true?) Now $A + B = (U_1 + V_1) + (U_2 + V_2) = (U_1 + U_2) + (V_1 + V_2)$. Since $\mathscr{S}$ and $\mathscr{T}$ are vector spaces, $U_1 + U_2 \in \mathscr{S}$ and $V_1 + V_2 \in \mathscr{T}$, therefore $\mathscr{S} + \mathscr{T}$ is closed under addition. Closure under scalar multiplication follows similarly.

11. a. Def. of $\mathbf{0}$; Def. of Y; Assoc. of $+$; Comm. of $+$; Def. of $-U$; Comm. of $+$; Def. of $\mathbf{0}$. b. Def. of additive inverse; Result just established; Assoc. of $+$; Def. of $-0U$; Def. of $\mathbf{0}$.

12. Suppose $rU = \mathbf{0}$, then either $r = 0$ or $r \neq 0$. Show that if $r \neq 0$, then $U = \mathbf{0}$.

13. If $r \neq s$, is $rU \neq sU$?

14. You want a chain of equalities yielding $Z = \mathbf{0}$. Start with

$$Z = Z + \mathbf{0} = Z + (U + (-U)) = \cdots.$$

§3 Page 49

1. a. $(0, 3, 3)$ is in the span if there exist $x, y \in R$ such that $(0, 3, 3) = x(-2, 1, -9) + y(2, -4, 6)$. This equation gives $-2x + 2y = 0$, $x - 4y = 3$, and $-9x + 6y = 3$. These equations have the common solution $x = y = -1$. Thus $(0, 3, 3)$ is in the span.

b. Here the equations are $-2x + 2y = 1$, $x - 4y = 0$, and $-9x + 6y = 1$. The first two equations are satisfied only by $y = 1/6$ and $x = 2/3$, but these values do not satisfy the third equation. Thus $(1, 0, 1)$ is not in the span.

c. $\mathbf{0}$ is in the span of any set. d. Not in the span.

e. $x = 1$, $y = 3/2$. f. $x = 2/3$, $y = 7/6$.

2. a. Suppose $t^2 = x(t^3 - t + 1) + y(3t^2 + 2t) + zt^3$. Then collecting terms in t gives, $(x + z)t^3 + (3y - 1)t^2 + (2y - x)t + x = \mathbf{0}$. Therefore, $x + z = 0$, $3y - 1 = 0$, $2y - x = 0$, and $x = 0$. Since these equations imply $y = 1/3$ and $y = 0$, there is no solution for x, y, and z, and the vector t^2 is not in the span.

b. With x, y, z as above, $x = -1$, $y = 0$, and $z = 1$.

c. $x = 4$, $y = 2$, $z = 1$. d. Not in the span.

e. $x = -1$, $y = 1$, $z = 2$.

3. a. It spans; $(a, b, c) = [b - c](1, 1, 0) + [b - a](0, 1, 1) + [a - b + c](1, 1, 1)$.

b. It spans with $(a, b, c) = a(1, 0, 0) + 0(4, 2, 0) + b(0, 1, 0) + (0, 0, 0) + c(0, 0, 1)$.

c. It does not span. From the vector equation $(a, b, c) = x(1, 1, 0) + y(2, 0, -1) + z(4, 4, 0) + w(5, 3, -1)$ can be obtained the two equations $-y - w = c$ and $2y + 2w = a - b$. Therefore (a, b, c) is in the span only if $a - b + 2c = 0$.

d. It does not span. (a, b, c) is in the span only if $a - 2b + c = 0$.

4. a. "U is a linear combination of V and W," or "U is in the span of V and W" of "U is linearly dependent on V and W."
b. "$\mathscr{S}$ is contained in the span of U and V."
c. "$\mathscr{S}$ is spanned by U and V."
d. "The set of all linear combinations of the vectors $U_1, \ldots, U_k$" or "the span of $U_1, \ldots, U_k$."

5. a. Your proof must hold for any finite subset of $R[t]$, including a set such as $\{t + 5t^6, t^2 - 1\}$.

6. Because addition of vectors is associative.

7. $\{(a, b), (c, d)\}$ is linearly dependent if and only if there exist x and y, not both zero, such that $x(a, b) + y(c, d) = (0, 0)$. This vector equation yields the system of linear equations

$$ax + cy = 0$$
$$bx + dy = 0.$$

Eliminating one variable from each equation yields the system

$$(ad - bc)x = 0$$
$$(ad - bc)y = 0,$$

which has $x = y = 0$ as the only solution if and only if $ad - bc \neq 0$.

8. a. $a \neq 1$ and $b \neq 2$. b. $2a - b - 4 \neq 0$. c. $ab \neq 0$.

9. a. You cannot choose U_1 or U_n to be the "some vector," you must use U_i where i is any integer between 1 and n.
c. Besides the fact that the statement in part a is rather cumbersome to write out, it does not imply that $\{\mathbf{0}\}$ is linearly dependent.

10. a. Independent. b. Dependent. c. Dependent. d. Dependent, e. Independent. f. Independent. g. Independent.

11. a. The vector equation $x(2, -4) + y(1, 3) + z(-6, 3) = \mathbf{0}$ has many solutions. They can be found, in this case, by letting z have an arbitrary value and solving for x and y in terms of z. Each solution is of the form $x = 3k$, $y = 6k$, $z = 2k$ for some real number k.
b. $(5, 0) = a(2, -4) + b(1, 3) + c(-6, -3)$ with $a = 3k + 3/2$, $b = 6k + 2$, and $c = 2k$, for any $k \in R$. That is, $(5, 0) = (5, 0) + \mathbf{0}$, and $(5, 0) = \frac{3}{2}(2, -4) + 2(1, 3)$.
For $(0, -20)$, $a = 3k + 2$, $b = 6k - 4$, and $c = 2k$, $k \in R$.
d. $\mathscr{L}(S) \subset \mathscr{V}_2$ is immediate, and part c gives $\mathscr{V}_2 \subset \mathscr{L}(S)$.

12. a. $3k(6, 2, -3) + 5k(-2, -4, 1) - 2k(r, -7, -2) = \mathbf{0}$ for all $k \in R$.
b. If $(0, 0, 1)$ is in the span, then there must be scalars x, y, z such that $6x - 2y + 4z = 0$ and $-3x + y - 2z = 1$.
c. Let x, y, z be the scalars needed in the linear combination. For $\mathbf{0}$, see part a. For $(0, -1, 0)$; $x = 3k + 1/10$, $y = 5k + 3/10$, and $z = -2k$, $k \in R$, For $(-2, 0, -1)$; $x = 3k + 2/5$, $y = 5k + 1/5$, and $z = -2k$, $k \in R$. For $(6, 2, -3)$; $x = 3k + 1$, $y = 5k$, and $z = -2k$, $k \in R$.

13. In problems such as this, you may not be able to complete a proof at first, but you should be able to start. Here you are given two conditions and asked to prove a third. Start by writing out exactly what these conditions mean. In this situation, this entails writing out expressions involving linear combinations and making statements about the coefficients. From these expressions, you should be able to complete a proof, but without them a proof is impossible.

14. The same remarks apply here as were given for problem 13.

15. Note that for any n vectors, $\mathbf{0} = 0U_1 + \cdots + 0U_n$.

16. The first principle of mathematical induction states that if S is a set of integers which are all greater than or equal to b, and S satisfies:

1. the integer b is in S,
2. whenever n is in S, then $n + 1$ is also in S, then S contains all integers greater than or equal to b.

There is a nice analogy for the first principle of induction. Suppose an infinite collection of dominos is set up so that whenever one falls over, it knocks the next one over. The principle is then equivalent to saying that if the first domino is pushed over, they all fall over.

For this problem, we are asked to prove that the set $S = \{k | r(U_1 + \cdots + U_k) = rU_1 + \cdots + rU_k\}$ includes all positive integers. (Here $b = 1$, although it would be possible to take $b = 2$.) Thus there are two parts to the proof:

1. Prove that the statement, $r(U_1 + \cdots + U_k) = rU_1 + \cdots + rU_k$ is true when $k = 1$.
2. Prove that if the statement is true for $k = n$, then it is true for $k = n + 1$. That is, prove that $r(U_1 + \cdots + U_n) = rU_1 + \cdots + rU_n$ implies

$$r(U_1 + \cdots + U_n + U_{n+1}) = rU_1 + \cdots + rU_n + rU_{n+1}.$$

For the second part, notice that

$$r(U_1 + \cdots + U_n + U_{n+1}) = r([U_1 + \cdots + U_n] + U_{n+1}).$$

§4 Page 57

1. Show that $\{1, i\}$ is a basis.

3. The dimension of a subspace must be an integer.

4. a. No; $\dim \mathscr{V}_3 = 3$. b. Yes. Show that the set is linearly independent, then use problem 1 and Corollary 2.8.
c. No; by Corollary 2.9. d. No; $10(-9) - (-15)(6) = 0$.
e. No; it cannot span. f. Yes; $2(4) - 1(-3) = 11 \neq 0$.

5. a. True. b. False. A counterexample: $0(1, 4) = \mathbf{0}$, but the set $\{(1, 4)\}$ is not linearly dependent. c. False.

6. a. Let $\dim \mathscr{V} = n$ and $S = \{U_1, \ldots, U_n\}$ be a spanning set for $\mathscr{V}$. Suppose

$a_1U_1 + \cdots + a_nU_n = \mathbf{0}$ and show that if one of the scalars is nonzero, then $\mathscr{V}$ is spanned by $n - 1$ vectors. This implies that $\dim \mathscr{V} \leq n - 1 < n$.
b. Let $\dim \mathscr{V} = n$ and $S = \{U_1, \ldots, U_n\}$ be linearly independent in $\mathscr{V}$. Suppose S does not span $\mathscr{V}$ and show that this implies the existence of a linearly independent set with $n + 1$ vectors within $\mathscr{V}_n$.

7. a. 2. b. 3. c. 2

8. Consider the set $\left\{\begin{pmatrix}1 & 0\\0 & 0\end{pmatrix}, \begin{pmatrix}0 & 1\\0 & 0\end{pmatrix}, \begin{pmatrix}0 & 0\\1 & 0\end{pmatrix}, \begin{pmatrix}0 & 0\\0 & 1\end{pmatrix}\right\}$.

9. a. Add any vector (a, b) such that $4b + 7a \neq 0$.
b. Add any vector (a, b, c) such that $2a - 5b - 3c \neq 0$.
c. Add any vector $a + bt + ct^2$ such that $a - 4c \neq 0$.
d. Add any vector $a + bt + ct^2$ such that $a + b - 5c \neq 0$.

10. If $\mathscr{S} \cap \mathscr{T}$ has $S = \{U_1, \ldots, U_k\}$ as a basis, then $\dim(\mathscr{S} \cap \mathscr{T}) = k$. $\mathscr{S} \cap \mathscr{T}$ is a subspace of $\mathscr{S}$, therefore S can be extended to a basis of $\mathscr{S}$. Call this basis $\{U_1, \ldots, U_k, V_{k+1}, \ldots, V_n\}$, so $\dim \mathscr{S} = n$. Similarly S can be extended to a basis $\{U_1, \ldots, U_k, W_{k+1}, \ldots, W_m\}$ of $\mathscr{T}$, and $\dim \mathscr{T} = m$.

You should now be able to show that the set $\{U_1, \ldots, U_k, V_{k+1}, \ldots, V_n, W_{k+1}, \ldots, W_m\}$ is a basis for $\mathscr{S} + \mathscr{T}$. Notice that if $a_1U_1 + \cdots + a_kU_k + b_{k+1}V_{k+1} + \cdots + b_nV_n + c_{k+1}W_{k+1} + \cdots + c_mW_m = \mathbf{0}$, then the vector $a_1U_1 + \cdots + a_kU_k + b_{k+1}V_{k+1} + \cdots + b_nV_n = -(c_{k+1}W_{k+1} + \cdots + c_mW_m)$ is in both $\mathscr{S}$ and $\mathscr{T}$.

11. A plane through the origin in 3-space can be viewed as a 2-dimensional subspace of $\mathscr{V}_3$.

§5 Page 64

1. a. You need to find a and b such that $(2, -5) = a(6, -5) + b(2, 5)$. The solution is $a = -b = 1/2$, so $(2, -5)\colon (1/2, -1/2)_{\{(6,-5),(2,5)\}}$.
b. $(1/10, -17/20)_{\{(3,1),(-2,6)\}}$. c. $(2, -5)_{\{E_i\}}$.
d. $(0, 1)_{\{(1,0),(2,-5)\}}$.

2. a. $(3/40, 11/40)_{\{(6,-5),(2,5)\}}$. b. $(2/5, 1/10)_{\{(3,1),(-2,6)\}}$.
c. $(1, 1)_{\{E_i\}}$. d. $(7/5, -1/5)_{\{(1,0),(2,-5)\}}$.

3. a. $(0, 1)_B$. b. $(1, -4)_B$. c. $(-6, 27)_B$. d. $(8, -30)_B$.
e. $(-5, 27)_B$.

4. a. $t^2 + 1\colon (0, 0, 1, 0)_B$. d. $t^2\colon (-1, 1, 1, -1)_B$.
b. $t^2\colon (1, 0, 0, 0)_B$. e. $t^3 - t^2\colon (2, -1, -1, 1)_B$.
c. $4\colon (4, -4, 0, 4)_B$. f. $t^2 - t\colon (0, 0, 1, -1)_B$.

5. a. $(0, -1)_B$. b. $(-3/5, -1/5)_B$. c. $(1, 2)_B$. d. $(-2, -1)_B$.
e. $(-3, -2)_B$.

6. a. $(5, -8, 5)_{B_1}$. b. $(0, -1/2, 0)_{B_2}$. c. $(-3, 0, 5)_{B_3}$. d. $(2, 2, 1)_{B_4}$.

7. $B = \{(2/3, 5/21), (1, 1/7)\}$.

8. $B = \{(1, 4), (1/2, 5/2)\}$.

9. If $B = \{U, V, W\}$, then $t^2 = U + W$, $t = V + W$, and $1 = U + V$, so $B = \{\frac{1}{2}t^2 - \frac{1}{2}t + \frac{1}{2}, -\frac{1}{2}t^2 + \frac{1}{2}t + \frac{1}{2}, \frac{1}{2}t^2 + \frac{1}{2}t - \frac{1}{2}\}$.

10. $B = \{4 + 2t, 2 + 3t\}$.

11. (Notice that T is not defined using coordinates.) To establish condition 2 in the definition of isomorphism, suppose $T(a, b) = T(c, d)$. Then $a + bt = c + dt$ and $a = c$, $b = d$. Therefore, $T(a, b) = T(c, d)$ implies $(a, b) = (c, d)$. For the other part of condition 2, suppose $W \in R_2[t]$, say $W = e + ft$. Then $e, f \in R$, so there is a vector $(e, f) \in \mathscr{V}_2$ and $T(e, f) = W$. Thus condition 2 holds.

12. If $\mathscr{S}$ and $\mathscr{T}$ are two such subspaces, then there exist nonzero vectors $U, V \in \mathscr{V}_2$ such that $\mathscr{S} = \{tU | t \in R\}$ and $\mathscr{T} = \{tV | t \in R\}$.

13. a. The coordinates of (4, 7) with respect to B are 1 and 5.
b. The vector $2t^2 - 2$ has coordinates -2, 1, and 1 with respect to the basis $\{t + 2, 3t^2, 4 - t^2\}$.

14. a. Although $T(a, b) = T(c, d)$ implies $(a, b) = (c, d)$, not every vector of $\mathscr{V}_3$ corresponds to a vector in $\mathscr{V}_2$, e.g. consider (0, 0, 1). Thus condition 2 does not hold.

Condition 3 holds for $T(U + V) = T((a, b) + (c, d)) = T(a + c, b + d) = (a + c, b + d, 0) = (a, b, 0) + (c, d, 0) = T(a, b) + T(c, d) = T(U) + T(V)$.

b. For condition 2, suppose $W = (x, y) \in \mathscr{V}_2$, then $(x, 0, 2x - y)$ is a vector in $\mathscr{V}_3$ and $T(x, 0, 2x - y) = (x, y)$. Therefore, half the condition holds. But if $T(a, b, c) = T(d, e, f)$, then (a, b, c) need not equal (d, e, f). For $T(a, b, c) = T(d, e, f)$ implies $a + b = d + e$ and $2a - c = 2d - f$. And these equations have many solutions besides $a = d$, $b = e$, $c = f$. For example $a = d + 1$, $b = e - 1$, $c = f + 2$, thus $T(4, 6, 3) = T(5, 5, 5)$ and condition 2 fails to hold. T satisfies condition 4 since $T(rU) = T(r(a, b, c)) = T(ra, rb, rc) = (ra + rb, 2(ra) - rc) = r(a + b, 2a - c) = rT(a, b, c) = rT(U)$.

15. a. $2 + t$ and $4 + 6t$.

16. Suppose T is a correspondence from $\mathscr{V}_2$ to $\mathscr{V}_3$ satisfying conditions 1, 3, and 4 in the definition of isomorphism (the map in problem 14a is an example.) Show that T fails to satisfy condition 2 by showing that the set $\{T(V) | V \in \mathscr{V}_2\}$ cannot contain three linearly independent vectors.

§6 Page 70

1. $(a, b) \in \mathscr{L}\{(1, 7)\} \cap \mathscr{L}\{(2, 3)\}$ if $(a, b) = x(1, 7)$ and $(a, b) = y(2, 3)$ for some $x, y \in R$. Thus $x(1, 7) = y(2, 3)$, but $\{(1, 7), (2, 3)\}$ is linearly independent, so $x = y = 0$ and the sum is direct. Since $\{(1, 7), (2, 3)\}$ must span $\mathscr{V}_2$, for any $W \in \mathscr{V}_2$, there exist $a, b \in R$ such that $W = a(1, 7) + b(2, 3)$. Now $a(1, 7)$

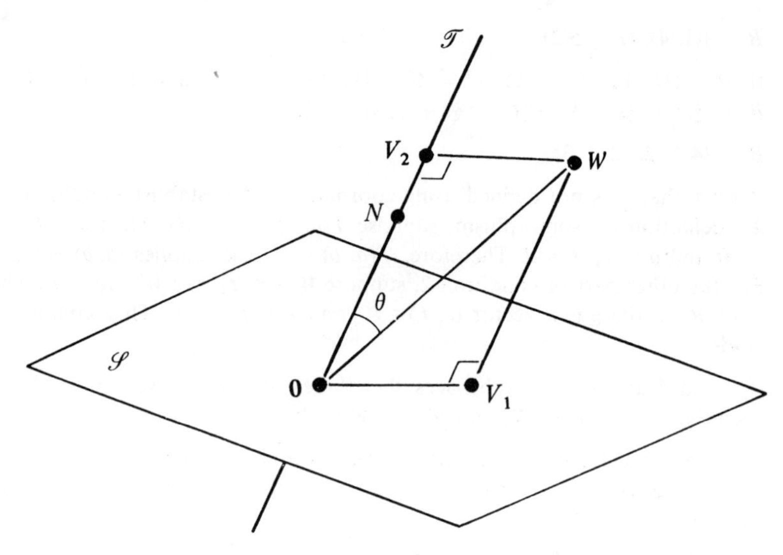

Figure A1

$\in \mathscr{L}\{(1, 7)\}$ and $b(2, 3) \in \mathscr{L}\{(2, 3)\}$ so $\mathscr{V}_2$ is contained in the sum and

$$\mathscr{V}_2 = \mathscr{L}\{(1, 7)\} \oplus \mathscr{L}\{(2, 3)\}.$$

2. No.

3. a. If $(a, b, c) \in \mathscr{S} \cap \mathscr{T}$, then there exist $r, s, t, u \in R$ such that

$$(a, b, c) = r(1, 0, 0) + s(0, 1, 0) = t(0, 0, 1) + u(1, 1, 1).$$

Therefore $(r, s, 0) = (u, u, t + u)$ and $\mathscr{S} \cap \mathscr{T} = \mathscr{L}\{(1, 1, 0)\}$.
b. Direct sum. c. Direct sum. d. $\mathscr{S} \cap \mathscr{T} = \mathscr{L}\{(4, 1, 1, 4)\}$.

5. If $W = \mathbf{0}$, it is immediate, so assume $W \neq \mathbf{0}$. Let θ be the angle between W and N, see Figure A1. Then $\cos\theta = \pm \|V_2\|/\|W\|$ with the sign depending on whether θ is acute or obtuse. And $V_2 = \pm\|V_2\|(N/\|N\|)$ with the sign depending on whether W and N are on the same or opposite sides of the plane $\mathscr{S}$, i.e., on whether θ is acute or obtuse.

6. a. $\mathscr{V}_3 = \mathscr{S} + \mathscr{T}$, the sum is not direct. b. $\mathscr{V}_3 \neq \mathscr{S} + \mathscr{T}$.
c. $\mathscr{V}_3 = \mathscr{S} \oplus \mathscr{T}$. d. $\mathscr{V}_3 \neq \mathscr{S} + \mathscr{T}$.

7. a. $U = (4a, a, 3a)$ and $V = (2b + c, b, b + 2c)$ for some $a, b, c \in R$. $U = (-20, -5, -15)$ and $V = (16, 4, 20)$.
b. $U = (1, -8, 6)$ and $V = (-1, 2, -1)$.

8. $U = (a, b, a + b)$ and $V = (c, d, 3c - d)$ for some $a, b, c, d \in R$. Since the sum $\mathscr{S} + \mathscr{T}$ is not direct, there are many solutions. For example, taking $d = 0$ gives $U = (2, 1, 3)$ and $V = (-1, 0, -3)$.

9. Choose a basis for $\mathscr{S}$ and extend it to a basis for $\mathscr{V}$.

10. Two possiblities are $R_4[t] \oplus \mathscr{L}\{t^4, t^5, \ldots, t^n, \ldots\}$ and

$$\mathscr{L}\{1, t^2, t^4, \ldots, t^{2n}, \ldots\} \oplus \mathscr{L}\{t, t^3, t^5, \ldots, t^{2n+1}, \ldots\}.$$

Review Problems Page 71

1. a, b, and c are false. d is true but incomplete and therefore not very useful. e and f are false.

2. b, d, g, h, i, and p are meaningful uses of the notation. s is meaningful if $U = V = \mathbf{0}$. The other expressions are meaningless.

3. b. Note that showing S is linearly dependent amounts to showing that $\mathbf{0}$ may be expressed as a linear combination of the vectors in S in an infinite number of ways. $t = (-1/4 - 4k)(2t + 2) + (1/2 + 5k)(3t + 1) + k(3 - 7t)$ for any $k \in R$.
c. Use the fact that $2t + 2 = \frac{5}{4}(3t + 1) + \frac{1}{4}(3 - 7t)$. Compare with problem 14, on page 51.

4. Only three of the statements are true.
a. A counterexample for this statement is given by $\{\mathbf{0}, t\}$ in $\mathscr{V} = R_2[t]$ or $\{(1, 3), (2, 6)\}$ in $\mathscr{V} = \mathscr{V}_2$.

5. b. If $U = (a, 2a + 4b, b)$ and $V = (c, 3d, 5d)$, then there are many solutions to the equations $a + c = 2$, $2a + 4b + 3d = 9$, and $b + 5d = 6$. One solution is given by $a = b = c = d = 1$.

7. a. For $n = 3$, $k = 1$, choose $\mathscr{V} = \mathscr{V}_3$, $\{U_1, U_2, U_3\} = \{E_i\}$, and $\mathscr{S} = \mathscr{L}\{(4, 1, 7)\}$.

Chapter 3

§1 Page 81

1. a.
$$\begin{aligned} 3y \quad\quad &= 0 \\ x + y + 4z &= 3 \\ x \quad\quad + 7z &= 0. \end{aligned}$$
b.
$$\begin{aligned} 6x + 7y &= 2 \\ 9x - 2y &= 3. \end{aligned}$$
c.
$$\begin{aligned} x + 2y &= 2 \\ 3x + y &= 1 \\ 6x + y &= -4. \end{aligned}$$

2. a. $3y = 0$, $x + y + z = 0$, $z = 0$, $-5y = 0$.
b. $3x - y + 6z = 0$, $4x + 2y - z = 0$.
c. Suppose $ae^x + be^{2x} + ce^{5x} = \mathbf{0}(x)$ for all $x \in [0, 1]$. Then $x = 0$ gives $a + b + c = 0$; $x = 1$ gives $ea + e^2b + e^5c = 0$, and so on.

3. $a_{11} = 2$, $a_{23} = 1$, $a_{13} = 0$, $a_{24} = -7$, and $a_{21} = 0$.

4. a. $A = \begin{pmatrix} 2 & -5 & 0 & 6 \\ 0 & 3 & 1 & -7 \end{pmatrix}$, $A^* = \begin{pmatrix} 2 & -5 & 0 & 6 & -6 \\ 0 & 3 & 1 & -7 & 8 \end{pmatrix}$.

b. $A = \begin{pmatrix} 2 & -1 \\ 1 & 1 \\ 5 & -1 \end{pmatrix}$, $A^* = \begin{pmatrix} 2 & -1 & 7 \\ 1 & 1 & 3 \\ 5 & -1 & 6 \end{pmatrix}$.

c. $A = \begin{pmatrix} 2 & -3 & 1 \\ 1 & 0 & -2 \end{pmatrix}$, $A^* = \begin{pmatrix} 2 & -3 & 1 & 0 \\ 1 & 0 & -2 & 0 \end{pmatrix}$.

5. a. $\begin{pmatrix} -5 \\ 6 \end{pmatrix}$. b. $\begin{pmatrix} 0 \\ -10 \\ 10 \\ -2 \end{pmatrix}$. c. $\begin{pmatrix} 6 \\ 3 \end{pmatrix}$.

6. a. $\begin{pmatrix} 2 & -5 & 0 & 6 \\ 0 & 3 & 1 & -7 \end{pmatrix}\begin{pmatrix} x_1 \\ x_2 \\ x_3 \\ x_4 \end{pmatrix} = \begin{pmatrix} -6 \\ 8 \end{pmatrix}$. b. $\begin{pmatrix} 2 & -1 \\ 1 & 1 \\ 5 & -1 \end{pmatrix}\begin{pmatrix} x \\ y \end{pmatrix} = \begin{pmatrix} 7 \\ 3 \\ 6 \end{pmatrix}$.

c. $\begin{pmatrix} 2 & -3 & 1 \\ 1 & 0 & -2 \end{pmatrix}\begin{pmatrix} x_1 \\ x_2 \\ x_3 \end{pmatrix} = \begin{pmatrix} 0 \\ 0 \end{pmatrix}$.

7. a. In determining if $S \subset \mathscr{V}_n$ is linearly dependent, one equation is obtained for each component. A is 2×3, so $AX = \mathbf{0}$ has two equations and $n = 2$. Further there is one unknown for each vector. Since A has three columns, S must contain three vectors from $\mathscr{V}_2$. If the first equation is obtained from the first component, then $S = \{(5, 1), (2, 3), (7, 8)\}$.
b. $\{(2, 1), (4, 3)\}$.
c. $\{(2, 1, 2), (5, 3, 1), (7, 0, 0), (8, 2, 6)\}$.
d. $\{(5, 2, 1, 6), (1, 4, 0, 1), (3, 9, 7, 5)\}$.

8. a. $W = (3, 2)$, $U_1 = (2, 1)$, and $U_2 = (-1, 1)$.
b. $W = (0, 0, 0)$, $U_1 = (1, 2, 4)$, and $U_2 = (5, 3, 9)$.
c. $W = (4, 8, 6, 0)$, $U_1 = (9, 2, 1, 0)$, and $U_2 = (5, 1, 7, 2)$.

10. a. $\begin{pmatrix} 2 & -1 & 8 \\ 4 & 11 & 3 \end{pmatrix}$. b. $\begin{pmatrix} 2 \\ 8 \\ -6 \end{pmatrix}$. c. $\begin{pmatrix} 6 & 3 \\ 12 & 6 \\ 3 & 0 \end{pmatrix}$.

11. $\begin{pmatrix} 0 & 0 \\ 0 & 0 \\ 0 & 0 \end{pmatrix}$, $\begin{pmatrix} 0 \\ 0 \\ 0 \end{pmatrix}$, $(0, 0, 0, 0)$, and $\begin{pmatrix} 0 & 0 \\ 0 & 0 \end{pmatrix}$.

12. The zero in $AX = \mathbf{0}$ is $n \times 1$; the zero in $X = \mathbf{0}$ is $m \times 1$.

14. a. $\mathbf{0} \notin S$. b. A plane not through the origin.
c. S may be empty, or a plane or line not passing through the origin.

§2 Page 93

1. a. $\begin{pmatrix} 1 & 1 & 1 \\ 4 & 1 & 3 \end{pmatrix}\begin{pmatrix} 3 - 2k \\ 3 - k \\ 3k \end{pmatrix} = \begin{pmatrix} 3 - 2k + 3 - k + 3k \\ 12 - 8k + 3 - k + 9k \end{pmatrix} = \begin{pmatrix} 6 \\ 15 \end{pmatrix}$.

Therefore all the solutions satisfy this system.

b. $\begin{pmatrix} 3 & 0 & 2 \\ 0 & 3 & 1 \\ 1 & -2 & 0 \end{pmatrix}\begin{pmatrix} 3 - 2k \\ 3 - k \\ 3k \end{pmatrix} = \begin{pmatrix} 9 \\ 9 \\ -3 \end{pmatrix}$ holds for all values of k.

c. $\begin{pmatrix} 1 & 1 & -2 \\ 3 & 1 & 1 \end{pmatrix}\begin{pmatrix} 3 - 2k \\ 3 - k \\ 3k \end{pmatrix} = \begin{pmatrix} 6 - 9k \\ 12 - 4k \end{pmatrix} \neq \begin{pmatrix} 0 \\ 1 \end{pmatrix}$ for all k.

Therefore the solutions do not satisfy this system.

2. $A_1X = B_1$ and $A_2X = B_2$ are the same except for their hth equations which are $a_{h1}x_1 + \cdots + a_{hm}x_m = b_h$ and $(a_{h1} + ra_{k1})x_1 + \cdots + (a_{hm} + ra_{km})x_m = b_h + rb_k$, respectively. Therefore, it is necessary to show that if $A_1C = B_1$, then $X = C$ is a solution of the hth equation in $A_2X = B_2$ and if $A_2C = B_2$, then $X = C$ is a solution of the hth equation of $A_1X = B_1$.

3. a. $\begin{pmatrix} 1 & 0 & 0 \\ 0 & 1 & 0 \\ 0 & 0 & 1 \end{pmatrix}$. b. $\begin{pmatrix} 1 & 0 \\ 0 & 1 \\ 0 & 0 \end{pmatrix}$. c. $\begin{pmatrix} 0 & 1 & -2 & 0 & -1 \\ 0 & 0 & 0 & 1 & 3 \\ 0 & 0 & 0 & 0 & 0 \end{pmatrix}$. d. $\begin{pmatrix} 1 & 4 & 0 & 0 \\ 0 & 0 & 1 & 0 \\ 0 & 0 & 0 & 1 \end{pmatrix}$.

4. a. From the echelon form $\begin{pmatrix} 1 & 0 & 0 & 2 \\ 0 & 1 & 0 & 3 \\ 0 & 0 & 1 & 1 \end{pmatrix}$, $x = 2$, $y = 3$, and $z = 1$.

b. $\begin{pmatrix} 1 & 0 & 1 \\ 0 & 1 & 1 \\ 0 & 0 & 0 \end{pmatrix}$, gives $x = y = 1$. c. $\begin{pmatrix} 1 & 0 & -1 & 0 \\ 0 & 1 & 2 & 0 \\ 0 & 0 & 0 & 1 \end{pmatrix}$, no solution.

5. The only 2×2 matrices in echelon form are $\begin{pmatrix} 0 & 0 \\ 0 & 0 \end{pmatrix}$, $\begin{pmatrix} 0 & 1 \\ 0 & 0 \end{pmatrix}$, $\begin{pmatrix} 1 & 0 \\ 0 & 1 \end{pmatrix}$ and $\begin{pmatrix} 1 & k \\ 0 & 0 \end{pmatrix}$, where k is any real number.

6. Call the span $\mathscr{S}$. Then $(x, y, z) \in \mathscr{S}$, if $(x, y, z) = a(1, 0, -4) + b(0, 1, 2) = (a, b, -4a + 2b)$. Therefore $\mathscr{S} = \{(x, y, z) | z = 2y - 4x\}$, i.e., $\mathscr{S}$ is the plane in 3-space with the equation $4x - 2y + z = 0$.

7. a. The echelon form is $\begin{pmatrix} 1 & 0 \\ 0 & 1 \\ 0 & 0 \end{pmatrix}$, so the row space is $\mathscr{V}_2$.

b. $\begin{pmatrix} 1 & 0 & -2 \\ 0 & 1 & 3 \\ 0 & 0 & 0 \end{pmatrix}$ is the echelon form, therefore U is in the row space if $U = a(1, 0, -2) + b(0, 1, 3)$ and the row space is the set $\{(x, y, z) | 2x - 3y + z = 0\}$.

c. $\begin{pmatrix} 1 & 0 & 1 \\ 0 & 1 & 5 \end{pmatrix}$, $\{(x, y, z) | z = x + 5y\}$.

d. $\mathscr{V}_3$. e. $\begin{pmatrix} 1 & 0 & 0 & 1 \\ 0 & 1 & 0 & 2 \\ 0 & 0 & 1 & 3 \end{pmatrix}$, $\{(x, y, z, w) | w = x + 2y + 3z\}$.

f. $\begin{pmatrix} 1 & 0 & 2 & 2 \\ 0 & 1 & 0 & 6 \end{pmatrix}$, $\{(x, y, z, w) | z = 2x \text{ and } w = 2x + 6y\}$.

8. a. 2. b. 2. c. 2. d. 3. e. 3. f. 2.

11. a. The rank of $\begin{pmatrix} 6 & 9 \\ 4 & 6 \end{pmatrix}$ is 1, so the vectors are linearly dependent.

b. Rank is 2. c. Rank is 2. d. Rank is 3.

12. "A is an n by m matrix."

13. Let $U_1, \ldots, U_n$ be the rows of A and $U_1, \ldots, U_h + rU_k, \ldots, U_n$ the rows of B. Then it is necessary to show

$$\mathscr{L}\{U_1, \ldots, U_k, \ldots, U_h, \ldots, U_n\} \subset \mathscr{L}\{U_1, \ldots, U_k, \ldots, U_h + rU_k, \ldots, U_n\}$$

and

$$\mathscr{L}\{U_1, \ldots, U_k, \ldots, U_h + rU_k, \ldots, U_n\} \subset \mathscr{L}\{U_1, \ldots, U_k, \ldots, U_h, \ldots, U_n\}.$$

14. The idea is to find a solution for one system which does not satisfy the other by setting some of the unknowns equal to 0. Show that it is sufficient to consider the following two cases.

Case 1: e_{kh} is the leading entry of 1 in the kth row of E, and the leading 1 in the kth row of F is f_{kj} with $j > h$.

Case 2: Neither e_{kh} nor f_{kh} is a leading entry of 1.

In the first case, show that x_j must be a parameter in the solutions of $EX = \mathbf{0}$, but when $x_{j+1} = \cdots = x_m = 0$ in a solution for $FX = \mathbf{0}$, then $x_j = 0$.

§3 Page 100

1. $x = 3$, $y = -2$.

2. $x = 1 + 2t$, $y = t$, $t \in R$.

3. $x = 1/2$, $y = 1/3$.

4. Inconsistent, it is equivalent to

$$\begin{aligned} x - 3y &= 0. \\ 0 &= 1. \end{aligned}$$

5. Equivalent to

$$\begin{aligned} x - \tfrac{3}{2}y &= -2 \\ z &= -3, \end{aligned}$$

so $x = 3t - 2$, $y = 2t$, and $z = -3$, for any $t \in R$.

6. Inconsistent. **7.** $x = 3 + 2s - t$, $y = s$, $z = t$, $s, t \in R$.

8. $x = y = z = 1$. **9.** $x = -2t$, $y = 3t - 2$, $z = t$, $w = 4$, $t \in R$.

10. The system is equivalent to

$$\begin{aligned} x - \tfrac{1}{7}z + w &= 0 \\ y - \tfrac{5}{7}z + w &= 0, \end{aligned}$$

therefore the solutions are given by $x = s - t$, $y = 5s - t$, $z = 7s$, $w = t$, for $s, t \in R$.

11. a. Independent. b, c, and d are dependent.

12. a. $\begin{pmatrix} 2 & 3 \\ 1 & 6 \end{pmatrix}$, b. $\begin{pmatrix} 4 & 1 \\ 2 & 3 \\ 5 & 7 \end{pmatrix}$. c. $(2, 5, 6)$. d. $\begin{pmatrix} 5 & 2 & 3 \\ 2 & 5 & 4 \\ 6 & 0 & 2 \end{pmatrix}$.

e. $\begin{pmatrix} 3 & 1 & 4 & 1 \\ 1 & 0 & 2 & 3 \\ 7 & 2 & 0 & 2 \end{pmatrix}$.

13. a. 2. b. 2. c. 1. d. 2. e. 3.

15. From the definition of the span of a set, obtain a system of n linear equations in n unknowns. Use the fact that $\{V_1, \ldots, V_n\}$ is linearly independent to show that such a system is consistent.

§4 Page 105

1. a. 2; a point in the plane. b. 1; a line in the plane.
c. 2; a line through the origin in 3-space.
d. 1; a plane through the origin in 3-space.
e. 1; a hyperplane in 4-space. f. 3; the origin in 3-space.

2. a. In a line. b. The planes coincide.
c. At the origin. d. In a line.

3. Rank A = rank A^* = 3; the planes intersect in a point.
Rank A = rank A^* = 2; the planes intersect in a line.
(Rank A = rank A^* = 1 is excluded since the planes are distinct.)
Rank A = 1, rank A^* = 2; three parallel planes.
Rank A = 2, rank A^* = 3; either two planes are parallel and intersect the third in a pair of parallel lines, or each pair of planes intersects in one of three parallel lines.

4. a. Set $\mathscr{S} = \{(2a + 3b, 3a + 5b, 4a + 5b) | a, b \in R\}$. Then $\mathscr{S} = \mathscr{L}\{(2, 3, 4), (3, 5, 5)\}$ is the row space of $A = \begin{pmatrix} 2 & 3 & 4 \\ 3 & 5 & 5 \end{pmatrix}$. Since A is row-equivalent to $\begin{pmatrix} 1 & 0 & 5 \\ 0 & 1 & -2 \end{pmatrix}$, $\mathscr{S} = \mathscr{L}\{(1, 0, 5), (0, 1, -2)\}$. That is, $(x, y, z) \in \mathscr{S}$ if $(x, y, z) = x(1, 0, 5) + y(0, 1, -2)$. This gives the equation $z = 5x - 2y$.
b. $\begin{pmatrix} 1 & 2 & -1 \\ 0 & -3 & 2 \end{pmatrix}$ is row-equivalent to $\begin{pmatrix} 1 & 0 & 1/3 \\ 0 & 1 & -2/3 \end{pmatrix}$, and an equation is $x - 2y - 3z = 0$.
c. $x - 3y + z = 0$. Notice that although the space is defined using 3 parameters, only two are necessary.
d. $4x - 2y + 3z - w = 0$. e. $x + y + z + w = 0$.

5. a. $(x, y, z, w) = (4, 0, 0, 2) + t(-3, 2, 1, 0)$, $t \in R$. This is the line with direction numbers $(-3, 2, 1, 0)$ which passes through the point $(4, 0, 0, 2)$.
b. $(x, y, z, w) = (-3, 1, 0, 0) + t(-1, -2, 1, 0) + s(-2, 0, 0, 1)$, $t, s \in R$. This is the plane in 4-space, parallel to $\mathscr{L}\{(-1, -2, 1, 0), (-2, 0, 0, 1)\}$ and passing through the point $(-3, 1, 0, 0)$.
c. $(x_1, x_2, x_3, x_4, x_5) = (1, 2, 0, 1, 0) + t(-3, -1, 1, 0, 0) + s(0, -2, 0, 0, 1)$, $t, s \in R$. The plane in 5-space through the point $(1, 2, 0, 1, 0)$ and parallel to the plane $\mathscr{L}\{(-3, -1, 1, 0, 0), (0, -2, 0, 0, 1)\}$.

6. a. An arbitrary point on the line is given by $t(P - Q) + P = (1 + t)P - tQ$. If $S = (x_j)$, $P = (p_j)$ and $Q = (q_j)$, show that $x_j = (1 + t)p_j - tq_j$, for $1 \le j \le n$, satisfies an arbitrary equation in the system $AX = B$.
b. Recall that both $S \subset \{U + V | V \in \mathscr{S}\}$ and $\{U + V | V \in \mathscr{S}\} \subset S$ must be established. For the first containment, notice that, for any $W \in S$, $W = U + (W - U)$.

8. a. $S = \{(-3, -7k, -1 - 4k, k) | k \in R\}$. This is the line with vector equation $P = t(-7, -4, 1) + (-3, -4, 0)$, $t \in R$.
b. $\mathscr{S}$ is the line with equation $P = t(-7, -4, 1)$, $t \in R$.

9. a. $S = \{(4 - 2a + b, 3 - 3a + 2b, a, b) | a, b \in R\}$;

$$\mathscr{S} = \mathscr{L}\{(-2, -3, 1, 0), (1, 2, 0, 1)\}.$$

b. $S = \{(4, 2, 0, 0) + (b - 2a, 2b - 3a, a, b) | a, b \in R\}$.

c. S is a plane.

§5 Page 115

1. 2. **2.** −5. **3.** 2. **4.** 1. **5.** 0. **6.** 0. **7.** 2. **8.** 0.

9. a. A parallelogram, since A is the origin and $D = B + C$. Area = 13.

b. A parallelogram since $B - A = D - C$. The area is the same as the area of the parallelogram determined by $B - A$ and $C - A$. The area is 34.

c. Area = 9.

10. a. 1. b. 30.

12. Suppose the hth row of A is multiplied by r, $A = (a_{ij})$ and $B = (b_{ij})$. Expand B along the hth row where $b_{hj} = ra_{hj}$ and $B_{hj} = A_{hj}$. The corollary can also be proved using problem 11.

13. The statement is immediate for 1×1 matrices, therefore assume the property holds for all $(n - 1) \times (n - 1)$ matrices, $n \geq 2$. To complete the proof by induction, it must be shown that if $A = (a_{ij})$ is an $n \times n$ matrix, then $|A^T| = |A|$.

Let $A^T = (b_{ij})$, so $b_{ij} = a_{ji}$, then

$$\begin{aligned} |A^T| &= b_{11}B_{11} + b_{12}B_{12} + \cdots + b_{1n}B_{1n} \\ &= a_{11}B_{11} + a_{21}B_{12} + \cdots + a_{n1}B_{1n}. \end{aligned}$$

The cofactor B_{1j} of b_{1j} is $(-1)^{1+j}M_{1j}$, where M_{1j} is the minor of b_{1j}. The minor M_{1j} is the determinant of the $(n - 1) \times (n - 1)$ matrix obtained from A^T by deleting the 1st row and jth column. But if the 1st column and jth row of A are deleted, the transpose of M_{1j} is obtained. Since these matrices are $(n - 1) \times (n - 1)$, the induction assumption implies that M_{1j} is equal to the minor of a_{j1} in A. That is, $B_{1j} = (-1)^{1+j}M_{1j} = (-1)^{j+1}M_{1j} = A_{j1}$. Therefore the above expansion becomes $|A^T| = a_{11}A_{11} + a_{21}A_{21} + \cdots + a_{n1}A_{n1}$, which is the expansion of $|A|$ along the first column.

Now the statement is true for all 1×1 matrices, and if it is true for all $(n - 1) \times (n - 1)$ matrices $(n \geq 2)$, it is true for all $n \times n$ matrices. Therefore the proof is complete by induction.

14. a. $\left(\begin{vmatrix} 0 & 0 \\ 1 & 0 \end{vmatrix}, -\begin{vmatrix} 1 & 0 \\ 0 & 0 \end{vmatrix}, \begin{vmatrix} 1 & 0 \\ 0 & 1 \end{vmatrix}\right) = (0, 0, 1).$ b. $(0, 0, -1)$.

Note that parts a and b show that the cross product is not commutative.

c. $\mathbf{0}$. d. $(-2, -4, 8)$. e. $\mathbf{0}$. f. $(0, -19, 0)$.

15. a. If the matrix is called $A = (a_{ij})$, then

$$A_{11}a_{11} = \begin{vmatrix} b_2 & b_3 \\ c_2 & c_3 \end{vmatrix} E_1 = [b_2c_3 - b_3c_2](1, 0, 0).$$

16. $A \circ (B \times C)$ is the volume of the parallelepiped determined by A, B, and C. Since there is no volume, A, B, and C must lie in a plane that passes through the origin.

17. a. (18, −5, −16) and (22, 8, −4).
b. The cross product is not an associative operation.
c. The expression is not defined because the order in which the operations are to be performed has not been indicated.

§6 Page 119

1. a. 13. b. 15. c. 0. d. 1. e. −144. f. 96.

3. a. The determinant is −11, so the set is linearly independent. b. The determinant is −5. c. The determinant is 0.

4. The set yields a lower triangular matrix with all the elements on the main diagonal nonzero.

5. a. Follow the pattern set in the proof of Theorem 3.12, that interchanging rows changes the sign of a determinant, and the proof that $|A^T| = |A|$, problem 13, page 115.

6. $$|A| = a_{11}\begin{vmatrix} 1 & a_{12}/a_{11} & \cdots & a_{1n}/a_{11} \\ 0 & a_{22} - a_{21}(a_{12}/a_{11}) & \cdots & a_{2n} - a_{21}(a_{1n}/a_{11}) \\ & & \vdots & \\ 0 & a_{n2} - a_{n1}(a_{12}/a_{11}) & \cdots & a_{nn} - a_{n1}(a_{1n}/a_{11}) \end{vmatrix}$$

7. a. $$\begin{vmatrix} 3 & 7 & 2 \\ 5 & 1 & 4 \\ 9 & 6 & 8 \end{vmatrix} = \frac{1}{3}\begin{vmatrix} 3\cdot 1 - 5\cdot 7 & 3\cdot 4 - 5\cdot 2 \\ 3\cdot 6 - 9\cdot 7 & 3\cdot 8 - 9\cdot 2 \end{vmatrix} = \frac{1}{3}\begin{vmatrix} -32 & 2 \\ -27 & 6 \end{vmatrix} = -34$$

b. $$\begin{vmatrix} 5 & 4 & 2 \\ 2 & 7 & 2 \\ 3 & 2 & 3 \end{vmatrix} = \frac{1}{5}\begin{vmatrix} 5\cdot 7 - 2\cdot 4 & 5\cdot 2 - 2\cdot 2 \\ 5\cdot 2 - 3\cdot 4 & 5\cdot 3 - 3\cdot 2 \end{vmatrix} = 51.$$

c. $$\begin{vmatrix} 2 & 1 & 3 & 2 \\ 1 & 3 & 2 & 1 \\ 2 & 2 & 1 & 1 \\ 3 & 1 & 2 & 1 \end{vmatrix} = \frac{1}{2^2}\begin{vmatrix} 2\cdot 3 - 1\cdot 1 & 2\cdot 2 - 1\cdot 3 & 2\cdot 1 - 1\cdot 2 \\ 2\cdot 2 - 2\cdot 1 & 2\cdot 1 - 2\cdot 3 & 2\cdot 1 - 2\cdot 3 \\ 2\cdot 1 - 3\cdot 1 & 2\cdot 2 - 3\cdot 3 & 2\cdot 1 - 3\cdot 2 \end{vmatrix}$$

$$= \frac{1}{4}\begin{vmatrix} 5 & 1 & 0 \\ 2 & -4 & -2 \\ -1 & -5 & -4 \end{vmatrix} = \frac{1}{4}\cdot\frac{1}{5}\begin{vmatrix} -22 & -10 \\ -24 & -20 \end{vmatrix} = 10.$$

§7 Page 126

1. a. $\begin{pmatrix} 1 & 0 \\ 0 & 1 \\ 0 & 0 \end{pmatrix}$. b. $\begin{pmatrix} 1 & 0 & 0 \\ 0 & 1 & 0 \\ 0 & 0 & 0 \end{pmatrix}$. c. $\begin{pmatrix} 1 \\ 0 \\ 0 \end{pmatrix}$. d. $\begin{pmatrix} 1 & 0 & 0 & 0 & 0 \\ 0 & 1 & 0 & 0 & 0 \\ 0 & 0 & 1 & 0 & 0 \end{pmatrix}$.

e. $\begin{pmatrix} 1 & 0 \\ 0 & 1 \end{pmatrix}$.

2. a. $\mathscr{M}_{2\times 1}$, rank 2. c. $\mathscr{M}_{3\times 1}$, rank 3.

b. $\mathscr{L}\left\{\begin{pmatrix} 1 \\ 0 \\ 2 \end{pmatrix}, \begin{pmatrix} 0 \\ 1 \\ 3 \end{pmatrix}\right\}$, rank 2. d. $\mathscr{L}\left\{\begin{pmatrix} 1 \\ 0 \\ 2/3 \end{pmatrix}, \begin{pmatrix} 0 \\ 1 \\ -1/3 \end{pmatrix}\right\}$, rank 2.

4. a. The row space of $A = \mathscr{L}\{(1, 0, 11/3), (0, 1, -1/3)\}$ and the row space of $A^T = \mathscr{L}\{(1, 0, -1), (0, 1, 1)\}$.

b. The row space of $A = \mathscr{L}\{(1, 0, 1), (0, 1, -2)\}$ and the row space of $A^T = \mathscr{L}\{(1, 0, -3), (0, 1, 1)\}$.

5. "A transpose" or "the transpose of A."

6. If $U = (a, b, c)$ and $V = (d, e, f)$, then the components of $U \times V$ are plus or minus the determinants of the 2×2 submatrices of

$$\begin{pmatrix} a & b & c \\ d & e & f \end{pmatrix}.$$

7. That is, if $x_1U_1 + \cdots + x_nU_n = W$ has a solution for all $W \in \mathscr{V}_n$, show that $x_1U_1 + \cdots + x_nU_n = \mathbf{0}$ implies $x_1 = \cdots = x_n = 0$.

Review Problems Page 126

2. a. The two equations in two unknowns would be obtained in determining if $\{4 + t, 5 + 3t\}$ (or $\{1 + 4t, 3 + 5t\}$) is linearly independent.
b. $S = \{2 + 9t + t^2, 8 + 3t + 7t^2\} \subset R_3[t]$.
c. $S = \{3 + 2t, 1 - t, 4 + 6t\} \subset R_2[t]$.

5. a. $\begin{vmatrix} 1 & 0 & x \\ 0 & 1 & y \\ 1 & 1 & 1 \end{vmatrix} = 1 - y - x.$ So an equation is $y = 1 - x$.
b. $y = 9x - 23$. c. $3y - 8x + 37 = 0$.

7. a. $5x + 6y - 8z = 0$. b. $3x - y = 0$.
c. $3x + 8y + 2z = 0$. d. $x + z = 0$.

9. If $AX = B$ is a system of n linear equations in n unknowns, then there is a unique solution if and only if the determinant of A is nonzero. Since $|A|$ may equal any real number, it is reasonable to expect that a given $n \times n$ matrix has nonzero determinant. That is, it is the exceptional case when the value of the determinant of an $n \times n$ matrix is 0, or any other particular number.

11. a. $A\begin{pmatrix} 1 \\ 1 \end{pmatrix} = \begin{pmatrix} 5 \\ 5 \\ 8 \end{pmatrix}$, $\begin{pmatrix} 5 \\ 10 \\ -1 \end{pmatrix}$, and $\begin{pmatrix} 2 \\ 1 \\ 5 \end{pmatrix}$, respectively.
b. $\begin{pmatrix} 0 \\ 0 \end{pmatrix}$, $\begin{pmatrix} 0 \\ 0 \end{pmatrix}$, and $\begin{pmatrix} 0 \\ 1 \end{pmatrix}$, respectively.
c. The requirement is that, for each $B \in \mathscr{M}_{n \times 1}$, there exists one and only one $X \in \mathscr{M}_{m \times 1}$, such that $AX = B$. What does this say about A when $B = \mathbf{0}$? Is there always an X when $m < n$?

Chapter 4

§1 Page 135

1. a, c, e, and g are not linear. b, d, and f are linear maps.

2. Remember that an if and only if statement requires a proof for both implications.

3. "T is a map from $\mathscr{V}$ to $\mathscr{W}$," or "T sends $\mathscr{V}$ into $\mathscr{W}$."

4. a. Suppose $T(a, b) = (-1, 5)$. Then $2a - b = -1$ and $3b - 4a = 5$. These equations have a unique solution, and $(1, 3)$ is the only preimage.
b. $\{(t, -t, 3t) | t \in R\}$. c. No preimage.
d. The problem is to determine which vectors in $R_3[t]$ are sent to t by the map T. That is, for what values of a, b, c, if any, is $T(a + bt + ct^2) = t$? This happens only if $a + b = 1$ and $c - a = 0$. Therefore any vector in $\{r + (1 - r)t + rt^2 | r \in R\}$ is a preimage of t under the map T.
e. $\{\frac{1}{2}t^2 + k | k \in R\}$. f. $\{3 + k - kt + kt^2 | k \in R\}$.

5. a. For each ordered 4-tuple W, there is at least one ordered triple V such that $T(V) = W$.
b. For each 2×2 matrix A, there is a 2×3 matrix B, such that $T(B) = A$.
c. For each polynomial P of degree less than 2, there is a polynomial Q of degree less than 3, such that $T(Q) = P$.

6. a. Onto but not 1–1. For example, $T\begin{pmatrix} 0 & k & k \\ k & k & 0 \end{pmatrix} = \begin{pmatrix} 0 & 0 \\ 0 & 0 \end{pmatrix}$, for any scalar k.
b. Neither 1–1 nor onto. c. 1–1, but not onto.
d. 1–1 and onto. e. Neither. f. 1–1, but not onto.

7. a. $(14, -8, -6)$. b. $(4, 10, 14)$. c. $(1, 4, -5)$.
d. $(2, 5, 7)$. e. $(x, y, -x - y)$. f. $(0, 0, 0)$.

8. a. Suppose $\mathscr{V} = \mathscr{S} \oplus \mathscr{T}$ and $V = U + W$, with $U \in \mathscr{S}$, $W \in \mathscr{T}$. Then $P_1(V) = U$. If $r \in R$, then $rV = r(U + W) = rU + rW$, and $rU \in \mathscr{S}$, $rW \in \mathscr{T}$. (Why?) So $P_1(rV) = rU = rP_1(V)$, and P_1 preserves scalar multiplication. The proof that a projection map preserves addition is similar.

9. a. If $\mathscr{S} = \mathscr{L}\{(1, 0)\}$, then $\mathscr{T} = \mathscr{L}\{(0, 1)\}$ is perpendicular to $\mathscr{S}$ and $\mathscr{E}_2 = \mathscr{S} \oplus \mathscr{T}$. Thus P_1 is the desired projection. Since $(a, b) = (a, 0) + (0, b)$, $P_1(a, b) = (a, 0)$.
b. $P_1(a, b) = (a/10 + 3b/10, 3a/10 + 9b/10)$.

10. a. $T(a, b) = (a/2 - \sqrt{3}b/2, \sqrt{3}a/2 + b/2)$, $T(1, 0) = (1/2, \sqrt{3}/2)$.
c. To show that a rotation is 1–1 and onto, let (a, b) be any vector in the codomain, and suppose $T(x, y) = (a, b)$. That is,

$$(x \cos \theta - y \sin \theta, x \sin \theta + y \cos \theta) = (a, b).$$

This vector equation yields a system of linear equations. Consider the coefficient matrix of this system and show that, for any $a, b \in R$, there is a unique solution for x and y.

11. A translation is linear only if $h = k = 0$. Every translation is both one to one and onto.

12. The reflection is linear, 1–1, and onto.

13. Show that $T(U) = T(\mathbf{0})$.

14. This is proved for $n = 2$ in problem 2. The induction assumption is that it is true for any linear combination of $n - 1$ terms, $n \geq 3$.

§2 Page 144

1. a. $\mathscr{I}_T = \{(3a - 2c, b + c)|a, b, c \in R\} = \mathscr{L}\{(3, 0), (0, 1), (-2, 1)\}$
$= \mathscr{V}_2$.
$\mathscr{N}_T = \{(a, b, c)|3a - 2c = 0, b + c = 0\} = \{t(2, -3, 3)|t \in R\}$
$= \mathscr{L}\{(2, -3, 3)\}$.
b. $\mathscr{I}_T = \{(x, y, z)|5x + 7y - 6z = 0\}$; $\mathscr{N}_T = \{\mathbf{0}\}$.
c. $\mathscr{I}_T = R$; $\mathscr{N}_T = \{a - ai|a \in R\} = \mathscr{L}\{1 - i\}$.
d. $\mathscr{I}_T = \mathscr{L}\{1 + t^2, t + 2t^2, 2 - t\} = \mathscr{L}\{1 + t^2, 2 - t\}$.
$\mathscr{N}_T = \{a + bt + ct^2|b = c, a = -2c\} = \mathscr{L}\{2 - t - t^2\}$.

2. a. $\{(4t - 2(5t), 3(5t) - 6t)|t \in R\} = \{t(2, -3)|t \in R\} = \mathscr{L}\{(2, -3)\}$.
b. $\{\mathbf{0}\}$. c. $\{(2, -3)\}$. d. $\mathscr{L}\{(2, -3)\}$. e. $\{(-4, 6)\}$.
f. If it is parallel to $\mathscr{L}\{(1, 2)\}$, i.e., the line with equation $y = 2x$.

3. a. $\mathscr{L}\{(1, 0, -2), (0, 1, 1)\} = \{(x, y, z)|2x - y + z = 0\}$.
b. $\mathscr{L}\{(1, 0, 1/2), (0, 1, 5/6)\} = \{(x, y, z)|3x + 5y - 6z = 0\}$.
c. $\mathbf{0}$. d. $\mathscr{V}_2$.

4. a. $(a, b, c) \in \mathscr{N}_T$ if $2a + b = 0$ and $a + b + c = 0$. A homogeneous system of 2 equations in 3 unknowns has nontrivial solutions, so T is not 1–1.
b. One to one. c. $T(a, -2a, 0, 0) = \mathbf{0}$, for all $a \in R$.
d. One to one. d. Not one to one.

5. Each is a map from a 3-dimensional space to a 2-dimensional space.

6. $\mathscr{I}_P = \mathscr{T}$ and $\mathscr{N}_P = \mathscr{S}$.

7. a. $\mathscr{N}_T$. b. Why is $T^{-1}[\mathscr{T}]$ nonempty? For closure under addition and scalar multiplication, is it sufficient to take $U, V \in T^{-1}[\mathscr{T}]$ and show that $rU + sV \in T^{-1}[\mathscr{T}]$ for any $r, s \in R$?
c. $T^{-1}[\{W\}]$ must contain at least one vector.

8. $(x, y) = [x - y](4, 3) + [2y - \frac{3}{2}x](2, 2)$, therefore $T(x, y) = (4x - 3y, 2x - y)$.

9. $(x, y) = [x - 3y](4, 1) + [4y - x](3, 1)$, so $T(x, y) = (10y - 2x, 3x - 11y, 7y - 2x)$.

10. a. $(2, 4) = \frac{2}{3}(3, 6)$, but $T(2, 4) \neq \frac{2}{3}T(3, 6)$. Therefore any extension of T to all $\mathscr{V}_2$ will fail to be linear.
b. Yes, but not uniquely, for now $T(2, 4) = \frac{3}{2}T(3, 6)$. An extension can be obtained by extending $\{(2, 4)\}$ to a basis for $\mathscr{V}_2$, say to $\{(2, 4), (5, 7)\}$, and then assigning an image in $\mathscr{V}_3$ to the second basis vector, say $T(5, 7) = (0, 3, 8)$. For these choices, T may be extended linearly to obtain the map

$$T(x, y) = (\tfrac{5}{3}y - \tfrac{7}{3}x, 2x - y, \tfrac{7}{3}y - \tfrac{5}{3}x).$$

11. Observe that since $T(1, 0) \in \mathscr{V}_2$, there exist $a, c \in R$ such that $T(1, 0) = (a, c)$.

13. a. $\mathscr{N}_T = \{(x, y, z)|2x + y - 3z = 0\}$.
b. Recall that a plane parallel to $\mathscr{N}_T$ has the Cartesian equation $2x + y - 3z = k$ for some $k \in R$; or that for any fixed vector $U \in \mathscr{V}_3$, $\{V + U|V \in \mathscr{N}_T\}$ is a plane parallel to $\mathscr{N}_T$.
d. $\mathscr{S}$ could be any subspace of dimension 1 which is not contained in $\mathscr{N}_T$.

§3 Page 153

1. a. $[T_1 + T_2](a, b) = (4a, 2a + 5b, 2a - 12b)$.
b. $[5T](a, b) = (15a - 5b, 10a + 20b, 5a - 40b)$.
c. $[T_3 - 3T_2](a, b) = (2b - 2a, -2a - 3b, 4b)$.
d. $\mathbf{0}_{\text{Hom}(\mathscr{V}_2, \mathscr{V}_3)}$.

2. a. $x_1T_1 + x_2T_2 + x_3T_3 + x_4T_4 = \mathbf{0}$ yields the equations $x_1 + x_2 + x_3 = 0$, $x_1 + x_3 = 0$, $x_1 + x_2 = 0$, and $x_2 + x_4 = 0$. This system has only the trivial solution, so the vectors are linearly independent. b. $2T_1 - T_2 - T_3 = \mathbf{0}$. c. $T_1 - T_2 + T_3 - T_4 = \mathbf{0}$. d. $T_1 + T_2 - 2T_3 = \mathbf{0}$.

3. "Let T be a linear map from $\mathscr{V}$ to $\mathscr{W}$."

4. b. Since we cannot multiply vectors, the first zero must be the zero map in $\text{Hom}(\mathscr{V}, \mathscr{W})$, the second is the zero vector in $\mathscr{V}$, and the fourth is the zero vector in $\mathscr{W}$.

5. a. If $T \in \text{Hom}(\mathscr{V}, \mathscr{W})$ and $r \in R$, then it is necessary to show that $rT: \mathscr{V} \to \mathscr{W}$, and that the map rT is linear.
b. Why is it necessary to show that $[(a + b)T](V) = [aT + bT](V)$ for all $V \in \mathscr{V}$? Where is a distributive law in $\mathscr{W}$ used?

6. a. $(0, 3)$. b. $(0, -6)$. c. $(-9, 5)$. d. $(-12, 0)$.
e. $(4, -3, 0, 1)_B$. f. $(0, 0, 0, 0)_B$. g. $(4, 0, 1, 0)_B$.
h. $(2, 1, 3, -5)_B$.

7. a. $a_{12}T_{12} + a_{22}T_{22} + a_{32}T_{32}$. b. $a_{31}T_{31} + a_{32}T_{32}$.
c. $a_{11}T_{11} + a_{21}T_{21} + a_{31}T_{31} + a_{12}T_{12} + a_{22}T_{22} + a_{32}T_{32}$.

8. a. $\sum_{i=1}^{2}\sum_{j=1}^{2} r_{ij}T_{ij}$. b. $\sum_{j=1}^{4} a_{4j}T_{4j}$. c. $\sum_{i=1}^{3}\sum_{j=1}^{2} r_i a_{ij}T_{ij}$.

9. a. $T_{12}(x, y) = (y, y, 0)$. b. $T_{22}(x, y) = (x - 3y, x - 3y, 0)$.
c. $T_{23}(4, 7) = (-17, 0, 0)$. d. $T_{11}(5, 9) = (9, 9, 9)$

10. a. $T(3, 1) = (2, 13, 3)$: $(3, 10, -11)_{B_2}$, therefore $T(3, 1) = [3T_{11} + 10T_{12} - 11T_{13}](3, 1)$. And $T(1, 0) = (1, 4, 0)$: $(0, 4, -3)_{B_2}$, so $T(1, 0) = [0T_{21} + 4T_{22} - 3T_{23}](1, 0)$. Therefore $T = 3T_{11} + 10T_{12} - 11T_{12} + 0T_{21} + 4T_{22} - 3T_{23}$, and T: $(3, 10, -11, 0, 4, -3)_B$. You should be obtaining the coordinates of vectors with respect to B_2 by inspection.
b. T: $(8, -8, 12, 2, -2, 5)_B$. c. T: $(2, 1, 2, 1, 0, 0)_B$.
d. T: $(6, -2, 3, 2, -2, 3)_B$.

11. a. $T_{11}(a + bt) = a$. $T_{13}(a + bt) = at^2$. $T_{21}(a + bt) = b$.
b. $[2T_{11} + 3T_{21} - 4T_{13}](a + bt) = 2a + 3b - 4at^2$.
c. $T = T_{12} + T_{23}$. d. $T = 2T_{21} + T_{12} - 4T_{22} + 5T_{13}$.
e. $(0, 0, 1, 0, 0, 1)_B$. f. $(3, -3, 7, -1, 4, 5)_B$.
g. $(0, 0, 0, 0, 0, 0)_B$.

12. a. $(0, 0)$. b. $(1, 2)$. c. $(-5, -10)$. d. You must first find the coordinates of $(3, 4, 5)$ with respect to B_1. $T_{22}(3, 4, 5) = -2(1, 2) = (-2, -4)$.
e. $T_{12}(1, 0, 0) = (4, 8)$. f. T: $(3, 2, 2, -2, 9, -1)_B$.
g. T: $(7, 0, -10, 3, -5, 4)_B$.

13. a. $0 \in R$, $(0, 0) \in \mathscr{V}_2$, $(0, 0, 0, 0) \in \mathscr{V}_4$, and the map in $\text{Hom}(\mathscr{V}_2, \mathscr{V}_4)$ that sends (x, y) to $(0, 0, 0, 0)$ for all $(x, y) \in \mathscr{V}_2$.

§4 Page 164

1. a. $\begin{pmatrix} 1 & 0 \\ 0 & 1 \end{pmatrix}$. b. $\begin{pmatrix} 0 & 0 \\ 1 & 0 \end{pmatrix}$. c. $\begin{pmatrix} 0 & 0 & 0 \\ 0 & 0 & 1 \\ 0 & 0 & 0 \end{pmatrix}$. d. $\begin{pmatrix} 2 & -1 \\ 3 & 4 \\ 1 & 3 \end{pmatrix}$.

2. a. $\begin{pmatrix} 1 & 0 \\ 0 & 1 \\ 0 & 0 \end{pmatrix}$. b. $\begin{pmatrix} 0 & 0 \\ 0 & 0 \\ 1 & 1 \end{pmatrix}$. c. $\begin{pmatrix} 0 & 0 \\ 0 & 0 \\ 0 & 0 \end{pmatrix}$. d. $\begin{pmatrix} 3 & -1 \\ -3 & 4 \\ 7 & 5 \end{pmatrix}$.

3. a. $\begin{pmatrix} 0 & 0 \\ 0 & 0 \\ 0 & 0 \end{pmatrix}$. b. $\begin{pmatrix} 0 & 0 \\ 1 & 0 \\ 0 & 0 \end{pmatrix}$. c. $\begin{pmatrix} 18 & 20 \\ -3 & -3 \\ -7 & -8 \end{pmatrix}$.

4. a. $\begin{pmatrix} 1 & 3 & 0 \\ 0 & 1 & -2 \\ 1 & 0 & 6 \end{pmatrix}$. b. $\begin{pmatrix} 7 & -3 & 4 \\ 5 & -4 & 1 \\ -8 & 9 & 1 \end{pmatrix}$. c. $\begin{pmatrix} 1 & 0 & 1 \\ 0 & 1 & 1 \\ 0 & 0 & 0 \end{pmatrix}$.

d. and e. $\begin{pmatrix} 1 & 0 & 0 \\ 0 & 1 & 0 \\ 0 & 0 & 0 \end{pmatrix}$.

5. $A = \begin{pmatrix} -3 & 1 & 2 \\ -3 & -1 & 4 \\ 1 & 0 & -1 \end{pmatrix}$. a. $A\begin{pmatrix} 0 \\ 1 \\ 0 \end{pmatrix} = \begin{pmatrix} 1 \\ -1 \\ 0 \end{pmatrix}$, so $T(0, 1, 0) = (1, -1, 0)$.

b. $(-1, 7, -1)$. c. $(0, 0, 0)$.

6. $A = \begin{pmatrix} -3 & -4 \\ 5 & 7 \end{pmatrix}$. a. $(1, 1)$: $(1, 0)_{\{(1,1),(1,0)\}}$, $A\begin{pmatrix} 1 \\ 0 \end{pmatrix} = \begin{pmatrix} -3 \\ 5 \end{pmatrix}$, and $-3(1, 1) + 5(1, 0) = (2, -3)$, so $T(1, 1) = (2, -3)$.

b. $A\begin{pmatrix} 0 \\ 4 \end{pmatrix} = \begin{pmatrix} -16 \\ 28 \end{pmatrix}$ and $T(4, 0) = (12, -16)$.

c. $A\begin{pmatrix} 2 \\ 1 \end{pmatrix} = \begin{pmatrix} -10 \\ 17 \end{pmatrix}$ and $T(3, 2) = (7, -10)$.

d. $A\begin{pmatrix} 1 \\ -1 \end{pmatrix} = \begin{pmatrix} 1 \\ -2 \end{pmatrix}$ and $T(0, 1) = (-1, 1)$.

7. The matrix of T with respect to B_1 and B_2 is $A = \begin{pmatrix} 2 & 0 & -1 \\ 3 & 1 & 0 \\ -1 & -3 & 1 \end{pmatrix}$.

a. $\begin{pmatrix} 2 & 0 & -1 \\ 3 & 1 & 0 \\ -1 & -3 & 1 \end{pmatrix}\begin{pmatrix} 0 \\ 1 \\ 0 \end{pmatrix} = \begin{pmatrix} 0 \\ 1 \\ -3 \end{pmatrix}$, so $T(t) = t - 3t^2$.

b. $A(1, 0, 1)^T = (1, 3, 0)^T$, so $T(1 + t^2) = 1 + 3t$.

c. $A(-4, 3, 0)^T = (-8, -9, -5)^T$, so $T(3t - 4) = -8 - 9t - 5t^2$.

d. $A(0, 0, 0)^T = (0, 0, 0)^T$, so $T(\mathbf{0}) = \mathbf{0}$.

8. a. $T(V_3) = \mathbf{0}$. b. $T(V_4) = 7W_2$. c. $T(V_1)$ is a linear combination of two vectors in B_2. d. $\mathscr{I}_T \subset \mathscr{L}\{W_1, W_m\}$. e. $T(V_1) = a_{11}W_1, \ldots, T(V_n) = a_{nn}W_n$ provided $n \leq m$. What if $n > m$?

9. a. T is onto. b. T is 1–1. c. T is not onto. d. T is onto, but not 1–1.

10. There are an infinite number of answers. If $B_1 = \{E_i\}$, then B_2 must be $\{(3-4), (-1, 1)\}$.

11. If $B_1 = \{V_1, V_2, V_3\}$, then V_3 must be in the null space of T. One possible answer is $B_1 = \{(1, 0, 0), (0, 1, 0), (1, 1, 1)\}$ and $B_2 = \{(-3, -3, 1), (1, -1, 0), (1, 0, 0)\}$. Check that B_1 and B_2 are in fact bases and that they give the desired result.

12. If $B_1 = \{V_1, V_2, V_3, V_4\}$ and $B_2 = \{W_1, W_2, W_3\}$, then $V_3, V_4 \in \mathscr{N}_T$ and $T(V_1) = W_1 - W_3$, $T(V_2) = W_2 + W_3$.

13. $\begin{pmatrix} -5 & 7 & 2 \\ 26 & -33 & -3 \end{pmatrix}$, $\begin{pmatrix} 0 & -3 & -3 \\ -4 & 12 & 12 \end{pmatrix}$, $\begin{pmatrix} -5 & 4 & -1 \\ 22 & -21 & 9 \end{pmatrix}$, $\begin{pmatrix} -10 & 5 & -5 \\ 40 & -30 & 30 \end{pmatrix}$.

14. Let $B_1 = \{V_1, \ldots, V_n\}$ and $B_2 = \{W_1, \ldots, W_m\}$. If $\varphi(T) = A = (a_{ij})$, then $T(V_j) = \sum_{i=1}^m a_{ij} W_i$. To show that φ preserves scalar multiplication, notice that $[rT](V_j) = r[T(V_j)] = r\sum_{i=1}^m a_{ij}W_i = \sum_{i=1}^m ra_{ij}W_i$. Therefore $[rT](V_j){:}(ra_{1j}, \ldots, ra_{mj})_{B_2}$ and $\varphi(rT) = (ra_{ij})$. But $(ra_{ij}) = r(a_{ij}) = rA = r\varphi(T)$.

15. a. $A = \begin{pmatrix} 1 & -1 \\ 4 & 1 \\ 2 & 3 \end{pmatrix}$.

16. If T: $(b_{11}, b_{12}, b_{13}, b_{21}, b_{22}, b_{23})_B$, then $T = \sum_{h=1}^2 \sum_{k=1}^3 b_{hk} T_{hk}$. Therefore $T(V_j) = \sum_{h=1}^2 \sum_{k=1}^3 b_{hk} T_{hk}(V_j) = \sum_{k=1}^3 b_{jk} W_{jk}$.

Hence $T(V_j)$: $(b_{j1}, b_{j2}, b_{j3})_{B_2}$, and $\begin{pmatrix} b_{11} & b_{21} \\ b_{12} & b_{22} \\ b_{13} & b_{23} \end{pmatrix} = \begin{pmatrix} a_{11} & a_{12} \\ a_{21} & a_{22} \\ a_{31} & a_{32} \end{pmatrix}$.

§5 Page 174

1. a. $S \circ T(a, b) = (4b - 2a, 8a - 4b)$, $T \circ S(x, y) = (2y - 6x, 12x)$.
b. $S \circ T(a, b, c) = (a - b + c, -b, a + c)$, $T \circ S(x, y) = (2y, x + y)$.
c. $S \circ T$ is undefined, $T \circ S(a, b, c) = (2a, 2a + 3b, 4a - 4b, b)$.
d. $S \circ T(a + bt) = -b - 6at^2$, $T \circ S$ is undefined.

2. a. $T^2(a, b) = (4a, b - 3a)$.
b. $T^3(a, b) = (8a, b - 7a)$.
c. $T^n(a, b) = (2^n a, b - [2^n - 1]a)$.
d. $[T^2 - 2T](a, b) = (0, -a - b)$.
e. $[T + 3I](a, b) = (5a, 4b - a)$.

5. Since by definition, $S \circ T: \mathscr{V} \to \mathscr{W}$, it need only be shown that if $U, V \in \mathscr{V}$ and $a, b \in R$, then $[S \circ T](aU + bV) = a([S \circ T](U)) + b([S \circ T](V))$.

7. b. When multiplication is commutative, $a(b + c) = (b + c)a$.

8. If $S \in \text{Hom}(\mathscr{V}_2)$, then there exist scalars x, y, z, and w such that $S(a, b) = (xa + yb, za + wb)$ (problem 11, page 145), and since a linear map is determined by its action on a basis, $S \circ T = T \circ S$ if and only if $S \circ T(E_i) = T \circ S(E_i)$, $i = 1, 2$, where $\{E_i\}$ is the standard basis for $\mathscr{V}_2$.
a. $\{S | S(a, b) = (xa, wb) \text{ for some } x, w \in R\}$. b. Same as a.

c. $\{S(a, b) = (xa + yb, za + wb) | x + 2y - w = 0, 3y + z = 0\}$.
d. $\{S(a, b) = (xa + yb, za + wb) | x + y - w = 0, y + z = 0\}$.

9. b. For any map T, $T \circ I = I \circ T$.

10. a. Not invertible.
b. Solving the system $a - 2b = x$, $b - a = y$ for a and b, gives $T^{-1}(x, y) = (-x - 2y, -x - y)$.
c. Suppose $T^{-1}(x + yt + zt^2) = (a, b, c)$, then $a + c = x$, $b = y$, and $-a - b = z$. So $T^{-1}(x + yt + zt^2) = (-y - z, y, x + y + z)$.
d. $T^{-1}(x, y) = y - x + [3x - 2y]t$.
e. A map from $R_2[t]$ to $R_3[t]$ is not invertible.
f. $T^{-1}(x, y, z) = (-3x + 2y - 2z, 6x - 3y + 4z, -2x + y - z)$.

11. a. Assume S_1 and S_2 are inverses for T and consider the fact that
$$S_1 = S_1 \circ I = S_1 \circ [T \circ S_2].$$

13. The vector space cannot be finite dimensional in either a or b.

15. If $V = [T \circ S](W)$, consider $S(V)$. Where do you need problem 14?

16. a. $S \circ T(a, b) = (17a + 6b, 14a + 5b)$.
b. $S^{-1}(a, b) = (a - b, 4b - 3a)$. $T^{-1}(a, b) = (2a - b, 3b - 5a)$. $(S \circ T)^{-1}(a, b) = (5a - 6b, 17b - 14a)$.

§6 Page 186

1. a. $\begin{pmatrix} -1 & 8 \\ 18 & 12 \end{pmatrix}$. b. (6). c. $\begin{pmatrix} 3 & 3 \\ 11 & 1 \\ 12 & 3 \end{pmatrix}$. d. $\begin{pmatrix} 28 & 14 & -21 \\ -8 & -4 & 6 \\ 12 & 6 & -9 \end{pmatrix}$.

e. $\begin{pmatrix} 0 & 0 \\ 0 & 0 \end{pmatrix}$. f. $\begin{pmatrix} 19 & 15 & 16 \\ 30 & -8 & 4 \\ 5 & 1 & 3 \end{pmatrix}$.

3. a. $\begin{pmatrix} 3/2 & -2 \\ -1/2 & 1 \end{pmatrix}$. b. $\begin{pmatrix} 2/11 & 1/11 \\ -3/11 & 4/11 \end{pmatrix}$. c. $\begin{pmatrix} -3/2 & 1 \\ 1 & -1/2 \end{pmatrix}$.

d. $\frac{1}{4}\begin{pmatrix} 2 & -4 & 2 \\ -1 & 6 & -3 \\ -3 & 6 & -1 \end{pmatrix}$. e. $\begin{pmatrix} 1 & 0 & 0 \\ 0 & 0 & 1 \\ 0 & 1 & 0 \end{pmatrix}$. f. $\frac{1}{25}\begin{pmatrix} 1 & -2 & 8 \\ 12 & 1 & -4 \\ -3 & 6 & 1 \end{pmatrix}$.

4. a. Show that the product $B^{-1}A^{-1}$ is the inverse of AB.
b. By induction on k for $k \geq 2$. Use the fact that $(A_1A_2 \cdots A_kA_{k+1})^{-1} = ((A_1A_2 \cdots A_k)A_{k+1})^{-1}$.

5. For Theorem 4.18 ($\Rightarrow$) Suppose T is nonsingular, then T^{-1} exists. Let $\varphi(T^{-1}) = B$ and show that $B = A^{-1}$.
($\Leftarrow$) Suppose A^{-1} exists. Then there exists S such that $\varphi(S) = A^{-1}$ (Why?) Show that $S = T^{-1}$.

6. a. $\begin{pmatrix} x \\ y \end{pmatrix} = \begin{pmatrix} 2 & -3 \\ 1 & 2 \end{pmatrix}^{-1}\begin{pmatrix} 5 \\ 4 \end{pmatrix} = \frac{1}{7}\begin{pmatrix} 2 & 3 \\ -1 & 2 \end{pmatrix}\begin{pmatrix} 5 \\ 4 \end{pmatrix} = \begin{pmatrix} 22/7 \\ 3/7 \end{pmatrix}$.
b. $x = 14$, $y = 3$, $z = 13$. c. $a = 9/8$, $b = -3/8$, $c = 13/16$, $d = 1/2$.

7. a. $\begin{pmatrix} 3 & 2 & -1 \\ 6 & 4 & -1 \\ -4 & -3 & 1 \end{pmatrix}\begin{pmatrix} 1 \\ 0 \\ 0 \end{pmatrix} = \begin{pmatrix} 3 \\ 6 \\ -4 \end{pmatrix}$, therefore $(1, 0, 0)\colon (3, 6, -4)_B$.

b. $(0, 3, -1)_B$. c. $(-3, 2, 1)_B$. d. $(0, 0, 2)_B$.

8. a. $\begin{pmatrix} -8 & -9 & 21/2 \\ -2 & -2 & 5/2 \\ 9 & 10 & -23/2 \end{pmatrix}\begin{pmatrix} 4 \\ -1 \\ 2 \end{pmatrix} = \begin{pmatrix} -2 \\ -1 \\ 3 \end{pmatrix}$, so $(4, -1, 2)\colon (-2, -1, 3)_B$.

b. $(0, 1, 0)_B$. c. $(-9, -2, 10)_B$. d. $(21, 5, -23)_B$.

9. $A = \begin{pmatrix} 2 & -3 \\ -3 & 3 \end{pmatrix}$ and $A^{-1} = \begin{pmatrix} -1 & -1 \\ -1 & -2/3 \end{pmatrix}$. For a. $\begin{pmatrix} -1 & -1 \\ -1 & -2/3 \end{pmatrix}\begin{pmatrix} 1 \\ 0 \end{pmatrix} = \begin{pmatrix} -1 \\ -2/3 \end{pmatrix}$, therefore $T^{-1}(0, 1) = (-1, -2/3)$. b. $(-1, 1)$. c. $(-4, -3)$.
d. $(-x - y, -x - \frac{2}{3}y)$.

10. The matrices of T and T^{-1} with respect to the standard basis for $\mathscr{V}_3$ are $\begin{pmatrix} 2 & 1 & 3 \\ 2 & 2 & 1 \\ 4 & 2 & 4 \end{pmatrix}$ and $\begin{pmatrix} -3/2 & -1/2 & 5/4 \\ 1 & 1 & -1 \\ 1 & 0 & -1/2 \end{pmatrix}$ respectively.

a. $(10, -8, -4)$. b. $(1, -2, 3)$.
c. $T^{-1}(x, y, z) = (-\frac{3}{2}x - \frac{1}{2}y + \frac{5}{4}z, x + y - z, x - \frac{1}{2}z)$.

§7 Page 194

1. a. $\begin{pmatrix} 1 & 0 \\ -5/2 & 1 \end{pmatrix}$. b. $\begin{pmatrix} 0 & 1 & 0 \\ 1 & 0 & 0 \\ 0 & 0 & 1 \end{pmatrix}$. c. $\begin{pmatrix} 1/2 & 0 \\ 0 & 1 \end{pmatrix}$. d. $\begin{pmatrix} 1 & 0 & 0 \\ 0 & 0 & 1 \\ 0 & 1 & 0 \end{pmatrix}$.

e. $\begin{pmatrix} 1 & 0 & 0 \\ -2 & 1 & 0 \\ 0 & 0 & 1 \end{pmatrix}$. f. $\begin{pmatrix} 1 & 0 & 0 \\ 0 & 1 & 0 \\ 0 & -3 & 1 \end{pmatrix}$. g. $\begin{pmatrix} 1 & 0 & 0 \\ 0 & 1 & 0 \\ 0 & 0 & 1/4 \end{pmatrix}$.

2. a. $E_1 = \begin{pmatrix} 1/2 & 0 \\ 0 & 1 \end{pmatrix}$, $E_2 = \begin{pmatrix} 1 & 0 \\ -5 & 1 \end{pmatrix}$, $E_3 = \begin{pmatrix} 1 & 0 \\ 0 & -1/6 \end{pmatrix}$, $E_4 = \begin{pmatrix} 1 & -3 \\ 0 & 1 \end{pmatrix}$.

c. $E_1^{-1} - \begin{pmatrix} 2 & 0 \\ 0 & 1 \end{pmatrix}$, $E_2^{-1} = \begin{pmatrix} 1 & 0 \\ 5 & 1 \end{pmatrix}$, $E_3^{-1} = \begin{pmatrix} 1 & 0 \\ 0 & -6 \end{pmatrix}$, $E_4^{-1} = \begin{pmatrix} 1 & 3 \\ 0 & 1 \end{pmatrix}$.

d. $\begin{pmatrix} 2 & 6 \\ 5 & 9 \end{pmatrix}$.

3. There are many possible answers for each matrix.

a. $\begin{pmatrix} 1 & 0 \\ 0 & 5 \end{pmatrix}\begin{pmatrix} 0 & 1 \\ 1 & 0 \end{pmatrix}\begin{pmatrix} 1 & 0 \\ 0 & 3 \end{pmatrix}\begin{pmatrix} 1 & -2 \\ 0 & 1 \end{pmatrix}$. b. $\begin{pmatrix} 1 & 0 \\ 3 & 1 \end{pmatrix}\begin{pmatrix} 1 & 0 \\ 0 & -4 \end{pmatrix}\begin{pmatrix} 1 & 4 \\ 0 & 1 \end{pmatrix}$.

c. $\begin{pmatrix} 2 & 0 & 0 \\ 0 & 1 & 0 \\ 0 & 0 & 1 \end{pmatrix}\begin{pmatrix} 1 & 0 & 0 \\ 0 & 1 & 0 \\ 3 & 0 & 1 \end{pmatrix}\begin{pmatrix} 1 & 0 & 0 \\ 0 & 0 & 1 \\ 0 & 1 & 0 \end{pmatrix}\begin{pmatrix} 1 & 2 & 0 \\ 0 & 1 & 0 \\ 0 & 0 & 1 \end{pmatrix}\begin{pmatrix} 1 & 0 & 0 \\ 0 & 1 & 0 \\ 0 & 1 & 1 \end{pmatrix}\begin{pmatrix} 1 & 0 & 3 \\ 0 & 1 & 0 \\ 0 & 0 & 1 \end{pmatrix}$.

d. $\begin{pmatrix} 0 & 0 & 1 \\ 0 & 1 & 0 \\ 1 & 0 & 0 \end{pmatrix}\begin{pmatrix} 1 & 0 & 0 \\ 0 & 1 & 0 \\ 2 & 0 & 1 \end{pmatrix}\begin{pmatrix} 1 & 0 & 0 \\ 0 & 0 & 1 \\ 0 & 1 & 0 \end{pmatrix}\begin{pmatrix} 1 & 0 & 0 \\ 0 & 1 & 0 \\ 0 & 1 & 1 \end{pmatrix}\begin{pmatrix} 1 & 0 & 0 \\ 0 & 1 & 0 \\ 0 & 0 & 2 \end{pmatrix}\begin{pmatrix} 1 & 0 & 1 \\ 0 & 1 & 0 \\ 0 & 0 & 1 \end{pmatrix}$.

4. a. $\begin{pmatrix} 3 & 1 & 4 & 1 & 0 & 0 \\ 1 & 0 & 2 & 0 & 1 & 0 \\ 2 & 5 & 0 & 0 & 0 & 1 \end{pmatrix} \to \begin{pmatrix} 1 & 0 & 2 & 0 & 1 & 0 \\ 3 & 1 & 4 & 1 & 0 & 0 \\ 2 & 5 & 0 & 0 & 0 & 1 \end{pmatrix} \to$

$$\begin{pmatrix} 1 & 0 & 2 & 0 & 1 & 0 \\ 0 & 1 & -2 & 1 & -3 & 0 \\ 0 & 5 & -4 & 0 & -2 & 1 \end{pmatrix} \to \begin{pmatrix} 1 & 0 & 2 & 0 & 1 & 0 \\ 0 & 1 & -2 & 1 & -3 & 0 \\ 0 & 0 & 6 & -5 & 13 & 1 \end{pmatrix} \to$$

$$\begin{pmatrix} 1 & 0 & 2 & 0 & 1 & 0 \\ 0 & 1 & -2 & 1 & -3 & 0 \\ 0 & 0 & 1 & -5/6 & 13/6 & 1/6 \end{pmatrix} \to \begin{pmatrix} 1 & 0 & 0 & 5/3 & -10/3 & -1/3 \\ 0 & 1 & 0 & -2/3 & 4/3 & 1/3 \\ 0 & 0 & 1 & -5/6 & 13/6 & 1/6 \end{pmatrix}.$$

Therefore $\begin{pmatrix} 3 & 1 & 4 \\ 1 & 0 & 2 \\ 2 & 5 & 0 \end{pmatrix}^{-1} = \begin{pmatrix} 5/3 & -10/3 & -1/3 \\ -2/3 & 4/3 & 1/3 \\ -5/6 & 13/6 & 1/6 \end{pmatrix}$. b. $\begin{pmatrix} 2 & -1 \\ -5/2 & 3/2 \end{pmatrix}$.

c. $\begin{pmatrix} -3/8 & 1 & 1/8 & -17/12 \\ -1/2 & 0 & 1/2 & -1/3 \\ 3/16 & -1/2 & -1/16 & 21/24 \\ 3/8 & 0 & -1/8 & 1/12 \end{pmatrix}$. d. $\begin{pmatrix} 9/2 & 8 & -15/2 \\ -3 & -5 & 5 \\ 2 & 3 & -3 \end{pmatrix}$.

5. Consider the product $\begin{pmatrix} -2 & 1 \\ 10 & -5 \end{pmatrix}\begin{pmatrix} 2 & 3 \\ 4 & 6 \end{pmatrix}$.

6. Let φ be the isomorphism $\varphi\colon \mathrm{Hom}(\mathscr{V}_n) \to \mathscr{M}_{n\times n}$ with $\varphi(S) = A$ and $\varphi(T) = B$. Show that $T = \mathbf{0}$ so that $B = \varphi(\mathbf{0}) = \mathbf{0}$.

7. a. Since A is row-equivalent to B, there exists a sequence of elementary row operations that transforms A to B. If this sequence of operations is performed using the elementary matrices $E_1, \ldots, E_k$, show that $Q = E_k \cdots E_2E_1$ is nonsingular and satisfies the equation $B = QA$.

b. Suppose $Q_1A = Q_2A$, if A is nonsingular, then A^{-1} exists.

c. $\begin{pmatrix} 1 & 0 \\ -2 & 1 \end{pmatrix}\begin{pmatrix} 1 & 2 \\ 2 & 4 \end{pmatrix} = \begin{pmatrix} 1 & 2 \\ 0 & 0 \end{pmatrix} = \begin{pmatrix} 0 & 1/2 \\ 1 & -1/2 \end{pmatrix}\begin{pmatrix} 1 & 2 \\ 2 & 4 \end{pmatrix}$.

8. a. $\begin{pmatrix} 3/2 & -2 \\ -1/2 & 1 \end{pmatrix}$. b. $\begin{pmatrix} 5 & -4 & 0 \\ -1 & 1 & 0 \\ -13 & 10 & 1 \end{pmatrix}$. c. $\begin{pmatrix} -5 & 2 & 0 \\ 3 & -1 & 0 \\ 2 & -2 & 1 \end{pmatrix}$.

d. A^{-1}.

9. a. $\begin{pmatrix} 1 & 0 & 0 \\ 0 & 1/2 & 0 \\ 0 & 0 & 1 \end{pmatrix}$. b. $\begin{pmatrix} 1 & -2 & 0 \\ 0 & 1 & 0 \\ 0 & 0 & 1 \end{pmatrix}$. c. $\begin{pmatrix} 0 & 1 & 0 \\ 1 & 0 & 0 \\ 0 & 0 & 1 \end{pmatrix}$. d. $\begin{pmatrix} 1 & 0 & 0 \\ 0 & 1 & 3 \\ 0 & 0 & 1 \end{pmatrix}$.

e. $\begin{pmatrix} 1 & 0 & 0 \\ 0 & 0 & 1 \\ 0 & 1 & 0 \end{pmatrix}$.

10. See problem 7a.

11. a. $\begin{pmatrix} 3 & -4/3 & 14/3 \\ -2 & 1 & -4 \\ 0 & 0 & 1 \end{pmatrix}$. b. $\begin{pmatrix} 3 & -5 \\ -1 & 2 \end{pmatrix}$.

12. a. With Q as in 8a, b. With Q as in 8c, c. With P as in 11b,

$P = \begin{pmatrix} 1 & 0 & -8 \\ 0 & 1 & 2 \\ 0 & 0 & 1 \end{pmatrix}$. $P = \begin{pmatrix} 1 & 0 & 18 \\ 0 & 1 & -11 \\ 0 & 0 & 1 \end{pmatrix}$. $Q = \begin{pmatrix} 1 & 0 & 0 \\ 0 & 1 & 0 \\ -5 & 6 & 1 \end{pmatrix}$.

Review Problems Page 196

1. a. $(2, 5)\colon (-1, 2)_{\{(0,1),(1,3)\}}$ and $\begin{pmatrix} 2 & 3 \\ 1 & 4 \\ 5 & 1 \end{pmatrix}\begin{pmatrix} -1 \\ 2 \end{pmatrix} = \begin{pmatrix} 4 \\ 7 \\ -3 \end{pmatrix}$, so $T(2, 5) = 4(-2, 1, 0) + 7(3, 1, -4) - 3(0, -2, 3) = (13, 17, -37)$.

b. $(6, 5, -13)$. c. $(9a - b, 26a - 7b, -46a + 11b)$.

2. a. T is the identity map.
b. $T(a, b, c) = (6a - 6b - 4c, -6a + 10b + 3c, 4a - 2b + c)$.

3. a. Show that I is an isomorphism.
b. If $T: \mathscr{V} \to \mathscr{W}$ is an isomorphism, show that T^{-1} exists and is an isomorphism.
c. Show that the composition of two isomorphisms is an isomorphism.

4. a. $A = \begin{pmatrix} 2 & -3 \\ -1 & 1 \end{pmatrix}$. b. The rank of A is $2 \neq 1$. d. $B_1 = \{(5, 2), (3, 4)\}$, $B_2 = \{E_i\}$.

6. a. $\mathscr{N}_T = \mathscr{L}\{(3, -8, 2)\}$.
b. There are many possible answers. B_1 can be taken to be $\{(1, 0, 0), (0, 1, 0), (3, -8, 2)\}$, which is clearly a basis for $\mathscr{V}_3$. Then since $T(0, 1, 0) = (0, 1, 3)$, the basis B_2 must have the form $\{U, (0, 1, 3), V\}$, with $T(1, 0, 0) = 2U - (0, 1, 3) + 3V$. There are now an infinite number of choices for B_2 given by the solutions for this equation, satisfying the condition that U, $(0, 1, 3)$, V are linearly independent.

8. For the first part, suppose $A^2 = A$ and A is nonsingular. Show that this implies $A = I_n$.

The 2×2 matrix $\begin{pmatrix} a & b \\ c & d \end{pmatrix}$ satisfies these conditions if $a + d = 1$ and $bc = a - a^2$.

9. a. $\mathscr{I}_T = \{(x, y, z) | y = 3x\}$. b. $\mathscr{N}_T = \mathscr{L}\{(2, 1, -7)\}$.
c. The line parallel to $\mathscr{N}_T$ that passes through U has the vector equation $P = t(2, 1, -7) + U$, $t \in R$.

11. b. $T(a, b, c) \in \mathscr{S}$ if $(a + b) - (b + c) + (a - c) = 0$, or $a = c$. Therefore $T^{-1}[\mathscr{S}] = \{(a, b, c) | a = c\}$ and $T[T^{-1}[\mathscr{S}]] = \{T(a, b, a) | a, b \in R\} = \mathscr{L}\{(1, 1, 0)\} \neq \mathscr{S}$.
c. $T^{-1}[\mathscr{T}] = \{(a, b, c) | 4a - b - 5c = 0\}$. d. $\mathscr{V}_3$.

12. a. $k = -12$. b. No values. c. $k = 0, -2$.

13. Suppose $S \in \operatorname{Hom}(\mathscr{E}_2)$. We can show that if S is nonsingular, then S sends parallelograms to parallelograms; and if S is singular, then it collapses parallelograms into lines.

For $U, V \in \mathscr{E}_2$, write (U^T, V^T) for the matrix with U and V as columns. Then area $(\mathbf{0}, U, V, U + V) = |\det(U^T, V^T)|$ is the area of the parallelogram with vertices $\mathbf{0}$, U, V, and $U + V$.

If A is the matrix of S with respect to the standard basis, then $(S(U))^T = AU^T$. That is, in this case, coordinates are simply the components of the vector.

S sends the vertices $\mathbf{0}, U, V, U + V$ to $\mathbf{0} = S(\mathbf{0}), S(U), S(V), S(U) + S(V) = S(U + V)$. Now the area of the figure with these four vertices is given by

$$|\det((S(U))^T, S(V))^T)| = |\det(AU^T, AV^T)| = |\det A(U^T, V^T)|$$
$$= |\det A \det(U^T, V^T)| = |\det A| \text{ area } (\mathbf{0}, U, V, U + V).$$

So if det $A = 0$ (and S is singular), then the image of every parallelogram has no area.

14. a. $A^2 = 4A - 3I_2$. b. $A^2 = 7A$. c. $A^3 = 2A^2$.
d. $A^3 = 4A^2 - 9I_3$. e. $A^3 = 4A^2 - A - 6I_3$. f. $A^4 = \mathbf{0}$.

Chapter 5

§1 Page 207

1. For each part, except b, you should be able to find the coordinates of the vectors in B' with respect to B by inspection.

a. $\begin{pmatrix} 3 & 2 \\ -5 & 2 \end{pmatrix}$. b. $\begin{pmatrix} 2/9 & -1/9 \\ -1/9 & 5/9 \end{pmatrix}$. c. $\begin{pmatrix} 3 & 1 & 5 \\ -3 & 3 & -3 \\ 2 & -5 & 1 \end{pmatrix}$.

d. $\begin{pmatrix} 2 & 0 & 1 \\ 3 & -4 & 2 \\ 1 & 1 & 0 \end{pmatrix}$. e. $\begin{pmatrix} -1 & -1 & -2 \\ -3 & 2 & 1 \\ 4 & 2 & 0 \end{pmatrix}$. f. $\begin{pmatrix} 1 & 2 & 1 \\ 1 & -4 & -1 \\ 4 & 5 & 1 \end{pmatrix}$.

2. a. The transition matrix from B to $\{E_i\}$ is $\begin{pmatrix} 3 & 5 \\ 1 & 2 \end{pmatrix}^{-1} = \begin{pmatrix} 2 & -5 \\ -1 & 3 \end{pmatrix}$, and $\begin{pmatrix} 2 & -5 \\ -1 & 3 \end{pmatrix}\begin{pmatrix} 4 \\ 3 \end{pmatrix} = \begin{pmatrix} -7 \\ 5 \end{pmatrix}$, so $(4, 3)\colon (-7, 5)_B$.
b. $(-13, 8)_B$. c. $(4, -3)_B$. d. $(2a - 5b, 3b - a)_B$.

3. $\begin{pmatrix} 2 & 3 & 3 \\ -2 & 0 & 5 \\ 1 & 2 & 3 \end{pmatrix}^{-1} = \begin{pmatrix} -10 & -3 & 15 \\ 11 & 3 & -16 \\ -4 & -1 & 6 \end{pmatrix}$.
a. $(2, -2, 1)_B$. b. $(1, -1, 1)_B$. c. $(2, 0, 1)_B$.

4. a. For this choice, $(0, 1) = (2, 1) - 2(1, 0)$ and $(1, 4) = 4(2, 1) - 7(1, 0)$. If $V\colon (x_1, x_2)_B$ and $V\colon (x_1', x_2')_{B'}$, then

$$\begin{aligned} V &= x_1(2, 1) + x_2(1, 0) = x_1'(0, 1) + x_2'(1, 4) \\ &= x_1'[(2, 1) - 2(1, 0)] + x_2'[4(2, 1) - 7(1, 0)] \\ &= [x_1' + 4x_2'](2, 1) + [-2x_1' - 7x_2'](1, 0). \end{aligned}$$

Therefore $x_1 = x_1' + 4x_2'$ and $x_2 = -2x_1' - 7x_2'$ or

$$X = \begin{pmatrix} x_1 \\ x_2 \end{pmatrix} = \begin{pmatrix} 1 & 4 \\ -2 & -7 \end{pmatrix}\begin{pmatrix} x_1' \\ x_2' \end{pmatrix} = PX'.$$

5. a. To prove $(AB)^T = B^TA^T$, let $A = (a_{ij})$, $B = (b_{ij})$, $A^T = (c_{ij})$, $B^T = (d_{ij})$, $AB = (e_{ij})$, $(AB)^T = (f_{ij})$, and $B^TA^T = (g_{ij})$. Then express f_{ij} and g_{ij} in terms of the elements in A and B to show that $f_{ij} = g_{ij}$.

6. a. The transformation changes length, angle, and area. See Figure A2.
b. The transformation is called a "shear." The points on one axis are fixed, so not all lengths are changed. See Figure A3.
c. The transformation is a "dilation." Lengths are doubled, but angles remain unchanged.

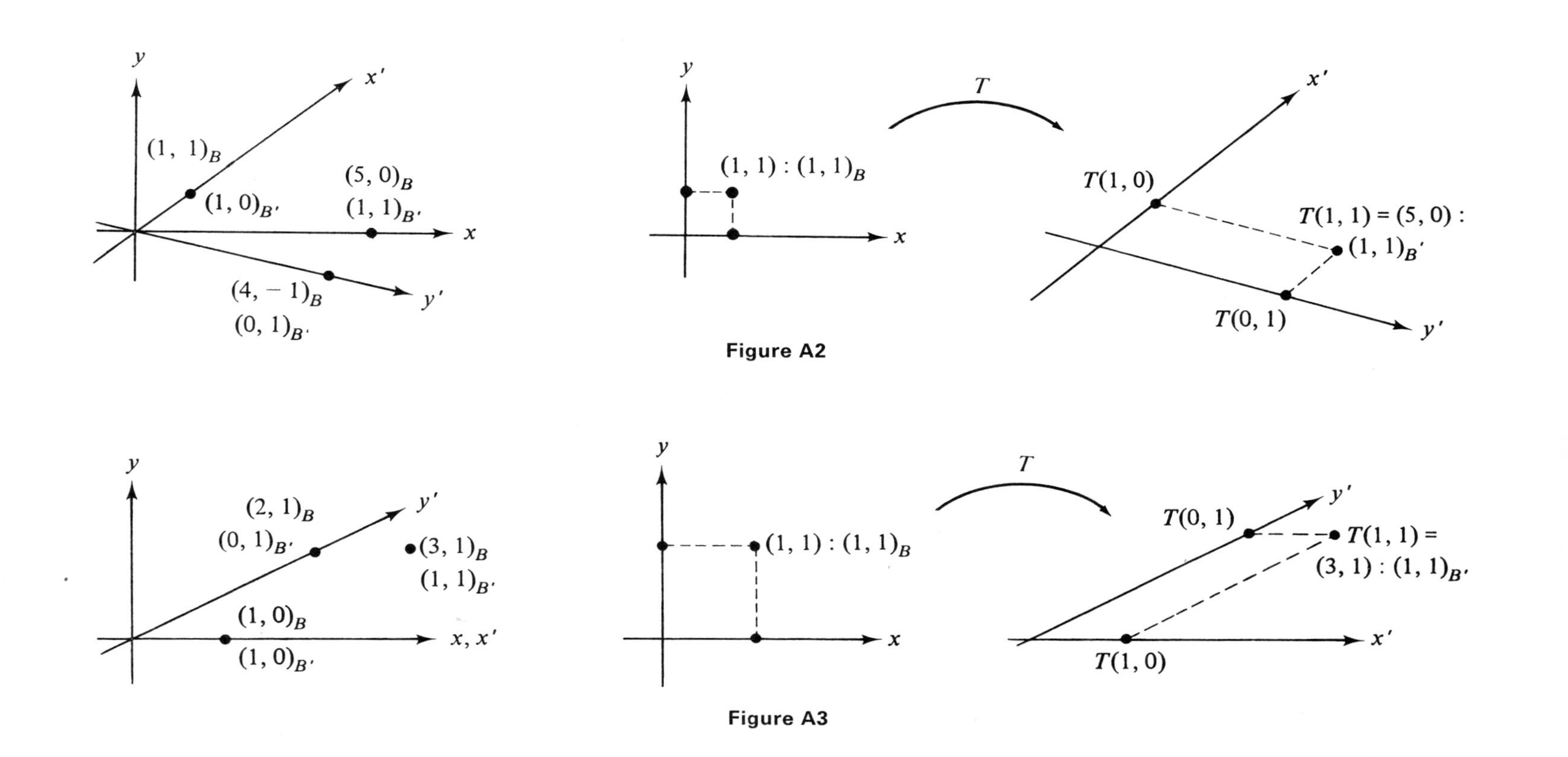

Figure A2

Figure A3

d. The transformation is a rotation through 45°, length, angle, and area remain unchanged by a rotation.
e. The transformation is a reflection in the y axis. Length and area are unchanged but the sense of angles is reversed.
f. This transformation can be regarded as either a rotation through 180° or a reflection in the origin.

8. Suppose P is the transition matrix from B to B', where B and B' are bases for $\mathscr{V}$ and dim $\mathscr{V} = n$. Then the columns of P are coordinates of n independent vectors with respect to B. Use an isomorphism between $\mathscr{V}$ and $\mathscr{M}_{n\times 1}$ to show that the n columns of P are linearly independent.

§2 Page 213

1. $A = \begin{pmatrix} 2 & -1 \\ 3 & -4 \\ 1 & 1 \end{pmatrix}$, $A' = \begin{pmatrix} -6 & -5 \\ -4 & -3 \\ 11 & 8 \end{pmatrix}$, $P = \begin{pmatrix} 2 & 1 \\ 3 & 2 \end{pmatrix}$, $Q = \begin{pmatrix} 1 & 1 & 1 \\ 1 & 0 & 0 \\ 1 & 0 & 1 \end{pmatrix}$.

2. $A = \begin{pmatrix} 5 & -3 \\ 2 & 1 \end{pmatrix}$, $A' = \begin{pmatrix} -1 & -18 \\ 1 & 7 \end{pmatrix}$, $P = \begin{pmatrix} 1 & 3 \\ 1 & 4 \end{pmatrix}$.

3. If P is nonsingular, P is the product of elementary matrices, $P = E_1E_2 \cdots E_k$, and $AP = AE_1E_2 \cdots E_k$ is a matrix obtained from A with k elementary column operations.

4. a. Suppose $Q = I_2$, and $P = \begin{pmatrix} 2 & 1 \\ 3 & 4 \end{pmatrix}^{-1}$.

5. a. If $T(a, b, c) = (a + b + c, -3a + b + 5c, 2a + b)$, then the matrix of T with respect to $\{(1, 0, 0), (0, 1, 0), (1, -2, 1)\}$ and $\{(1, -3, 2), (1, 1, 1), (1, 0, 0)\}$ is the desired matrix.

7. If $I_n = P^{-1}AP$, then $PI_nP^{-1} = A$.

12. $T(U_j') = r_jU_j'$.

§3 Page 221

1. a. Reflexive and transitive, but not symmetric.
b. Symmetric, but neither reflexive nor transitive.
c. The relation is an equivalence relation.
d. Symmetric and transitive, but not reflexive.
e. Reflexive and symmetric, but not transitive.

2. Suppose a is not related to anything. Consider problem 1d.

3. a. For symmetry: if $a \sim b$, then $a - b = 5c$, for some $c \in Z$. Therefore $b - a = 5(-c)$. Since $-c$ is an integer, $b \sim a$. This equivalence realtion is usually denoted $a \equiv b \bmod 5$, read "a is congruent to b modulo 5."
b. Five. For example the equivalence class with representative 7 is

$$\begin{aligned}[7] &= \{p \in Z \mid 7 - p = 5k \text{ for some } k \in Z\} \\ &= \{7 - 5k \mid k \in Z\} = \{2 + 5h \mid h \in Z\} = [2].\end{aligned}$$

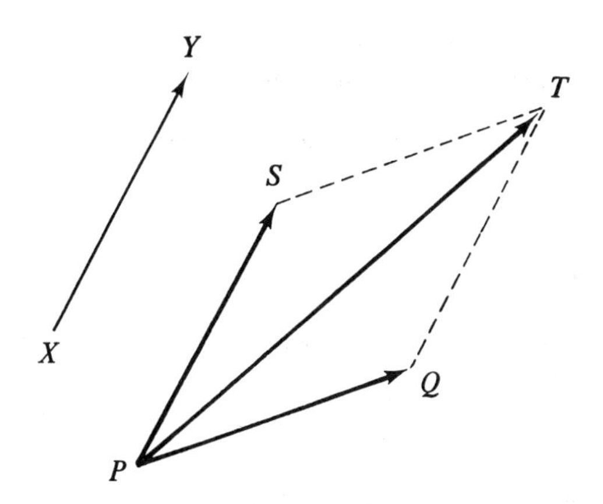

Figure A4

d. The property of having a certain remainder when divided by 5.

4. a. For transitivity, suppose $a \sim b$ and $b \sim c$, then there exist rational numbers r, s such that $a - b = r$ and $b - c = s$. The rationals are closed under addition, so $r + s$ is rational. Since $a - c = r + s$, $a \sim c$ and $\sim$ is transitive.
b. $[1] = [2/3] = \{r | r \text{ is rational}\}$, $[\sqrt{2}] = \{\sqrt{2} + r | r \text{ is rational}\}$.
c. The number is uncountably infinite.

5. b. Given $\overrightarrow{PQ}$ and any point $X \in E^2$, there exists a point Y such that $\overrightarrow{PQ} \sim \overrightarrow{XY}$. Therefore drawing all the line segments in $[\overrightarrow{PQ}]$ would completely fill the plane. So only representatives for $[\overrightarrow{PQ}]$ can be drawn.
c. The set of all arrows starting at the origin is a good set of canonical forms. Compare with Example 2, page 33.
d. It is possible to define addition and scalar multiplication for equivalence classes of directed line segments in such a way that the set of equivalence classes becomes a vector space. For example, $[\overrightarrow{PQ}] + [\overrightarrow{XY}]$ can be defined to be $[\overrightarrow{PT}]$ where $\overrightarrow{XY} \sim \overrightarrow{PS}$ and P, Q, S, T are vertices of a parallelogram as in Figure A4. In this vector space any arrow is a representative of some vector, but the vectors are not arrows.

6. b. The lines with equations $y = 2x + b$, where $b \in R$ is arbitrary.

7. a. $\mathcal{N}_T$. b. All planes with normal $(2, -1, -4)$.

8. a. Row-equivalence is a special case of equivalence with $P = I_m$.
b. Rank because of the reason for part a.
c. The (row) echelon form.
d. An infinite number, since $\begin{pmatrix} 0 & 0 \\ 0 & 0 \end{pmatrix}$, $\begin{pmatrix} 0 & 1 \\ 0 & 0 \end{pmatrix}$, $\begin{pmatrix} 1 & 0 \\ 0 & 1 \end{pmatrix}$, and $\begin{pmatrix} 1 & x \\ 0 & 0 \end{pmatrix}$, for each $x \in R$, belong to different equivalence classes.

9. A few are area, perimenter, being isoceles or having a certain value for the lengths of two sides and the included angle.

10. Write the matrix as $\begin{pmatrix} I_k & \mathbf{0}_1 \\ \mathbf{0}_2 & \mathbf{0}_3 \end{pmatrix}$, then $\mathbf{0}_1$ is $k \times (m - k)$, $\mathbf{0}_2$ is $(n - k) \times k$, and $\mathbf{0}_3$ is $(n - k) \times (m - k)$.

§4 Page 230

1. For each part, there are many choices for P and the diagonal entries in D might be in any order.

a. $P = \begin{pmatrix} 1 & -1 \\ -1 & 5 \end{pmatrix}$, $D = \begin{pmatrix} 2 & 0 \\ 0 & -2 \end{pmatrix}$. b. $P = \begin{pmatrix} 3 & 1 \\ -8 & -1 \end{pmatrix}$, $D = \begin{pmatrix} 4 & 0 \\ 0 & -1 \end{pmatrix}$.

c. $P = \begin{pmatrix} 0 & 2 & 9 \\ 1 & 4 & -7 \\ 0 & 1 & -3 \end{pmatrix}$, $D = \text{diag}(3, 5, 0)$.

d. $P = \begin{pmatrix} 2 & 4 & 0 \\ -1 & -5 & 1 \\ 0 & 18 & 2 \end{pmatrix}$, $D = \text{diag}(2, -4, 10)$.

2. a. There are no real characteristic roots, therefore it is not diagonalizable. b. The characteristic equation has three distinct real roots since the discriminant of the quadratic factor is 28. c. $(a, b, c, d)^T$ is a characteristic vector for 1, if $4a - 3c + d = 0$ and this equation has three independent solutions. d. There are only two independent characteristic vectors.

3. All vectors of the form $(h, k, h + 2k)^T$, with $h^2 + k^2 \neq 0$, are characteristic vectors for 2 and $(k, -2k, 3k)$, with $k \neq 0$, are characteristic vectors for -4.

If $P = \begin{pmatrix} 1 & 0 & 1 \\ 0 & 1 & -2 \\ 1 & 2 & 3 \end{pmatrix}$, then $P^{-1}AP = \text{diag}(2, 2, -4)$.

4. $(2k, k, -k)^T$, with $k \neq 0$, are the characteristic vectors for 2, and $(k, k, k)^T$, $k \neq 0$, are the characteristic vectors for 1.

5. $A\mathbf{0} = \lambda\mathbf{0}$ for all matrices A and all real numbers λ.

6. Not necessarily, consider $\begin{pmatrix} 2 & 0 \\ 1 & 2 \end{pmatrix}$.

7. The matrix $A - \lambda I_n$ is singular for each characteristic value λ, therefore the system of linear equations $(A - \lambda I_n)X = \mathbf{0}$ has at most $n - 1$ independent equations and n unknowns.

8. a. 2, 4, and 3, for subtracting any of these numbers from the main diagonal yields a singular matrix. b. 5 and 0; for 0, note that two columns are equal. c. 7 and -2; subtracting -2 yields two equal rows.

9. a. Every polynomial equation of odd degree has at least one real root.

b. $\begin{pmatrix} 0 & 0 & 0 \\ 1 & 0 & 0 \\ 1 & 1 & 0 \end{pmatrix}$ and $\begin{pmatrix} 1 & 0 & 0 \\ 0 & 0 & 1 \\ 0 & -1 & 0 \end{pmatrix}$ give two different types of examples.

10. a. The diagonal entries. b. They are invariant under similarity.

11. a. T is singular. b. $\mathcal{N}_T$. c. $T(U_i) = a_i U_i$.

§5 Page 237

1. a. $-7 - \lambda + \lambda^2$. b. $4 + 13\lambda - 2\lambda^2 - \lambda^3$. c. $-11 + 2\lambda + \lambda^2$.
d. $-5\lambda - \lambda^3$.

2. a. $2 - 5\lambda + \lambda^2$. b. $1 + \lambda^2$. c. $-1 + \lambda^2$. d. $-23 - 3\lambda + \lambda^2$.

3. a. $s_{n-1} = \sum_{i=1}^{n} A_{ii}$, where A_{ii} is the cofactor of a_{ii}.
b. The characteristic polynomial of A is $|A| - s_2\lambda + (tr\ A)\lambda^2 - \lambda^3$.
i. $-2 + 6\lambda + 2\lambda - \lambda^3$. ii. $35\lambda - \lambda^3$. iii. $1 - 3\lambda + 3\lambda^2 - \lambda^3$.

4. You should be able to find two matrices with the same characteristic polynomial but different ranks.

5. a. The characteristic polynomial is $2 - 4\lambda + 3\lambda^2 - \lambda^3$. $\ell = \mathscr{L}\{(1, 0, 1)\}$.
b. The plane has the equation $x + z = 0$.

7. b. $\mathcal{N}_T$.

8. The maps in a, b, and e are diagonalizable.

9. b. $\mathscr{S}(1) = \mathscr{L}\{(1, -1, -1)\}$ and $\mathscr{S}(3) = \mathscr{L}\{(1, 0, 1), (0, 1, 0)\}$. Since $\{(1, -1, -1), (1, 0, 1), (0, 1, 0)\}$ is a basis for $\mathscr{V}_3$, $\mathscr{V}_3 = \mathscr{S}(1) \oplus \mathscr{S}(3)$. This also implies that $\mathscr{S}(1) \cap \mathscr{S}(2) = \{\mathbf{0}\}$, but to prove it directly, suppose $U \in \mathscr{S}(1) \cap \mathscr{S}(3)$. Then $U = (a, -a, -a)$ and $U = (b, c, b)$ for some $a, b, c \in R$. But then $a = b = -a = c$ implies $U = \mathbf{0}$ and $\mathscr{S}(1) \cap \mathscr{S}(3) = \{\mathbf{0}\}$. Thus the sum is direct.

10. Any plane π parallel to $\mathscr{T}$ can be written as $\{U + P | U \in \mathscr{T}\}$ for some fixed $P \in \mathscr{S}(1)$.

11. a. $T(U) = (1, 2, 2)$, $T^2(U) = (-1, 3, 7)$. b. $\begin{pmatrix} 0 & 0 & 4 \\ 1 & 0 & -8 \\ 0 & 1 & 5 \end{pmatrix}$.

c. $T^2(U) \in \mathscr{L}\{U, T(U)\}$. d. $\begin{pmatrix} 2 & 0 & 0 \\ 1 & 2 & 0 \\ 0 & 0 & 1 \end{pmatrix}$. Notice that the characteristic values for T are on the main diagonal.

Review Problems Page 238

1. a. $\begin{pmatrix} 1 & 0 & 0 \\ 0 & 1 & 0 \end{pmatrix}$. b. There are many answers. If $B_1' = \{(1, 0, 0), (0, 1, 0), (1, -2, -1)\}$ then $B_2' = \{(2, 1), (-1, 2)\}$.

2. b. For consistent systems, choose the equations exhibiting the solutions using the system with augmented matrix in echelon form. What about inconsistent systems?

3. c. $\begin{pmatrix} 1 & 0 & 0 \\ 0 & 1 & 0 \\ 0 & 0 & x \end{pmatrix}$ for each x in R.

5. The discriminant of the characteristic equation is $-4 \sin^2 \theta$. Therefore a rotation in the plane has characteristic values only if $\theta = n\pi$, where n is an integer. There are two cases, n even or n odd.

6. b, c, and d are diagonalizable.

7. Define the sum by $\mathscr{S} + \mathscr{T} + \mathscr{U} = \{U + V + W | U \in \mathscr{S},\ V \in \mathscr{T},\ W \in \mathscr{U}\}$. Two equivalent definitions for the direct sum are: $\mathscr{S} + \mathscr{T} + \mathscr{U}$ is a direct sum if $\mathscr{S} \cap (\mathscr{T} + \mathscr{U}) = \mathscr{T} \cap (\mathscr{S} + \mathscr{U}) = \mathscr{U} \cap (\mathscr{S} + \mathscr{T}) = \{\mathbf{0}\}$. Or $\mathscr{S} + \mathscr{T} + \mathscr{U}$ is a direct sum if for every vector $V \in \mathscr{V}$, there exist unique vectors $X \in \mathscr{S}$, $Y \in \mathscr{T}$, $Z \in \mathscr{U}$ such that $V = X + Y + Z$.

8. a. There are three distinct characteristic values.
b. $\mathscr{S}(2) = \mathscr{L}\{(0, 0, 1, 0), (1, 0, 0, 1)\}$, $\mathscr{S}(-2) = \mathscr{L}\{(1, 0, 3, -3)\}$, and $\mathscr{S}(-3) = \mathscr{L}\{(0, 1, 0, 0)\}$.
c. Follows since $\{(0, 0, 1, 0), (1, 0, 0, 1), (1, 0, 3, -3), (0, 1, 0, 0)\}$ is a basis for $\mathscr{V}_4$.

9. a. $8 - 4\lambda + 2\lambda^2 - \lambda^3$. d. $\begin{pmatrix} 0 & 0 & 8 \\ 1 & 0 & -4 \\ 0 & 1 & 2 \end{pmatrix}$.

12. a. $T(a + bt) = -33a - 5b + (203a + 31b)t$.
b. $B' = \{5 - 29t, 1 - 7t\}$.

Chapter 6

§1 Page 246

1. a. 2 b. 0. c. 13. d. -4.

2. a. i. 1. ii. 10. iii. 1. iv. 0. v. 8. vi. -14.
b. For linearity in the first variable, let $r, s \in R$ and $U, V, W \in \mathscr{V}_2$. Suppose $U:(a_1, a_2)_B$, $V:(b_1, b_2)_B$, and $W:(c_1, c_2)_B$, then $(rU + sV):(ra_1 + sb_1, ra_2 + sb_2)_B$ (Why?). Using this fact, $\langle rU + sV, W\rangle_B = (ra_1 + sb_1)c_1 + (ra_2 + sb_2)c_2$ and $r\langle U, W\rangle_B + s\langle V, W\rangle_B = r(a_1c_1 + a_2c_2) + s(b_1c_1 + b_2c_2)$.

4. a. 1. b. 0. c. 9. d. -14.

5. Recall that $0V = \mathbf{0}$.

6. a. 5, 0, -5. b. Since b is not symmetric, it is necessary to show that $b(U, V)$ is linear in each variable.

7. a. 2, sin 1. b. $b(rf + sg, h) = (rf(1) + sg(1))h(1) = (rf(1))h(1) + sg(1)h(1) = rb(f, h) + sb(g, h)$, fill in the reasons. Since b is symmetric, this implies b is bilinear. c. Consider $f(x) = \sin \pi x$.

8. a. 8, 0.

10. *a.* $x = z = 1$, $y = 0$.
b. $x = 13/100$, $y = -11/100$, $z = 17/100$.
c. $x = 3$, $y = -1$, $x = 2$.

11. f is symmetric and bilinear but it should be easy to find values for x, y, and z for which f is not positive definite.

§2 Page 254

1. a. $1/\sqrt{3}$. b. $\sqrt{13/3}$. c. $1/\sqrt{2}$.

2. a. $\langle U, V\rangle^2 = 25$, $\|U\|^2 = 10$ and $\|V\|^2 = 5$.
b. $\langle U, V\rangle^2 = 1/9$, $\|U\|^2 = 1/3$ and $\|V\|^2 = 2/3$.

4. The direction ($\Leftarrow$) is not difficult. For ($\Rightarrow$), consider the case $\langle U, V\rangle = \|U\|\|V\|$ with $U, V \neq \mathbf{0}$. The problem is to find nonzero scalars a and b such that $aU + bV = \mathbf{0}$. If such scalars exist, then $0 = \langle aU + bV, V\rangle = \|V\|(a\|U\| + b\|V\|)$. Therefore a possible choice is $a = \|V\|$ and $b = -\|U\|$. Consider $\langle aU + bV, aU + bV\rangle$ for this choice.

5. The second vector (x, y) must satisfy $(x, y) \circ (3/5, 4/5) = 0$ and $x^2 + y^2 = 1$. Solutions: $(4/5, -3/5)$ and $(-4/5, 3/5)$.

6. a. There are two vectors which complete an orthonormal basis, $\pm(1/\sqrt{2}, 0, 1/\sqrt{2})$. Note that these are the two possible cross products of the given vectors.
b. $\pm (1/3\sqrt{2})(4, 1, -1)$. c. $\pm(3/7, -6/7, 2/7)$.

7. Prove that $\{V | \langle V, U\rangle = 0 \text{ for all } U \in S\}$ is a subspace of $\mathscr{V}$.

8. a. Use $\langle\ ,\ \rangle_B$. b. Since $(2, -7)$: $(2, -3)_B$, $\|(2, -7)\| = \sqrt{13}$.

9. b. $\{(1/\sqrt{6})(1, 0, 2, 1), (1/\sqrt{14})(2, 3, -1, 0), (1/\sqrt{22})(-3, 2, 0, 3)\}$.

10. a. Use the identity $2 \sin A \sin B = \cos(A - B) - \cos(A + B)$. Why isn't the integral zero when $n = m$?
b. $\frac{1}{2} - (1/\pi)\sum_{n=1}^{\infty}(1/n)\sin 2n\pi x$. Notice that this series does not equal $f(x)$ when $x = 0$ or $x = 1$; at these points the series equals $\frac{1}{2}(f(0) + f(1))$.

§3 Page 260

1. a. $V_1 = (1, 0, 0)$ and $V_2 = (0, 0, 1)$ for $W_2 = (3, 0, 5) - [(3, 0, 5) \circ (1, 0, 0)](1, 0, 0) = (0, 0, 5)$.
b. $\{(2/\sqrt{13}, 3/\sqrt{13}), (-3/\sqrt{13}, 2/\sqrt{13})\}$.
c. $\{(1/\sqrt{3}, -1/\sqrt{3}, 1/\sqrt{3}), (0, 1/\sqrt{2}, 1/\sqrt{2})\}$.
d. $\{(1/2, 1/2, 1/2, 1/2), (1/\sqrt{6}, 0, -2/\sqrt{6}, 1/\sqrt{6}), (-2/\sqrt{30}, -3/\sqrt{30}, 1/\sqrt{30}, 4/\sqrt{30})\}$.

2. $\{1, 2\sqrt{3}t - \sqrt{3}, 6\sqrt{5}t^2 - 6\sqrt{5}t + \sqrt{5}\}$.

3. a. $\{(1/2, 0), (1/\sqrt{28}, 2/\sqrt{7})\}$.
b. Check that your set is orthonormal.

4. $\{(1, 0, 0), (2, 1, 0), (-2/\sqrt{3}, -1/\sqrt{3}, 1/\sqrt{3})\}$.

6. a. $\mathscr{S}^\perp = \{(x, y, z) | 2x + y + 4z = 0, x + 3y + z = 0\}$
$= \mathscr{L}\{(11, -2, -5)\}$.
b. $\mathscr{L}\{(3, -1)\}$. c. $\{(x, y, z) | 2x - y + 3z = 0\}$.
d. $\{(x, y, z) | x = 2z\}$. e. $\mathscr{L}\{(-2, -4, 4, 7)\}$.

8. b. $\mathscr{L}\{(0, 1, 0)\}$ is invariant under T. $\mathscr{L}\{(0, 1, 0)\}^T = \{(x, y, z) | y = 0\}$.

9. $\mathscr{S}^\perp = \mathscr{L}\{(4, -1)\}$. $\mathscr{E}_2 = \mathscr{S} + \mathscr{S}^\perp$ because $\{(1, 4), (4, -1)\}$ spans $\mathscr{E}_2$. The sum is direct, that is, $\mathscr{S} \cap \mathscr{S}^\perp = \{\mathbf{0}\}$, because $\{(1, 4), (4, -1)\}$ is linearly independent.

10. $\mathscr{S}^\perp = \mathscr{L}\{(0, 1, 0, 0), (4, 0, 8, -5)\}$.

§4 Page 269

1. a. Since the matrix of T with respect to $\{E_i\}$ is $\begin{pmatrix} 1/2 & \sqrt{3}/2 \\ \sqrt{3}/2 & -1/2 \end{pmatrix}$, T is a reflection. The line reflected in is $y = (1\sqrt{3})x$.
b. The rotation through the angle $-\pi/3$.
c. The rotation through φ where $\cos \varphi = -4/5$ and $\sin \varphi = -3/5$.
d. The identity map, $\varphi = 0$. e. Reflection in the line $y = -x$.

3. You need an expression for the inner product in terms of the norm. Consider $\|U - V\|^2$.

4. a. $\begin{pmatrix} 1/3 & 2/3 & -2/3 \\ -2/3 & 2/3 & 1/3 \\ 2/3 & 1/3 & 2/3 \end{pmatrix}$. b. $\begin{pmatrix} 1/\sqrt{3} & 1/\sqrt{3} & 1/\sqrt{3} \\ -1/\sqrt{2} & 0 & 1/\sqrt{2} \\ 1/\sqrt{6} & -2/\sqrt{6} & 1/\sqrt{6} \end{pmatrix}$.

5. a. The composition of the reflection of $\mathscr{E}_3$ in the xy plane and the rotation with the z axis as the axis of rotation and angle $-\pi/4$.
b. Not orthogonal.
c. The rotation with axis $\mathscr{L}\{(1 - \sqrt{2}, -1, 0)\}$ and angle π.
d. The rotation about the y axis with angle $\pi/4$.
e. Not orthogonal.

6. Use the definition of orthogonal for one, and problem 2 for the other.

7. a. The determinant is 1, so T is a rotation. (2, 1, 0) is a characteristic vector for 1, so the axis of the rotation is $\mathscr{L}\{(2, 1, 0)\}$. The orthogonal complement of the axis is the plane $\mathscr{S} = \mathscr{L}\{(1, -2, 0), (0, 0, 1)\}$; and the restriction $T_{\mathscr{S}}$ rotates $\mathscr{S}$ through the angle φ satisfying $\cos \varphi = (0, 0, 1) \circ T(0, 0, 1) = 2/7$. The direction of this rotation can be determined by examining the matrix of T with respect to the orthonormal basis $\{(2/\sqrt{5}, 1/\sqrt{5}, 0), (1/\sqrt{5}, -2/\sqrt{5}, 0), (0, 0, 1)\}$.
b. The determinant is -1, so -1 is a characteristic value for T. The charac-

teristic polynomial for T is $-1 + \lambda + \lambda^2 - \lambda^3 = -(\lambda + 1)(\lambda^2 - 2\lambda + 1)$, so the other two characteristic values are both 1. Therefore T is the reflection in the plane through the origin which has a characteristic vector corresponding to -1 as normal. $(1, -2, 3)$ is such a vector, so T is the reflection of $\mathscr{E}_3$ in the plane $x - 2y + 3z = 0$.

c. The determinant is -1, so -1 is a characteristic value for T. The characteristic polynomial is as in part b. Therefore T is a reflection. $T(1, -1, 1) = -(1, -1, 1)$, hence T reflects $\mathscr{E}_3$ in the plane with equation $x - y + z = 0$.

d. The determinant is -1 and the characteristic polynomial of T is $-1 + \frac{2}{3}\lambda + \frac{2}{3}\lambda^2 - \lambda^3 = -(\lambda + 1)(\lambda^2 - 5\lambda/3 + 1)$. Since the discriminant of the second degree factor is negative, there is only one real characteristic root. Thus T is the composition of a reflection and a rotation. A characteristic vector corresponding to -1 is $(1, -3, 1)$. Therefore the reflection is in the plane $x - 3y + z = 0$ and the axis of the rotation is $\mathscr{L}\{(1, -3, 1)\}$. $(1, 0, -1) \in \mathscr{L}\{(1, -3, 1)\}^{\perp}$ so the angle of the rotation φ, disregarding its sense, is the angle between $(1, 0, -1)$ and $T(1, 0, -1) = (4/3, 1/3, -1/3)$. Thus $\cos\varphi = 5/6$.

e. T is the rotation with axis $\mathscr{L}\{(2, -2, 1)\}$ and angle π.

f. T is the rotation with axis $\mathscr{L}\{(1, 1, 0)\}$ and angle φ satisfying $\cos\varphi = -7/9$.

9. Suppose T is the reflection of $\mathscr{E}_2$ in the line ℓ with equation $P = tU + A$, $t \in R$. Consider Figure A5.

10. b. An orthonormal basis $\{U_1, U_2, U_3, U_4\}$ can be found for $\mathscr{E}_4$ for which the matrix of T is either

$$\begin{pmatrix} 1 & 0 & 0 & 0 \\ 0 & 1 & 0 & 0 \\ 0 & 0 & a & c \\ 0 & 0 & b & d \end{pmatrix}, \quad \begin{pmatrix} 1 & 0 & 0 & 0 \\ 0 & 1 & 0 & 0 \\ 0 & 0 & 1 & 0 \\ 0 & 0 & 0 & -1 \end{pmatrix}, \quad \begin{pmatrix} 1 & 0 & 0 & 0 \\ 0 & -1 & 0 & 0 \\ 0 & 0 & a & c \\ 0 & 0 & b & d \end{pmatrix}, \quad \text{or} \quad \begin{pmatrix} -1 & 0 & 0 & 0 \\ 0 & -1 & 0 & 0 \\ 0 & 0 & a & c \\ 0 & 0 & b & d \end{pmatrix}$$

where $\begin{pmatrix} a & c \\ b & d \end{pmatrix} = \begin{pmatrix} \cos\varphi & -\sin\varphi \\ \sin\varphi & \cos\varphi \end{pmatrix}$ for some angle φ. The first matrix gives a rotation in each plane parallel to $\mathscr{L}\{U_3, U_4\}$, that is, a rotation about the plane $\mathscr{L}\{U_1, U_2\}$, and the second matrix gives a reflection of E_4 in the hyperplane $\mathscr{L}\{U_1, U_2, U_3\}$.

11. This may be proved without using the property that complex roots come in conjugate pairs. Combine the facts $\det A^{-1} = 1$ and $\det(I_n - A) = (-1)^n \det(A - I_n)$ to show that $\det(A - I_n) = 0$.

§5 Page 275

1. a. $(7, 7i - 3)$. b. $(2 + 4i, 2i - 14, 10)$.

c. $\begin{pmatrix} 8i - 6 \\ 8i \end{pmatrix}$. d. $\begin{pmatrix} 1 - i & 8 - 6i & 1 + 4i \\ 2 + 5i & 13 - i & 4 + 13i \\ i & 1 + 3i & i - 1 \end{pmatrix}$.

2. a. $\text{Det}\begin{pmatrix} 1 + i & 2 \\ 3i & 3 + 2i \end{pmatrix} = 1 - i \neq 0$, so the set is linearly independent. Since the dimension of $\mathscr{V}_2(C)$ is 2, the set is a basis.

b. $(1 + i, 3i)$: $(4, -3)_B$, $(4i, 7i - 4)$: $(1 - i, 2i - 1)_B$.

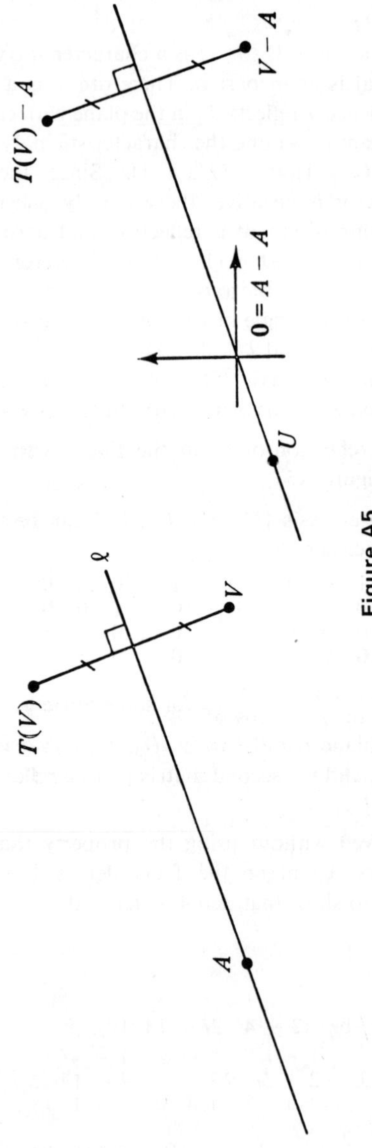

Figure A5

3. b. For every $w \in C$, $[1 + i - iw](i, 1 - i, 2) + [2 + (1 - 3)w](2, 1, -i) + w(5 - 2i, 4, -1 - i) = (3 + i, 4, 2)$.

6. a. $3 + 4 = 2, 2 + 3 = 0, 3 + 3 = 1$.
b. $2 \cdot 4 = 3, 3 \cdot 4 = 2, 4^2 = 1, 2^6 = 4$.
c. $-1 = 4, -4 = 1, -3 = 2$. d. $2^{-1} = 3, 3^{-1} = 2, 4^{-1} = 4$.

7. $2^2 = 0$ in Z_4. Why does this imply that Z_4 is not a field?

8. a. $\mathscr{L}\{(1, 4)\} = \{(0, 0), (1, 4), (2, 3), (3, 2), (4, 1)\}$.
b. No. c. 25. d. No.

9. a. If $U = (z, w)$ and $a \in C$, then $T(aU) = T(a(z, w)) = T(az, aw) = (aw, -az) = a(w, -z) = aT(U)$. Similarly $T(U + V) = T(U) + T(V)$.
b. $\lambda^2 + 1$. c. If z is any nonzero complex number, then $z(1, i)$ are characteristic vectors for the characteristic value i, and $z(1, -i)$ are characteristic vectors for $-i$.
d. The matrix of T with respect to the basis $\{(1, i), (1, -i)\}$ is diag$(i, -i)$.

10. b. $\lambda^2 - (3 + i)\lambda + 2 + 6i$.
c. The discriminant is $-18i$. To find $\sqrt{-18i}$, solve $(a + bi)^2 = -18i$ for a and b. If z is any nonzero complex number, $z(i - 2, i)$ are characteristic vectors for the characteristic value $2i$, and $z(1 + i, -2i)$ are characteristic vectors for $3 - i$.
d. The matrix of T with respect to the basis $\{(i - 2, i), (1 + i, -2i)\}$ is diag$(2i, 3 - i)$.

13. b. Dim $\mathscr{V} = 4$. Notice that $\mathscr{V} \neq \mathscr{V}_2(C)$.

14. b. Vectors such as $(\sqrt{2}, 0)$ and $(1, e)$ are in $\mathscr{V}$ but they are not in $\mathscr{V}_2(Q)$.
c. No. d. If $(a/b)1 + (c/d)\sqrt{2} = 0$ with $a, b, c, d \in Z$ and $c \neq 0$, then $\sqrt{2} = -ad/bc$ but $\sqrt{2} \notin Q$. e. No since $\pi \notin Q$. f. The dimension is at least 3. In fact, $\mathscr{V}$ is an infinite-dimensional vector space even though the vectors are ordered pairs of real numbers.

15. Suppose there exists a set $P \subset C$ closed under addition and multiplication. Show that either $i \in P$ or $-i \in P$ leads to both 1 and -1 being in P.

§6 Page 285

1. a. $4i - 1$. b. $-4 - 11i$.

2. a. $\{(x, y, z) | -ix + 2y + (1 - i)z = 0\}$. b. $\mathscr{L}\{(i, 2i, -1)\}$.

4. a. Use $\langle\ ,\ \rangle_B$. b. $(4i, 2)$: $(1 + i, -2)_B$, so $\|(4i, 2)\| = \sqrt{6}$.

5. a. Neither. b. Both. c. Symmetric. d. Hermitian. e. Neither.
f. Neither.

7. a. The characteristic equation for A is $\lambda^2 + 9 = 0$, so A has no real characteristic values.
b. If $P = \begin{pmatrix} 2 & 1 - i \\ 3 - 3i & 3 \end{pmatrix}$, then $P^{-1}AP = \text{diag}(3i, -3i)$.

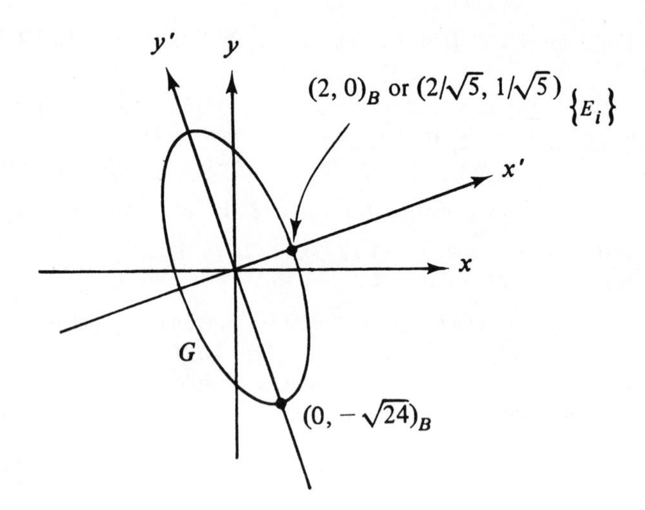

Figure A6

8. a. $\begin{pmatrix} 2 & 1+i \\ 1-i & 3 \end{pmatrix}$.

b. $\lambda = 1, 4$. If $P = \begin{pmatrix} 1+i & 1 \\ -1 & 1-i \end{pmatrix}$, then $P^{-1}AP = \text{diag}(1, 4)$.

9. Yes. Examples may be found with real or complex entries.

10. a. $T(a, b) = (a/\sqrt{2} - b/\sqrt{2}, a/\sqrt{2} + b/\sqrt{2})$.

b. $B = \{(1/\sqrt{2}, 1/\sqrt{2}), (-1/\sqrt{2}, 1/\sqrt{2})\}$.

c. The curve is a hyperbola with vertices at the points with coordinates $(\pm 3, 0)_B$ or $(\pm 3/\sqrt{2}, \pm 3/\sqrt{2})_{\{E_i\}}$ and asymptotes given by $y' = \pm\sqrt{2}x'$ in terms of coordinates with respect to B.

11. a. If $B = \{(2/\sqrt{5}, 1/\sqrt{5}), (1/\sqrt{5}, -2/\sqrt{5})\}$, then $G = \{V \in \mathscr{E}_2 | V{:}(x', y')_B$ and $6x'^2 + y'^2 = 24\}$.

b. Figure A6.

Review Problems Page 286

1. $r > 1$.

2. a. $\{(1/\sqrt{2}, 0), (3/\sqrt{2}, \sqrt{2})\}$. b. $\{(1/\sqrt{3}, 0), (-1/\sqrt{15}, \sqrt{3/5})\}$.

7. a. To see that $\mathscr{S}$ need not equal $(\mathscr{S}^\perp)^\perp$, consider the inner product space of continuous functions $\mathscr{F}$ with the integral inner product. Let $\mathscr{S} = R[x]$, the set of all polynomial functions. Then using the Maclaurin series expansion for $\sin x$, it can be shown that the sine function is in $(\mathscr{S}^\perp)^\perp$, but it is not in $\mathscr{S}$.

8. a. $T^*(x, y) = (2x + 5y, 3x - y)$. b. A^T; $\bar{A}^T$.

9. a. $\left(\begin{array}{c|c} A_1A_2 + B_1C_2 & A_1B_2 + B_1D_2 \\ \hline C_1A_2 + D_1C_2 & C_1B_2 + D_1D_2 \end{array}\right)$.

b. $E = -A^{-1}B$, $F = D - B^TA^{-1}B$.

c. $\det\begin{pmatrix} A & B \\ B^T & D \end{pmatrix} = \det A \det(D - B^TA^{-1}B)$.

10. a. -277. b. 85. c. 162.

Chapter 7

§1 Page 297

1. a. $(x_1 \quad x_2)\begin{pmatrix} 1 & 3 \\ 3 & -5 \end{pmatrix}\begin{pmatrix} x_1 \\ x_2 \end{pmatrix}$. b. $(x_1 \quad x_2)\begin{pmatrix} 4 & 7/2 \\ 7/2 & 0 \end{pmatrix}\begin{pmatrix} x_1 \\ x_2 \end{pmatrix}$.

c. $(x \quad y \quad z)\begin{pmatrix} 1 & 5/2 & 0 \\ 5/2 & -1 & 3 \\ 0 & 3 & 2 \end{pmatrix}\begin{pmatrix} x \\ y \\ z \end{pmatrix}$. d. $(x \quad y \quad z)\begin{pmatrix} 5 & 0 & -1 \\ 0 & 4 & 1/2 \\ -1 & 1/2 & 0 \end{pmatrix}\begin{pmatrix} x \\ y \\ z \end{pmatrix}$.

2. $A = \begin{pmatrix} 1 & x \\ y & 7 \end{pmatrix}$ where $x + y = -5$.

3. a. $3a^2 + 2ab + 3b^2$. b. $(1/25)(99a^2 + 14ab + 51b^2)$. This is obtained using $(a, b)\colon(\frac{1}{5}(3a + 4b), \frac{1}{5}(4a - 3b))_B$. c. $4a^2 + 2b^2$. d. $a^2 + b^2$.

5. When $n = 2$, the sum $\sum_{j=1}^n \sum_{i=1}^n a_{ij}x_ix_j$ is

$$\sum_{j=1}^2 a_{1j}x_1x_j + \sum_{j=1}^2 a_{2j}x_2x_j = a_{11}x_1x_1 + a_{12}x_1x_2 + a_{21}x_2x_1 + a_{22}x_2x_2.$$

Three other ways to write the sum are

$$\sum_{i=1}^n \sum_{j=1}^n a_{ij}x_ix_j, \qquad \sum_{i=1}^n x_i \sum_{j=1}^n a_{ij}x_j, \qquad \sum_{j=1}^n x_j \sum_{i=1}^n a_{ij}x_i.$$

7. a. If $b(U, V) = X^TAY$ with X and Y the coordinates of U and V with respect to B, then $b(U, V) = 21x_1y_1 + 3x_1y_2 + 21x_2y_1 + 30x_2y_2$. For example,

$$a_{12} = (4, 1) \circ T(1, -2) = (4, 1) \circ (4, -3) = 3.$$

b. $14x_1y_1 + 7x_1y_2 + 5x_2y_1 + 3x_2y_2$.

9. a. $q(x, y) = 4xy - 5y^2$. b. $q(x, y, z) = 3x^2 - 2xz + 2y^2 + 8yz$.

10. a. $x_1y_1 - 4x_1y_2 - 4x_2y_1 + 7x_2y_2$.

b. $x_1y_1 + x_1y_2 + x_2y_1 - x_2y_2 - 6x_2y_3 - 6x_3y_2$.

12. Consider Example 4 on page 296.

13. a. It should be a homogeneous 1st degree polynomial in the coordinates of an arbitrary vector. Thus a *linear form* on $\mathscr{V}$ with basis B is an expression of the form $a_1x_1 + \cdots + a_nx_n$ where $x_1, \ldots, x_n$ are the coordinates of an arbitrary vector with respect to B.

b. If $U\colon(x_1, \ldots, x_n)_B$ and T is defined by $T(U) = \sum_{i=1}^n a_ix_i$ for $U \in \mathscr{V}$, then $T \in \operatorname{Hom}(\mathscr{V}, R)$ and $(a_1, \ldots, a_n)\begin{pmatrix} x_1 \\ \vdots \\ x_n \end{pmatrix} = AX$ is the matrix expression for T in terms of coordinates with respect to B.

c. Show that if W is held fixed, then $T(U) = b(U, W)$ and $S(U) = b(W, U)$ are expressed as linear forms.

§2 Page 304

1. a. $A = \begin{pmatrix} 1 & 8 \\ 8 & -11 \end{pmatrix}$. b. $A' = \begin{pmatrix} 1 & 0 \\ 0 & -1 \end{pmatrix}$.
c. $B' = \{(2/5, 1/5), (-1/5\sqrt{3}, 2/5\sqrt{3})\}$.
d. $q(a, b) = x_1^2 - x_2^2$ if $(a, b)\colon (x_1, x_2)_{B'}$.

2. b. The characteristic values are 9, 18, 0; $A' = \text{diag}(1, 1, 0)$.
c. $B' = \{(1/9, 2/9, -2/9), (\sqrt{2}/9, \sqrt{2}/18, \sqrt{2}/9), (-2/3, 2/3, 1/3)\}$.
d. $q(a, b, c) = x_1^2 + x_2^2$ if $(a, b, c)\colon (x_1, x_2, x_3)_{B'}$.

3. a. $A = \begin{pmatrix} 14 & 8 \\ 8 & 4 \end{pmatrix}$. b. $A' = P^T A P = \begin{pmatrix} -4 & 0 \\ 0 & 2 \end{pmatrix}$.
c. $q(a + bt) = -4x_1^2 + 2x_2^2$ if $a + bt\colon (x_1, x_2)_{B'}$.

5. Consider letting $X = E_i^T$ and $Y = E_j^T$, E_i and E_j from $\{E_i\}$.

6. a. $r = 2$, $s = 0$, diag$(1, -1)$. b. $r = 2$, $s = -2$, diag$(-1, -1)$.
c. det $A = -3$, so $r = 3$, tr $A = 6$, so $s = 1$, diag$(1, 1, -1)$.
d. The characteristic polynomial of A is $-13 + 5\lambda - 3\lambda^2 - \lambda^3$.
Det $A = -13$, so $r = 3$. Tr $A = -3$, so the roots might be all negative or only one might be negative. But the second symmetric function s_2 of the roots is -5. If all the roots were negative, s_2 would be positive so the canonical form is diag$(1, 1, -1)$.

7. a. The matrix $\begin{pmatrix} 2 & 6 \\ 6 & 18 \end{pmatrix}$ is singular.
b. The characteristic polynomial for the matrix A of q with respect to $\{E_i\}$ is $8 - 21\lambda + 11\lambda^2 - \lambda^3$. This polynomial is positive for all values of $\lambda \leq 0$, therefore all three characteristic roots are positive and A is congruent to I_3.

§3 Page 311

1. If P is a transition matrix, what are the possible signs of det P?

2. Figure A7.

3. a. A line has the equation $P = tU + V$ for some $U, V \in \mathscr{E}_n$, $t \in R$. Show that $T(P)$ lies on some line ℓ for all $t \in R$ and that if $W \in \ell$, then $T(tU + V) = W$ for some $t \in R$.
b. That is, does $\|T(U) - T(V)\| = \|U - V\|$ for all $U, V \in \mathscr{E}_n$?

4. Suppose A, B, and C are collinear with B between A and C. Show that if $T(B)$ is not between $T(A)$ and $T(C)$, a contradiction results.

5. a. Given two lines ℓ_1 and ℓ_2 find a translation which sends ℓ_1 to a line through the origin. Show that there exists a rotation that takes the translated line to a line parallel to ℓ_2 and then translate this line onto ℓ_2.

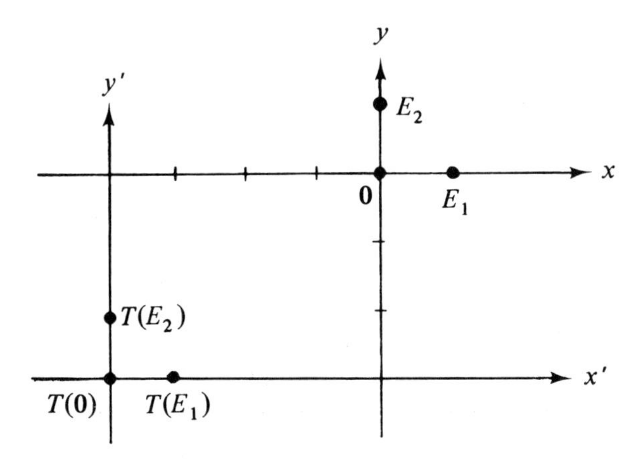

Figure A7

b. Use the normal to obtain the rotation. In this case you must show that your map sends the first hyperplane onto the second. (Why wasn't this necessary in a?)

6. Use the fact that $\mathbf{0}$, U, V, $U + V$, and $\mathbf{0}$, $T(U)$, $T(V)$, $T(U) + T(V)$ determine parallelograms to show that $T(U + V) = T(U) + T(V)$.

7. Show that two spheres are congruent if and only if they have the same radius. Recall that X is on the sphere with center C and radius r if $\|X - C\| = r$.

§4 Page 320

1. a. $A = \begin{pmatrix} 3 & -2 \\ -2 & 0 \end{pmatrix}$, $\quad B = (0, 7)$, $\quad \hat{A} = \begin{pmatrix} 3 & -2 & 0 \\ -2 & 0 & 7/2 \\ 0 & 7/2 & 8 \end{pmatrix}$.

   b. $A = \begin{pmatrix} 0 & 4 \\ 4 & 0 \end{pmatrix}$, $\quad B = (-4, 6)$, $\quad \hat{A} = \begin{pmatrix} 0 & 4 & -2 \\ 4 & 0 & 3 \\ -2 & 3 & 0 \end{pmatrix}$.

2. The characteristic values of A are 26 and 13 with corresponding characteristic vectors (3, 2) and (−2, 3). Therefore the rotation given by

$$\begin{pmatrix} x \\ y \end{pmatrix} = \begin{pmatrix} 3/\sqrt{13} & -2/\sqrt{13} \\ 2/\sqrt{13} & 3/\sqrt{13} \end{pmatrix} \begin{pmatrix} x' \\ y' \end{pmatrix}$$

yields $26x'^2 - 13y'^2 - 26 = 0$ or $x'^2 - y'^2/2 = 1$. See Figure A8.

3. The rigid motion given by

$$\begin{pmatrix} x \\ y \\ 1 \end{pmatrix} = \begin{pmatrix} 4/5 & 3/5 & -1/5 \\ -3/5 & 4/5 & -18/5 \\ 0 & 0 & 1 \end{pmatrix} \begin{pmatrix} x'' \\ y'' \\ 1 \end{pmatrix}$$

yields $x''^2 = 4py''$ with $p = 2$. Let x', y' be coordinates with respect to $B' = \{(4/5, -3/5), (3/5, 4/5)\}$, then B'' is obtained from B' by the translation $(x'', y'') = (x' - 2, y' + 3)$. See Figure A9.

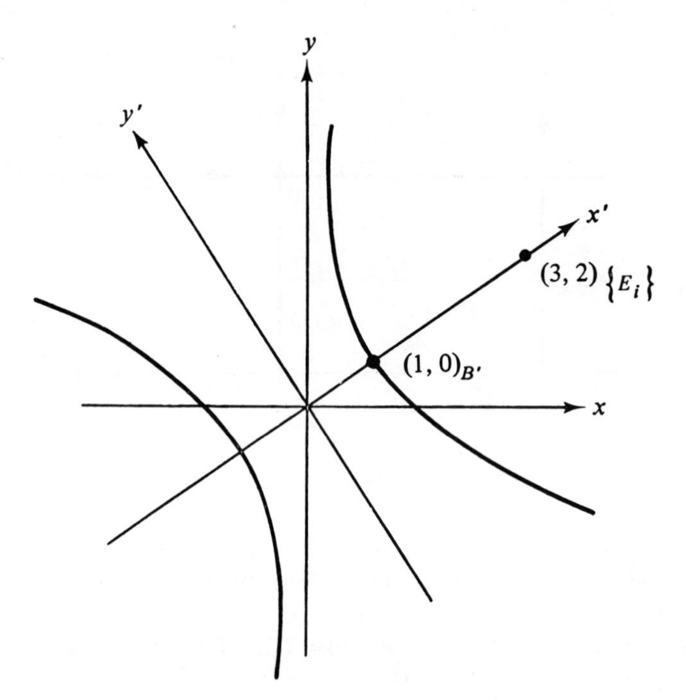

Figure A8

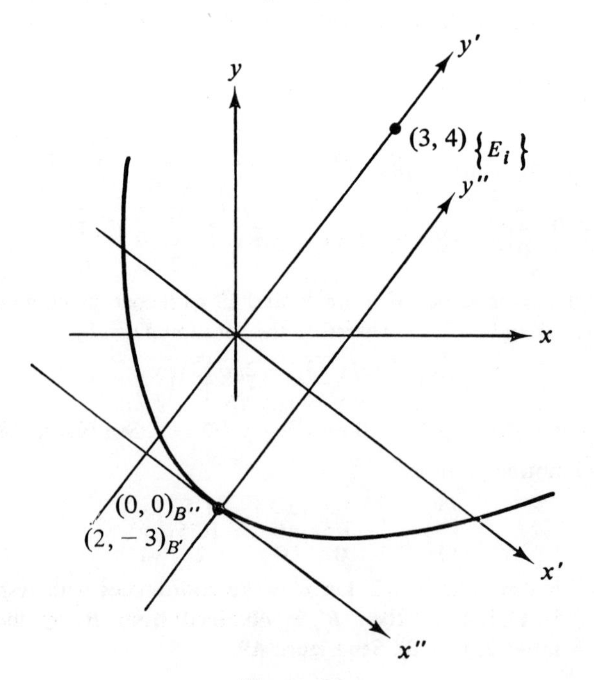

Figure A9

4. $\hat{X}^T\hat{A}\hat{X} = (x \quad y \quad 1)\begin{pmatrix} 5 & 0 & -5 \\ 0 & -3 & -6 \\ -5 & -6 & -7 \end{pmatrix}\begin{pmatrix} x \\ y \\ 1 \end{pmatrix}$. If $\hat{P} = \begin{pmatrix} 1 & 0 & 1 \\ 0 & 1 & -2 \\ 0 & 0 & 1 \end{pmatrix}$ and $\hat{X} = \hat{P}\hat{X}'$, then $\hat{X}'^T(\hat{P}^T\hat{A}\hat{P})\hat{X}' = 5x'^2 - 3y'^2$.

5. a. Intersecting lines. b. Parabola. c. Ellipse, (0, 0) satisfies the equation. d. No graph (imaginary parallel lines). e. Hyperbola.

6. Probably, but since A is a 2×2 matrix, $\det(rA) = r^2 \det A$, so the sign of the determinant is unchanged if $r \neq 0$.

7. b. The equation is $-7t^2 + 28t - 21 = 0$. The values $t = 1, 3$ give $(-1, 1)$ and $(1, -3)$ as the points of intersection.

8. a. $B' = \{(1/\sqrt{2}, 1/\sqrt{2}), (-1/\sqrt{2}, 1/\sqrt{2})\}$. The canonical form is $x'^2 = 9$. See Figure A10(a).
b. $B' = \{(1/\sqrt{10}, -3/\sqrt{10}), (3/\sqrt{10}, 1/\sqrt{10})\}$. The canonical form is $x'^2/3^2 + y'^2/(\sqrt{3})^2 = 1$. See Figure A10(b).

§5 Page 329

1. a. $\begin{pmatrix} x' \\ y' \\ z' \end{pmatrix} = \begin{pmatrix} 0 & 1 & 0 \\ 0 & 0 & 1 \\ 1 & 0 & 0 \end{pmatrix}\begin{pmatrix} x \\ y \\ z \end{pmatrix}$, and multiplication by -1 yields $x'^2/4 + y'^2/9 - z'^2 = -1$.

b. $\begin{pmatrix} x \\ y \\ z \end{pmatrix} = \begin{pmatrix} 1 & 0 & 0 \\ 0 & 0 & -1 \\ 0 & 1 & 0 \end{pmatrix}\begin{pmatrix} x' \\ y' \\ z' \end{pmatrix}$, yields $x'^2/5 - y'^2 = 1$.

c. The translation $(x', y', z') = (x - 2, y, z + 1)$ yields $4z'^2 + 2x' + 3y' = 0$, and $4x''^2 - \sqrt{13}y'' = 0$ or $x''^2 = 4py''$, with $p = \sqrt{13}/16$, results from

$$\begin{pmatrix} x'' \\ y'' \\ z'' \end{pmatrix} = \begin{pmatrix} 0 & 0 & 1 \\ -2/\sqrt{13} & -3/\sqrt{13} & 0 \\ 3/\sqrt{13} & -2/\sqrt{13} & 0 \end{pmatrix}\begin{pmatrix} x' \\ y' \\ z' \end{pmatrix}$$

d. $(x', y', z') = (x + 2, y + 1, z + 3)$, $x'^2/2 + y'^2 + z'^2/3 = -1$.

e. The rotation

$$\begin{pmatrix} x \\ y \\ z \\ 1 \end{pmatrix} = \begin{pmatrix} 1/3 & 2/3 & -2/3 & 0 \\ -2/3 & -1/3 & -2/3 & 0 \\ -2/3 & 2/3 & 1/3 & 0 \\ 0 & 0 & 0 & 1 \end{pmatrix}\begin{pmatrix} x' \\ y' \\ z' \\ 1 \end{pmatrix}$$

yields $9x'^2 - 9y'^2 = 18z'$ or $x'^2 - y'^2 = 2z'$.

2. The plane $y = mx + k$, $m, k \in R$, intersects the paraboloid $x^2/a^2 - y^2/b^2 = 2z$ in the curve given by

$$(1/a^2 - m^2/b^2)x^2 - 2mx/b^2 + k^2/b^2 = 2z, \qquad y = mx + k.$$

If $1/a^2 \neq m^2/b^2$, then there is a translation under which the equation becomes $x'^2 = 4pz'$, $y' = m'x' + k'$. This is the equation of a parabola in the given plane. If $1/a^2 = m^2/b^2$, the section is a line which might be viewed as a degenerate parabola.

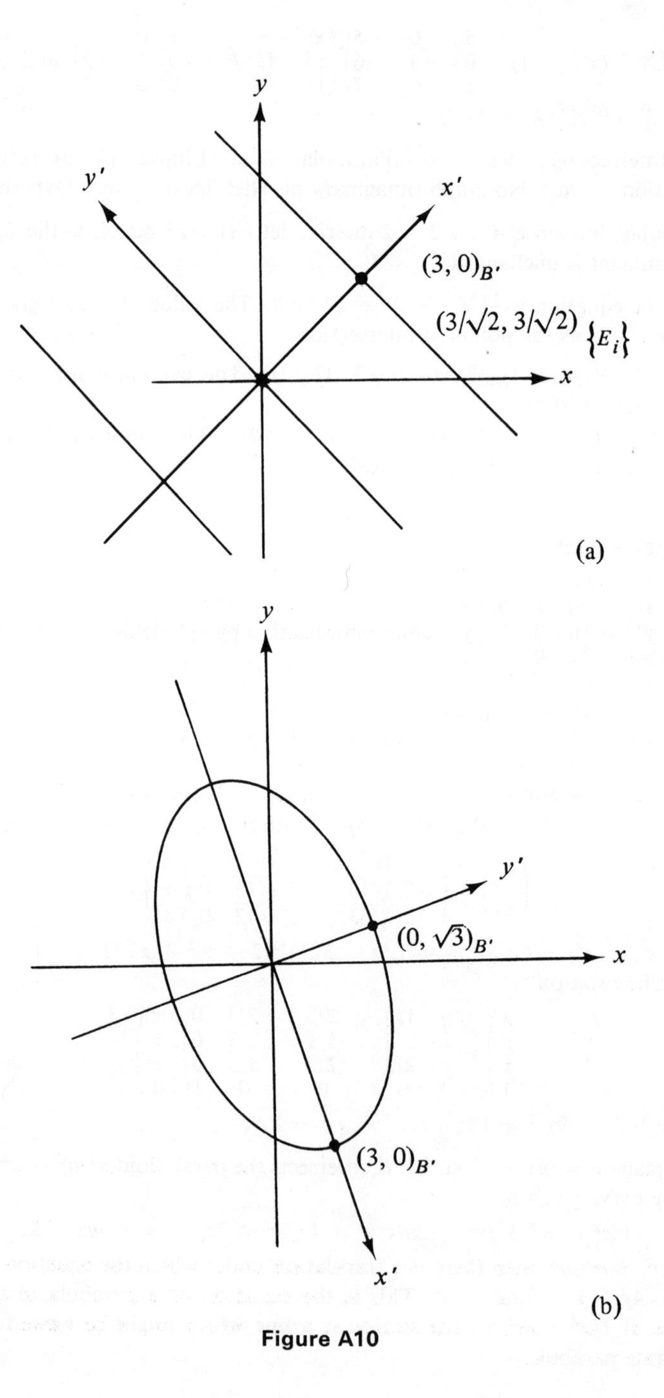

Figure A10

4. Intersecting lines, parallel lines, and concident lines.

6. a. $x'^2/8 + y'^2/8 - z'^2/4 = -1$. b. $x'^2/\sqrt{6} + y'^2/2 - z'^2/2 = 0$.

7. a. Parabolic cylinder. b. Hyperboloid of 2 sheets.
c. Point. d. Intersecting planes.

8. Consider a section of the two surfaces in the plane given by $y = mx$. It is sufficient to show that as x approaches infinity, the difference between the z coordinates approaches zero.

§6 Page 338

1. a. $(-2, -2, 2)$ and $(-25/4, 9/4, -13/2)$. The diametral plane is given by $x + 7y - z + 1 = 0$.
b. $(1, 0, -1)$ is a single point of intersection.
c. $(-\sqrt{2}/2, -1/2, 0)$, two conincident points of intersection. The diametral plane is given by $2x + 2\sqrt{2}y + (4\sqrt{2} - 2)z + 2\sqrt{2} = 0$.

2. Suppose the line passes through the point $\hat{U}$, then if $\hat{U}$ is not on the plane with equation $4x + y - 2z = 5$, the line intersects the surface in a single point. If $\hat{U}$ is on this plane but not on the line with equations $4x + y - 2z = 5$, $7x - 7z = 15$, then the line does not intersect the surface. If $\hat{U}$ is on the line given above, then the line through it with direction $(1, -2, 1)$ is a ruling. (Does this give an entire plane of rulings?)

4. a. For $\hat{A} = \text{diag}(1/5, 1/3, -1/2, 0)$ and $\hat{U} = (5, 3, 4, 1)^T$ we must find $\hat{K} = (\alpha, \beta, \gamma, 0)^T$ such that $\hat{K}^T\hat{A}\hat{K} = \hat{K}^T\hat{A}\hat{U} = 0$. These equations can be combined to yield $(3\gamma - 4\beta)^2 = 0$, $\alpha = 2\gamma - \beta$, therefore $\hat{K}$ may be taken as $(5, 3, 4, 0)^T$ as expected.
b. $\hat{K} = (1, \sqrt{8}, 4 - \sqrt{2}, 0)^T$ or $(1, -\sqrt{8}, 4 + \sqrt{2}, 0)^T$.
c. $\hat{K} = (1, 1, 1, 0)^T$ or $(9, 2, -12, 0)^T$.
d. $\hat{K} = (0, 1, 2, 0)^T$.

5. a. There are no centers. b. A plane, $x + y + z = -1$, of centers.
c. $(1, 1, 3)$. d. A line, $X = t(-3, 2, 1) + (3, -2, 0)$, of centers.

6. a. $\hat{K}^T\hat{A}\hat{U} = 0$, for then the line either intersects the surface in 2 coincident points ($\hat{K}^T\hat{A}\hat{K} \neq 0$) or it is a ruling ($\hat{K}^T\hat{A}\hat{K} = 0$).
b. If $\hat{U}$ is a center, then $\hat{K}^T\hat{A}\hat{K} = 0$ for every direction.
c. The point $\hat{X}$ is on the tangent plane if $\hat{X} - \hat{U}$ is the direction of a tangent or a ruling ($\hat{X} \neq \hat{U}$). Therefore $(\hat{X} - \hat{U})^T\hat{A}\hat{U} = 0$. But $\hat{U}$ is on the surface, so the equation is $\hat{X}^T\hat{A}\hat{U} = 0$. Is the converse clear? That if $\hat{V}^T\hat{A}\hat{U} = 0$, then $\hat{V}$ is a point on the tangent plane.

7. a. $z = 1$. b. $(1, 1, -1)$ is a singular point—the center or vertex of the cone. c. $x - 8y - 6z + 3 = 0$. d. $21x - 25y + 4z + 38 = 0$.

8. If $\hat{A} = \text{diag}(1/4, 1, 1/4, -1)$ and $\hat{U} = (2, 1, 2, 1)^T$, then we must find $\hat{K}$ such that $(2\hat{K}^T\hat{A}\hat{U})^2 - 4(\hat{K}^T\hat{A}\hat{K})(\hat{U}^T\hat{A}\hat{U}) = 0$. (Why?) The tangent lines are given by $X = t(1, 0, 1) + (2, 1, 2)$ and $X = t(9, 2, 1) + (2, 1, 2)$.

Review Problems Page 339

1. a. $A = \begin{pmatrix} 3 & -5/2 \\ -5/2 & 2 \end{pmatrix}$, $B = (5, -7)$. b. $A = \begin{pmatrix} 1 & 2 \\ 2 & 0 \end{pmatrix}$, $B = (0, 6)$.
c. $A = \begin{pmatrix} 4 & 2 & 0 \\ 2 & -1 & 5 \\ 0 & 5 & 0 \end{pmatrix}$, $B = (4, 0, -9)$. d. $A = \begin{pmatrix} 0 & 3/2 & -2 \\ 3/2 & 3 & 1 \\ -2 & 1 & 0 \end{pmatrix}$, $B =$ $(-5, 1, -2)$.

3. a. An inner product. b. Not symmetric. c. Not positive definite.

4. a. -2; diag$(-1, -1)$. b. 0; diag$(1, -1)$. c. 2; diag$(1, 1, 0)$.

5. a. $x'^2/16 - y'^2/4 = 1$; $x' = x + 2$, $y' = y - 3$.
b. $x'^2 - y'^2 = 0$; $x' = x - 5$, $y' = y + 2$.
c. $y' = 5x'^2$; $x = \frac{1}{5}(3x' - 4y')$, $y = \frac{1}{5}(4x' + 3y')$.
d. $x'^2 = 4\sqrt{2}$; $x = (1/\sqrt{2})(x' + y')$, $y = (1/\sqrt{2})(y' - x')$.
e. $x'^2/5 + y'^2/10 = 1$; $x = (1/\sqrt{5})(2x' + y' - 4)$, $y = (1/\sqrt{5})(-x' + 2y' - 3)$.

7. a. $(4, -2)$. b. $\{(x, y) | 3x + 6y + 1 = 0\}$. c. $(1, 1)$.

8. a. Hyperboloid of 2 sheets. b. Parallel planes. c. Hyperbolic cylinder.

9. Ellipsoid or hyperboloid. See problem 9, page 127.

10. a. Any plane containing the z axis.
b. Any plane parallel to the z axis.
c. Any plane parallel to the z axis but not parallel to either trace of the surface in the xy plane.

11. Each line intersects the surface in a single point.

12. $\hat{P} = \begin{pmatrix} 4/5 & -3/5 & 11/5 \\ 3/5 & 4/5 & 2/5 \\ 0 & 0 & 1 \end{pmatrix}$. $x'^2/8 + y'^2/2 = 1$.

Chapter 8

§1 Page 347

1. a. $\mathscr{S}(U, T) = \mathscr{L}\{(1, 0, 0), (3, 1, 1), (10, 8, 6)\}$ and $a(x) = x^3 - 8x^2 + 18x - 12 = (x - 2)(x^2 - 6x + 6)$.
b. $\mathscr{S}(U, T) = \mathscr{L}\{U\}$; $a(x) = x - 2$.
c. $\mathscr{S}(U, T) = \mathscr{L}\{U, (-6, 0, 0), (-18, -6, -6)\}$; $a(x)$ as in part a.
d. $\mathscr{L}(U, T) = \mathscr{L}\{(0, 1, 0), (1, 3, 1)\}$: $a(x) = x^2 - 6x + 6$.

2. a. $\mathscr{S}(U, T) = \mathscr{L}\{(-1, 0, 1), (0, 1, 2), (3, 7, 8)\}$ and $a(x) = x^3 - 8x^2 + 16x - 9 = (x - 1)(x^2 - 7x + 9)$.
b. $\mathscr{S}(U, T) = \mathscr{L}\{U\}$; $a(x) = x - 1$.

c. $\mathscr{S}(U, T) = \mathscr{L}\{(1, 1, 1), (3, 6, 6)\}$; $a(x) = x^2 - 7x + 9$.
d. $x^3 - 8x^2 + 16x - 9$.

4. b. $$\begin{pmatrix} 0 & -3 & & & 0 \\ 1 & 2 & & & \\ & & -1 & & \\ & & & 0 & -1 \\ 0 & & & 1 & 1 \end{pmatrix}.$$

5. For the two block case, suppose $A_1 \in \mathscr{M}_{r \times r}$ and $A_2 \in \mathscr{M}_{s \times s}$, and use the fact that

$$\begin{pmatrix} A_1 & 0 \\ 0 & A_2 \end{pmatrix} = \begin{pmatrix} A_1 & 0 \\ 0 & I_s \end{pmatrix}\begin{pmatrix} I_r & 0 \\ 0 & A_2 \end{pmatrix}.$$

6. a. $\mathscr{L}\{(1, -3, 1)\}$. b. $\{(x, y, z) | x - 3y + z = 0\}$.
c. $\{\mathbf{0}\}$. d. $\mathscr{V}_3$

7. a. $\mathscr{L}\{(0, 1, 0)\}$. b. $\mathscr{L}\{(0, 1, 0), (1, 0, 1)\}$. c. $\mathscr{L}\{(2, 1, -2)\}$.
d. $\mathscr{L}\{(0, 1, 0), (2, 1, -2)\}$. e. $\mathscr{V}_3$, for $p(T) = \mathbf{0}$.

8. It is the zero map.

9. a. $\mathscr{L}\{(1, 0, 2)\}$, b. $\mathscr{L}\{(1, 0, 2)\}$, compare with 7a and b.
c. $\mathscr{L}\{(1, 0, 1), (0, 1, 0)\}$. d. $\{\mathbf{0}\}$.
e. Use problem 8 with parts a and c.

§2 Page 358

1. Consider $x^2 - x \in Z_2[x]$ or in general the polynomial $(x - c_1)(x - c_2) \cdots (x - c_h)$ when $F = \{c_1, \ldots, c_h\}$.

4. a. $q(x) = 3x^2 - x + 2$, $r(x) = x + 3$. c. $q(x) = x^3 - 2$, $r(x) = x$.
b. $q(x) = 0(x)$, $r(x) = x^2 + 2x + 1$. d. $q(x) = x^3 - 3x + 4$, $r(x) = 0(x)$.

5. Suppose $a(x) = b(x)q_1(x) + r_1(x)$ with deg $r_1 <$ deg b and $a(x) = b(x)q_2(x) + r_2(x)$ with deg $r_2 <$ deg b. Show that $q_1(x) \neq q_2(x)$ and $r_1(x) \neq r_2(x)$ leads to a contradiction involving degree.

6. Use the unique factorization theorem.

7. a. g.c.d.: $(x - 1)^3(x + 2)^2$; l.c.m.: $x(x - 1)^4(x + 2)^4(x^2 + 5)$.
b. g.c.d.: $(x - 4)(x - 1)$; l.c.m.: $(x^2 - 16)(x - 4)^2(x^3 - 1)$.

8. The last equation implies that $r_k(x)$ is a g.c.d. of $r_k(x)$ and $r_{k-1}(x)$, then the next to the last equation implies that $r_k(x)$ is a g.c.d. of $r_{k-1}(x)$ and $r_{k-2}(x)$.

9. a. $x^2 + 1$ is a g.c.d. The polynomials obtained are:
$q_1(x) = 2x$, $r_1(x) = x^3 - 2x^2 + x - 2$; $q_2(x) = x + 1$, $r_2(x) = x^2 + 1$; $q_3(x) = x - 2$, $r_3(x) = 0(x)$.
b. Relatively prime. $q_1(x) = 2$, $r_1(x) = x^3 - 4x + 6$: $q_2(x) = x + 2$, $r_2(x) = x^2 + x - 4$; $q_3(x) = x - 1$, $r_3(x) = x + 2$; $q_4(x) = x - 1$, $r_4(x) = -2$; $q_5(x) = -\frac{1}{2}(x + 2)$, $r_5(x) = 0(x)$.
c. $2x + 1$ is a g.c.d. $q_1(x) = x + 1$, $r_1(x) = 2x^3 - x^2 + x + 1$; $q_2(x) = 3$, $r_2(x) = 2x^2 + x$; $q_3(x) = x - 1$, $r_3(x) = 2x + 1$; $q_4(x) = x$.

10. Suppose c/d is in lowest terms, that is c and d are relatively prime as integers, and consider $d^n p(c/d)$.

11. a. $\pm 6, \pm 3, \pm 3/2, \pm 2, \pm 1/2, \pm 1$. c. $\pm 8, \pm 4, \pm 2, \pm 1$.
b. ± 1. d. $\pm 1/9, \pm 1/3, \pm 1$.

12. Consider $x^2 + x + 1$.

13. Why is it only necessary to prove that well ordering implies the first principle of induction? Given A such that $1 \in A$ and $k \in A$ implies $k + 1 \in A$, consider the set B of all positive integers not in A.

17. c. $z - 4 = a - b$, $x^2 + y^2 + (z - 4)^2 - 4 = (z - 3)a + (4 - z)b$, and $4x^2 + 4y^2 - z^2 = (z + 4)b - za$.
d. A curve in 3-space cannot be described by a single equation.

§3 Page 366

1. a. $x^3 - 5x^2 + 7x - 3 = (x - 1)^2(x - 3)$. b. $x - 3$. c. $(x - 1)^2$.
d. $x - 1$.

2. a. $x^2 + x + 2$. c. If $T \in \mathrm{Hom}(\mathscr{V}_2(C))$, then T is diagonalizable.

3. a. $(x + 4)(x - 3)$. b. $(x + 2)^2$. c. $x^3 - 3x^2 + 3x - 1$.
d. $(x^2 + 8x + 16)(x + 5)$. e. $(x^2 + x + 1)(x + 2)$.
f. $x^4 + 6x^2 + 9$. g. $x^2 - x - 2$.

4. 3a and 3g.

5. a. $\mathscr{N}_{T+4I} \oplus \mathscr{N}_{T-3I} = \mathscr{S}(-4) \oplus \mathscr{S}(3) = \mathscr{L}\{(1, -1)\} \oplus \mathscr{L}\{(1, 6)\}$.
b. $\mathscr{N}_{(T+2I)^2} = \mathscr{V}_2$. c. $\mathscr{N}_{(T-I)^3} = \mathscr{V}_3$.
d. $\mathscr{N}_{(T+4I)^2} \oplus \mathscr{N}_{T+5I} = \mathscr{L}\{(1, 0, 0), (0, 1, 0)\} \oplus \mathscr{L}\{(0, 0, 1)\}$.
e. $\mathscr{N}_{T^2+T+I} \oplus \mathscr{N}_{T+2I} = \mathscr{L}\{(1, 0, -1), (0, 1, 0)\} \oplus \mathscr{L}\{(1, 0, -2)\}$.
f. $\mathscr{N}_{(T^2+3I)^2} = \mathscr{V}_4$.
g. $\mathscr{N}_{T-2I} \oplus \mathscr{N}_{T+I} = \{(x, y, z) | 5x + 10y - 2z = 0\} \oplus \mathscr{L}\{(1, -1, -1)\}$.

6. $\begin{pmatrix} 0 & -1 & 0 \\ 1 & -1 & 0 \\ 0 & 0 & -2 \end{pmatrix}$. **7.** c. $\begin{pmatrix} 0 & -7 \\ 1 & 5 \end{pmatrix}$.

8. a. $(2, -1, -5) = (T - I)E_1$. b. $(3, 3, 3)$. c. $(4, 2, -2)$.

15. Define $T_i = s_i(T) \circ q_i(T)$, then $I = T_1 + \cdots + T_k$ and therefore each T_i is a projection. $T_i \circ T_j = \mathbf{0}$ for $i \neq j$ follows from the fact that $p_h(x) | q_g(x)$ when $h \neq g$. $m(x) = p_i^{e_i}(x)q_i(x)$ implies $\mathscr{N}_{p_i^{e_i}(T)} \subset T_i[\mathscr{V}]$, and it remains to obtain the reverse containment.

§4 Page 376

1. a. (5). b. $\begin{pmatrix} 0 & -25 \\ 1 & 10 \end{pmatrix}$. c. $\begin{pmatrix} 0 & 0 & 125 \\ 1 & 0 & -75 \\ 0 & 1 & 15 \end{pmatrix}$. d. $\begin{pmatrix} 0 & -4 \\ 1 & 3 \end{pmatrix}$.

e. $\begin{pmatrix} 0 & 0 & 0 & -16 \\ 1 & 0 & 0 & 24 \\ 0 & 1 & 0 & -17 \\ 0 & 0 & 1 & 6 \end{pmatrix}$.

2. a. Show that if $B = \{U_1, \ldots, U_h\}$, then $B = \{U_1, T(U_1), \ldots, T^{h-1}(U_1)\}$ and if $q(x)$ is not the minimum polynomial of T, then B is linearly dependent.
b. Expand $\det(C(q) - \lambda I_h)$ along the last column and show that the cofactor of the element in the kh position (occupied by $-a_{k-1}$) is $(-1)^{k+h}(-\lambda)^{k-1}(1)^{k-1}$ when $k < h$ and $(-1)^{h+h}(-\lambda)^{h-1}$ when $k = h$.

3. a. Since the minimum polynomial is $(x - 2)^3$, $\mathscr{V}_3 = \mathscr{S}(U_1, T)$ with U_1 any vector such that $(T - 2I)^2 U_1 \neq \mathbf{0}$. A possible cyclic basis is $\{(0, 0, 1), (-1, 1, 2), (-6, 5, 4)\}$, which yields the matrix

$$\begin{pmatrix} 0 & 0 & 8 \\ 1 & 0 & -12 \\ 0 & 1 & 6 \end{pmatrix}.$$

b. The minimum polynomial is $(x - 2)^2$. For $U_1 \notin \mathscr{S}(2)$, choose $U_2 \in \mathscr{S}(2)$ such that $U_2 \notin \mathscr{S}(U_1, T)$. For example, $B = \{(1, 0, 0), (-2, 2, -4), (1, -1, 0)\}$ yields the matrix

$$\begin{pmatrix} 0 & -4 & 0 \\ 1 & 4 & 0 \\ 0 & 0 & 2 \end{pmatrix}.$$

4. b. Suppose $U_1, \ldots, U_{i-1}$ have been chosen for $1 < i \leq h$ and $\mathscr{V} = \mathscr{S}(U_1, T) \oplus \cdots \oplus \mathscr{S}(U_{i-1}, T)$. Then a maximal integer r_i exists such that $p^{r_i - 1}(T)W \notin \mathscr{S}$ and $p^{r_i}(T)W \in \mathscr{S}$, for some $W \in \mathscr{V}$. Show that $p^{r_i - 1}(T)V = \mathbf{0}$ for all $V \in B$ leads to a contradiction.

5. a. $\operatorname{diag}(C(p^2), C(p^2), C(p^2))$, $\operatorname{diag}(C(p^2), C(p^2), C(p), C(p))$, $\operatorname{diag}(C(p^2), -3I_4)$.
b. $4I_6$. c. $\operatorname{diag}(C(p^2), C(p))$. d. $C(p^3)$.
e. $\operatorname{diag}(C(p^3), C(p^3))$, $\operatorname{diag}(C(p^3), C(p^2), C(p))$, $\operatorname{diag}(C(p^3), I_3)$.

6. a. $(T - cI)^r U = \mathbf{0}$ implies $(T - cI)((T - cI)^{r-1} U) = \mathbf{0}$.

7. a. There are three linearly independent characteristic vectors for the characteristic value -1, therefore the $(x + 1)$-elementary divisors are $(x + 1)^2$, $x + 1$, $x + 1$.
b. There are two linearly independent characteristic vectors for -1 so the $(x + 1)$-elementary divisors are $(x + 1)^2$, $(x + 1)^2$.

8. a. $$\begin{bmatrix} 0 & -1 & & & & 0 \\ 1 & -1 & & & & \\ & & 0 & -1 & & \\ & & 1 & -1 & & \\ & & & & 0 & -1 \\ 0 & & & & 1 & -1 \end{bmatrix}.$$

b. $$\begin{bmatrix} 0 & 0 & -27 & & & & 0 \\ 1 & 0 & -27 & & & & \\ 0 & 1 & -9 & & & & \\ & & & 0 & -9 & & \\ & & & 1 & -6 & & \\ & & & & & 0 & -9 \\ 0 & & & & & 1 & -6 \end{bmatrix}.$$

c. $$\begin{bmatrix} 0 & 0 & 0 & -49 & & 0 \\ 1 & 0 & 0 & 56 & & \\ 0 & 1 & 0 & -30 & & \\ 0 & 0 & 1 & 8 & & \\ & & & & 0 & -7 \\ 0 & & & & 1 & 4 \end{bmatrix}.$$

9. Suppose the given matrix is the matrix of $T \in \operatorname{Hom}(\mathscr{V}_4)$ with respect to the standard basis.
a. The minimum polynomial of T is $(x + 1)^2(x - 1)^2$. Therefore

diag($C((x+1)^2)$, $C((x-1)^2)$) is the matrix of T with respect to some basis. This matrix is similar to the given matrix since both are matrices for T.
b. The minimum polynomial of T is $(x-2)^2(x+4)$ and T has only one characteristic vector for the characteristic value -4. Therefore T has only one $(x-4)$-elementary divisor and because of dimension restrictions, there are two $(x-2)$-elementary divisors, $(x-2)^2$ and $x-2$. Thus diag($C((x-2)^2)$, $C(x-2)$, $C(x-4)$) is a diagonal block matrix similar to the given matrix.

§5 Page 386

1. a. $$D_1 = \begin{pmatrix} 0 & 0 & 27 & & 0 \\ 1 & 0 & -27 & & \\ 0 & 1 & 9 & & \\ & & & 3 & \\ 0 & & & & 3 \end{pmatrix}, \qquad D_2 = \begin{pmatrix} 3 & 0 & 0 & & 0 \\ 1 & 3 & 0 & & \\ 0 & 1 & 3 & & \\ & & & 3 & \\ 0 & & & & 3 \end{pmatrix}.$$

b. $$D_1 = \begin{pmatrix} 0 & 0 & 0 & -16 & & 0 \\ 1 & 0 & 0 & 0 & & \\ 0 & 1 & 0 & -8 & & \\ 0 & 0 & 1 & 0 & & \\ & & & & 0 & -4 \\ 0 & & & & 1 & -4 \end{pmatrix}, \qquad D_2 = \begin{pmatrix} 0 & -4 & & & & 0 \\ 1 & 0 & & & & \\ & 1 & 0 & -4 & & \\ & & 1 & 0 & & \\ & & & & -2 & 0 \\ 0 & & & & 1 & -2 \end{pmatrix}.$$

c. $$D_1 = \begin{pmatrix} 0 & 0 & 0 & -256 & & 0 \\ 1 & 0 & 0 & 0 & & \\ 0 & 1 & 0 & -32 & & \\ 0 & 0 & 1 & 0 & & \\ & & & & 0 & -16 \\ 0 & & & & 1 & 0 \end{pmatrix}, \qquad D_2 = \begin{pmatrix} 0 & -16 & 0 & 0 & & 0 \\ 1 & 0 & 0 & 0 & & \\ 0 & 1 & 0 & -16 & & \\ 0 & 0 & 1 & 0 & & \\ & & & & 0 & -16 \\ 0 & & & & 1 & 0 \end{pmatrix}.$$

d. $$D_1 = \begin{pmatrix} 0 & 0 & 0 & 0 & 0 & -8 \\ 1 & 0 & 0 & 0 & 0 & -24 \\ 0 & 1 & 0 & 0 & 0 & -36 \\ 0 & 0 & 1 & 0 & 0 & -32 \\ 0 & 0 & 0 & 1 & 0 & -18 \\ 0 & 0 & 0 & 0 & 1 & -6 \end{pmatrix}, \qquad D_2 = \begin{pmatrix} 0 & -2 & 0 & 0 & 0 & 0 \\ 1 & -2 & 0 & 0 & 0 & 0 \\ 0 & 1 & 0 & -2 & 0 & 0 \\ 0 & 0 & 1 & -2 & 0 & 0 \\ 0 & 0 & 0 & 1 & 0 & -2 \\ 0 & 0 & 0 & 0 & 1 & -2 \end{pmatrix}.$$

2. a. No change, $D_3 = D_1$ and $D_4 = D_2$.
b. $(x-2i)^2$, $(x+2i)^2$, $(x+2)^2$;

$$D_3 = \begin{pmatrix} 0 & 4 & & & & 0 \\ 1 & 4i & & & & \\ & & 0 & 4 & & \\ & & 1 & -4i & & \\ & & & & 0 & -4 \\ 0 & & & & 1 & -4 \end{pmatrix}, \qquad D_4 = \begin{pmatrix} 2i & 0 & & & & 0 \\ 1 & 2i & & & & \\ & & -2i & 0 & & \\ & & 1 & -2i & & \\ & & & & -2 & 0 \\ 0 & & & & 1 & -2 \end{pmatrix}.$$

c. $(x-4i)^2$, $x-4i$, $(x+4i)^2$, $x+4i$.

$$D_3 = \begin{pmatrix} 0 & 16 & & & & 0 \\ 1 & 8i & & & & \\ & & 4i & & & \\ & & & 0 & 16 & \\ & & & 1 & -8i & \\ 0 & & & & & -4i \end{pmatrix},$$

$$D_4 = \begin{pmatrix} 4i & 0 & & & & 0 \\ 1 & 4i & & & & \\ & & 4i & & & \\ & & & -4i & 0 & \\ & & & 1 & -4i & \\ 0 & & & & & -4i \end{pmatrix}.$$

d. $(x + 1 - i)^3$ and $(x + 1 + i)^3$ are the elementary divisors.

$$D_3 = \begin{pmatrix} 0 & 0 & 2+2i & & & 0 \\ 1 & 0 & 6i & & & \\ 0 & 1 & 3i-3 & & & \\ & & & 0 & 0 & 2-2i \\ & & & 1 & 0 & -6i \\ 0 & & & 0 & 1 & -3-3i \end{pmatrix},$$

$$D_4 = \begin{pmatrix} i-1 & 0 & 0 & & & 0 \\ 1 & i-1 & 0 & & & \\ 0 & 1 & i-1 & & & \\ & & & -i-1 & 0 & 0 \\ & & & 1 & -i-1 & \\ 0 & & & 0 & 1 & -i-1 \end{pmatrix}.$$

3. b. $B_c = \{(1, 0, 0, 0), (0, 1, 0, 0), (-1, -2, 0, 1), (0, 1, -1, -1)\}$.

4. The minimum polynomial of A is $x^2 + 9$, i.e., $A^2 + 9I_4 = \mathbf{0}$. A classical form over R is diag $\left\{\begin{pmatrix} 0 & -9 \\ 1 & 0 \end{pmatrix}, \begin{pmatrix} 0 & -9 \\ 1 & 0 \end{pmatrix}\right\}$ and a Jordan form over C is diag$(3i, 3i, -3i, -3i)$.

6. a. The characteristic polynomial is $-(x - 2)^3$. Since there is only one linearly independent characteristic vector for 2, the minimum polynomial is $(x - 2)^3$, and the Jordan form is

$$\begin{pmatrix} 2 & 0 & 0 \\ 1 & 2 & 0 \\ 0 & 1 & 2 \end{pmatrix}.$$

b. The characteristic polynomial is $-(x - 2)^3$, but there are two linearly independent characteristic vectors for 2. So the minimum polynomial is $(x - 2)^2$ and a Jordan form is

$$\begin{pmatrix} 2 & 0 & 0 \\ 1 & 2 & 0 \\ 0 & 0 & 2 \end{pmatrix}.$$

c. The characteristic polynomial is $-(x - 1)(x - i)(x + i)$ and this must be the negative of the minimum polynomial. A Jordan form is diag$(1, i, -i)$.

7. a. There are 11 possibilities from $(x - 2)^6$ to $x - 2, x - 2, x - 2, x - 2, x - 2, x - 2$.

b. The first and last (as in part a) and $(x - 2)^2, x - 2, x - 2, x - 2, x - 2$.

8. Show by induction that

$$\begin{pmatrix} a & 0 \\ 1 & a \end{pmatrix}^n = \begin{pmatrix} a^n & 0 \\ na^{n-1} & a^n \end{pmatrix}.$$

9. Let k denote the number of independent characteristic vectors for 2.

a. $(x - 2)^2, (x - 2)^2, k = 2$.

b. $(x-2)^3, x-2, k=2$.
c. $(x-2)^2, x-2, x-2, k=3$.

10. b. $B_c = \{(0, 0, 1, -1), (0, 1, 0, 0), (0, 0, -1, 0), (-1, -2, 0, -1)\}$.
c. Let $p_1(x) = x - \sqrt{2}i$ and $p_2(x) = x + \sqrt{2}i$, $B_c = \{U_1, \ldots, U_4\}$. If $W = (0, 0, 1, -1)$, $U_1 = p_2^2(T)W$ and $U_3 = p_1^2(T)W$, then $B_c = \{(0, 2\sqrt{2}i, -5, 4), (-1, -2, -\sqrt{2}i, -1), (0, -2\sqrt{2}i, -5, 4), (-1, -2, \sqrt{2}i, -1)\}$.

11. b. There are five possible sets: x^4; x^3, x; x^2, x^2; x^2, x, x; and x, x, x, x. The corresponding Jordan forms are:

$$\begin{pmatrix}0&0&0&0\\1&0&0&0\\0&1&0&0\\0&0&1&0\end{pmatrix}, \begin{pmatrix}0&0&0&0\\1&0&0&0\\0&1&0&0\\0&0&0&0\end{pmatrix}, \begin{pmatrix}0&0&0&0\\1&0&0&0\\0&0&0&0\\0&0&1&0\end{pmatrix}, \begin{pmatrix}0&0&0&0\\1&0&0&0\\0&0&0&0\\0&0&0&0\end{pmatrix}, \begin{pmatrix}0&0&0&0\\0&0&0&0\\0&0&0&0\\0&0&0&0\end{pmatrix}.$$

12. The elementary divisors for the given matrix are:
a. x^3. b. x^2, x. c. x^2, x^2.

13. a. If $B_c = \{(1, 0), (1, 1)\}$, the matrix of T with respect to B_c is $\begin{pmatrix}4&0\\1&4\end{pmatrix}$. $T_n(a, b) = (a - b, a - b)$ is the map with matrix $\begin{pmatrix}0&0\\1&0\end{pmatrix}$ with respect to B_c, and $T_d(a, b) = (4a, 4b)$ has $\begin{pmatrix}4&0\\0&4\end{pmatrix}$ as its matrix with respect to B_c. Thus $T = T_n + T_d$. Notice that this answer does not depend on the choice of basis B_c. Will this always be true?
b. The matrix of T with respect to $B_c = \{(0, 0, 1), (-1, 1, -2), (0, 1, -1)\}$ is $\begin{pmatrix}2&0&0\\1&2&0\\0&0&3\end{pmatrix}$. If $\begin{pmatrix}0&0&0\\1&0&0\\0&0&0\end{pmatrix}$ is the matrix of T_n with respect to B_c, and diag $(2, 2, 3)$ is the matrix of T_d with respect to B_c, then $T = T_n + T_d$, and $T_n(a, b, c) = (a - b - c, b - a + c, 2a - 2b - 2c)$ and $T_d(a, b, c) = (2a, a + 3b, 2c - a - b)$.

14. There are two approaches. Either show that a matrix with all characteristic roots zero is nilpotent, or show that if $A = (a_{ij})$ and $A^2 = (b_{ij})$, then $a_{ij} = 0$ for $i \le j$ implies $b_{ij} = 0$ for $i - 1 \le j$ and continue by induction.

§6 Page 401

1. a. $P = \begin{pmatrix}2&3\\1&2\end{pmatrix}$, $P^{-1}AP = \begin{pmatrix}-2&0\\0&-3\end{pmatrix}$, $x' = -13e^{-2t}$, $y' = 9e^{-3t}$; $X = P(-13e^{-2t}, 9e^{-3t})^T$.
b. $P = \begin{pmatrix}5&-2\\8&-3\end{pmatrix}$, $P^{-1}AP = \begin{pmatrix}1&0\\1&1\end{pmatrix}$; $x' = 4e^t$, $(D_ty' - y' = 4e^t)e^{-t}$ and $y'(0) = 9$ yields $y' = (4t + 9)e^t$; $X = P(4e^t, (4t + 9)e^t)^T$.
c. $P = \begin{pmatrix}2&1&-1\\1&1&2\\1&1&3\end{pmatrix}$, $P^{-1}AP = \operatorname{diag}(1, 1, -1)$; $X = P(5e^t, -6e^t, e^{-t})^T$.

d. $P = \begin{pmatrix} 2 & 1 & 2 \\ 1 & 1 & 1 \\ 0 & 1 & 1 \end{pmatrix}$, $P^{-1}AP = \begin{pmatrix} 2 & 0 & 0 \\ 1 & 2 & 0 \\ 0 & 1 & 2 \end{pmatrix}$; $X = PX'$, $x' = e^{2t}$, $y' = (t + 3)e^{2t}$, $z' = (t^2/2 + 3t - 2)e^{2t}$.

2. a. $P = \begin{pmatrix} 2 & 1 \\ 1 & 1 \end{pmatrix}$, $e^A = P^{-1}e^{\text{diag}(1,3)}\,P = \begin{pmatrix} 2e - e^3 & e - e^3 \\ 2e^3 - 2e & 2e^3 - e \end{pmatrix}$. $e^{tA} = P^{-1}\,\text{diag}(e^t, e^{3t})P$.

b. $e^A = e^3\begin{pmatrix} 1 & 0 \\ 1 & 1 \end{pmatrix}$, $e^{tA} = e^{3t}\begin{pmatrix} 1 & 0 \\ t & 1 \end{pmatrix}$.

c. $e^A = \begin{pmatrix} e^4 & 0 & 0 \\ 0 & e^5 & 0 \\ 0 & e^5 & e^5 \end{pmatrix}$, $e^{tA} = \begin{pmatrix} e^{4t} & 0 & 0 \\ 0 & e^{5t} & 0 \\ 0 & te^{5t} & e^{5t} \end{pmatrix}$.

d. $e^A = e^3\begin{pmatrix} 1 & 0 & 0 \\ 1 & 1 & 0 \\ 1/2 & 1 & 1 \end{pmatrix}$, $e^{tA} = e^{3t}\begin{pmatrix} 1 & 0 & 0 \\ t & 1 & 0 \\ t^2/2 & t & 1 \end{pmatrix}$.

4. a. $A^T = C(p)$ with $p(\lambda) = \lambda^n + a_{n-1}\lambda^{n-1} + \cdots + a_1\lambda + a_0$.

5. a. $x = 7e^{2t} - 5e^{3t}$. b. $x = (6t + 7)e^{4t}$. c. $x = e^t + 2e^{-t} - e^{-2t}$.

6. a. diag(1, 0). b. Does not exist. c. **0**, characteristic values are 1/2 and 1/3. d. Does not exist.

e. $\begin{pmatrix} 0 & 0 & 0 \\ 6 & 1 & 0 \\ -2 & 0 & 1 \end{pmatrix}$. f. $\begin{pmatrix} 0 & 0 & 0 & 0 \\ 3/4 & 1 & 0 & 0 \\ 0 & 0 & 0 & 0 \\ 5/2 & 0 & -12 & 1 \end{pmatrix}$. g. $\begin{pmatrix} -1 & 8 & -6 \\ -1 & 5 & -3 \\ -1 & 4 & -2 \end{pmatrix}$.

7. b. For the converse, consider diag(1/4, 3).

9. Both limits exist.

Review Problems Page 402

1. d. $\begin{pmatrix} 0 & -1 & 0 \\ 1 & -1 & 0 \\ 0 & 0 & 1 \end{pmatrix}$. e. No.

3. a. $(x - 2)^2(x + 1)^2$. b. Why is it also $(x - 2)^2(x + 1)^2$?

c. $\text{diag}\left(\begin{pmatrix} 0 & -4 \\ 1 & 4 \end{pmatrix}, \begin{pmatrix} -1 & 0 \\ 1 & -2 \end{pmatrix}\right)$. d. $\text{diag}\left(\begin{pmatrix} 2 & 0 \\ 1 & 2 \end{pmatrix}, \begin{pmatrix} -1 & 0 \\ 1 & -1 \end{pmatrix}\right)$.

4. c. Notice that if $T(V) = \lambda V$ and $p(x) = \sum_{i=0}^{k} a_i x^i$, then $p(T)V = \sum_{i=0}^{k} a_i\lambda^i V = p(\lambda)V$.

5. a. $\begin{pmatrix} 0 & 0 & -8 \\ 1 & 0 & -12 \\ 0 & 1 & -6 \end{pmatrix}$, $\begin{pmatrix} 3 & 0 & 0 \\ 1 & 3 & 0 \\ 0 & 1 & 3 \end{pmatrix}$. b. $\begin{pmatrix} 0 & 0 & 0 & -1 \\ 1 & 0 & 0 & -2 \\ 0 & 1 & 0 & -3 \\ 0 & 0 & 1 & -2 \end{pmatrix}$, $\begin{pmatrix} 0 & -1 & 0 & 0 \\ 1 & -1 & 0 & 0 \\ 0 & 1 & 0 & -1 \\ 0 & 0 & 1 & -1 \end{pmatrix}$.

6. a. $\text{diag}\left(\begin{pmatrix} 1 & 0 \\ 1 & 1 \end{pmatrix}, \begin{pmatrix} 2 & 0 \\ 1 & 2 \end{pmatrix}\right)$. b. $\text{diag}\left(\begin{pmatrix} 1 & 0 \\ 1 & 1 \end{pmatrix}, (2), (2)\right)$.

7. a. $(x + 4)^3$, $(x + 1)^2$; one independent vector each for -4 and -1.
b. $(x + 4)^3$, $(x + 4)^2$, $(x + 1)^2$; 2 for -4 and 1 for -1.

$(x+4)^3$, $x+4$, $(x+1)^2$, $x+1$; 2 for -4 and 2 for -1.
$(x+4)^3$, $(x+1)^2$, $(x+1)^2$; 1 for -4 and 2 for -1.
$(x+4)^3$, $(x+1)^2$, $x+1$, $x+1$; 1 for -4 and 3 for -1.

8. a. $\begin{pmatrix} 0 & -4 & & & & 0 \\ 1 & 0 & & & & \\ & 1 & 0 & -4 & & \\ & & 1 & 0 & & \\ & & & 1 & 0 & -4 \\ 0 & & & & 1 & 0 \end{pmatrix}$. b. $\begin{pmatrix} 2i & 0 & 0 & & & 0 \\ 1 & 2i & 0 & & & \\ 0 & 1 & 2i & & & \\ & & & -2i & 0 & 0 \\ & & & 1 & -2i & 0 \\ 0 & & & 0 & 1 & -2i \end{pmatrix}$.

9. a. $\begin{pmatrix} 3 & 0 & 0 \\ 1 & 3 & 0 \\ 0 & 1 & 3 \end{pmatrix}$. b. $\begin{pmatrix} 3 & 0 & 0 & & 0 \\ 1 & 3 & 0 & & \\ 0 & 1 & 3 & & \\ & & & 3 & 0 \\ 0 & & & 1 & 3 \end{pmatrix}$, $\begin{pmatrix} 3 & 0 & 0 & & 0 \\ 1 & 3 & 0 & & \\ 0 & 1 & 3 & & \\ & & & 3 & 0 \\ 0 & & & 0 & 3 \end{pmatrix}$.

11. a. subdiag(1, 1, 0, 0, 0). b. subdiag(1, 0, 1, 0, 1, 0).
c. subdiag(1, 1, 0, 1, 0, 1) or subdiag(1, 1, 0, 1, 1, 0).

12. a. $X = \begin{pmatrix} 1 & 2 & 1 \\ 1 & 1 & 1 \\ 0 & 1 & 1 \end{pmatrix} \begin{pmatrix} e^{2t} & 0 & 0 \\ 0 & e^{-t} & 0 \\ 0 & 0 & e^{-3t} \end{pmatrix} \begin{pmatrix} 0 & 1 & -1 \\ 1 & -1 & 0 \\ -1 & 1 & 1 \end{pmatrix} \begin{pmatrix} 1 \\ -2 \\ 1 \end{pmatrix}$.

13. Consider problem 8, page 387.

Appendix

§1 Page 411

1. a. -1. b. 0 c. -1. d. 1.
2. a. 1. b. 0. c. -1. d. 0.
7. Yes.

§2 Page 417

1. The even permutations in $\mathscr{P}_3$ are 1, 2, 3; 3, 1, 2; and 2, 3, 1. The odd permutations are 1, 3, 2; 3, 2, 1; and 2, 1, 3.

2. a. 3, 4, 2, 1, 5. b. 4, 1, 3, 5, 2. c. σ. d. 2, 3, 5, 4, 1.

3. I, σ, σ^2.

6. a. $(x_1 - x_2)(x_1 - x_3)(x_1 - x_4)(x_2 - x_3)(x_2 - x_4)(x_3 - x_4)$.
b. For σ: $(x_4 - x_2)(x_4 - x_1)(x_4 - x_3)(x_2 - x_1)(x_2 - x_3)(x_1 - x_3)$
$= (-1)^4 P(x_1, x_2, x_3, x_4)$, so sgn $\sigma = 1$. Sgn $\tau = (-1)^3 = -1$.

7. $n!$

8. b. If σ is odd, $\tau \circ \sigma$ is even. Set $\tau \circ \sigma = \mu$, then $\sigma = \tau^{-1} \circ \mu = \tau \circ \sigma$.

9. a. 10; +1. b. 15; -1. c. 6; +1. d. 9; -1.

§3 Page 426

4. If $AZ = I_n$ and $Z = (Z_1, \ldots, Z_n)$, then $AZ_j = E_j$ for $1 \le j \le n$.

7. b. $3\sqrt{61}$. c. $\frac{1}{2}\|(A - B) \times (C - B)\|$. Would $\frac{1}{2}\|(A - B) \times (B - C)\|$ also give the area? d. $11\sqrt{11}/2$.

8. The only property it satisfies is closure.

9. a. For $U, V, W \in \mathscr{E}_4$, let $U \times V \times W$ be the vector satisfying $U \times V \times W \circ Y = \det(U^T, V^T, W^T, Y^T)$ for all $Y \in \mathscr{V}_4$. Such a vector exists since the map $Y \to \det(U^T, V^T, W^T, Y^T)$ is linear. The value of $U \times V \times W$ may be obtained by computing $U \times V \times W \circ E_i$ for $1 \le i \le 4$.
c. $(0, 0, 0, 1)$. d. $(3, 4, -6, -6)$.

Symbol Index†

Vector Spaces

Matrix Notation

Maps

† Numbers in parentheses refer to problem numbers.

Vectors

Miscellaneous

Subject Index†

† Numbers in parentheses refer to problem numbers.

T

U

V

W

Z